THE OXFORD A

General Editor: Frank Kermode

SAMUEL TAYLOR COLERIDGE, poet, critic, and thinker, was born in Devon in 1772 and died at Highgate in 1834. As a radical young poet in the years following the French Revolution, he collaborated with Wordsworth in *Lyrical Ballads* (1798). He was by turns poet, dramatist, political journalist, essayist, and public lecturer, and brought his literary and philosophical interests together in a remarkable work of criticism, the *Biographica Literaria* (1817). Chronic ill health and addiction to opium led him to take up residence with a surgeon in 1816, and in the Gillman household he spent his last years peacefully and productively, publishing among other things an enlarged edition of his periodical *The Friend* (1818), a volume of meditations, the *Aids to Reflection* (1825), and a treatise on political theory, *On the Constitution of the Church and State* (1829).

H. J. JACKSON, associate professor of English at Scarborough College in the University of Toronto, is co-editor of the volumes of *Shorter Works and Fragments* in the Bollingen edition of Coleridge's *Collected Works*.

FRANK KERMODE, retired King Edward VII Professor of English Literature at Cambridge, is the author of many books, including *Romantic Image*, *The Sense of an Ending*, *The Classic*, *The Genesis of Secrecy*, *Forms of Attention*, and *History and Value*; he is also co-editor with John Hollander of *The Oxford Anthology of English Literature*.

THE OXFORD AUTHORS

SAMUEL TAYLOR COLERIDGE

EDITED BY

H. J. JACKSON

Oxford New York
OXFORD UNIVERSITY PRESS

Oxford University Press, Walton Street, Oxford OX2 6DP
Oxford New York Toronto
Delhi Bombay Calcutta Madras Karachi
Kuala Lumpur Singapore Hong Kong Tokyo
Nairobi Dar es Salaam Cape Town
Melbourne Auckland Madrid
and associated companies in
Berlin Ibadan

Oxford is a trade mark of Oxford University Press

First published 1985

British Library Cataloguing in Publication Data
Coleridge, Samuel Taylor
Samuel Taylor Coleridge.—(The Oxford authors)
I. Title II. Jackson, H. J.
828'.709 PR4471
ISBN 0-19-281383-8

Library of Congress Cataloging in Publication Data
Coleridge, Samuel Taylor, 1772-1834.
Samuel Taylor Coleridge.
(The Oxford authors)
Bibliography: p. Includes indexes.
I. Jackson, H. J. II. Title III. Series.
PR4472.J3 1985 821'.7 84-9641
ISBN 0-19-281383-8

5 7 9 10 8 6 4

Printed in Great Britain by
Biddles Ltd
Guildford and King's Lynn

CONTENTS

An asterisk (*) indicates an incomplete item

PROSE

INTRODUCTION

THIS is an exciting time to be beginning the study of Coleridge. Although he first became known to his contemporaries as a poet of democratic principles, by the end of his life it was rather as a talker and thinker that he was famous. He influenced his own and a later generation both through his published prose works and through personal contact, for many people interested in literature or philosophy had attended his public lectures or sought him out at home in Highgate. Towards the end of the nineteenth century, as memories of the man faded and his ideas, never exactly in fashion, went increasingly out of it, his reputation was sustained by his poetry, eventually available in new editions by James Dykes Campbell and E. H. Coleridge. But in the twentieth century, following the appearance of the Shawcross edition of the *Biographia Literaria* with related aesthetic essays (1907), Coleridge again found an interested and respectful audience for his ideas, and has with justice been claimed as a precursor by virtually every school of literary criticism, practical, historical, or deconstructionist. Now, with all his works being freshly edited and made more easily accessible than ever before, our sense of the complex personality underlying them may be—or be in a position to become—more vivid than it was to his contemporaries or even to his closest friends, thanks to the publication of such private documents as notebooks and marginalia, or comments in books, complete and properly annotated for the first time. As they are gradually assessed and assimilated by critical readers, the editions of the seventies and eighties will be the basis of a new image of Coleridge.

Readers will want to make their own observations and come to their own conclusions about the works represented in this volume, with the biographical Chronology, the headnotes, and the annotation as guides in matters of fact; but it may be helpful to bear in mind some of the prominent features of Coleridge's career. Coleridge's was a polymathic, synthesizing mind. He drives commentators to catalogues as they try to sum him up, for he read in several languages (Latin, German, Greek, Italian; and, with less ease and pleasure, French, Spanish, and Portuguese), he wrote more than competently in many fields, and he sought constantly to articulate the one comprehensive pattern in which all his steadily increasing knowledge could find a place. One recognizes this impulse and these habits of mind in the unusual, sometimes unstable combination of philosophic seriousness, psychological acuteness, and metaphoric brilliance in his best and characteristic works—the *Biographia*, the notebooks, 'Christabel'. But such works were not easy to produce. Throughout his life, Coleridge suffered from ill health. It made him a sickly and in some ways an over-protected child, and it led directly to his addiction to opium from about the age of thirty. After some periods of suicidal depression and unsuccessful attempts to break the habit by strength of will, he settled as a permanent paying guest in the house of a surgeon, James Gillman, who helped him to bring his dependence upon opium under control, although Coleridge was never able to give up the drug.

Throughout his life, he was harassed by financial embarrassment: there was never enough money. Coleridge had been one of a family of thirteen; by cutting short his studies at Cambridge, he sacrificed to religious and political conviction the chance of a steady professional life; he married young, and soon had a family to support; and when he separated from his wife he had somehow to find money for two establishments. The consequence was that he had to take up whatever money-making schemes suggested themselves—journalism, lectures, translations, or other booksellers' projects—with the exception of reviewing, to which he had moral objections. Then, of course, he could not settle down to write the great work that he dreamt of; he was perpetually unsettled and interrupted by the pressing needs of the day. Furthermore, for the first forty-four years of his life he was physically rootless, or perhaps it would be closer to the truth to say that it was his repeated experience, having thought himself settled in one place, to have to tear himself away and start again elsewhere. The changes of place corresponded generally to shifting friendships, too, for although Coleridge attracted many devoted friends, he never found the unqualified love that he wanted: in the myth of the mariner 'alone on a wide wide sea' he forecast the sense of isolation that was with him for most of his life.

A counterweight to the distresses of his life was religious conviction. Although the orthodox practice of his childhood was deflected while he was at Cambridge, when he became and for some years remained a Unitarian, the fundamental assumptions of his faith were unshaken, and his belief in the importance of spiritual religion, in opposition to a prevalent materialism, can be seen as the basis of many of his original contributions in poetry and criticism, political theory and philosophy, as well as in religious thought. Coleridge's faith was the fountain of his best work, but it did not make him at all narrowly dogmatic. In fact, as a motto for his life and work one could do worse than take a phrase that appears in both verse and prose, 'What if': 'And what if all of animated nature / Be but organic harps diversely framed'; 'And what if Monads of the infinite mind?' The enquiring spirit is everywhere to be found.

Contents

The contents of this anthology have been selected with the reader of literature in mind. There is poetry from every period of Coleridge's life, but chiefly from the fruitful early years. The modern reader, knowing Coleridge primarily by the great poems of the imagination, as they are generally considered, the 'Ancient Mariner', 'Christabel', and 'Kubla Khan', may be surprised by the prominence of political subjects in the early verse; but then it was a turbulent time in which, as Carl Woodring has observed, even a poem about a mad mother would be recognized instantly as a statement against the war and against the current ministry. The range of modes and models in Coleridge's verse was not so typical, and still strikes us as extraordinary. He wrote ballads, odes, sonnets, and lyrics; he imitated Shakespeare, Milton, Gray, Cowper, and many lesser figures; he turned most of the popular modes—the sentimental, the reflective, the sublime, the satirical—successfully to his own ends. And he continued to write and to experiment with verse to the very end of his life.

In this volume, the poetry is followed by the complete text of Coleridge's major critical work, the *Biographia Literaria*, unabridged so that readers may decide for themselves the important questions of its coherence and value. Published at the mid-point of Coleridge's career, *Biographia Literaria* analyses the sources of false taste in his day and, after exposing the lack of clear standards for literary judgement, it sets about establishing and exercising criteria of literary worth. It does much more besides, using the popular form of the autobiography as a vehicle for subjects that Coleridge knew would never be popular themselves. It is an eccentric, difficult, and altogether remarkable book. Its central position in this collection is deliberate: the *Biographia* illuminates the poetry, and the extracts from other prose works that follow it have been chosen chiefly because they take up or take further ideas raised in the *Biographia*. Coleridge is a hard man to take excerpts from, partly because he could hold a complex concept in suspension while discussing a bit of it, so that in extracting the bit one loses the whole; and partly because flatter passages are often preliminaries to the high points which cannot stand alone. Although selection is unavoidable in such a collection as this, the *Biographia* at least can be seen as a whole, and other passages in the context that it provides.

Coleridge is to many readers a more attractive figure in his letters, notebooks, and marginalia, where one finds him at ease, curious, whimsical, and spontaneous, than when he appears to be straining over an argument in more formal public writings. For that reason, his private papers have been given considerable prominence here, but since it is possible to offer only a fraction of the available material, extracts have again been chosen largely for their literary interest, as supporting the reading of the poetry and the *Biographia*.

Annotation

Notes are by policy kept to a minimum in this series. In the case of Coleridge, difficult though he sometimes is, the policy is wise for more than practical reasons. The reader trying to follow the development of thoughts, especially in the first or second reading of a work, is liable to lose his place mentally if he has to turn to the back of the book for a note. The most conspicuous aspect of the annotation of this volume is the translation of phrases and lines in foreign languages in square brackets as part of the text itself. (Where the translation is taken from a published source such as the Loeb Classical Library or the Bollingen edition of Coleridge, that source is also indicated within the square brackets.) Phrases which Coleridge himself goes on to translate as part of his text are not retranslated, and titles are not normally translated, since the works concerned are to be found under their foreign titles and Coleridge usually gives enough of the title to allow the book to be traced easily.

The most conspicuously *absent* kind of annotation is the recording of sources for quotations or ideas. Coleridge's sources have for the most part been identified, and advanced readers may wish to trace them through fully annotated scholarly editions, such as are listed in the guide to Further Reading at the end of the volume. Coleridge did not, however, tend to use quotations as allusions evoking the whole of the work from which they came. The reader who is trying to follow the argument will generally find that the quotations express ideas as part of a connected argument, and that a note that provides a line reference to *Paradise Lost* is distracting and unprofitable, whereas sustained and careful attention to the sense of the words will reveal Coleridge's meaning. The words themselves, if they can be found in the *Oxford English Dictionary* or another good English dictionary, have not been glossed here unless they are words that one might not expect to find in an English dictionary, or that one might not find easily—words such as 'perdue', 'ne plus ultra', and 'rifacciamento'. Occasionally, when Coleridge uses an ordinary word in an unusual way, an endnote draws attention to the fact. But it is worth observing that the *OED* is in some ways the best companion for Coleridge studies, for an essential part of Coleridge's method was the restoring, redefining, and coining of words, and a historical dictionary enables the reader to assess and appreciate his usage.

For proper names, too, it is assumed that the reader has access to a standard one-volume encyclopaedia or similar reference work. As a rule, Coleridge himself tells his readers as much as they need to know for the purposes of his argument about the famous figures to whom he refers (Plato, for instance), and further information is often not needed. I have, however, tried to identify in an appropriate way contemporary figures such as Erasmus Darwin and William Roscoe, household names to the first readers of the *Biographia*, but not to us. Finally, it is hoped that the Index to the Prose at the end of the volume will fill some of the gaps in the annotation by permitting cross-reference within Coleridge's own work.

ACKNOWLEDGEMENTS

The gradual publication of Coleridge's writings in modern editions has been the occasion of much scholarly co-operation. I wish to thank Oxford University Press for granting permission to use its editions of Coleridge's *Collected Letters* and *Complete Poetical Works* as the basis of those sections in this collection, and Princeton University Press and Routledge & Kegan Paul Ltd. for similar permission with regard to the *Notebooks* and *Collected Works*. On a more personal level, I have been happy to be able once again to take advantage of the generous spirits of Kathleen Coburn, J. R. de J. Jackson, and J. C. C. Mays. George Whalley, who had with characteristic selflessness allowed me access to the unpublished marginalia, died as this volume was in preparation, to the sorrow of Coleridgeans and humanists everywhere.

CHRONOLOGY

1772 Birth of Samuel Taylor Coleridge at Ottery St Mary, Devon, 21 October.

1781 Death of father.

1782 Sent to Christ's Hospital School; Charles Lamb a schoolmate, James Boyer a memorable master.

1789 Fall of the Bastille 14 July, beginning of the French Revolution.

1791 Enters Jesus College, Cambridge, with scholarships.

1793 Execution of the King (January) and Queen (October) of France. Declaration of war between France and England (February). Reign of Terror begins in March, ending July 1794 with execution of Robespierre. Coleridge publishes his first poem but, fearing disgrace for debts, enlists in a company of dragoons (as Silas Tomkyn Comberbache) in December.

1794 Returns to Cambridge in April. In June, on a walking tour, meets Robert Southey at Oxford, and plans with him to establish a Utopian community ('pantisocracy') in America; becomes engaged to Sara Fricker, the sister of Southey's fiancée; leaves Cambridge in December, without a degree.

1795 Lectures with Southey at Bristol (January to June), but pantisocracy is abandoned. Meets Wordsworth (September?). Marries Sara Fricker 4 October.

1796 Publishes a political newspaper, *The Watchman*, in ten numbers (March to May). Publishes *Poems on Various Subjects*. Birth of Hartley Coleridge 19 September. Moves with family to Nether Stowey, near Thomas Poole (December).

1797 William and Dorothy Wordsworth rent Alfoxden House, not far from Stowey (July); *The Rime of the Ancient Mariner* begun (November).

1798 Accepts an annuity of £150 from Thomas and Josiah Wedgwood (reduced to £75 in November 1812). Birth of Berkeley Coleridge 14 May. *Lyrical Ballads* published anonymously (September). Coleridge and the Wordsworths set out for Germany.

1799 Wordsworths at Ratzeburg, Coleridge at Göttingen and travelling. Death of Berkeley Coleridge 10 February. Return to England in July. On a walking tour with Wordsworth, Coleridge meets Sara Hutchinson, whom he is to love hopelessly for many years. In London as a regular contributor to the *Morning Post* December 1799 to April 1800.

1800 Birth of Derwent Coleridge 14 September.

1801	*Lyrical Ballads* (1800) published January, with Preface. Coleridge an occasional contributor to the *Morning Post* September 1801 to August 1803.
1802	Napoleon made life consul (May); French invasion of Switzerland (October). Wordsworth marries Mary Hutchinson (October). Founding of the *Edinburgh Review* (October). Birth of Sara Coleridge 23 December.
1804	Napoleon made emperor (May). Spain declares war on Britain (December). Coleridge leaves England in April to live in Malta and Italy, acting for part of the time as Public Secretary in Malta; on return (August 1806) resolved to separate from his wife.
1808	Lectures on poetry at the Royal Institution; fitful contributor to the *Courier* (until 1814).
1809–10	Dictates *The Friend*, in 28 numbers, to Sara Hutchinson during a prolonged domestication with the Wordsworths at Allan Bank, Grasmere. Serious break with Wordsworth (October 1810); partial reconciliation May 1812. Goes to live with John Morgan and his wife and sister-in-law, at first in London.
1811–12	Three series of lectures on literature.
1813	Coleridge's tragedy, *Remorse*, has a successful run at Drury Lane. Lectures at Bristol October to November.
1814	Lectures at Bristol (April). Wordsworth's *Excursion* published.
1815	Waterloo (18 June); restoration of Louis XVIII. Wordsworth's *Poems* (1815) and *The White Doe of Rylstone*. *Biographia Literaria* dictated to John Morgan at Calne, Wiltshire.
1816	Accepted as patient and housemate by James Gillman, surgeon, at Highgate. *Christabel*, *Kubla Khan*, *and The Pains of Sleep* published in May; *The Statesman's Manual* in December. Composes 'Theory of Life' (pub. 1848).
1817	Publication of *A Lay Sermon*, *Biographia Literaria*, *Sibylline Leaves*, and a play, *Zapolya*.
1818–19	Lectures on poetry and drama (January to March 1818) and alternately on literature and on the history of philosophy (December 1818 to March 1819). Bankruptcy of Coleridge's publisher, Rest Fenner.
1825	Address to the Royal Society of Literature 'On the Prometheus of Aeschylus'. *Aids to Reflection* published.
1828	Rhine tour with William Wordsworth and his daughter Dora. Publication of Coleridge's *Poetical Works* (3 vols.).
1829	Second edition of *Poetical Works*. Publication of *On the Constitution of the Church and State* (December).
1834	*Poetical Works* (3rd edn.). Death of Coleridge, 25 July.

NOTE ON THE TEXT

THE fact that the texts selected for different parts of this anthology came from various sources and required varying degrees of editorial interference is in part a reflection of the transitional state of Coleridge studies. It has produced unavoidable inconsistencies in the volume as a whole, although it is hoped that the rationale of each part (given in the headnote that identifies the copy-text) will make for consistency for that part. In general, it has seemed desirable to remove as many as possible of the mechanical barriers that stand in the way of comprehension. The prose texts published by different houses and often under difficult circumstances during Coleridge's lifetime have been modernized in conformity with the guidelines of the Oxford Authors series: specifically, the use of capitals, small capitals, and italics to indicate degrees of emphasis has been eliminated, and spellings have been reduced to consistency, although some older forms have been retained when they were not ambiguous. The punctuation of the originals has been retained except in the very few cases (generally involving only the position of a comma) in which it produced ambiguity. For other texts, however, the policy of the series has been relaxed. The printers of the 1834 *Poetical Works* produced a text which, though not rigorously consistent, presents no serious difficulty to the modern reader, and it has been possible to retain the accidentals of that text. Finally, the informal writings now published from Coleridge's manuscripts—the letters, notebooks, and marginalia—owe so much of their vigour and spontaneity to practices in spelling and capitalization that are certainly Coleridge's and not the printer's that they had to be respected. The following minor changes have, however, been made consistently throughout the volume, even in the texts from manuscript. Proper names have been given in their modern spelling, for ease of reference: thus, the playwright whose name is spelt five different ways in the original text of the *Biographia* is referred to consistently as Shakespeare. Obvious printer's errors have been corrected silently, and raised letters, as in 'M^{r}', have been lowered. The contracted letters that were used in the setting of Greek in Coleridge's day have been expanded to *ου* and *στ*, but the eccentric diacritical marks have not been corrected. Possessives have been standardized to 'its', 'theirs', 'hers' from the various forms in which they appeared. Quotation marks have been made to conform with modern practice. And in the case of what was originally manuscript material, no record is given of insertions or deletions.

The degree sign (°) indicates a note at the end of the book.

POEMS°

Genevieve

Maid of my Love, sweet Genevieve!
In Beauty's light you glide along:
Your eye is like the star of eve,
And sweet your Voice, as Seraph's song.
Yet not your heavenly Beauty gives
This heart with passion soft to glow:
Within your soul a Voice there lives!
It bids you hear the tale of Woe.
When sinking low the Sufferer wan
Beholds no hand outstretcht to save, 10
Fair, as the bosom of the Swan
That rises graceful o'er the wave,
I've seen your breast with pity heave,
And therefore love I you, sweet Genevieve!

Epitaph on an Infant

Ere Sin could blight or Sorrow fade,
Death came with friendly care;
The opening bud to Heaven conveyed,
And bade it blossom there.

Monody on the Death of Chatterton°

O what a wonder seems the fear of death,
Seeing how gladly we all sink to sleep,
Babes, Children, Youths, and Men,
Night following night for threescore years and ten!
But doubly strange, where life is but a breath
To sigh and pant with, up Want's rugged steep.

Away, Grim Phantom! Scorpion King, away!
Reserve thy terrors and thy stings display
For coward Wealth and Guilt in robes of State!

Lo! by the grave I stand of one, for whom 10
A prodigal Nature and a niggard Doom
(That all bestowing, this withholding all,)
Made each chance knell from distant spire or dome
Sound like a seeking Mother's anxious call,
Return, poor Child! Home, weary Truant, home!

Thee, Chatterton! these unblest stones protect
From want, and the bleak freezings of neglect.
Too long before the vexing Storm-blast driven
Here hast thou found repose! beneath this sod!
Thou! O vain word! thou dwell'st not with the clod! 20
Amid the shining Host of the Forgiven
Thou at the throne of Mercy and thy God
The triumph of redeeming Love dost hymn
(Believe it, O my Soul!) to harps of Seraphim.

Yet oft, perforce ('tis suffering Nature's call)
I weep, that heaven-born Genius so should fall;
And oft, in Fancy's saddest hour, my soul
Averted shudders at the poisoned bowl.
Now groans my sickening heart, as still I view
  Thy corse of livid hue; 30
Now Indignation checks the feeble sigh,
Or flashes through the tear that glistens in mine eye!

Is this the land of song-ennobled line?
Is this the land, where Genius ne'er in vain
  Pour'd forth his lofty strain?
Ah me! yet Spenser, gentlest bard divine,
Beneath chill Disappointment's shade,
His weary limbs in lonely anguish laid.
  And o'er her darling dead
  Pity hopeless hung her head, 40
While 'mid the pelting of that merciless storm,'
Sunk to the cold earth Otway's famished form!

Sublime of thought, and confident of fame,
From vales where Avon winds the Minstrel came.
  Light-hearted youth! aye, as he hastes along,
  He meditates the future song,

How dauntless Aella fray'd the Dacyan foe;°
　　And while the numbers flowing strong
　　In eddies whirl, in surges throng,
Exulting in the spirits' genial throe 50
In tides of power his life-blood seems to flow.

And now his cheeks with deeper ardors flame,
His eyes have glorious meanings, that declare
More than the light of outward day shines there,
A holier triumph and a sterner aim!
Wings grow within him; and he soars above
Or Bard's or Minstrel's lay of war or love.
Friend to the friendless, to the Sufferer health,
He hears the widow's prayer, the good man's praise;
To scenes of bliss transmutes his fancied wealth, 60
And young and old shall now see happy days.
On many a waste he bids trim Gardens rise,
Gives the blue sky to many a prisoner's eyes;
And now in wrath he grasps the patriot steel,
And her own iron rod he makes Oppression feel.
Sweet Flower of Hope! free Nature's genial child!
That didst so fair disclose thy early bloom,
Filling the wide air with a rich perfume!
For thee in vain all heavenly aspects smil'd;
From the hard world brief respite could they win— 70
The frost nipp'd sharp without, the canker prey'd within!
Ah! where are fled the charms of vernal Grace,
And Joy's wild gleams that lighten'd o'er thy face?
Youth of tumultuous soul, and haggard eye!
Thy wasted form, thy hurried steps I view,
On thy wan forehead starts the lethal dew,
And oh! the anguish of that shuddering sigh!

　Such were the struggles of the gloomy hour,
　　When Care, of wither'd brow,
　Prepared the poison's death-cold power: 80
Already to thy lips was raised the bowl,
　When near thee stood Affection meek
　(Her bosom bare, and wildly pale her cheek)
Thy sullen gaze she bade thee roll
On scenes that well might melt thy soul;

Thy native cot she flash'd upon thy view,
Thy native cot, where still, at close of day,
Peace smiling sate, and listen'd to thy lay;
Thy sister's shrieks she bade thee hear,
And mark thy mother's thrilling tear; 90
    See, see her breast's convulsive throe,
    Her silent agony of woe!
Ah! dash the poison'd chalice from thy hand!

And thou had'st dashed it, at her soft command,
But that Despair and Indignation rose,
And told again the story of thy woes;
Told the keen insult of the unfeeling heart,
The dread dependence on the low-born mind;
Told every pang, with which thy soul must smart,
Neglect, and grinning Scorn, and Want combined! 100
Recoiling quick, thou bad'st the friend of pain
Roll the black tide of Death through every freezing vein!

O Spirit blest!
Whether the Eternal's throne around,
Amidst the blaze of Seraphim,
Thou pourest forth the grateful hymn;
Or soaring thro' the blest domain
Enrapturest Angels with thy strain,—
Grant me, like thee, the lyre to sound,
Like thee with fire divine to glow;— 110
But ah! when rage the waves of woe,
Grant me with firmer breast to meet their hate,
And soar beyond the storm with upright eye elate!

Ye woods! that wave o'er Avon's rocky steep,
To Fancy's ear sweet is your murmuring deep!
For here she loves the cypress wreath to weave
Watching, with wistful eye, the saddening tints of eve.
Here, far from men, amid this pathless grove,
In solemn thought the Minstrel wont to rove,
Like star-beam on the slow sequester'd tide 120
Lone-glittering, through the high tree branching wide.
And here, in Inspiration's eager hour,
When most the big soul feels the mastering power,

These wilds, these caverns roaming o'er,
Round which the screaming sea-gulls soar,
With wild unequal steps he passed along,
Oft pouring on the winds a broken song:
Anon, upon some rough rock's fearful brow
Would pause abrupt—and gaze upon the waves below.

Poor Chatterton! he sorrows for thy fate 130
Who would have praised and loved thee, ere too late.
Poor Chatterton! farewell! of darkest hues
This chaplet cast I on thy unshaped tomb;
But dare no longer on the sad theme muse,
Lest kindred woes persuade a kindred doom:
For oh! big gall-drops, shook from Folly's wing,
Have blackened the fair promise of my spring;
And the stern Fate transpierced with viewless dart
The last pale Hope that shiver'd at my heart!

Hence, gloomy thoughts! no more my soul shall dwell 140
On joys that were! No more endure to weigh
The shame and anguish of the evil day,
Wisely forgetful! O'er the ocean swell
Sublime of Hope I seek the cottaged dell
Where Virtue calm with careless step may stray;
And, dancing to the moon-light roundelay,
The wizard Passions weave an holy spell!

O Chatterton! that thou wert yet alive!
Sure thou would'st spread the canvass to the gale,
And love with us the tinkling team to drive 150
O'er peaceful Freedom's undivided dale;
And we, at sober eve, would round thee throng,
Would hang, enraptured, on thy stately song,
And greet with smiles the young-eyed Poesy
All deftly mask'd, as hoar Antiquity.
Alas, vain Phantasies! the fleeting brood
Of Woe self-solac'd in her dreamy mood!
Yet will I love to follow the sweet dream,
Where Susquehana pours his untamed stream;
And on some hill, whose forest-frowning side 160
Waves o'er the murmurs of his calmer tide,

Will raise a solemn Cenotaph to thee,
Sweet Harper of time-shrouded Minstrelsy!
And there, soothed sadly by the dirgeful wind,
Muse on the sore ills I had left behind.

Sonnet

TO THE RIVER OTTER

Dear native brook! wild streamlet of the West!
How many various-fated years have past,
What happy, and what mournful hours, since last
I skimmed the smooth thin stone along thy breast,
Numbering its light leaps! yet so deep imprest
Sink the sweet scenes of childhood, that mine eyes
I never shut amid the sunny ray,
But straight with all their tints thy waters rise,
Thy crossing plank, thy marge with willows grey,
And bedded sand that, vein'd with various dyes, 10
Gleamed through thy bright transparence! On my way,
Visions of Childhood! oft have ye beguil'd
Lone manhood's cares, yet waking fondest sighs:
Ah! that once more I were a careless child!

Songs of the Pixies

The Pixies, in the superstition of Devonshire, are a race of beings invisibly small, and harmless or friendly to man. At a small distance from a village in that county, half way up a wood-covered hill, is an excavation called the Pixies' Parlour. The roots of old trees form its ceiling; and on its sides are innumerable cyphers, among which the author discovered his own and those of his brothers, cut by the hand of their childhood. At the foot of the hill flows the river Otter.

To this place the Author, during the Summer months of the year 1793, conducted a party of young ladies; one of whom, of stature elegantly small, and of complexion colourless yet clear, was proclaimed the Faery Queen. On which occasion the following Irregular Ode was written.

I

Whom the untaught Shepherds call
  Pixies in their madrigal,
Fancy's children, here we dwell:
  Welcome, Ladies! to our cell.
Here the wren of softest note
  Builds its nest and warbles well;
Here the blackbird strains his throat;
  Welcome, Ladies! to our cell.

II

When fades the moon to shadowy-pale,
And scuds the cloud before the gale, 10
Ere the Morn, all gem-bedight,
Hath streak'd the East with rosy light,
We sip the furze-flower's fragrant dews
Clad in robes of rainbow hues:
Or sport amid the shooting gleams
To the tune of distant-tinkling teams,
While lusty Labour scouting sorrow
Bids the Dame a glad good-morrow,
Who jogs the accustomed road along,
And paces cheery to her cheering song. 20

III

But not our filmy pinion
  We scorch amid the blaze of day,
  When Noontide's fiery-tressed minion
    Flashes the fervid ray.
  Aye from the sultry heat
  We to the cave retreat
O'ercanopied by huge roots intertwined
With wildest texture, blackened o'er with age:
Round them their mantle green the ivies bind,
    Beneath whose foliage pale 30
    Fanned by the unfrequent gale
We shield us from the Tyrant's mid-day rage.

IV

Thither, while the murmuring throng
Of wild-bees hum their drowsy song,
By Indolence and Fancy brought,
A youthful Bard, 'unknown to Fame,'
Wooes the Queen of Solemn Thought,
And heaves the gentle misery of a sigh
Gazing with tearful eye,
As round our sandy grot appear
Many a rudely sculptured name
To pensive Memory dear!
Weaving gay dreams of sunny-tinctured hue
We glance before his view:
O'er his hush'd soul our soothing witcheries shed
And twine the future garland round his head.

V

When Evening's dusky car
Crowned with her dewy star
Steals o'er the fading sky in shadowy flight;
On leaves of aspen trees 50
We tremble to the breeze
Veiled from the grosser ken of mortal sight.
Or, haply, at the visionary hour,
Along our wildly-bowered sequestered walk,
We listen to the enamoured rustic's talk;
Heave with the heavings of the maiden's breast,
Where young-eyed Loves have hid their turtle nest;
Or guide of soul-subduing power
The glance, that from the half-confessing eye
Darts the fond question or the soft reply. 60

VI

Or through the mystic ringlets of the vale
We flash our faery feet in gamesome prank;
Or, silent-sandal'd, pay our defter court,
Circling the Spirit of the Western Gale,
Where wearied with his flower-caressing sport,
Supine he slumbers on a violet bank;

Then with quaint music hymn the parting gleam
By lonely Otter's sleep-persuading stream;
Or where his wave with loud unquiet song
Dash'd o'er the rocky channel froths along; 70
Or where, his silver waters smooth'd to rest,
The tall tree's shadow sleeps upon his breast.

VII

Hence thou lingerer, Light!
Eve saddens into Night.
Mother of wildly-working dreams! we view
The sombre hours, that round thee stand
With down-cast eyes (a duteous band!)
Their dark robes dripping with the heavy dew.
Sorceress of the ebon throne!
Thy power the Pixies own, 80
When round thy raven brow
Heaven's lucent roses glow,
And clouds in watery colours drest
Float in light drapery o'er thy sable vest:
What time the pale moon sheds a softer day
Mellowing the woods beneath its pensive beam:
For mid the quivering light 'tis ours to play,
Aye dancing to the cadence of the stream.

VIII

Welcome, Ladies! to the cell
Where the blameless Pixies dwell: 90
But thou, sweet Nymph! proclaimed our Faery Queen,
With what obeisance meet
Thy presence shall we greet?
For lo! attendant on thy steps are seen
Graceful Ease in artless stole,
And white-robed Purity of soul,
With Honour's softer mien;
Mirth of the loosely-flowing hair,
And meek-eyed Pity eloquently fair,
Whose tearful cheeks are lovely to the view, 100
As snow-drop wet with dew.

IX

Unboastful Maid! though now the Lily pale
Transparent grace thy beauties meek;
Yet ere again along the impurpling vale,
The purpling vale and elfin-haunted grove,
Young Zephyr his fresh flowers profusely throws,
We'll tinge with livelier hues thy cheek;
And, haply, from the nectar-breathing Rose
Extract a Blush for Love!

To a Young Ass

Its Mother being tethered near it

Poor little Foal of an oppressed Race!
I love the languid Patience of thy face:
And oft with gentle hand I give thee bread,
And clap thy ragged Coat, and pat thy head.
But what thy dulled Spirits hath dismayed,
That never thou dost sport along the glade?
And (most unlike the nature of things young)
That earthward still thy moveless head is hung?
Do thy prophetic Fears anticipate,
Meek Child of Misery! thy future fate? 10
The starving meal, and all the thousand aches
'Which patient Merit of the Unworthy takes'?
Or is thy sad heart thrilled with filial pain
To see thy wretched Mother's shorten'd Chain?
And truly, very piteous is her Lot—
Chained to a Log within a narrow spot,
Where the close-eaten Grass is scarcely seen,
While sweet around her waves the tempting Green!
Poor Ass! thy master should have learnt to show
Pity—best taught by fellowship of Woe! 20
For much I fear me that He lives like thee,
Half famished in a land of Luxury!
How askingly its footsteps hither bend,
It seems to say, 'And have I then one Friend?'
Innocent Foal! thou poor despised Forlorn!
I hail thee Brother—spite of the fool's scorn!

And fain would take thee with me, in the Dell
Of Peace and mild Equality to dwell,
Where Toil shall call the charmer Health his bride,
And Laughter tickle Plenty's ribless side! 30
How thou wouldst toss thy heels in gamesome play,
And frisk about, as lamb or kitten gay!
Yea! and more musically sweet to me
Thy dissonant harsh bray of joy would be,
Than warbled melodies that soothe to rest
The aching of pale Fashion's vacant breast!

Sonnets on Eminent Characters°

BURKE

As late I lay in slumber's shadowy vale,
With wetted cheek and in a mourner's guise,
I saw the sainted form of Freedom rise:
She spake! not sadder moans the autumnal gale—
'Great Son of Genius! sweet to me thy name,
Ere in an evil hour with altered voice
Thou bad'st Oppression's hireling crew rejoice
Blasting with wizard spell my laurelled fame.
Yet never, Burke! thou drank'st Corruption's bowl!
Thee stormy Pity and the cherished lure 10
Of Pomp, and proud Precipitance of soul
Wildered with meteor fires. Ah Spirit pure!
That error's mist had left thy purged eye:
So might I clasp thee with a Mother's joy!'

PRIESTLEY

Though roused by that dark Vizir Riot rude
Have driven our Priestley o'er the Ocean swell;
Though Superstition and her wolfish brood
Bay his mild radiance, impotent and fell;
Calm in his halls of brightness he shall dwell!
For lo! Religion at his strong behest
Starts with mild anger from the Papal spell,
And flings to Earth her tinsel-glittering vest,

Her mitred state and cumbrous pomp unholy;
And Justice wakes to bid the Oppressor wail 10
Insulting aye the wrongs of patient Folly;
And from her dark retreat by Wisdom won
Meek Nature slowly lifts her matron veil
To smile with fondness on her gazing son!

PITT°

Not always should the tear's ambrosial dew
Roll its soft anguish down thy furrow'd cheek!
Not always heaven-breath'd tones of suppliance meek
Beseem thee, MERCY! Yon dark Scowler view,
Who with proud words of dear-lov'd Freedom came—
More blasting than the mildew from the South!
And kiss'd his country with Iscariot mouth
(Ah! foul apostate from his Father's fame!)
Then fix'd her on the cross of deep distress,
And at safe distance marks the thirsty lance 10
Pierce her big side! But O! if some strange trance
The eye-lids of thy stern-brow'd Sister° press,
Seize, MERCY! thou more terrible the brand,
And hurl her thunderbolts with fiercer hand!

TO THE REV. W. L. BOWLES°

My heart has thanked thee, Bowles! for those soft strains
Whose sadness soothes me, like the murmuring
Of wild-bees in the sunny showers of spring!
For hence not callous to the mourner's pains
Through Youth's gay prime and thornless paths I went:
And when the mightier throes of mind began,
And drove me forth, a thought-bewildered man,
Their mild and manliest melancholy lent
A mingled charm, such as the pang consigned
To slumber, though the big tear it renewed; 10
Bidding a strange mysterious Pleasure brood
Over the wavy and tumultuous mind,
As the great Spirit erst with plastic sweep
Mov'd on the darkness of the unformed deep.

Religious Musings

A desultory Poem, written on the Christmas Eve of 1794

ARGUMENT°

Introduction. Person of Christ. His prayer on the Cross. The process of his Doctrines on the mind of the Individual. Character of the Elect. Superstition. Digression to the present War. Origin and Uses of Government and Property. The present State of Society. The French Revolution. Millennium. Universal Redemption. Conclusion.

This is the time, when most divine to hear,
The voice of adoration rouses me,
As with a Cherub's trump: and high upborne,
Yea, mingling with the choir, I seem to view
The vision of the heavenly multitude,
Who hymned the song of peace o'er Bethlehem's fields!
Yet thou more bright than all the angel blaze,
That harbingered thy birth, Thou, Man of Woes!
Despised Galilean! For the great
Invisible (by symbols only seen) 10
With a peculiar and surpassing light
Shines from the visage of the oppressed good man,
When heedless of himself the scourged Saint
Mourns for the oppressor. Fair the vernal mead,
Fair the high grove, the sea, the sun, the stars;
True impress each of their creating Sire!
Yet nor high grove, nor many-coloured mead,
Nor the green Ocean with his thousand isles,
Nor the starred azure, nor the sovran sun,
E'er with such majesty of portraiture 20
Imaged the supreme beauty uncreate,
As thou, meek Saviour! at the fearful hour
When thy insulted anguish winged the prayer
Harped by Archangels, when they sing of mercy!
Which when the Almighty heard from forth his throne
Diviner light filled Heaven with ecstasy!
Heaven's hymnings paused: and Hell her yawning mouth
Closed a brief moment.

Lovely was the death
Of Him whose life was Love! Holy with power

He on the thought-benighted Sceptic beamed 30
Manifest Godhead, melting into day
What floating mists of dark idolatry
Broke and misshaped the omnipresent Sire:
And first by Fear uncharmed the drowsed Soul.
Till of its nobler nature it 'gan feel
Dim recollections; and thence soared to Hope,
Strong to believe whate'er of mystic good
The Eternal dooms for his immortal sons.
From Hope and firmer Faith to perfect Love
Attracted and absorbed: and centered there 40
God only to behold, and know, and feel,
Till by exclusive consciousness of God
All self-annihilated it shall make
God its identity: God all in all!
We and our Father one!

And blest are they,
Who in this fleshly World, the elect of Heaven,
Their strong eye darting through the deeds of men,
Adore with steadfast unpresuming gaze
Him Nature's essence, mind, and energy!
And gazing, trembling, patiently ascend 50
Treading beneath their feet all visible things
As steps, that upward to their Father's throne
Lead gradual—else nor glorified nor loved.
They nor contempt embosom nor revenge:
For they dare know of what may seem deform
The Supreme Fair sole operant: in whose sight
All things are pure, his strong controlling Love
Alike from all educing perfect good.
Theirs too celestial courage, inly armed—
Dwarfing Earth's giant brood, what time they muse 60
On their great Father, great beyond compare!
And marching onwards view high o'er their heads
His waving banners of Omnipotence.

Who the Creator love, created might
Dread not: within their tents no terrors walk.
For they are holy things before the Lord
Aye unprofaned, though Earth should league with Hell;
God's altar grasping with an eager hand

Fear, the wild-visaged, pale, eye-starting wretch,
Sure-refug'd hears his hot pursuing fiends 70
Yell at vain distance. Soon refreshed from Heaven
He calms the throb and tempest of his heart.
His countenance settles; a soft solemn bliss
Swims in his eye—his swimming eye upraised:
And Faith's whole armour glitters on his limbs!
And thus transfigured with a dreadless awe,
A solemn hush of soul, meek he beholds
All things of terrible seeming: yea, unmoved
Views e'en the immitigable ministers
That shower down vengeance on these latter days. 80
For kindling with intenser Deity
From the celestial Mercy-seat they come,
And at the renovating wells of Love
Have fill'd their vials with salutary wrath,
To sickly Nature more medicinal
Than what soft balm the weeping good man pours
Into the lone despoiléd traveller's wounds!

Thus from the Elect, regenerate through faith,
Pass the dark Passions and what thirsty Cares
Drink up the Spirit, and the dim regards 90
Self-centre. Lo they vanish! or acquire
New names, new features—by supernal grace
Enrobed with Light, and naturalized in Heaven.
As when a shepherd on a vernal morn
Through some thick fog creeps timorous with slow foot,
Darkling he fixes on the immediate road
His downward eye: all else of fairest kind
Hid or deformed. But lo! the bursting Sun!
Touched by the enchantment of that sudden beam
Straight the black vapour melteth, and in globes 100
Of dewy glitter gems each plant and tree;
On every leaf, on every blade it hangs!
Dance glad the new-born intermingling rays,
And wide around the landscape streams with glory!

There is one Mind, one omnipresent Mind,
Omnific. His most holy name is Love.
Truth of subliming import! with the which

Who feeds and saturates his constant soul,
He from his small particular orbit flies
With blest outstarting! From himself he flies, 110
Stands in the sun, and with no partial gaze
Views all creation; and he loves it all,
And blesses it, and calls it very good!
This is indeed to dwell with the most High!
Cherubs and rapture-trembling Seraphim
Can press no nearer to the Almighty's Throne.
But that we roam unconscious, or with hearts
Unfeeling of our universal Sire,
And that in his vast family no Cain
Injures uninjured (in her best-aimed blow 120
Victorious murder a blind suicide)
Haply for this some younger Angel now
Looks down on human nature: and, behold!
A sea of blood bestrewed with wrecks, where mad
Embattling interests on each other rush
With unhelmed rage!

'Tis the sublime of man,
Our noontide majesty, to know ourselves
Parts and proportions of one wondrous whole!
This fraternises man, this constitutes
Our charities and bearings. But 'tis God 130
Diffused through all, that doth make all one whole;
This the worst superstition, him except
Aught to desire, Supreme Reality!
The plenitude and permanence of bliss!
O Fiends of Superstition! not that oft
The erring priest hath stained with brother's blood
Your grisly idols, not for this may wrath
Thunder against you from the Holy One!
But o'er some plain that steameth to the sun,
Peopled with death; or where more hideous Trade 140
Loud-laughing packs his bales of human anguish;
I will raise up a mourning, O ye Fiends!
And curse your spells, that film the eye of Faith,
Hiding the present God; whose presence lost,
The moral world's cohesion, we become
An anarchy of Spirits! Toy-bewitched,

Made blind by lusts, disherited of soul,
No common centre Man, no common sire
Knoweth! A sordid solitary thing,
Mid countless brethren with a lonely heart 150
Through courts and cities the smooth savage roams
Feeling himself, his own low self the whole;
When he by sacred sympathy might make
The whole one self! Self, that no alien knows!
Self, far diffused as Fancy's wing can travel!
Self, spreading still! Oblivious of its own,
Yet all of all possessing! This is Faith!
This the Messiah's destined victory!

But first offences needs must come! Even now°
(Black Hell laughs horrible—to hear the scoff!) 160
Thee to defend, meek Galilean! Thee
And thy mild laws of Love unutterable,
Mistrust and enmity have burst the bands
Of social peace: and listening treachery lurks
With pious fraud to snare a brother's life;
And childless widows o'er the groaning land
Wail numberless; and orphans weep for bread
Thee to defend, dear Saviour of mankind!
Thee, Lamb of God! Thee, blameless Prince of peace!
From all sides rush the thirsty brood of War,— 170
Austria, and that foul Woman of the North,
The lustful murderess of her wedded lord!°
And he, connatural mind! whom (in their songs
So bards of elder time had haply feigned)
Some Fury fondled in her hate to man,
Bidding her serpent hair in mazy surge
Lick his young face, and at his mouth imbreathe
Horrible sympathy! And leagued with these
Each petty German princeling, nursed in gore!
Soul-hardened barterers of human blood! 180
Death's prime slave-merchants! Scorpion-whips of Fate!
Nor least in savagery of holy zeal,
Apt for the yoke, the race degenerate,
Whom Britain erst had blushed to call her sons!
Thee to defend the Moloch priest prefers
The prayer of hate, and bellows to the herd,

That Deity, accomplice Deity
In the fierce jealousy of wakened wrath
Will go forth with our armies and our fleets
To scatter the red ruin on their foes! 190
O blasphemy! to mingle fiendish deeds
With blessedness!

Lord of unsleeping Love,
From everlasting Thou! We shall not die.
These, even these, in mercy didst thou form,
Teachers of Good through Evil, by brief wrong
Making Truth lovely, and her future might
Magnetic o'er the fixed untrembling heart.
In the primeval age a dateless while
The vacant Shepherd wandered with his flock,
Pitching his tent where'er the green grass waved. 200
But soon Imagination conjured up
A host of new desires: with busy aim,
Each for himself, Earth's eager children toiled.
So Property began, twy-streaming fount,
Whence Vice and Virtue flow, honey and gall.
Hence the soft couch, and many-coloured robe,
The timbrel, and arch'd dome and costly feast,
With all the inventive arts, that nursed the soul
To forms of beauty, and by sensual wants
Unsensualised the mind, which in the means 210
Learnt to forget the grossness of the end,
Best pleasured with its own activity.
And hence Disease that withers manhood's arm,
The daggered Envy, spirit-quenching Want,
Warriors, and Lords, and Priests—all the sore ills
That vex and desolate our mortal life.
Wide-wasting ills! yet each the immediate source
Of mightier good. Their keen necessities
To ceaseless action goading human thought
Have made Earth's reasoning animal her Lord; 220
And the pale-featured Sage's trembling hand
Strong as a host of armed Deities,
Such as the blind Ionian fabled erst.°

From avarice thus, from luxury and war
Sprang heavenly science; and from science freedom

O'er waken'd realms Philosophers and Bards
Spread in concentric circles: they whose souls,
Conscious of their high dignities from God,
Brook not wealth's rivalry! and they who long
Enamoured with the charms of order hate 230
The unseemly disproportion: and whoe'er
Turn with mild sorrow from the victor's car
And the low puppetry of thrones, to muse
On that blest triumph, when the patriot Sage°
Called the red lightnings from the o'er-rushing cloud
And dashed the beauteous terrors on the earth
Smiling majestic. Such a phalanx ne'er
Measured firm paces to the calming sound
Of Spartan flute! These on the fated day,
When, stung to rage by pity, eloquent men 240
Have roused with pealing voice the unnumbered tribes
That toil and groan and bleed, hungry and blind,—
These hush'd awhile with patient eye serene
Shall watch the mad careering of the storm;
Then o'er the wild and wavy chaos rush
And tame the outrageous mass, with plastic might
Moulding confusion to such perfect forms,
As erst were wont,—bright visions of the day!—
To float before them, when, the summer noon,
Beneath some arch'd romantic rock reclined 250
They felt the sea breeze lift their youthful locks;
Or in the month of blossoms, at mild eve,
Wandering with desultory feet inhaled
The wafted perfumes, and the flocks and woods
And many-tinted streams and setting sun
With all his gorgeous company of clouds
Ecstatic gazed! then homeward as they strayed
Cast the sad eye to earth, and inly mused
Why there was misery in a world so fair.
Ah! far removed from all that glads the sense, 260
From all that softens or ennobles Man,
The wretched Many! Bent beneath their loads
They gape at pageant Power, nor recognise
Their cots' transmuted plunder! From the tree
Of Knowledge, ere the vernal sap had risen
Rudely disbranched! Blest Society!

Fitliest depictured by some sun-scorched waste,
Where oft majestic through the tainted noon
The Simoom sails, before whose purple pomp°
Who falls not prostrate dies! And where by night, 270
Fast by each precious fountain on green herbs
The lion couches: or hyaena dips
Deep in the lucid stream his bloody jaws;
Or serpent plants his vast moon-glittering bulk,
Caught in whose monstrous twine Behemoth yells,°
His bones loud-crashing!

O ye numberless,
Whom foul oppression's ruffian gluttony
Drives from life's plenteous feast! O thou poor wretch
Who nursed in darkness and made wild by want,
Roamest for prey, yea thy unnatural hand 280
Dost lift to deeds of blood! O pale-eyed form,
The victim of seduction, doomed to know
Polluted nights and days of blasphemy;
Who in loathed orgies with lewd wassailers
Must gaily laugh, while thy remembered home
Gnaws like a viper at thy secret heart!
O aged women! ye who weekly catch
The morsel tossed by law-forced charity,
And die so slowly, that none call it murder!
O loathly suppliants! ye, that unreceived 290
Totter heart-broken from the closing gates
Of the full Lazar-house: or, gazing, stand°
Sick with despair! O ye to glory's field
Forced or ensnared, who, as ye gasp in death,
Bleed with new wounds beneath the vulture's beak!
O thou poor widow, who in dreams dost view
Thy husband's mangled corse, and from short doze
Start'st with a shriek; or in thy half-thatched cot
Waked by the wintry night-storm, wet and cold,
Cow'rst o'er thy screaming baby! Rest awhile 300
Children of wretchedness! More groans must rise,
More blood must stream, or ere your wrongs be full.
Yet is the day of retribution nigh:
The Lamb of God hath opened the fifth seal:°

And upward rush on swiftest wing of fire
The innumerable multitude of Wrongs
By man on man inflicted! Rest awhile,
Children of wretchedness! The hour is nigh;
And lo! the great, the rich, the mighty Men,
The Kings and the chief Captains of the World, 310
With all that fixed on high like stars of Heaven
Shot baleful influence, shall be cast to earth,
Vile and down-trodden, as the untimely fruit
Shook from the fig-tree by a sudden storm.
Even now the storm begins: each gentle name,°
Faith and meek Piety, with fearful joy
Tremble far-off—for lo! the giant Frenzy
Uprooting empires with his whirlwind arm
Mocketh high Heaven; burst hideous from the cell
Where the old Hag, unconquerable, huge, 320
Creation's eyeless drudge, black ruin, sits
Nursing the impatient earthquake.

O return!
Pure Faith! meek Piety! The abhorred Form°
Whose scarlet robe was stiff with earthly pomp,
Who drank iniquity in cups of gold,
Whose names were many and all blasphemous,
Hath met the horrible judgment! Whence that cry?
The mighty army of foul Spirits shrieked
Disherited of earth! For she hath fallen
On whose black front was written Mystery; 330
She that reeled heavily, whose wine was blood;
She that worked whoredom with the Demon Power,
And from the dark embrace all evil things
Brought forth and nurtured: mitred atheism!
And patient Folly who on bended knee
Gives back the steel that stabbed him; and pale Fear
Haunted by ghastlier shapings than surround
Moon-blasted Madness when he yells at midnight!
Return pure Faith! return meek Piety!
The kingdoms of the world are yours: each heart 340
Self-governed, the vast family of Love
Raised from the common earth by common toil
Enjoy the equal produce. Such delights

As float to earth, permitted visitants!
When in some hour of solemn jubilee
The massy gates of Paradise are thrown
Wide open, and forth come in fragments wild
Sweet echoes of unearthly melodies,
And odours snatched from beds of amaranth,
And they, that from the crystal river of life 350
Spring up on freshened wing, ambrosial gales!
The favoured good man in his lonely walk
Perceives them, and his silent spirit drinks
Strange bliss which he shall recognise in heaven.
And such delights, such strange beatitudes
Seize on my young anticipating heart
When that blest future rushes on my view!
For in his own and in his Father's might
The Saviour comes! While as the Thousand Years°
Lead up their mystic dance, the Desert shouts! 360
Old Ocean claps his hands! The mighty Dead
Rise to new life, whoe'er from earliest time
With conscious zeal had urged Love's wondrous plan,
Coadjutors of God. To Milton's trump
The high groves of the renovated Earth
Unbosom their glad echoes: inly hushed,
Adoring Newton his serener eye
Raises to heaven: and he of mortal kind
Wisest, he first who marked the ideal tribes
Up the fine fibres through the sentient brain.° 370
Lo! Priestley there, patriot, and saint, and sage,
Him, full of years, from his loved native land
Statesmen blood stained and priests idolatrous
By dark lies maddening the blind multitude
Drove with vain hate. Calm, pitying he retired,
And mused expectant on these promised years.

O Years! the blest pre-eminence of Saints!
Ye sweep athwart my gaze, so heavenly bright,
The wings that veil the adoring Seraph's eyes,
What time they bend before the Jasper Throne° 380
Reflect no lovelier hues! Yet ye depart,
And all beyond is darkness! Heights most strange,
Whence Fancy falls, fluttering her idle wing.

For who of woman born may paint the hour,
When seized in his mid course, the Sun shall wane
Making noon ghastly! Who of woman born
May image in the workings of his thought,
How the black-visaged, red-eyed Fiend outstretched
Beneath the unsteady feet of Nature groans,
In feverous slumbers—destined then to wake, 390
When fiery whirlwinds thunder his dread name
And Angels shout, Destruction! How his arm
The last great Spirit lifting high in air
Shall swear by Him, the ever-living One,
Time is no more!

Believe thou, O my soul,
Life is a vision shadowy of Truth;
And vice, and anguish, and the wormy grave,
Shapes of a dream! The veiling clouds retire,
And lo! the Throne of the redeeming God
Forth flashing unimaginable day 400
Wraps in one blaze earth, heaven, and deepest hell.

Contemplant Spirits! ye that hover o'er
With untired gaze the immeasurable fount
Ebullient with creative Deity!
And ye of plastic power, that interfused
Roll through the grosser and material mass
In organizing surge! Holies of God!
(And what if Monads of the infinite mind?)
I haply journeying my immortal course
Shall sometime join your mystic choir. Till then 410
I discipline my young and novice thought
In ministeries of heart-stirring song,
And aye on Meditation's heaven-ward wing
Soaring aloft I breathe the empyreal air
Of Love, omnific, omnipresent Love,
Whose day-spring rises glorious in my soul
As the great Sun, when he his influence
Sheds on the frost-bound waters—The glad stream
Flows to the ray and warbles as it flows.

To an Infant

Ah! cease thy tears and sobs, my little Life!
I did but snatch away the unclasped knife:°
Some safer toy will soon arrest thine eye,
And to quick laughter change this peevish cry!
Poor stumbler on the rocky coast of woe,
Tutor'd by pain each source of pain to know!
Alike the foodful fruit and scorching fire
Awake thy eager grasp and young desire;
Alike the Good, the Ill offend thy sight,
And rouse the stormy sense of shrill affright! 10
Untaught, yet wise! mid all thy brief alarms
Thou closely clingest to thy Mother's arms,
Nestling thy little face in that fond breast
Whose anxious heavings lull thee to thy rest!
Man's breathing Miniature! thou mak'st me sigh—
A Babe art thou—and such a Thing am I!
To anger rapid and as soon appeased,
For trifles mourning and by trifles pleased,
Break Friendship's mirror with a tetchy blow,
Yet snatch what coals of fire on Pleasure's altar glow! 20

O thou that rearest with celestial aim
The future Seraph in my mortal frame,
Thrice holy Faith! whatever thorns I meet
As on I totter with unpractised feet,
Still let me stretch my arms and cling to thee,
Meek nurse of souls through their long infancy!

Lines°

Written at Shurton Bars, near Bridgewater, September 1795,
in Answer to a Letter from Bristol

Good verse *most* good, and bad verse then seems better
Received from absent friend by way of Letter.
For what so sweet can laboured lays impart
As one rude rhyme warm from a friendly heart?—Anon.

Nor travels my meandering eye
The starry wilderness on high;
  Nor now with curious sight
I mark the glow-worm, as I pass,
Move with 'green radiance' through the grass,
  An EMERALD of Light.

O ever present to my view!
My wafted spirit is with you,
  And soothes your boding fears:
I see you all oppressed with gloom 10
Sit lonely in that cheerless room—
  Ah me! You are in tears!

Beloved Woman! did you fly
Chilled Friendship's dark disliking eye,
  Or Mirth's untimely din?
With cruel weight these trifles press
A temper sore with tenderness,
  When aches the Void within.

But why with sable wand unblessed
Should Fancy rouse within my breast 20
  Dim-visag'd shapes of Dread?
Untenanting its beauteous clay
My SARA's soul has winged its way,
  And hovers round my head!

I felt it prompt the tender Dream,
When slowly sunk the day's last gleam;
  You roused each gentler sense
As sighing o'er the Blossom's bloom
Meek Evening wakes its soft perfume
  With viewless influence. 30

And hark, my Love! The sea-breeze moans
Through yon reft house! O'er rolling stones
  In bold ambitious sweep
The onward-surging tides supply
The silence of the cloudless sky
  With mimic thunders deep.

Dark reddening from the channell'd Isle*
(Where stands one solitary pile
  Unslated by the blast)
The Watchfire, like a sullen star 40
Twinkles to many a dozing Tar
  Rude cradled on the mast.

Even there—beneath that light-house tower—
In the tumultuous evil hour
  Ere Peace with SARA came,
Time was, I should have thought it sweet
To count the echoings of my feet,
  And watch the storm-vexed flame.

And there in black soul-jaundiced fit
A sad gloom-pampered Man to sit, 50
  And listen to the roar:
When mountain Surges bellowing deep
With an uncouth monster leap
  Plunged foaming on the shore.

Then by the Lightning's blaze to mark
Some toiling tempest-shattered bark;
  Her vain distress-guns hear;
And when a second sheet of light
Flash'd o'er the blackness of the night—
  To see *no* Vessel there! 60

But Fancy now more gaily sings;
Or if awhile she droop her wings,
  As skylarks 'mid the corn,
On summer fields she grounds her breast:
The oblivious Poppy o'er her nest
  Nods, till returning morn.

O mark those smiling tears, that swell
The opened Rose! From heaven they fell,
  And with the sun-beam blend.
Blest visitations from above, 70
Such are the tender woes of Love
  Fostering the heart, they bend!

* The Holmes in the Bristol Channel.

When stormy Midnight howling round
Beats on our roof with clattering sound,
To me your arms you'll stretch:
Great God! you'll say—To us so kind,
O shelter from this loud bleak wind
The houseless, friendless wretch!

The tears that tremble down your cheek,
Shall bathe my kisses chaste and meek 80
In Pity's dew divine;
And from your heart the sighs that steal
Shall make your rising bosom feel
The answering swell of mine!

How oft, my Love! with shapings sweet
I paint the moment, we shall meet!
With eager speed I dart—
I seize you in the vacant air,
And fancy, with a Husband's care
I press you to my heart! 90

'Tis said, in Summer's evening hour
Flashes the golden-coloured flower
A fair electric flame:
And so shall flash my love-charged eye
When all the heart's big ecstasy
Shoots rapid through the frame!

The Eolian Harp°

Composed at Clevedon, Somersetshire

My pensive Sara! thy soft cheek reclined
Thus on mine arm, most soothing sweet it is
To sit beside our cot, our cot o'ergrown
With white-flowered jasmin, and the broad-leav'd myrtle,
(Meet emblems they of Innocence and Love!)
And watch the clouds, that late were rich with light,
Slow saddening round, and mark the star of eve
Serenely brilliant (such should wisdom be)

Shine opposite! How exquisite the scents
Snatched from yon bean-field! and the world so hushed! 10
The stilly murmur of the distant sea
Tells us of silence.

And that simplest lute,
Placed length-ways in the clasping casement, hark!
How by the desultory breeze caressed,
Like some coy maid half yielding to her lover,
It pours such sweet upbraiding, as must needs
Tempt to repeat the wrong! And now, its strings
Boldlier swept, the long sequacious notes
Over delicious surges sink and rise,
Such a soft floating witchery of sound 20
As twilight Elfins make, when they at eve
Voyage on gentle gales from Fairy-Land,
Where Melodies round honey-dropping flowers,
Footless and wild, like birds of Paradise,
Nor pause, nor perch, hovering on untamed wing!
O the one life within us and abroad,
Which meets all motion and becomes its soul,
A light in sound, a sound-like power in light,
Rhythm in all thought, and joyance every where—
Methinks, it should have been impossible 30
Not to love all things in a world so filled;
Where the breeze warbles, and the mute still air
Is Music slumbering on her instrument.

And thus, my love! as on the midway slope
Of yonder hill I stretch my limbs at noon,
Whilst through my half-closed eye-lids I behold
The sunbeams dance, like diamonds, on the main,
And tranquil muse upon tranquillity;
Full many a thought uncall'd and undetain'd,
And many idle flitting phantasies, 40
Traverse my indolent and passive brain,
As wild and various as the random gales
That swell and flutter on this subject lute!

And what if all of animated nature
Be but organic harps diversely framed,

That tremble into thought, as o'er them sweeps
Plastic and vast, one intellectual breeze,
At once the Soul of each, and God of All?

But thy more serious eye a mild reproof
Darts, O beloved woman! nor such thoughts 50
Dim and unhallowed dost thou not reject,
And biddest me walk humbly with my God.
Meek Daughter in the family of Christ!
Well hast thou said and holily dispraised
These shapings of the unregenerate mind;
Bubbles that glitter as they rise and break
On vain Philosophy's aye-babbling spring.
For never guiltless may I speak of him,
The Incomprehensible! save when with awe
I praise him, and with Faith that inly feels; 60
Who with his saving mercies healed me,
A sinful and most miserable man,
Wildered and dark, and gave me to possess
Peace, and this cot, and thee, heart-honour'd Maid!

Reflections on Having Left a Place of Retirement

Sermoni propriora [fitter for discourse].—Hor.

Low was our pretty Cot: our tallest rose
Peeped at the chamber-window. We could hear
At silent noon, and eve, and early morn,
The sea's faint murmur. In the open air
Our myrtles blossomed; and across the porch
Thick jasmins twined: the little landscape round
Was green and woody, and refreshed the eye.
It was a spot which you might aptly call
The Valley of Seclusion! Once I saw
(Hallowing his Sabbath-day by quietness) 10
A wealthy son of commerce saunter by,
Bristowa's citizen: methought, it calmed°
His thirst of idle gold, and made him muse
With wiser feelings: for he paused, and looked
With a pleased sadness, and gazed all around,

Then eyed our Cottage, and gazed round again,
And sighed, and said, it was a Blessed Place.
And we were blessed. Oft with patient ear
Long-listening to the viewless sky-lark's note
(Viewless, or haply for a moment seen 20
Gleaming on sunny wings) in whispered tones
I've said to my beloved, 'Such, sweet girl!
The inobtrusive song of happiness,
Unearthly minstrelsy! then only heard
When the soul seeks to hear; when all is hushed,
And the heart listens!'

But the time, when first
From that low dell, steep up the stony mount
I climbed with perilous toil and reached the top,
Oh! what a goodly scene! Here the bleak mount,
The bare bleak mountain speckled thin with sheep; 30
Grey clouds, that shadowing spot the sunny fields;
And river, now with bushy rocks o'er-browed,
Now winding bright and full, with naked banks;
And seats, and lawns, the Abbey and the wood,
And cots, and hamlets, and faint city-spire;
The Channel there, the Islands and white sails,
Dim coasts, and cloud-like hills, and shoreless Ocean—
It seemed like Omnipresence! God, methought,
Had built him there a temple: the whole World
Seemed imaged in its vast circumference, 40
No wish profaned my overwhelmed heart.
Blest hour! It was a luxury,—to be!

Ah! quiet dell! dear cot, and mount sublime!
I was constrained to quit you. Was it right,
While my unnumbered brethren toiled and bled,
That I should dream away the entrusted hours
On rose-leaf beds, pampering the coward heart
With feelings all too delicate for use?
Sweet is the tear that from some Howard's eye°
Drops on the cheek of one he lifts from earth: 50
And he that works me good with unmoved face,
Does it but half: he chills me while he aids,
My benefactor, not my brother man!

Yet even this, this cold beneficence
Praise, praise it, O my Soul! oft as thou scann'st
The sluggard Pity's vision-weaving tribe!
Who sigh for wretchedness, yet shun the wretched,
Nursing in some delicious solitude
Their slothful loves and dainty sympathies!
I therefore go, and join head, heart, and hand, 60
Active and firm, to fight the bloodless fight
Of science, freedom, and the truth in Christ.

Yet oft when after honourable toil
Rests the tired mind, and waking loves to dream,
My spirit shall revisit thee, dear Cot!
Thy jasmin and thy window-peeping rose,
And myrtles fearless of the mild sea-air.
And I shall sigh fond wishes—sweet abode!
Ah!—had none greater! And that all had such!
It might be so—but the time is not yet. 70
Speed it, O Father! Let thy kingdom come!

Ode to the Departing Year

Ἰοὺ, ἰού, ὢ ὢ κακά.
Ὑπ' αὖ με δεινὸς ὀρθομαντείας πόνος
Στροβεῖ, ταράσσων φροιμίοις ἐφημίοις
. . . . . .
Τὸ μέλλον ἥξει. Καὶ σύ μ' ἐν τάχει παρὼν
Ἄγαν ἀληθόμαντιν οἰκτείρας ἐρεῖς.

[(Cassandra speaks:) Ha, ha! Oh, oh, the agony! Once more the dreadful throes of true prophecy whirl and distract me with their ill-boding onset. . . . What is to come, will come. Soon thou, present here thyself, shalt of thy pity pronounce me all too true a prophetess. (LCL)]—Aeschyl. *Agam.*

ARGUMENT

The Ode commences with an address to the Divine Providence, that regulates into one vast harmony all the events of time, however calamitous some of them may appear to mortals. The second Strophe calls on men to suspend their private joys and sorrows, and devote them for a while to the cause of human nature in general. The first Epode speaks of the Empress of Russia, who died of an apoplexy on the 17th of November 1796; having just concluded a subsidiary treaty with the Kings combined against France. The first and second Antistrophe describe the Image of the Departing Year, etc., as in a vision. The second Epode prophesies, in anguish of spirit, the downfall of this country.

I

Spirit who sweepest the wild harp of Time!
It is most hard, with an untroubled ear
Thy dark inwoven harmonies to hear!
Yet, mine eye fixed on Heaven's unchanging clime
Long had I listened, free from mortal fear,
With inward stillness, and bowed mind;
When lo! its folds far waving on the wind,
I saw the train of the departing Year!
Starting from my silent sadness
Then with no unholy madness, 10
Ere yet the entered cloud foreclosed my sight,
I raised the impetuous song, and solemnized his flight.

II

Hither, from the recent tomb,
From the prison's direr gloom,
From distemper's midnight anguish;
And thence, where poverty doth waste and languish;
Or where, his two bright torches blending,
Love illumines manhood's maze;
Or where o'er cradled infants bending
Hope has fixed her wishful gaze; 20
Hither, in perplexed dance,
Ye Woes! ye young-eyed Joys! advance!

By Time's wild harp, and by the hand
Whose indefatigable sweep
Raises its fateful strings from sleep,
I bid you haste, a mix'd tumultuous band!
From every private bower,
And each domestic hearth,
Haste for one solemn hour;
And with a loud and yet a louder voice, 30
O'er Nature struggling in portentous birth,
Weep and rejoice!

Still echoes the dread name that o'er the earth
Let slip the storm, and woke the brood of Hell:
    And now advance in stately jubilee
  Justice and Truth! They too have heard thy spell,
    They too obey thy name, divinest Liberty!

III

I marked Ambition in his war-array!
    I heard the mailed Monarch's troublous cry—
'Ah! wherefore does the Northern Conqueress stay!° 40
Groans not her chariot on its onward way?'
      Fly, mailed Monarch, fly!
    Stunned by Death's twice mortal mace,
    No more on murder's lurid face
The insatiate hag shall gloat with drunken eye!
    Manes of the unnumbered slain!
    Ye that gasped on Warsaw's plain!
  Ye that erst at Ismail's tower,
When human ruin choked the streams,
  Fell in conquest's glutted hour, 50
Mid women's shrieks and infants' screams!
  Spirits of the uncoffin'd slain,
    Sudden blasts of triumph swelling,
  Oft, at night, in misty train,
    Rush around her narrow dwelling!
  The exterminating fiend is fled—
    (Foul her life, and dark her doom)
  Mighty armies of the dead
    Dance, like death-fires, round her tomb!
  Then with prophetic song relate, 60
  Each some tyrant-murderer's fate!

IV

Departing Year! 'twas on no earthly shore
  My soul beheld thy vision! Where alone,
  Voiceless and stern, before the cloudy throne,
Aye Memory sits: thy robe inscribed with gore,
With many an unimaginable groan
  Thou storied'st thy sad hours! Silence ensued,
  Deep silence o'er the ethereal multitude,
Whose locks with wreaths, whose wreaths with glories shone.

Then, his eye wild ardours glancing, 70
From the choired gods advancing,
The Spirit of the Earth made reverence meet,
And stood up, beautiful, before the cloudy seat.

V

Throughout the blissful throng,
Hush'd were harp and song:
Till wheeling round the throne the Lampads seven,°
(The mystic Words of Heaven)
Permissive signal make:
The fervent Spirit bow'd, then spread his wings and spake!
'Thou in stormy blackness throning 80
Love and uncreated Light,
By the Earth's unsolaced groaning,
Seize thy terrors, Arm of might!
By peace with proffer'd insult scared,
Masked hate and envying scorn!
By years of havoc yet unborn!
And hunger's bosom to the frost-winds bared!
But chief by Afric's wrongs,
Strange, horrible, and foul!
By what deep guilt belongs 90
To the deaf Synod, "full of gifts and lies!"
By wealth's insensate laugh! by torture's howl!
Avenger, rise!
For ever shall the thankless Island scowl,
Her quiver full, and with unbroken bow?
Speak! from thy storm-black Heaven O speak aloud!
And on the darkling foe
Open thine eye of fire from some uncertain cloud!
O dart the flash! O rise and deal the blow!
The Past to thee, to thee the Future cries! 100
Hark! how wide Nature joins her groans below!
Rise, God of Nature! rise.'

VI

The voice had ceased, the vision fled;
Yet still I gasped and reeled with dread.
And ever, when the dream of night
Renews the phantom to my sight,

Cold sweat-drops gather on my limbs;
My ears throb hot; my eye-balls start;
My brain with horrid tumult swims;
Wild is the tempest of my heart; 110
And my thick and struggling breath
Imitates the toil of death!
No stranger agony confounds
The soldier on the war-field spread,
When all foredone with toil and wounds,
Death-like he dozes among heaps of dead!
(The strife is o'er, the daylight fled,
And the night-wind clamours hoarse!
See! the starting wretch's head
Lies pillowed on a brother's corse!) 120

VII

Not yet enslaved, not wholly vile,
O Albion! O my mother Isle!
Thy valleys, fair as Eden's bowers,
Glitter green with sunny showers;
Thy grassy uplands' gentle swells
Echo to the bleat of flocks;
(Those grassy hills, those glittering dells
Proudly ramparted with rocks)
And Ocean mid his uproar wild
Speaks safety to his island-child! 130
Hence for many a fearless age
Has social Quiet lov'd thy shore;
Nor ever proud Invader's rage
Or sack'd thy towers, or stain'd thy fields with gore.

VIII

Abandoned of Heaven! mad avarice thy guide,
At cowardly distance, yet kindling with pride—
Mid thy herds and thy corn-fields secure hast thou stood,
And join'd the wild yelling of famine and blood!
The nations curse thee! They with eager wondering
Shall hear Destruction, like a vulture, scream! 140
Strange-eyed Destruction! who with many a dream
Of central fires through nether seas up-thundering
Soothes her fierce solitude; yet as she lies

By livid fount, or red volcanic stream,
  If ever to her lidless dragon-eyes,
  O Albion! thy predestined ruins rise,
The fiend-hag on her perilous couch doth leap,
Muttering distemper'd triumph in her charmed sleep.

IX

Away, my soul, away!
    In vain, in vain the birds of warning sing— 150
And hark! I hear the famished brood of prey
Flap their lank pennons on the groaning wind!
    Away, my soul, away!
    I unpartaking of the evil thing,
      With daily prayer and daily toil
      Soliciting for food my scanty soil,
    Have wailed my country with a loud Lament.
Now I recentre my immortal mind
    In the deep sabbath of meek self-content;
Cleansed from the vaporous passions that bedim 160
God's Image, sister of the Seraphim.

To the Rev. George Coleridge°

Of Ottery St. Mary, Devon

With some Poems

Notus in fratres animi paterni [known for his fatherly spirit towards his brothers].
—Hor. *Carm.* lib. II. 2.

A blessed lot hath he, who having passed
His youth and early manhood in the stir
And turmoil of the world, retreats at length,
With cares that move, not agitate the heart,
To the same dwelling where his father dwelt;
And haply views his tottering little ones
Embrace those aged knees and climb that lap,
On which first kneeling his own infancy
Lisped its brief prayer. Such, O my earliest Friend!
Thy lot, and such thy brothers too enjoy. 10

At distance did ye climb life's upland road,
Yet cheered and cheering: now fraternal love
Hath drawn you to one centre. Be your days
Holy, and blest and blessing may ye live!

To me the Eternal Wisdom hath dispensed
A different fortune and more different mind—
Me from the spot where first I sprang to light
Too soon transplanted, ere my soul had fixed
Its first domestic loves; and hence through life
Chasing chance-started friendships. A brief while 20
Some have preserved me from life's pelting ills;
But, like a tree with leaves of feeble stem,
If the clouds lasted, and a sudden breeze
Ruffled the boughs, they on my head at once
Dropped the collected shower; and some most false,
False and fair foliaged as the Manchineel,°
Have tempted me to slumber in their shade
E'en mid the storm; then breathing subtlest damps,
Mixed their own venom with the rain from Heaven,
That I woke poisoned! But, all praise to Him 30
Who gives us all things, more have yielded me
Permanent shelter; and beside one friend,
Beneath the impervious covert of one oak,
I've raised a lowly shed, and know the names
Of husband and of father; not unhearing
Of that divine and nightly-whispering voice,
Which from my childhood to maturer years
Spake to me of predestinated wreaths,
Bright with no fading colours!

Yet at times
My soul is sad, that I have roamed through life 40
Still most a stranger, most with naked heart
At mine own home and birth-place: chiefly then,
When I remember thee, my earliest friend!
Thee, who didst watch my boyhood and my youth;
Didst trace my wanderings with a father's eye;
And boding evil yet still hoping good,
Rebuk'd each fault, and over all my woes
Sorrowed in silence! He who counts alone

The beatings of the solitary heart,
That being knows, how I have loved thee ever, 50
Loved as a brother, as a son revered thee!
Oh! 'tis to me an ever new delight,
To talk of thee and thine: or when the blast
Of the shrill winter, rattling our rude sash,
Endears the cleanly hearth and social bowl;
Or when as now, on some delicious eve,
We in our sweet sequestered orchard-plot
Sit on the tree crooked earth-ward; whose old boughs,
That hang above us in an arborous roof,
Stirred by the faint gale of departing May, 60
Send their loose blossoms slanting o'er our heads!

Nor dost not thou sometimes recall those hours,
When with the joy of hope thou gav'st thine ear
To my wild firstling-lays. Since then my song
Hath sounded deeper notes, such as beseem
Or that sad wisdom folly leaves behind,
Or such as, tuned to these tumultuous times,
Cope with the tempest's swell!

These various strains,
Which I have framed in many a various mood,
Accept, my brother! and (for some perchance 70
Will strike discordant on thy milder mind)
If aught of error or intemperate truth
Should meet thine ear, think thou that riper age
Will calm it down, and let thy love forgive it!

This Lime-Tree Bower My Prison

[Addressed to Charles Lamb, of the India House, London]

In the June of 1797 some long-expected Friends paid a visit to the author's cottage; and on the morning of their arrival, he met with an accident, which disabled him from walking during the whole time of their stay. One evening, when they had left him for a few hours, he composed the following lines in the garden-bower.

Well, they are gone, and here must I remain,
This lime-tree bower my prison! I have lost
Beauties and feelings, such as would have been
Most sweet to my remembrance even when age
Had dimmed mine eyes to blindness! They, meanwhile,
Friends, whom I never more may meet again,
On springy heath, along the hill-top edge,
Wander in gladness, and wind down, perchance,
To that still roaring dell, of which I told;
The roaring dell, o'erwooded, narrow, deep, 10
And only speckled by the mid-day sun;
Where its slim trunk the ash from rock to rock
Flings arching like a bridge;—that branchless ash,
Unsunned and damp, whose few poor yellow leaves
Ne'er tremble in the gale, yet tremble still,
Fanned by the water-fall! and there my friends
Behold the dark green file of long lank weeds,
That all at once (a most fantastic sight!)
Still nod and drip beneath the dripping edge
Of the blue clay-stone.

Now, my friends emerge 20
Beneath the wide wide Heaven—and view again
The many-steepled tract magnificent
Of hilly fields and meadows, and the sea,
With some fair bark, perhaps, whose sails light up
The slip of smooth clear blue betwixt two Isles
Of purple shadow! Yes! they wander on
In gladness all; but thou, methinks, most glad,
My gentle-hearted Charles! for thou hast pined
And hungered after Nature, many a year,
In the great City pent, winning thy way 30
With sad yet patient soul, through evil and pain
And strange calamity! Ah! slowly sink
Behind the western ridge, thou glorious sun!
Shine in the slant beams of the sinking orb,
Ye purple heath-flowers! richlier burn, ye clouds!
Live in the yellow light, ye distant groves!
And kindle, thou blue ocean! So my Friend
Struck with deep joy may stand, as I have stood,
Silent with swimming sense; yea, gazing round

On the wide landscape, gaze till all doth seem 40
Less gross than bodily; and of such hues
As veil the Almighty Spirit, when yet he makes
Spirits perceive his presence.

A delight
Comes sudden on my heart, and I am glad
As I myself were there! Nor in this bower,
This little lime-tree bower, have I not marked
Much that has soothed me. Pale beneath the blaze
Hung the transparent foliage; and I watch'd
Some broad and sunny leaf, and loved to see
The shadow of the leaf and stem above 50
Dappling its sunshine! And that walnut-tree
Was richly tinged, and a deep radiance lay
Full on the ancient ivy, which usurps
Those fronting elms, and now, with blackest mass
Makes their dark branches gleam a lighter hue
Through the late twilight: and though now the bat
Wheels silent by, and not a swallow twitters,
Yet still the solitary humble bee
Sings in the bean-flower! Henceforth I shall know
That Nature ne'er deserts the wise and pure; 60
No plot so narrow, be but Nature there,
No waste so vacant, but may well employ
Each faculty of sense, and keep the heart
Awake to Love and Beauty! and sometimes
'Tis well to be bereft of promised good,
That we may lift the Soul, and contemplate
With lively joy the joys we cannot share.
My gentle-hearted Charles! when the last rook
Beat its straight path along the dusky air
Homewards, I blest it! deeming, its black wing 70
(Now a dim speck, now vanishing in light)
Had cross'd the mighty orb's dilated glory,
While thou stood'st gazing; or when all was still,
Flew creeking o'er thy head, and had a charm
For thee, my gentle-hearted Charles, to whom
No sound is dissonant which tells of Life.

The Wanderings of Cain

PREFATORY NOTE

A prose composition, one not in metre at least, seems *prima facie* [at first sight] to require explanation or apology. It was written in the year 1798, near Nether Stowey, in Somersetshire, at which place (*sanctum et amabile nomen* [holy and beloved name]! rich by so many associations and recollections) the author had taken up his residence in order to enjoy the society and close neighbourhood of a dear and honoured friend, T. Poole, Esq. The work was to have been written in concert with another [Wordsworth], whose name is too venerable within the precincts of genius to be unnecessarily brought into connection with such a trifle, and who was then residing at a small distance from Nether Stowey. The title and subject were suggested by myself, who likewise drew out the scheme and the contents for each of the three books or cantos, of which the work was to consist, and which, the reader is to be informed, was to have been finished in one night! My partner undertook the first canto: I the second: and which ever had *done first*, was to set about the third. Almost thirty years have passed by; yet at this moment I cannot without something more than a smile moot the question which of the two things was the more impracticable, for a mind so eminently original to compose another man's thoughts and fancies, or for a taste so austerely pure and simple to imitate the Death of Abel?° Methinks I see his grand and noble countenance as at the moment when having despatched my own portion of the task at full finger-speed, I hastened to him with my manuscript—that look of humorous despondency fixed on his almost blank sheet of paper, and then its silent mock-piteous admission of failure struggling with the sense of the exceeding ridiculousness of the whole scheme—which broke up in a laugh: and the Ancient Mariner was written instead.

Years afterward, however, the draft of the plan and proposed incidents, and the portion executed, obtained favour in the eyes of more than one person, whose judgment on a poetic work could not but have weighed with me, even though no parental partiality had been thrown into the same scale, as a make-weight: and I determined on commencing anew, and composing the whole in stanzas, and made some progress in realising this intention, when adverse gales drove my bark off the 'Fortunate Isles' of the Muses: and then other and more momentous interests prompted a different voyage, to firmer anchorage and a securer port. I have in vain tried to recover the lines from the palimpsest tablet of my memory: and I can only offer the introductory stanza, which had been committed to writing for the purpose of procuring a friend's judgment on the metre, as a specimen:—

Encinctured with a twine of leaves,
That leafy twine his only dress!
A lovely Boy was plucking fruits,
By moonlight, in a wilderness.

The moon was bright, the air was free,
And fruits and flowers together grew
On many a shrub and many a tree:
And all put on a gentle hue,
Hanging in the shadowy air
Like a picture rich and rare.
It was a climate where, they say,
The night is more belov'd than day.
But who that beauteous Boy beguil'd,
That beauteous Boy to linger here?
Alone, by night, a little child,
In place so silent and so wild—
Has he no friend, no loving mother near?

I have here given the birth, parentage, and premature decease of the 'Wanderings of Cain, a poem',—intreating, however, my Readers, not to think so meanly of my judgment as to suppose that I either regard or offer it as any excuse for the publication of the following fragment (and I may add, of one or two others in its neighbourhood) in its primitive crudity. But I should find still greater difficulty in forgiving myself were I to record pro *taedio* publico [for public *boredom*] a set of petty mishaps and annoyances which I myself wish to forget. I must be content therefore with assuring the friendly Reader, that the less he attributes its appearance to the Author's will, choice, or judgment, the nearer to the truth he will be.

S. T. COLERIDGE (1828)

THE WANDERINGS OF CAIN

CANTO II

'A little further, O my father, yet a little further, and we shall come into the open moonlight.' Their road was through a forest of fir-trees; at its entrance the trees stood at distances from each other, and the path was broad, and the moonlight and the moonlight shadows reposed upon it, and appeared quietly to inhabit that solitude. But soon the path winded and became narrow; the sun at high noon sometimes speckled, but never illuminated it, and now it was dark as a cavern.

'It is dark, O my father!' said Enos, 'but the path under our feet is smooth and soft, and we shall soon come out into the open moonlight.'

'Lead on, my child!' said Cain; 'guide me, little child!' And the innocent little child clasped a finger of the hand which had murdered the righteous Abel, and he guided his father. 'The fir branches drip upon thee, my son.' 'Yea, pleasantly, father, for I ran fast and eagerly to bring thee the pitcher and the cake, and my body is not yet cool.

How happy the squirrels are that feed on these fir-trees! they leap from bough to bough, and the old squirrels play round their young ones in the nest. I clomb a tree yesterday at noon, O my father, that I might play with them, but they leaped away from the branches, even to the slender twigs did they leap, and in a moment I beheld them on another tree. Why, O my father, would they not play with me? I would be good to them as thou art good to me: and I groaned to them even as thou groanest when thou givest me to eat, and when thou coverest me at evening, and as often as I stand at thy knee and thine eyes look at me?' Then Cain stopped, and stifling his groans he sank to the earth, and the child Enos stood in the darkness beside him.

And Cain lifted up his voice and cried bitterly, and said, 'The Mighty One that persecuteth me is on this side and on that; he pursueth my soul like the wind, like the sand-blast he passeth through me; he is around me even as the air! O that I might be utterly no more! I desire to die—yea, the things that never had life, neither move they upon the earth—behold! they seem precious to mine eyes. O that a man might live without the breath of his nostrils. So I might abide in darkness, and blackness, and an empty space! Yea, I would lie down, I would not rise, neither would I stir my limbs till I became as the rock in the den of the lion, on which the young lion resteth his head whilst he sleepeth. For the torrent that roareth far off hath a voice: and the clouds in heaven look terribly on me; the Mighty One who is against me speaketh in the wind of the cedar grove; and in silence am I dried up.' Then Enos spake to his father, 'Arise, my father, arise, we are but a little way from the place where I found the cake and the pitcher.' And Cain said, 'How knowest thou?' and the child answered—'Behold the bare rocks are a few of thy strides distant from the forest; and while even now thou wert lifting up thy voice, I heard the echo.' Then the child took hold of his father, as if he would raise him: and Cain being faint and feeble rose slowly on his knees and pressed himself against the trunk of a fir, and stood upright and followed the child.

The path was dark till within three strides' length of its termination, when it turned suddenly; the thick black trees formed a low arch, and the moonlight appeared for a moment like a dazzling portal. Enos ran before and stood in the open air; and when Cain, his father, emerged from the darkness, the child was affrighted. For the mighty limbs of Cain were wasted as by fire; his hair was as the matted curls on the bison's forehead, and so glared his fierce and sullen eye beneath: and the black abundant locks on either side, a rank and tangled mass, were stained and scorched, as though the grasp of a burning iron hand had striven to rend them; and his countenance told in a strange and terrible language of agonies that had been, and were, and were still to continue to be.

The scene around was desolate; as far as the eye could reach it was desolate: the bare rocks faced each other, and left a long and wide interval of thin white sand. You might wander on and look round and round, and peep into the crevices of the rocks and discover nothing that acknowledged the influence of the seasons. There was no spring, no summer, no autumn: and the winter's snow, that would have been lovely, fell not on these hot rocks and scorching sands. Never morning lark had poised himself over this desert; but the huge serpent often hissed there beneath the talons of the vulture, and the vulture screamed, his wings imprisoned within the coils of the serpent. The pointed and shattered summits of the ridges of the rocks made a rude mimicry of human concerns, and seemed to prophesy mutely of things that then were not; steeples, and battlements, and ships with naked masts. As far from the wood as a boy might sling a pebble of the brook, there was one rock by itself at a small distance from the main ridge. It had been precipitated there perhaps by the groan which the Earth uttered when our first father fell. Before you approached, it appeared to lie flat on the ground, but its base slanted from its point, and between its point and the sands a tall man might stand upright. It was here that Enos had found the pitcher and cake, and to this place he led his father. But ere they had reached the rock they beheld a human shape: his back was towards them, and they were advancing unperceived, when they heard him smite his breast and cry aloud, 'Woe is me! woe is me! I must never die again, and yet I am perishing with thirst and hunger.'

Pallid, as the reflection of the sheeted lightning on the heavy-sailing night-cloud, became the face of Cain; but the child Enos took hold of the shaggy skin, his father's robe, and raised his eyes to his father, and listening whispered, 'Ere yet I could speak, I am sure, O my father, that I heard that voice. Have not I often said that I remembered a sweet voice? O my father! this is it': and Cain trembled exceedingly. The voice was sweet indeed, but it was thin and querulous, like that of a feeble slave in misery, who despairs altogether, yet can not refrain himself from weeping and lamentation. And, behold! Enos glided forward, and creeping softly round the base of the rock, stood before the stranger, and looked up into his face. And the Shape shrieked, and turned round, and Cain beheld him, that his limbs and his face were those of his brother Abel whom he had killed! And Cain stood like one who struggles in his sleep because of the exceeding terribleness of a dream.

Thus as he stood in silence and darkness of soul, the Shape fell at his feet, and embraced his knees, and cried out with a bitter outcry, 'Thou eldest born of Adam, whom Eve, my mother, brought forth, cease to torment me! I was feeding my flocks in green pastures by the side of quiet rivers, and thou killedst me; and now I am in misery.' Then Cain closed his eyes, and hid them with his hands; and again he opened his eyes, and looked around him, and said to Enos, 'What beholdest thou? Didst thou hear a voice, my son?' 'Yes, my father, I beheld a man in unclean garments, and he uttered a sweet voice, full of lamentation.' Then Cain raised up the Shape that was like Abel, and said:—'The Creator of our father, who had respect unto thee, and unto thy offering, wherefore hath he forsaken thee?' Then the Shape shrieked a second time, and rent his garment, and his naked skin was like the white sands beneath their feet; and he shrieked yet a third time, and threw himself on his face upon the sand that was black with the shadow of the rock, and Cain and Enos sate beside him; the child by his right hand, and Cain by his left. They were all three under the rock, and within the shadow. The Shape that was like Abel raised himself up, and spake to the child: 'I know where the cold waters are, but I may not drink, wherefore didst thou then take away my pitcher?' But Cain said, 'Didst thou not find favour in the sight of the Lord thy God?' The Shape answered, 'The Lord is God of the living only, the dead have another God.' Then the child Enos lifted up his eyes and prayed; but Cain rejoiced secretly in his heart. 'Wretched shall they be all the days of their mortal life,' exclaimed the Shape, 'who sacrifice worthy and acceptable sacrifices to the God of the dead; but after death their toil ceaseth. Woe is me, for I was well beloved by the God of the living, and cruel wert thou, O my brother, who didst snatch me away from his power and his dominion.' Having uttered these words, he rose suddenly, and fled over the sands: and Cain said in his heart, 'The curse of the Lord is on me; but who is the God of the dead?' and he ran after the Shape, and the Shape fled shrieking over the sands, and the sands rose like white mists behind the steps of Cain, but the feet of him that was like Abel disturbed not the sands. He greatly outrun Cain, and turning short, he wheeled round, and came again to the rock where they had been sitting, and where Enos still stood; and the child caught hold of his garment as he passed by, and he fell upon the ground. And Cain stopped, and beholding him not, said, 'he has passed into the dark woods,' and he walked slowly back to the rocks; and when he reached it the child told him that he had caught hold of his garment as he passed by, and that the man had fallen upon the ground: and Cain once more sate beside him, and said, 'Abel, my brother, I would lament for thee, but that the spirit within me is withered, and burnt up with extreme agony. Now, I pray thee, by thy flocks, and by thy pastures, and by the quiet rivers which thou lovedst, that thou tell me all that thou knowest. Who is the God of the dead? where doth he make his dwelling? what sacrifices are acceptable unto him? for I have offered, but have not been received; I have prayed, and have not been heard; and how can I be afflicted more than I already am?' The Shape arose and answered, 'O that thou hadst had pity on me as I will have pity on thee. Follow me, Son of Adam! and bring thy child with thee!'

And they three passed over the white sands between the rocks, silent as the shadows.

The Rime of the Ancient Mariner°

In seven Parts

Facile credo, plures esse Naturas invisibiles quam visibiles in rerum universitate. Sed horum omnium familiam quis nobis enarrabit, et gradus et cognationes et discrimina et singulorum munera? Quid agunt? quae loca habitant? Harum rerum notitiam semper ambivit ingenium humanum, nunquam attigit. Juvat, interea, non diffiteor, quandoque in animo, tanquam in tabulâ, majoris et melioris mundi imaginem contemplari: ne mens assuefacta hodiernae vitae minutiis se contrahat nimis, et tota subsidat in pusillas cogitationes. Sed veritati interea invigilandum est, modusque servandus, ut certa ab incertis, diem a nocte, distinguamus. T. Burnet, *Archaeol. Phil.* p. 68.

[I can easily believe that there are more invisible creatures in the universe than visible ones. But who will tell us what family each belongs to, what their ranks and relationships are, and what their respective distinguishing characters may be? What do they do? Where do they live? Human wit has always circled around a knowledge of these things without ever attaining it. But I do not doubt that it is beneficial sometimes to contemplate in the mind, as in a picture, the image of a grander and better world; for if the mind grows used to the trivia of daily life, it may dwindle too much and decline altogether into worthless thoughts. Meanwhile, however, we must be on the watch for the truth, keeping a sense of proportion so that we can tell what is certain from what is uncertain and day from night.]

PART I

An ancient Mariner meeteth three gallants bidden to a wedding-feast, and detaineth one.

It is an ancient Mariner,
And he stoppeth one of three.
'By thy long grey beard and glittering eye,
Now wherefore stopp'st thou me?

The Bridegroom's doors are opened wide,
And I am next of kin;
The guests are met, the feast is set:
May'st hear the merry din.'

He holds him with his skinny hand,
'There was a ship,' quoth he. 10
'Hold off! unhand me, grey-beard loon!'
Eftsoons his hand dropt he.

The wedding guest is spellbound by the eye of the old seafaring man, and constrained to hear his tale.

He holds him with his glittering eye—
The wedding-guest stood still,
And listens like a three years' child:
The Mariner hath his will.

The wedding-guest sat on a stone:
He cannot choose but hear;
And thus spake on that ancient man,
The bright-eyed Mariner. 20

'The ship was cheered, the harbour cleared,
Merrily did we drop
Below the kirk, below the hill,
Below the lighthouse top.

The Mariner tells how the ship sailed southward with a good wind and fair weather, till it reached the line.

The sun came up upon the left,
Out of the sea came he!
And he shone bright, and on the right
Went down into the sea.

Higher and higher every day,
Till over the mast at noon—' 30
The Wedding-Guest here beat his breast,
For he heard the loud bassoon.

The wedding guest heareth the bridal music; but the mariner continueth his tale.

The bride hath paced into the hall,
Red as a rose is she;
Nodding their heads before her goes
The merry minstrelsy.

The Wedding-Guest he beat his breast,
Yet he cannot choose but hear;
And thus spake on that ancient man,
The bright-eyed Mariner. 40

The ship drawn by a storm toward the south pole.

And now the storm-blast came, and he
Was tyrannous and strong:
He struck with his o'ertaking wings,
And chased us south along.

With sloping masts and dipping prow,
As who pursued with yell and blow
Still treads the shadow of his foe,
And forward bends his head,
The ship drove fast, loud roared the blast,
And southward aye we fled. 50

And now there came both mist and snow,
And it grew wondrous cold:
And ice, mast-high, came floating by,
As green as emerald.

The land of ice, and of fearful sounds where no living thing was to be seen.

And through the drifts the snowy clifts
Did send a dismal sheen:
Nor shapes of men nor beasts we ken—
The ice was all between.

The ice was here, the ice was there,
The ice was all around: 60
It cracked and growled, and roared and howled,
Like noises in a swound!

Till a great sea-bird, called the Albatross, came through the snow-fog, and was received with great joy and hospitality.

At length did cross an Albatross,
Thorough the fog it came;
As if it had been a Christian soul,
We hailed it in God's name.

It ate the food it ne'er had eat,
And round and round it flew.
The ice did split with a thunder-fit;
The helmsman steered us through! 70

And lo! the Albatross proveth a bird of good omen, and followeth the ship as it returned northward through fog and floating ice.

And a good south wind sprung up behind;
The Albatross did follow,
And every day, for food or play,
Came to the mariner's hollo!

In mist or cloud, on mast or shroud,
It perched for vespers nine;
Whiles all the night, through fog-smoke white,
Glimmered the white moon-shine.

The ancient Mariner inhospitably killeth the pious bird of good omen.

'God save thee, ancient Mariner!
From the fiends, that plague thee thus!— 80
Why look'st thou so?'—With my cross-bow
I shot the Albatross.

PART II

The Sun now rose upon the right:
Out of the sea came he,
Still hid in mist, and on the left
Went down into the sea.

And the good south wind still blew behind,
But no sweet bird did follow,
Nor any day for food or play
Came to the mariners' hollo! 90

His shipmates cry out against the ancient Mariner, for killing the bird of good luck.

And I had done a hellish thing,
And it would work 'em woe:
For all averred, I had killed the bird
That made the breeze to blow.
Ah wretch! said they, the bird to slay,
That made the breeze to blow!

But when the fog cleared off, they justify the same, and thus make themselves accomplices in the crime.

Nor dim nor red, like God's own head,
The glorious Sun uprist:
Then all averred, I had killed the bird
That brought the fog and mist. 100
'Twas right, said they, such birds to slay,
That bring the fog and mist.

The fair breeze continues; the ship enters the Pacific Ocean, and sails northward, even till it reaches the Line.

The fair breeze blew, the white foam flew,
The furrow followed free;
We were the first that ever burst
Into that silent sea.

The ship hath been suddenly becalmed.

Down dropt the breeze, the sails dropt down,
'Twas sad as sad could be;
And we did speak only to break
The silence of the sea! 110

All in a hot and copper sky,
The bloody Sun, at noon,
Right up above the mast did stand,
No bigger than the Moon.

Day after day, day after day,
We stuck, nor breath nor motion;
As idle as a painted ship
Upon a painted ocean.

And the Albatross begins to be avenged.

Water, water, every where,
And all the boards did shrink; 120
Water, water, every where,
Nor any drop to drink.

The very deep did rot: O Christ!
That ever this should be!
Yea, slimy things did crawl with legs
Upon the slimy sea.

About, about, in reel and rout
The death-fires danced at night;
The water, like a witch's oils,
Burnt green, and blue and white. 130

A Spirit had followed them; one of the invisible inhabitants of this planet, neither departed souls nor angels; concerning whom the learned Jew, Josephus, and the Platonic Constantinopolitan, Michael Psellus, may be consulted. They are very numerous, and there is no climate or element without one or more.

And some in dreams assured were
Of the spirit that plagued us so;
Nine fathom deep he had followed us
From the land of mist and snow.

And every tongue, through utter drought,
Was withered at the root;
We could not speak, no more than if
We had been choked with soot.

The shipmates, in their sore distress, would fain throw the whole guilt on the ancient Mariner: in sign whereof they hang the dead sea-bird round his neck.

Ah! well a-day! what evil looks
Had I from old and young! 140
Instead of the cross, the Albatross
About my neck was hung.

PART III

There passed a weary time. Each throat
Was parched, and glazed each eye.
A weary time! a weary time!
How glazed each weary eye,
When looking westward, I beheld
A something in the sky.

The ancient Mariner beholdeth a sign in the element afar off.

At first it seemed a little speck,
And then it seemed a mist; 150
It moved and moved, and took at last
A certain shape, I wist.

A speck, a mist, a shape, I wist!
And still it neared and neared:
As if it dodged a water-sprite,
It plunged and tacked and veered.

At its nearer approach, it seemeth him to be a ship; and at a dear ransom he freeth his speech from the bonds of thirst.

With throats unslaked, with black lips baked,
We could nor laugh nor wail;
Through utter drought all dumb we stood!
I bit my arm, I sucked the blood, 160
And cried, A sail! a sail!

With throats unslaked, with black lips baked,
Agape they heard me call:
A flash of joy;
Gramercy! they for joy did grin,
And all at once their breath drew in,
As they were drinking all.

And horror follows. For can it be a ship that comes onward without wind or tide?

See! see! (I cried) she tacks no more!
Hither to work us weal;
Without a breeze, without a tide,
She steadies with upright keel! 170

The western wave was all a-flame.
The day was well nigh done!
Almost upon the western wave
Rested the broad bright Sun;
When that strange shape drove suddenly
Betwixt us and the Sun.

It seemeth him but the skeleton of a ship.

And straight the Sun was flecked with bars,
(Heaven's Mother send us grace!)
As if through a dungeon-grate he peered
With broad and burning face. 180

Alas! (thought I, and my heart beat loud)
How fast she nears and nears!
Are those her sails that glance in the Sun,
Like restless gossameres?

And its ribs are seen as bars on the face of the setting Sun. The spectre-woman and her death-mate, and no other on board the skeleton-ship.

Are those her ribs through which the Sun
Did peer, as through a grate?
And is that Woman all her crew?
Is that a Death? and are there two?
Is Death that woman's mate?

Like vessel, like crew!

Death and Life-in-death have diced for the ship's crew, and she (the latter) winneth the ancient Mariner.

Her lips were red, her looks were free, 190
Her locks were yellow as gold:
Her skin was as white as leprosy,
The Night-mare Life-in-Death was she,
Who thicks man's blood with cold.

The naked hulk alongside came,
And the twain were casting dice;
'The game is done! I've won! I've won!'
Quoth she, and whistles thrice.

No twilight within the courts of the sun.

The Sun's rim dips; the stars rush out:
At one stride comes the dark; 200
With far-heard whisper, o'er the sea,
Off shot the spectre-bark.

At the rising of the Moon,

We listened and looked sideways up!
Fear at my heart, as at a cup,
My life-blood seemed to sip!
The stars were dim, and thick the night,
The steersman's face by his lamp gleamed white;
From the sails the dew did drip—
Till clomb above the eastern bar
The horned Moon, with one bright star 210
Within the nether tip.

One after another,

One after one, by the star-dogged Moon,
Too quick for groan or sigh,
Each turned his face with a ghastly pang,
And cursed me with his eye.

His shipmates drop down dead.

Four times fifty living men,
(And I heard nor sigh nor groan)
With heavy thump, a lifeless lump,
They dropped down one by one.

But Life-in-Death begins her work on the ancient Mariner.

The souls did from their bodies fly,— 220
They fled to bliss or woe!
And every soul, it passed me by,
Like the whizz of my cross-bow!

PART IV

The wedding guest feareth that a Spirit is talking to him.

'I fear thee, ancient Mariner!
I fear thy skinny hand!
And thou art long, and lank, and brown,
As is the ribbed sea-sand.

I fear thee and thy glittering eye,
And thy skinny hand, so brown.'—
Fear not, fear not, thou Wedding-Guest! 230
This body dropt not down.

But the ancient Mariner assureth him of his bodily life, and proceedeth to relate his horrible penance.

Alone, alone, all, all alone,
Alone on a wide wide sea!
And never a saint took pity on
My soul in agony.

He despiseth the creatures of the calm,

The many men, so beautiful!
And they all dead did lie:
And a thousand thousand slimy things
Lived on; and so did I.

And envieth that they should live, and so many lie dead.

I looked upon the rotting sea, 240
And drew my eyes away;
I looked upon the rotting deck,
And there the dead men lay.

I looked to heaven, and tried to pray;
But or ever a prayer had gusht,
A wicked whisper came, and made
My heart as dry as dust.

I closed my lids, and kept them close,
And the balls like pulses beat;
For the sky and the sea, and the sea and the sky 250
Lay like a load on my weary eye,
And the dead were at my feet.

But the curse liveth for him in the eye of the dead men.

The cold sweat melted from their limbs,
Nor rot nor reek did they:
The look with which they looked on me
Had never passed away.

An orphan's curse would drag to hell
A spirit from on high;
But oh! more horrible than that
Is the curse in a dead man's eye! 260
Seven days, seven nights, I saw that curse,
And yet I could not die.

In his loneliness and fixedness he yearneth towards the journeying Moon, and the stars that still sojourn, yet still move onward; and every where the blue sky belongs to them, and is their appointed rest, and their native country and their own natural homes, which they enter unannounced, as lords that are certainly expected and yet there is a silent joy at their arrival.

The moving Moon went up the sky.
And no where did abide:
Softly she was going up,
And a star or two beside—

Her beams bemocked the sultry main,
Like April hoar-frost spread;
But where the ship's huge shadow lay,
The charmed water burnt alway 270
A still and awful red.

By the light of the Moon he beholdeth God's creatures of the great calm.

Beyond the shadow of the ship,
I watched the water-snakes:
They moved in tracks of shining white,
And when they reared, the elfish light
Fell off in hoary flakes.

Within the shadow of the ship
I watched their rich attire:
Blue, glossy green, and velvet black,
They coiled and swam; and every track 280
Was a flash of golden fire.

Their beauty and their happiness.

O happy living things! no tongue
Their beauty might declare:
A spring of love gushed from my heart,

He blesseth them in his heart.

And I blessed them unaware:
Sure my kind saint took pity on me,
And I blessed them unaware.

The spell begins to break.

The self-same moment I could pray;
And from my neck so free
The Albatross fell off, and sank 290
Like lead into the sea.

PART V

Oh sleep! it is a gentle thing,
Beloved from pole to pole!
To Mary Queen the praise be given!
She sent the gentle sleep from Heaven,
That slid into my soul.

By grace of the holy Mother, the ancient Mariner is refreshed with rain.

The silly buckets on the deck,
That had so long remained,
I dreamt that they were filled with dew;
And when I awoke, it rained. 300

My lips were wet, my throat was cold,
My garments all were dank;
Sure I had drunken in my dreams,
And still my body drank.

I moved, and could not feel my limbs:
I was so light—almost
I thought that I had died in sleep,
And was a blessed ghost.

He heareth sounds and seeth strange sights and commotions in the sky and the element.

And soon I heard a roaring wind:
It did not come anear; 310
But with its sound it shook the sails,
That were so thin and sere.

The upper air burst into life!
And a hundred fire-flags sheen,
To and fro they were hurried about!
And to and fro, and in and out,
The wan stars danced between.

And the coming wind did roar more loud,
And the sails did sigh like sedge;
And the rain poured down from one black cloud; 320
The Moon was at its edge.

The thick black cloud was cleft, and still
The Moon was at its side:
Like waters shot from some high crag,
The lightning fell with never a jag,
A river steep and wide.

The bodies of the ship's crew are inspired, and the ship moves on;

The loud wind never reached the ship,
Yet now the ship moved on!
Beneath the lightning and the moon
The dead men gave a groan. 330

They groaned, they stirred, they all uprose,
Nor spake, nor moved their eyes;
It had been strange, even in a dream,
To have seen those dead men rise.

The helmsman steered, the ship moved on;
Yet never a breeze up blew;
The mariners all 'gan work the ropes,
Where they were wont to do;
They raised their limbs like lifeless tools—
We were a ghastly crew. 340

The body of my brother's son
Stood by me, knee to knee:
The body and I pulled at one rope,
But he said nought to me.

But not by the souls of the men, nor by demons of earth or middle air, but by a blessed troop of angelic spirits, sent down by the invocation of the guardian saint.

'I fear thee, ancient Mariner!'
Be calm, thou Wedding-Guest!
'Twas not those souls that fled in pain,
Which to their corses came again,
But a troop of spirits blest:

For when it dawned—they dropped their arms, 350
And clustered round the mast;
Sweet sounds rose slowly through their mouths,
And from their bodies passed.

Around, around, flew each sweet sound,
Then darted to the Sun;
Slowly the sounds came back again,
Now mixed, now one by one.

Sometimes a-dropping from the sky
I heard the sky-lark sing;
Sometimes all little birds that are, 360
How they seemed to fill the sea and air
With their sweet jargoning!

And now 'twas like all instruments,
Now like a lonely flute;
And now it is an angel's song,
That makes the heavens be mute.

It ceased; yet still the sails made on
A pleasant noise till noon,
A noise like of a hidden brook
In the leafy month of June, 370
That to the sleeping woods all night
Singeth a quiet tune.

Till noon we quietly sailed on,
Yet never a breeze did breathe:
Slowly and smoothly went the ship,
Moved onward from beneath.

The lonesome spirit from the south-pole carries on the ship as far as the line, in obedience to the angelic troop, but still requireth vengeance.

Under the keel nine fathom deep,
From the land of mist and snow,
The spirit slid: and it was he
That made the ship to go. 380
The sails at noon left off their tune,
And the ship stood still also.

The Sun, right up above the mast,
Had fixed her to the ocean:
But in a minute she 'gan stir,
With a short uneasy motion—
Backwards and forwards half her length
With a short uneasy motion.

Then like a pawing horse let go,
She made a sudden bound: 390
It flung the blood into my head,
And I fell down in a swound.

The Polar Spirit's fellow-demons, the invisible inhabitants of the element, take part in his wrong; and two of them relate, one to the other, that penance long and heavy for the ancient Mariner hath been accorded to the Polar Spirit, who returneth southward.

How long in that same fit I lay,
I have not to declare;
But ere my living life returned,
I heard, and in my soul discerned
Two voices in the air.

'Is it he?' quoth one, 'Is this the man?
By him who died on cross,
With his cruel bow he laid full low 400
The harmless Albatross.

The spirit who bideth by himself
In the land of mist and snow,
He loved the bird that loved the man
Who shot him with his bow.'

The other was a softer voice,
As soft as honey-dew:
Quoth he, 'The man hath penance done,
And penance more will do.'

PART VI

FIRST VOICE

But tell me, tell me! speak again, 410
Thy soft response renewing—
What makes that ship drive on so fast?
What is the ocean doing?

SECOND VOICE

Still as a slave before his lord,
The ocean hath no blast;
His great bright eye most silently
Up to the Moon is cast—

If he may know which way to go;
For she guides him smooth or grim.
See, brother, see! how graciously 420
She looketh down on him.

FIRST VOICE

The Mariner hath been cast into a trance; for the angelic power causeth the vessel to drive northward faster than human life could endure.

But why drives on that ship so fast,
Without or wave or wind?

SECOND VOICE

The air is cut away before,
And closes from behind.

Fly, brother, fly! more high, more high!
Or we shall be belated:
For slow and slow that ship will go,
When the Mariner's trance is abated.

The supernatural motion is retarded; the Mariner awakes, and his penance begins anew.

I woke, and we were sailing on 430
As in a gentle weather:
'Twas night, calm night, the moon was high;
The dead men stood together.

All stood together on the deck,
For a charnel-dungeon fitter:
All fixed on me their stony eyes,
That in the Moon did glitter.

The pang, the curse, with which they died,
Had never passed away:
I could not draw my eyes from theirs, 440
Nor turn them up to pray.

The curse is finally expiated.

And now this spell was snapt: once more
I viewed the ocean green,
And looked far forth, yet little saw
Of what had else been seen—

Like one, that on a lonesome road
Doth walk in fear and dread,
And having once turned round walks on,
And turns no more his head;
Because he knows, a frightful fiend 450
Doth close behind him tread.

But soon there breathed a wind on me,
Nor sound nor motion made:
Its path was not upon the sea,
In ripple or in shade.

It raised my hair, it fanned my cheek
Like a meadow-gale of spring—
It mingled strangely with my fears,
Yet it felt like a welcoming.

Swiftly, swiftly flew the ship, 460
Yet she sailed softly too:
Sweetly, sweetly blew the breeze—
On me alone it blew.

And the ancient Mariner beholdeth his native country.

Oh! dream of joy! is this indeed
The light-house top I see?
Is this the hill? is this the kirk?
Is this mine own countree?

We drifted o'er the harbour-bar,
And I with sobs did pray—
O let me be awake, my God! 470
Or let me sleep alway.

The harbour-bay was clear as glass,
So smoothly it was strewn!
And on the bay the moonlight lay,
And the shadow of the moon.

The rock shone bright, the kirk no less,
That stands above the rock:
The moonlight steeped in silentness
The steady weathercock.

And the bay was white with silent light, 480
Till rising from the same,
Full many shapes, that shadows were,
In crimson colours came.

The angelic spirits leave the dead bodies,

And appear in their own forms of light.

A little distance from the prow
Those crimson shadows were:
I turned my eyes upon the deck—
Oh, Christ! what saw I there!

Each corse lay flat, lifeless and flat,
And, by the holy rood!
A man all light, a seraph-man, 490
On every corse there stood.

This seraph-band, each waved his hand:
It was a heavenly sight!
They stood as signals to the land,
Each one a lovely light;

This seraph-band, each waved his hand,
No voice did they impart—
No voice; but oh! the silence sank
Like music on my heart.

But soon I heard the dash of oars, 500
I heard the Pilot's cheer;
My head was turned perforce away,
And I saw a boat appear.

The Pilot and the Pilot's boy,
I heard them coming fast:
Dear Lord in Heaven! it was a joy
The dead men could not blast.

I saw a third—I heard his voice:
It is the Hermit good!
He singeth loud his godly hymns 510
That he makes in the wood.
He'll shrieve my soul, he'll wash away
The Albatross's blood.

PART VII

The Hermit of the wood,

This Hermit good lives in that wood
Which slopes down to the sea.
How loudly his sweet voice he rears!
He loves to talk with marineres
That come from a far countree.

He kneels at morn, and noon, and eve—
He hath a cushion plump: 520
It is the moss that wholly hides
The rotted old oak-stump.

The skiff-boat neared: I heard them talk,
'Why, this is strange, I trow!
Where are those lights so many and fair,
That signal made but now?'

Approacheth the ship with wonder.

'Strange, by my faith!' the Hermit said—
And they answered not our cheer!
The planks looked warped! and see those sails,
How thin they are and sere! 530
I never saw aught like to them,
Unless perchance it were

Brown skeletons of leaves that lag
My forest-brook along;
When the ivy-tod is heavy with snow,
And the owlet whoops to the wolf below,
That eats the she-wolf's young.'

'Dear Lord! it hath a fiendish look—
(The Pilot made reply)
I am a-feared'—'Push on, push on!' 540
Said the Hermit cheerily.

The boat came closer to the ship,
But I nor spake nor stirred;
The boat came close beneath the ship,
And straight a sound was heard.

The ship suddenly sinketh.

Under the water it rumbled on,
Still louder and more dread:
It reached the ship, it split the bay;
The ship went down like lead.

The ancient Mariner is saved in the Pilot's boat.

Stunned by that loud and dreadful sound, 550
Which sky and ocean smote,
Like one that hath been seven days drowned
My body lay afloat;
But swift as dreams, myself I found
Within the Pilot's boat.

Upon the whirl, where sank the ship,
The boat spun round and round;
And all was still, save that the hill
Was telling of the sound.

I moved my lips—the Pilot shrieked 560
And fell down in a fit;
The holy Hermit raised his eyes,
And prayed where he did sit.

I took the oars: the Pilot's boy,
Who now doth crazy go,
Laughed loud and long, and all the while
His eyes went to and fro.
'Ha! ha!' quoth he, 'full plain I see,
The Devil knows how to row.'

And now, all in my own countree, 570
I stood on the firm land!
The Hermit stepped forth from the boat,
And scarcely he could stand.

The ancient Mariner earnestly entreateth the Hermit to shrieve him; and the penance of life falls on him.

'O shrieve me, shrieve me, holy man!'
The Hermit crossed his brow.
'Say quick,' quoth he, 'I bid thee say—
What manner of man art thou?'

Forthwith this frame of mine was wrenched
With a woful agony,
Which forced me to begin my tale; 580
And then it left me free.

And ever and anon throughout his future life an agony constraineth him to travel from land to land;

Since then, at an uncertain hour,
That agony returns:
And till my ghastly tale is told,
This heart within me burns.

I pass, like night, from land to land;
I have strange power of speech;
That moment that his face I see,
I know the man that must hear me:
To him my tale I teach. 590

What loud uproar bursts from that door!
The wedding-guests are there:
But in the garden-bower the bride

And bride-maids singing are:
And hark the little vesper bell,
Which biddeth me to prayer!

O Wedding-Guest! this soul hath been
Alone on a wide wide sea:
So lonely 'twas, that God himself
Scarce seemed there to be. 600

O sweeter than the marriage-feast,
'Tis sweeter far to me,
To walk together to the kirk
With a goodly company!—

To walk together to the kirk,
And all together pray,
While each to his great Father bends,
Old men, and babes, and loving friends
And youths and maidens gay!

And to teach, by his own example, love and reverence to all things that God made and loveth.

Farewell, farewell! but this I tell 610
To thee, thou Wedding-Guest!
He prayeth well, who loveth well
Both man and bird and beast.

He prayeth best, who loveth best
All things both great and small;
For the dear God who loveth us,
He made and loveth all.

The Mariner, whose eye is bright,
Whose beard with age is hoar,
Is gone: and now the Wedding-Guest 620
Turned from the bridegroom's door.

He went like one that hath been stunned,
And is of sense forlorn:
A sadder and a wiser man,
He rose the morrow morn.

Christabel

PREFACE

The first part of the following poem was written in the year 1797, at Stowey, in the county of Somerset. The second part, after my return from Germany, in the year 1800, at Keswick, Cumberland. It is probable that if the poem had been finished at either of the former periods, or if even the first and second part had been published in the year 1800, the impression of its originality would have been much greater than I dare at present expect. But for this I have only my own indolence to blame. The dates are mentioned for the exclusive purpose of precluding charges of plagiarism or servile imitation from myself. For there is amongst us a set of critics, who seem to hold, that every possible thought and image is traditional; who have no notion that there are such things as fountains in the world, small as well as great; and who would therefore charitably derive every rill they behold flowing, from a perforation made in some other man's tank. I am confident, however, that as far as the present poem is concerned, the celebrated poets whose writings I might be suspected of having imitated,° either in particular passages, or in the tone and the spirit of the whole, would be among the first to vindicate me from the charge, and who, on any striking coincidence, would permit me to address them in this doggerel version of two monkish Latin hexameters.

'Tis mine and it is likewise yours;
But an if this will not do;
Let it be mine, good friend! for I
Am the poorer of the two.

I have only to add that the metre of the Christabel is not, properly speaking, irregular, though it may seem so from its being founded on a new principle: namely, that of counting in each line the accents, not the syllables. Though the latter may vary from seven to twelve, yet in each line the accents will be found to be only four. Nevertheless, this occasional variation in number of syllables is not introduced wantonly, or for the mere ends of convenience, but in correspondence with some transition, in the nature of the imagery or passion.

PART I

'Tis the middle of night by the castle clock,
And the owls have awakened the crowing cock;
Tu—whit!——Tu—whoo!
And hark, again! the crowing cock,
How drowsily it crew.

Sir Leoline, the Baron rich,
Hath a toothless mastiff bitch;
From her kennel beneath the rock
She maketh answer to the clock,
Four for the quarters, and twelve for the hour; 10
Ever and aye, by shine and shower,
Sixteen short howls, not over loud;
Some say, she sees my lady's shroud.

Is the night chilly and dark?
The night is chilly, but not dark.
The thin gray cloud is spread on high,
It covers but not hides the sky.
The moon is behind, and at the full;
And yet she looks both small and dull.
The night is chill, the cloud is gray: 20
'Tis a month before the month of May,
And the Spring comes slowly up this way.

The lovely lady, Christabel,
Whom her father loves so well,
What makes her in the wood so late,
A furlong from the castle gate?
She had dreams all yesternight
Of her own betrothed knight;
And she in the midnight wood will pray
For the weal of her lover that's far away. 30

She stole along, she nothing spoke,
The sighs she heaved were soft and low,
And naught was green upon the oak
But moss and rarest misletoe:
She kneels beneath the huge oak tree,
And in silence prayeth she.

The lady sprang up suddenly,
The lovely lady, Christabel!
It moaned as near, as near can be,
But what it is, she cannot tell.— 40
On the other side it seems to be,
Of the huge, broad-breasted, old oak tree.

The night is chill; the forest bare;
Is it the wind that moaneth bleak?
There is not wind enough in the air
To move away the ringlet curl
From the lovely lady's cheek—
There is not wind enough to twirl
The one red leaf, the last of its clan,
That dances as often as dance it can, 50
Hanging so light, and hanging so high,
On the topmost twig that looks up at the sky.

Hush, beating heart of Christabel!
Jesu, Maria, shield her well!
She folded her arms beneath her cloak,
And stole to the other side of the oak.
    What sees she there?

There she sees a damsel bright,
Drest in a silken robe of white,
That shadowy in the moonlight shone: 60
The neck that made that white robe wan,
Her stately neck, and arms were bare;
Her blue-veined feet unsandal'd were,
And wildly glittered here and there
The gems entangled in her hair.
I guess, 'twas frightful there to see
A lady so richly clad as she—
Beautiful exceedingly!

Mary mother, save me now!
(Said Christabel,) And who art thou? 70

The lady strange made answer meet,
And her voice was faint and sweet:—
Have pity on my sore distress,
I scarce can speak for weariness:
Stretch forth thy hand, and have no fear!
Said Christabel, How camest thou here?
And the lady, whose voice was faint and sweet,
Did thus pursue her answer meet:—

My sire is of a noble line,
And my name is Geraldine: 80
Five warriors seized me yestermorn,
Me, even me, a maid forlorn:
They choked my cries with force and fright,
And tied me on a palfrey white.
The palfrey was as fleet as wind,
And they rode furiously behind.
They spurred amain, their steeds were white:
And once we crossed the shade of night.
As sure as Heaven shall rescue me,
I have no thought what men they be; 90
Nor do I know how long it is
(For I have lain entranced I wis)
Since one, the tallest of the five,
Took me from the palfrey's back,
A weary woman, scarce alive.
Some muttered words his comrades spoke:
He placed me underneath this oak;
He swore they would return with haste;
Whither they went I cannot tell—
I thought I heard, some minutes past, 100
Sounds as of a castle bell.
Stretch forth thy hand (thus ended she),
And help a wretched maid to flee.

Then Christabel stretched forth her hand,
And comforted fair Geraldine:
O well, bright dame! may you command
The service of Sir Leoline;
And gladly our stout chivalry
Will he send forth and friends withal
To guide and guard you safe and free 110
Home to your noble father's hall.

She rose: and forth with steps they passed
That strove to be, and were not, fast.
Her gracious stars the lady blest,
And thus spake on sweet Christabel:
All our household are at rest,
The hall as silent as the cell;

Sir Leoline is weak in health,
And may not well awakened be,
But we will move as if in stealth, 120
And I beseech your courtesy,
This night, to share your couch with me.

They crossed the moat, and Christabel
Took the key that fitted well;
A little door she opened straight,
All in the middle of the gate;
The gate that was ironed within and without,
Where an army in battle array had marched out.
The lady sank, belike through pain,
And Christabel with might and main 130
Lifted her up, a weary weight,
Over the threshold of the gate:
Then the lady rose again,
And moved, as she were not in pain.

So free from danger, free from fear,
They crossed the court: right glad they were.
And Christabel devoutly cried
To the lady by her side,
Praise we the Virgin all divine
Who hath rescued thee from thy distress! 140
Alas, alas! said Geraldine,
I cannot speak for weariness.
So free from danger, free from fear,
They crossed the court: right glad they were.

Outside her kennel the mastiff old
Lay fast asleep, in moonshine cold.
The mastiff old did not awake,
Yet she an angry moan did make!
And what can ail the mastiff bitch?
Never till now she uttered yell 150
Beneath the eye of Christabel.
Perhaps it is the owlet's scritch:
For what can ail the mastiff bitch?

They passed the hall, that echoes still,
Pass as lightly as you will!
The brands were flat, the brands were dying,
Amid their own white ashes lying;
But when the lady passed, there came
A tongue of light, a fit of flame;
And Christabel saw the lady's eye, 160
And nothing else saw she thereby,
Save the boss of the shield of Sir Leoline tall,
Which hung in a murky old niche in the wall.
O softly tread, said Christabel,
My father seldom sleepeth well.

Sweet Christabel her feet doth bare,
And jealous of the listening air
They steal their way from stair to stair,
Now in glimmer, and now in gloom,
And now they pass the Baron's room, 170
As still as death with stifled breath!
And now have reached her chamber door;
And now doth Geraldine press down
The rushes of the chamber floor.

The moon shines dim in the open air,
And not a moonbeam enters here.
But they without its light can see
The chamber carved so curiously,
Carved with figures strange and sweet,
All made out of the carver's brain, 180
For a lady's chamber meet:
The lamp with twofold silver chain
Is fastened to an angel's feet.
The silver lamp burns dead and dim;
But Christabel the lamp will trim.
She trimmed the lamp, and made it bright,
And left it swinging to and fro,
While Geraldine, in wretched plight,
Sank down upon the floor below.

O weary lady, Geraldine, 190
I pray you, drink this cordial wine!

It is a wine of virtuous powers;
My mother made it of wild flowers.

And will your mother pity me,
Who am a maiden most forlorn?
Christabel answered—Woe is me!
She died the hour that I was born.
I have heard the grey-haired friar tell
How on her death-bed she did say,
That she should hear the castle-bell 200
Strike twelve upon my wedding-day.
O mother dear! that thou wert here!
I would, said Geraldine, she were!

But soon with altered voice, said she—
'Off, wandering mother! Peak and pine!
I have power to bid thee flee.'
Alas! what ails poor Geraldine?
Why stares she with unsettled eye?
Can she the bodiless dead espy?
And why with hollow voice cries she, 210
'Off, woman, off! this hour is mine—
Though thou her guardian spirit be,
Off, woman, off! 'tis given to me.'

Then Christabel knelt by the lady's side,
And raised to heaven her eyes so blue—
Alas! said she, this ghastly ride—
Dear lady! it hath wildered you!
The lady wiped her moist cold brow,
And faintly said, ''tis over now!'

Again the wild-flower wine she drank: 220
Her fair large eyes 'gan glitter bright,
And from the floor whereon she sank,
The lofty lady stood upright:
She was most beautiful to see,
Like a lady of a far countrée.

And thus the lofty lady spake—
'All they, who live in the upper sky,

Do love you, holy Christabel!
And you love them, and for their sake
And for the good which me befel, 230
Even I in my degree will try,
Fair maiden, to requite you well.
But now unrobe yourself; for I
Must pray, ere yet in bed I lie.'

Quoth Christabel, so let it be!
And as the lady bade, did she.
Her gentle limbs did she undress,
And lay down in her loveliness.

But through her brain of weal and woe
So many thoughts moved to and fro, 240
That vain it were her lids to close;
So half-way from the bed she rose,
And on her elbow did recline
To look at the lady Geraldine.

Beneath the lamp the lady bowed,
And slowly rolled her eyes around;
Then drawing in her breath aloud,
Like one that shuddered, she unbound
The cincture from beneath her breast:
Her silken robe, and inner vest, 250
Dropt to her feet, and full in view,
Behold! her bosom and half her side——
A sight to dream of, not to tell!
O shield her! shield sweet Christabel!

Yet Geraldine nor speaks nor stirs;
Ah! what a stricken look was hers!
Deep from within she seems half-way
To lift some weight with sick assay,
And eyes the maid and seeks delay;
Then suddenly, as one defied, 260
Collects herself in scorn and pride,
And lay down by the maiden's side!—
And in her arms the maid she took,
Ah well-a-day!

And with low voice and doleful look
These words did say:
In the touch of this bosom there worketh a spell,
Which is lord of thy utterance, Christabel!
Thou knowest to-night, and wilt know to-morrow,
This mark of my shame, this seal of my sorrow; 270
    But vainly thou warrest,
      For this is alone in
    Thy power to declare,
      That in the dim forest
    Thou heard'st a low moaning,
And found'st a bright lady, surpassingly fair;
And didst bring her home with thee in love and in charity,
To shield her and shelter her from the damp air.

THE CONCLUSION TO PART I

It was a lovely sight to see
The lady Christabel, when she 280
Was praying at the old oak tree.
    Amid the jagged shadows
    Of mossy leafless boughs,
    Kneeling in the moonlight,
    To make her gentle vows;
Her slender palms together prest,
Heaving sometimes on her breast;
Her face resigned to bliss or bale—
Her face, oh call it fair not pale,
And both blue eyes more bright than clear, 290
Each about to have a tear.

With open eyes (ah woe is me!)
Asleep, and dreaming fearfully,
Fearfully dreaming, yet I wis,
Dreaming that alone, which is—
O sorrow and shame! Can this be she,
The lady, who knelt at the old oak tree?
And lo! the worker of these harms,
That holds the maiden in her arms,
Seems to slumber still and mild, 300
As a mother with her child.

A star hath set, a star hath risen,
O Geraldine! since arms of thine
Have been the lovely lady's prison.
O Geraldine! one hour was thine—
Thou'st had thy will! By tairn and rill,
The night-birds all that hour were still.
But now they are jubilant anew,
From cliff and tower, tu—whoo! tu—whoo!
Tu—whoo! tu—whoo! from wood and fell! 310

And see! the lady Christabel
Gathers herself from out her trance;
Her limbs relax, her countenance
Grows sad and soft; the smooth thin lids
Close o'er her eyes; and tears she sheds—
Large tears that leave the lashes bright!
And oft the while she seems to smile
As infants at a sudden light!
Yea, she doth smile, and she doth weep,
Like a youthful hermitess, 320
Beauteous in a wilderness,
Who, praying always, prays in sleep.
And, if she move unquietly,
Perchance, 'tis but the blood so free
Comes back and tingles in her feet.
No doubt, she hath a vision sweet.
What if her guardian spirit 'twere,
What if she knew her mother near?
But this she knows, in joys and woes,
That saints will aid if men will call: 330
For the blue sky bends over all!

PART II

Each matin bell, the Baron saith,
Knells us back to a world of death.
These words Sir Leoline first said,
When he rose and found his lady dead:
These words Sir Leoline will say
Many a morn to his dying day!

And hence the custom and law began,
That still at dawn the sacristan,
Who duly pulls the heavy bell, 340
Five and forty beads must tell
Between each stroke—a warning knell,
Which not a soul can choose but hear
From Bratha Head to Wyndermere.

Saith Bracy the bard, So let it knell!
And let the drowsy sacristan
Still count as slowly as he can!
There is no lack of such, I ween,
As well fill up the space between.
In Langdale Pike and Witch's Lair, 350
And Dungeon-ghyll so foully rent,
With ropes of rock and bells of air
Three sinful sextons' ghosts are pent,
Who all give back, one after t'other,
The death-note to their living brother;
And oft too, by the knell offended,
Just as their one! two! three! is ended,
The devil mocks the doleful tale
With a merry peal from Borodale.

The air is still! through mist and cloud 360
That merry peal comes ringing loud;
And Geraldine shakes off her dread,
And rises lightly from the bed;
Puts on her silken vestments white,
And tricks her hair in lovely plight,
And nothing doubting of her spell
Awakens the lady Christabel.
'Sleep you, sweet lady Christabel?
I trust that you have rested well.'

And Christabel awoke and spied 370
The same who lay down by her side—
O rather say, the same whom she
Raised up beneath the old oak tree!
Nay, fairer yet! and yet more fair!
For she belike hath drunken deep
Of all the blessedness of sleep!

And while she spake, her looks, her air
Such gentle thankfulness declare,
That (so it seemed) her girded vests
Grew tight beneath her heaving breasts. 380
'Sure I have sinn'd!' said Christabel,
'Now heaven be praised if all be well!'
And in low faltering tones, yet sweet,
Did she the lofty lady greet
With such perplexity of mind
As dreams too lively leave behind.

So quickly she rose, and quickly arrayed
Her maiden limbs, and having prayed
That He, who on the cross did groan,
Might wash away her sins unknown, 390
She forthwith led fair Geraldine
To meet her sire, Sir Leoline.

The lovely maid and the lady tall
Are pacing both into the hall,
And pacing on through page and groom,
Enter the Baron's presence room.

The Baron rose, and while he prest
His gentle daughter to his breast,
With cheerful wonder in his eyes
The lady Geraldine espies, 400
And gave such welcome to the same,
As might beseem so bright a dame!

But when he heard the lady's tale,
And when she told her father's name,
Why waxed Sir Leoline so pale,
Murmuring o'er the name again,
Lord Roland de Vaux of Tryermaine?

Alas! they had been friends in youth;
But whispering tongues can poison truth;
And constancy lives in realms above; 410
And life is thorny; and youth is vain;
And to be wroth with one we love
Doth work like madness in the brain.

And thus it chanced, as I divine,
With Roland and Sir Leoline.
Each spake words of high disdain
And insult to his heart's best brother:
They parted—ne'er to meet again!
But never either found another
To free the hollow heart from paining— 420
They stood aloof, the scars remaining,
Like cliffs which had been rent asunder;
A dreary sea now flows between;—
But neither heat, nor frost, nor thunder,
Shall wholly do away, I ween,
The marks of that which once hath been.

Sir Leoline, a moment's space,
Stood gazing on the damsel's face:
And the youthful Lord of Tryermaine
Came back upon his heart again. 430

O then the Baron forgot his age,
His noble heart swelled high with rage;
He swore by the wounds in Jesu's side
He would proclaim it far and wide,
With trump and solemn heraldry,
That they who thus had wronged the dame,
Were base as spotted infamy!
'And if they dare deny the same,
My herald shall appoint a week,
And let the recreant traitors seek 440
My tourney court—that there and then
I may dislodge their reptile souls
From the bodies and forms of men!'
He spake: his eye in lightning rolls!
For the lady was ruthlessly seized; and he kenned
In the beautiful lady the child of his friend!

And now the tears were on his face,
And fondly in his arms he took
Fair Geraldine, who met the embrace,
Prolonging it with joyous look. 450
Which when she viewed, a vision fell
Upon the soul of Christabel,

The vision of fear, the touch and pain!
She shrunk and shuddered, and saw again—
(Ah, woe is me! Was it for thee,
Thou gentle maid! such sights to see?)
Again she saw that bosom old,
Again she felt that bosom cold,
And drew in her breath with a hissing sound:
Whereat the Knight turned wildly round, 460
And nothing saw, but his own sweet maid
With eyes upraised, as one that prayed.

The touch, the sight, had passed away,
And in its stead that vision blest,
Which comforted her after-rest
While in the lady's arms she lay,
Had put a rapture in her breast,
And on her lips and o'er her eyes
Spread smiles like light!
With new surprise,
'What ails then my beloved child?' 470
The Baron said—His daughter mild
Made answer, 'All will yet be well!'
I ween, she had no power to tell
Aught else: so mighty was the spell.

Yet he, who saw this Geraldine,
Had deemed her sure a thing divine:
Such sorrow with such grace she blended,
As if she feared she had offended
Sweet Christabel, that gentle maid!
And with such lowly tones she prayed, 480
She might be sent without delay
Home to her father's mansion.
'Nay!
Nay, by my soul!' said Leoline.
'Ho! Bracy the bard, the charge be thine!
Go thou, with music sweet and loud,
And take two steeds with trappings proud,
And take the youth whom thou lov'st best
To bear thy harp, and learn thy song,
And clothe you both in solemn vest,
And over the mountains haste along, 490

Lest wandering folk, that are abroad,
Detain you on the valley road.
And when he has crossed the Irthing flood,
My merry bard! he hastes, he hastes
Up Knorren Moor, through Halegarth Wood,
And reaches soon that castle good
Which stands and threatens Scotland's wastes.

Bard Bracy! bard Bracy! your horses are fleet,
Ye must ride up the hall, your music so sweet,
More loud than your horses' echoing feet! 500
And loud and loud to Lord Roland call,
Thy daughter is safe in Langdale hall!
Thy beautiful daughter is safe and free—
Sir Leoline greets thee thus through me!
He bids thee come without delay
With all thy numerous array
And take thy lovely daughter home:
And he will meet thee on the way
With all his numerous array
White with their panting palfreys' foam: 510
And, by mine honour! I will say,
That I repent me of the day
When I spake words of fierce disdain
To Roland de Vaux of Tryermaine!—
—For since that evil hour hath flown,
Many a summer's sun hath shone;
Yet ne'er found I a friend again
Like Roland de Vaux of Tryermaine.'

The lady fell, and clasped his knees,
Her face upraised, her eyes o'erflowing; 520
And Bracy replied, with faltering voice,
His gracious hail on all bestowing!—
'Thy words, thou sire of Christabel,
Are sweeter than my harp can tell;
Yet might I gain a boon of thee,
This day my journey should not be,
So strange a dream hath come to me;
That I had vowed with music loud
To clear yon wood from thing unblest,
Warned by a vision in my rest! 530

For in my sleep I saw that dove,
That gentle bird, whom thou dost love,
And call'st by thy own daughter's name—
Sir Leoline! I saw the same
Fluttering, and uttering fearful moan,
Among the green herbs in the forest alone.
Which when I saw and when I heard,
I wonder'd what might ail the bird;
For nothing near it could I see,
Save the grass and green herbs underneath the old tree. 540

And in my dream methought I went
To search out what might there be found;
And what the sweet bird's trouble meant,
That thus lay fluttering on the ground.
I went and peered, and could descry
No cause for her distressful cry;
But yet for her dear lady's sake
I stooped, methought, the dove to take,
When lo! I saw a bright green snake
Coiled around its wings and neck, 550
Green as the herbs on which it couched,
Close by the dove's its head it crouched;
And with the dove it heaves and stirs,
Swelling its neck as she swelled hers!
I woke; it was the midnight hour,
The clock was echoing in the tower;
But though my slumber was gone by,
This dream it would not pass away—
It seems to live upon my eye!
And thence I vowed this self-same day 560
With music strong and saintly song
To wander through the forest bare,
Lest aught unholy loiter there.'

Thus Bracy said: the Baron, the while,
Half-listening heard him with a smile;
Then turned to Lady Geraldine,
His eyes made up of wonder and love;
And said in courtly accents fine,
'Sweet maid, Lord Roland's beauteous dove,

With arms more strong than harp or song, 570
Thy sire and I will crush the snake!'
He kissed her forehead as he spake,
And Geraldine in maiden wise
Casting down her large bright eyes,
With blushing cheek and courtesy fine
She turned her from Sir Leoline;
Softly gathering up her train,
That o'er her right arm fell again;
And folded her arms across her chest,
And couched her head upon her breast, 580
And looked askance at Christabel—
Jesu, Maria, shield her well!

A snake's small eye blinks dull and shy;
And the lady's eyes they shrunk in her head,
Each shrunk up to a serpent's eye,
And with somewhat of malice, and more of dread,
At Christabel she looked askance!—
One moment—and the sight was fled!
But Christabel in dizzy trance
Stumbling on the unsteady ground 590
Shuddered aloud, with a hissing sound;
And Geraldine again turned round,
And like a thing, that sought relief,
Full of wonder and full of grief,
She rolled her large bright eyes divine
Wildly on Sir Leoline.

The maid, alas! her thoughts are gone,
She nothing sees—no sight but one!
The maid, devoid of guile and sin,
I know not how, in fearful wise, 600
So deeply had she drunken in
That look, those shrunken serpent eyes,
That all her features were resigned
To this sole image in her mind:
And passively did imitate
That look of dull and treacherous hate!
And thus she stood, in dizzy trance,
Still picturing that look askance

With forced unconscious sympathy
Full before her father's view—— 610
As far as such a look could be
In eyes so innocent and blue!
And when the trance was o'er, the maid
Paused awhile, and inly prayed:
Then falling at the Baron's feet,
'By my mother's soul do I entreat
That thou this woman send away!'
She said: and more she could not say:
For what she knew she could not tell,
O'er-mastered by the mighty spell. 620

Why is thy cheek so wan and wild,
Sir Leoline? Thy only child
Lies at thy feet, thy joy, thy pride,
So fair, so innocent, so mild;
The same, for whom thy lady died!
O by the pangs of her dear mother
Think thou no evil of thy child!
For her, and thee, and for no other,
She prayed the moment ere she died:
Prayed that the babe for whom she died, 630
Might prove her dear lord's joy and pride!
  That prayer her deadly pangs beguiled,
      Sir Leoline!
  And wouldst thou wrong thy only child,
      Her child and thine?

Within the Baron's heart and brain
If thoughts, like these, had any share,
They only swelled his rage and pain,
And did but work confusion there.
His heart was cleft with pain and rage, 640
His cheeks they quivered, his eyes were wild,
Dishonoured thus in his old age;
Dishonoured by his only child,
And all his hospitality
To the wrong'd daughter of his friend
By more than woman's jealousy
Brought thus to a disgraceful end—

He rolled his eye with stern regard
Upon the gentle minstrel bard,
And said in tones abrupt, austere— 650
'Why, Bracy! dost thou loiter here?
I bade thee hence!' The bard obeyed;
And turning from his own sweet maid,
The aged knight, Sir Leoline,
Led forth the lady Geraldine!

THE CONCLUSION TO PART II

A little child, a limber elf,
Singing, dancing to itself,
A fairy thing with red round cheeks,
That always finds, and never seeks,
Makes such a vision to the sight 660
As fills a father's eyes with light;
And pleasures flow in so thick and fast
Upon his heart, that he at last
Must needs express his love's excess
With words of unmeant bitterness.
Perhaps 'tis pretty to force together
Thoughts so all unlike each other;
To mutter and mock a broken charm,
To dally with wrong that does no harm.
Perhaps 'tis tender too and pretty 670
At each wild word to feel within
A sweet recoil of love and pity.
And what, if in a world of sin
(O sorrow and shame should this be true!)
Such giddiness of heart and brain
Comes seldom save from rage and pain,
So talks as it's most used to do.

Fire, Famine, and Slaughter

A War Eclogue°

The Scene a desolated Tract in La Vendée. FAMINE *is discovered lying on the ground; to her enter* FIRE *and* SLAUGHTER.

Fam. Sisters! sisters! who sent you here?
Slau. [*to Fire*]. I will whisper it in her ear.
Fire. No! no! no!
Spirits hear what spirits tell:
'Twill make a holiday in Hell.
No! no! no!
Myself, I named him once below,
And all the souls, that damned be,
Leaped up at once in anarchy,
Clapped their hands and danced for glee. 10
They no longer heeded me;
But laughed to hear Hell's burning rafters
Unwillingly re-echo laughters!
No! no! no!
Spirits hear what spirits tell:
'Twill make a holiday in Hell!
Fam. Whisper it, sister! so and so!
In a dark hint, soft and slow.
Slau. Letters four do form his name—
And who sent you?
Both. The same! the same! 20
Slau. He came by stealth, and unlocked my den,
And I have drunk the blood since then
Of thrice three hundred thousand men.
Both. Who bade you do't?
Slau. The same! the same!
Letters four do form his name.
He let me loose, and cried Halloo!
To him alone the praise is due.
Fam. Thanks, sister, thanks! the men have bled,
Their wives and their children faint for bread.
I stood in a swampy field of battle; 30
With bones and skulls I made a rattle,

To frighten the wolf and carrion-crow
And the homeless dog—but they would not go.
So off I flew: for how could I bear
To see them gorge their dainty fare?
I heard a groan and a peevish squall,
And through the chink of a cottage-wall—
Can you guess what I saw there?
Both. Whisper it, sister! in our ear.
Fam. A baby beat its dying mother: 40
I had starved the one and was starving the other!
Both. Who bade you do't?
Fam. The same! the same!
Letters four do form his name.
He let me loose, and cried, Halloo!
To him alone the praise is due.
Fire. Sisters! I from Ireland came!
Hedge and corn-fields all on flame,
I triumphed o'er the setting sun!
And all the while the work was done,
On as I strode with my huge strides, 50
I flung back my head and I held my sides,
It was so rare a piece of fun
To see the sweltered cattle run
With uncouth gallop through the night,
Scared by the red and noisy light!
By the light of his own blazing cot
Was many a naked rebel shot:
The house-stream met the flame and hissed,
While crash! fell in the roof, I wist,
On some of those old bed-rid nurses, 60
That deal in discontent and curses.
Both. Who bade you do't?
Fire. The same! the same!
Letters four do form his name.
He let me loose, and cried Halloo!
To him alone the praise is due.
All. He let us loose, and cried Halloo!
How shall we yield him honour due?
Fam. Wisdom comes with lack of food.
I'll gnaw, I'll gnaw the multitude,
Till the cup of rage o'erbrim: 70

They shall seize him and his brood—
  Slau. They shall tear him limb from limb!
  Fire. O thankless beldames and untrue!
And is this all that you can do
For him, who did so much for you?
Ninety months he, by my troth!
Hath richly catered for you both;
And in an hour would you repay
An eight years' work?—Away! away!
I alone am faithful! I 80
Cling to him everlastingly.

Frost at Midnight

The frost performs its secret ministry,
Unhelped by any wind. The owlet's cry
Came loud—and hark, again! loud as before.
The inmates of my cottage, all at rest,
Have left me to that solitude, which suits
Abstruser musings: save that at my side
My cradled infant slumbers peacefully.
'Tis calm indeed! so calm, that it disturbs
And vexes meditation with its strange
And extreme silentness. Sea, hill, and wood, 10
This populous village! Sea, and hill, and wood,
With all the numberless goings on of life,
Inaudible as dreams! the thin blue flame
Lies on my low burnt fire, and quivers not;
Only that film, which fluttered on the grate,°
Still flutters there, the sole unquiet thing.
Methinks, its motion in this hush of nature
Gives it dim sympathies with me who live,
Making it a companionable form,
Whose puny flaps and freaks the idling Spirit 20
By its own moods interprets, every where
Echo or mirror seeking of itself,
And makes a toy of Thought.

But O! how oft,
How oft, at school, with most believing mind,

Presageful, have I gazed upon the bars,
To watch that fluttering stranger! and as oft
With unclosed lids, already had I dreamt
Of my sweet birth-place, and the old church-tower,
Whose bells, the poor man's only music, rang
From morn to evening, all the hot Fair-day, 30
So sweetly, that they stirred and haunted me
With a wild pleasure, falling on mine ear
Most like articulate sounds of things to come!
So gazed I, till the soothing things I dreamt
Lulled me to sleep, and sleep prolonged my dreams!
And so I brooded all the following morn,
Awed by the stern preceptor's face, mine eye
Fixed with mock study on my swimming book:
Save if the door half opened, and I snatched
A hasty glance, and still my heart leaped up, 40
For still I hoped to see the *stranger's* face,
Townsman, or aunt, or sister more beloved,
My play-mate when we both were clothed alike!

Dear Babe, that sleepest cradled by my side,
Whose gentle breathings, heard in this deep calm,
Fill up the interspersed vacancies
And momentary pauses of the thought!
My babe so beautiful! it thrills my heart
With tender gladness, thus to look at thee,
And think that thou shalt learn far other lore 50
And in far other scenes! For I was reared
In the great city, pent 'mid cloisters dim,
And saw nought lovely but the sky and stars.
But thou, my babe! shalt wander like a breeze
By lakes and sandy shores, beneath the crags
Of ancient mountain, and beneath the clouds,
Which image in their bulk both lakes and shores
And mountain crags: so shalt thou see and hear
The lovely shapes and sounds intelligible
Of that eternal language, which thy God 60
Utters, who from eternity doth teach
Himself in all, and all things in himself.
Great universal Teacher! he shall mould
Thy spirit, and by giving make it ask.

Therefore all seasons shall be sweet to thee,
Whether the summer clothe the general earth
With greenness, or the redbreast sit and sing
Betwixt the tufts of snow on the bare branch
Of mossy apple-tree, while the nigh thatch
Smokes in the sun-thaw; whether the eave-drops fall 70
Heard only in the trances of the blast,
Or if the secret ministry of frost
Shall hang them up in silent icicles,
Quietly shining to the quiet Moon.

France: An Ode

ARGUMENT°

First Stanza. An invocation to those objects in Nature the contemplation of which had inspired the Poet with a devotional love of Liberty. *Second Stanza.* The exultation of the Poet at the commencement of the French Revolution, and his unqualified abhorrence of the Alliance against the Republic. *Third Stanza.* The blasphemies and horrors during the domination of the Terrorists regarded by the Poet as a transient storm, and as the natural consequence of the former despotism and of the foul superstition of Popery. Reason, indeed, began to suggest many apprehensions; yet still the Poet struggled to retain the hope that France would make conquests by no other means than by presenting to the observation of Europe a people more happy and better instructed than under other forms of Government. *Fourth Stanza.* Switzerland, and the Poet's recantation. *Fifth Stanza.* An address to Liberty, in which the Poet expresses his conviction that those feelings and that grand *ideal* of Freedom which the mind attains by its contemplation of its individual nature, and of the sublime surrounding objects (see Stanza the First) do not belong to men, as a society, nor can possibly be either gratified or realised, under any form of human government; but belong to the individual man, so far as he is pure, and inflamed with the love and adoration of God in Nature.

I

Ye Clouds! that far above me float and pause,
Whose pathless march no mortal may control!
Ye Ocean-Waves! that, wheresoe'er ye roll,
Yield homage only to eternal laws!

Ye Woods! that listen to the night-birds singing,
  Midway the smooth and perilous slope reclined,
Save when your own imperious branches swinging,
  Have made a solemn music of the wind!
Where, like a man beloved of God,
Through glooms, which never woodman trod, 10
    How oft, pursuing fancies holy,
My moonlight way o'er flowering weeds I wound,
    Inspired, beyond the guess of folly,
By each rude shape and wild unconquerable sound!
O ye loud Waves! and O ye Forests high!
  And O ye Clouds that far above me soared!
Thou rising Sun! thou blue rejoicing Sky!
  Yea, every thing that is and will be free!
  Bear witness for me, wheresoe'er ye be,
  With what deep worship I have still adored 20
    The spirit of divinest Liberty.

II

When France in wrath her giant-limbs upreared,
  And with that oath, which smote air, earth and sea,
  Stamped her strong foot and said she would be free,
Bear witness for me, how I hoped and feared!
With what a joy my lofty gratulation
  Unawed I sang, amid a slavish band:
And when to whelm the disenchanted nation,
  Like fiends embattled by a wizard's wand,
    The Monarchs marched in evil day, 30
    And Britain joined the dire array;
  Though dear her shores and circling ocean,
Though many friendships, many youthful loves
  Had swol'n the patriot emotion
And flung a magic light o'er all her hills and groves;
Yet still my voice, unaltered, sang defeat
  To all that braved the tyrant-quelling lance,
And shame too long delayed and vain retreat!
For ne'er, O Liberty! with partial aim
I dimmed thy light or damped thy holy flame; 40
  But blessed the paeans of delivered France,
And hung my head and wept at Britain's name.

III

'And what,' I said, 'though Blasphemy's loud scream
With that sweet music of deliverance strove!
Though all the fierce and drunken passions wove
A dance more wild than e'er was maniac's dream!
Ye storms, that round the dawning east assembled,
The Sun was rising, though ye hid his light!'
And when, to soothe my soul, that hoped and trembled,
The dissonance ceased, and all seemed calm and bright; 50
When France her front deep-scarr'd and gory
Concealed with clustering wreaths of glory;
When, insupportably advancing,
Her arm made mockery of the warrior's tramp;
While timid looks of fury glancing,
Domestic treason, crushed beneath her fatal stamp,
Writhed like a wounded dragon in his gore;
Then I reproached my fears that would not flee;
'And soon,' I said, 'shall Wisdom teach her lore
In the low huts of them that toil and groan! 60
And, conquering by her happiness alone,
Shall France compel the nations to be free,
Till Love and Joy look round, and call the Earth their own.'

IV

Forgive me, Freedom! O forgive those dreams!
I hear thy voice, I hear thy loud lament,
From bleak Helvetia's icy cavern sent—
I hear thy groans upon her blood-stained streams!
Heroes, that for your peaceful country perished,
And ye that, fleeing, spot your mountain-snows
With bleeding wounds; forgive me, that I cherished 70
One thought that ever blessed your cruel foes!
To scatter rage, and traitorous guilt,
Where Peace her jealous home had built;
A patriot-race to disinherit
Of all that made their stormy wilds so dear;
And with inexpiable spirit
To taint the bloodless freedom of the mountaineer—
O France, that mockest Heaven, adulterous, blind,
A patriot only in pernicious toils,

Are these thy boasts, Champion of human kind? 80
To mix with Kings in the low lust of sway,
Yell in the hunt, and share the murderous prey;
To insult the shrine of Liberty with spoils
From freemen torn; to tempt and to betray?

V

The Sensual and the Dark rebel in vain,
Slaves by their own compulsion! In mad game
They burst their manacles and wear the name
Of Freedom, graven on a heavier chain!
O Liberty! with profitless endeavour
Have I pursued thee, many a weary hour; 90
But thou nor swell'st the victor's strain, nor ever
Didst breathe thy soul in forms of human power.
Alike from all, howe'er they praise thee,
(Nor prayer, nor boastful name delays thee)
Alike from Priestcraft's harpy minions,
And factious Blasphemy's obscener slaves,
Thou speedest on thy subtle pinions,
The guide of homeless winds, and playmate of the waves!
And there I felt thee!—on that sea-cliff's verge,
Whose pines, scarce travelled by the breeze above, 100
Had made one murmur with the distant surge!
Yes, while I stood and gazed, my temples bare,
And shot my being through earth, sea and air,
Possessing all things with intensest love,
O Liberty! my spirit felt thee there.

Fears in Solitude

Written in April 1798, during the Alarm of an Invasion

A green and silent spot, amid the hills,
A small and silent dell! O'er stiller place
No singing sky-lark ever poised himself.
The hills are heathy, save that swelling slope,
Which hath a gay and gorgeous covering on,

All golden with the never-bloomless furze,
Which now blooms most profusely: but the dell,
Bathed by the mist, is fresh and delicate
As vernal corn-field, or the unripe flax,
When, through its half-transparent stalks, at eve, 10
The level sunshine glimmers with green light.
Oh! 'tis a quiet spirit-healing nook!
Which all, methinks, would love; but chiefly he,
The humble man, who, in his youthful years,
Knew just so much of folly, as had made
His early manhood more securely wise!
Here he might lie on fern or withered heath,
While from the singing lark (that sings unseen
The minstrelsy that solitude loves best),
And from the sun, and from the breezy air, 20
Sweet influences trembled o'er his frame;
And he, with many feelings, many thoughts,
Made up a meditative joy, and found
Religious meanings in the forms of nature!
And so, his senses gradually wrapt
In a half sleep, he dreams of better worlds,
And dreaming hears thee still, O singing lark,
That singest like an angel in the clouds!

My God! it is a melancholy thing
For such a man, who would full fain preserve 30
His soul in calmness, yet perforce must feel
For all his human brethren—O my God!
It weighs upon the heart, that he must think
What uproar and what strife may now be stirring
This way or that way o'er these silent hills—
Invasion, and the thunder and the shout,
And all the crash of onset; fear and rage,
And undetermined conflict—even now,
Even now, perchance, and in his native isle;
Carnage and groans beneath this blessed sun! 40
We have offended, Oh! my countrymen!
We have offended very grievously,
And been most tyrannous. From east to west
A groan of accusation pierces Heaven!
The wretched plead against us; multitudes

Countless and vehement, the sons of God,
Our brethren! Like a cloud that travels on,
Steamed up from Cairo's swamps of pestilence,
Even so, my countrymen! have we gone forth
And borne to distant tribes slavery and pangs, 50
And, deadlier far, our vices, whose deep taint
With slow perdition murders the whole man,
His body and his soul! Meanwhile, at home,
All individual dignity and power
Engulfed in courts, committees, institutions,
Associations and societies,
A vain, speech-mouthing, speech-reporting guild,
One benefit-club for mutual flattery,
We have drunk up, demure as at a grace,
Pollutions from the brimming cup of wealth; 60
Contemptuous of all honourable rule,
Yet bartering freedom and the poor man's life
For gold, as at a market! The sweet words
Of Christian promise, words that even yet
Might stem destruction, were they wisely preached,
Are muttered o'er by men, whose tones proclaim
How flat and wearisome they feel their trade:
Rank scoffers some, but most too indolent
To deem them falsehoods or to know their truth.
Oh! blasphemous! the Book of Life is made 70
A superstitious instrument, on which
We gabble o'er the oaths we mean to break;
For all must swear—all and in every place,
College and wharf, council and justice-court;
All, all must swear, the briber and the bribed,
Merchant and lawyer, senator and priest,
The rich, the poor, the old man and the young;
All, all make up one scheme of perjury,
That faith doth reel; the very name of God
Sounds like a juggler's charm; and, bold with joy, 80
Forth from his dark and lonely hiding-place,
(Portentous sight!) the owlet Atheism,
Sailing on obscene wings athwart the noon,
Drops his blue-fringed lids, and holds them close,
And hooting at the glorious sun in Heaven,
Cries out, 'Where is it?'

Thankless too for peace,
(Peace long preserved by fleets and perilous seas)
Secure from actual warfare, we have loved
To swell the war-whoop, passionate for war!
Alas! for ages ignorant of all 90
Its ghastlier workings, (famine or blue plague,
Battle, or siege, or flight through wintry snows,)
We, this whole people, have been clamorous
For war and bloodshed; animating sports,
The which we pay for as a thing to talk of,
Spectators and not combatants! No guess
Anticipative of a wrong unfelt,
No speculation on contingency,
However dim and vague, too vague and dim
To yield a justifying cause; and forth, 100
(Stuffed out with big preamble, holy names,
And adjurations of the God in Heaven,)
We send our mandates for the certain death
Of thousands and ten thousands! Boys and girls,
And women, that would groan to see a child
Pull off an insect's leg, all read of war,
The best amusement for our morning meal!
The poor wretch, who has learnt his only prayers
From curses, who knows scarcely words enough
To ask a blessing from his Heavenly Father, 110
Becomes a fluent phraseman, absolute
And technical in victories and defeats,
And all our dainty terms for fratricide;
Terms which we trundle smoothly o'er our tongues
Like mere abstractions, empty sounds to which
We join no feeling and attach no form!
As if the soldier died without a wound;
As if the fibres of this godlike frame
Were gored without a pang; as if the wretch,
Who fell in battle, doing bloody deeds, 120
Passed off to Heaven, translated and not killed;
As though he had no wife to pine for him,
No God to judge him! Therefore, evil days
Are coming on us, O my countrymen!
And what if all-avenging Providence,
Strong and retributive, should make us know

The meaning of our words, force us to feel
The desolation and the agony
Of our fierce doings?

Spare us yet awhile,
Father and God! O! spare us yet awhile! 130
Oh! let not English women drag their flight
Fainting beneath the burthen of their babes,
Of the sweet infants, that but yesterday
Laughed at the breast! Sons, brothers, husbands, all
Who ever gazed with fondness on the forms
Which grew up with you round the same fire-side,
And all who ever heard the sabbath-bells
Without the infidel's scorn, make yourselves pure!
Stand forth! be men! repel an impious foe,
Impious and false, a light yet cruel race, 140
Who laugh away all virtue, mingling mirth
With deeds of murder; and still promising
Freedom, themselves too sensual to be free,
Poison life's amities, and cheat the heart
Of faith and quiet hope, and all that soothes,
And all that lifts the spirit! Stand we forth;
Render them back upon the insulted ocean,
And let them toss as idly on its waves
As the vile sea-weed, which some mountain-blast
Swept from our shores! And oh! may we return 150
Not with a drunken triumph, but with fear,
Repenting of the wrongs with which we stung
So fierce a foe to frenzy!

I have told,
O Britons! O my brethren! I have told
Most bitter truth, but without bitterness.
Nor deem my zeal or factious or mistimed;
For never can true courage dwell with them,
Who, playing tricks with conscience, dare not look
At their own vices. We have been too long
Dupes of a deep delusion! Some, belike, 160
Groaning with restless enmity, expect
All change from change of constituted power;
As if a Government had been a robe,

On which our vice and wretchedness were tagged
Like fancy-points and fringes, with the robe
Pulled off at pleasure. Fondly these attach
A radical causation to a few
Poor drudges of chastising Providence,
Who borrow all their hues and qualities
From our own folly and rank wickedness, 170
Which gave them birth and nursed them. Others, meanwhile,
Dote with a mad idolatry; and all
Who will not fall before their images,
And yield them worship, they are enemies
Even of their country!

Such have I been deemed—
But, O dear Britain! O my Mother Isle!
Needs must thou prove a name most dear and holy
To me, a son, a brother, and a friend,
A husband, and a father! who revere
All bonds of natural love, and find them all 180
Within the limits of thy rocky shores.
O native Britain! O my Mother Isle!
How shouldst thou prove aught else but dear and holy
To me, who from thy lakes and mountain-hills,
Thy clouds, thy quiet dales, thy rocks and seas,
Have drunk in all my intellectual life,
All sweet sensations, all ennobling thoughts,
All adoration of the God in nature,
All lovely and all honourable things,
Whatever makes this mortal spirit feel 190
The joy and greatness of its future being?
There lives nor form nor feeling in my soul
Unborrowed from my country! O divine
And beauteous island! thou hast been my sole
And most magnificent temple, in the which
I walk with awe, and sing my stately songs,
Loving the God that made me!—

May my fears,
My filial fears, be vain! and may the vaunts
And menace of the vengeful enemy

Pass like the gust, that roared and died away 200
In the distant tree: which heard, and only heard
In this low dell, bowed not the delicate grass.

But now the gentle dew-fall sends abroad
The fruit-like perfume of the golden furze:
The light has left the summit of the hill,
Though still a sunny gleam lies beautiful,
Aslant the ivied beacon. Now farewell,
Farewell, awhile, O soft and silent spot!
On the green sheep-track, up the heathy hill,
Homeward I wind my way; and lo! recalled 210
From bodings that have well-nigh wearied me,
I find myself upon the brow, and pause
Startled! And after lonely sojourning
In such a quiet and surrounded nook,
This burst of prospect, here the shadowy main,
Dim tinted, there the mighty majesty
Of that huge amphitheatre of rich
And elmy fields, seems like society—
Conversing with the mind, and giving it
A livelier impulse and a dance of thought! 220
And now, beloved Stowey! I behold
Thy church-tower, and, methinks, the four huge elms
Clustering, which mark the mansion of my friend;
And close behind them, hidden from my view,
Is my own lowly cottage, where my babe
And my babe's mother dwell in peace! With light
And quickened footsteps thitherward I tend,
Remembering thee, O green and silent dell!
And grateful, that by nature's quietness
And solitary musings, all my heart 230
Is softened, and made worthy to indulge
Love, and the thoughts that yearn for human kind.

The Nightingale

A Conversation Poem. April 1798

No cloud, no relique of the sunken day
Distinguishes the West, no long thin slip
Of sullen light, no obscure trembling hues.
Come, we will rest on this old mossy bridge!
You see the glimmer of the stream beneath,
But hear no murmuring: it flows silently,
O'er its soft bed of verdure. All is still,
A balmy night! and though the stars be dim,
Yet let us think upon the vernal showers
That gladden the green earth, and we shall find 10
A pleasure in the dimness of the stars.
And hark! the Nightingale begins its song,
'Most musical, most melancholy' bird!°
A melancholy bird? Oh! idle thought!
In nature there is nothing melancholy.
But some night-wandering man, whose heart was pierced
With the remembrance of a grievous wrong,
Or slow distemper, or neglected love,
(And so, poor wretch! filled all things with himself,
And made all gentle sounds tell back the tale 20
Of his own sorrow) he, and such as he,
First named these notes a melancholy strain.
And many a poet echoes the conceit;
Poet who hath been building up the rhyme
When he had better far have stretched his limbs
Beside a brook in mossy forest-dell,
By sun or moon-light, to the influxes
Of shapes and sounds and shifting elements
Surrendering his whole spirit, of his song
And of his fame forgetful! so his fame 30
Should share in Nature's immortality,
A venerable thing! and so his song
Should make all Nature lovelier, and itself
Be loved like Nature! But 'twill not be so;
And youths and maidens most poetical,
Who lose the deepening twilights of the spring

In ball-rooms and hot theatres, they still
Full of meek sympathy must heave their sighs
O'er Philomela's pity-pleading strains.

My Friend, and thou, our Sister! we have learnt 40
A different lore: we may not thus profane
Nature's sweet voices, always full of love
And joyance! 'Tis the merry Nightingale
That crowds, and hurries, and precipitates
With fast thick warble his delicious notes,
As he were fearful that an April night
Would be too short for him to utter forth
His love-chant, and disburthen his full soul
Of all its music!

And I know a grove
Of large extent, hard by a castle huge, 50
Which the great lord inhabits not; and so
This grove is wild with tangling underwood,
And the trim walks are broken up, and grass,
Thin grass and king-cups grow within the paths.
But never elsewhere in one place I knew
So many nightingales; and far and near,
In wood and thicket, over the wide grove,
They answer and provoke each other's song,
With skirmish and capricious passagings,
And murmurs musical and swift jug jug, 60
And one low piping sound more sweet than all—
Stirring the air with such a harmony,
That should you close your eyes, you might almost
Forget it was not day! On moonlight bushes,
Whose dewy leaflets are but half-disclosed,
You may perchance behold them on the twigs,
Their bright, bright eyes, their eyes both bright and full,
Glistening, while many a glow-worm in the shade
Lights up her love-torch.

A most gentle Maid,
Who dwelleth in her hospitable home 70
Hard by the castle, and at latest eve
(Even like a Lady vowed and dedicate

To something more than Nature in the grove)
Glides through the pathways; she knows all their notes,
That gentle Maid! and oft a moment's space,
What time the moon was lost behind a cloud,
Hath heard a pause of silence; till the moon
Emerging, hath awakened earth and sky
With one sensation, and those wakeful birds
Have all burst forth in choral minstrelsy, 80
As if some sudden gale had swept at once
A hundred airy harps! And she hath watched
Many a nightingale perch giddily
On blossomy twig still swinging from the breeze,
And to that motion tune his wanton song
Like tipsy Joy that reels with tossing head.

Farewell, O Warbler! till to-morrow eve,
And you, my friends! farewell, a short farewell!
We have been loitering long and pleasantly,
And now for our dear homes.—That strain again! 90
Full fain it would delay me! My dear babe,
Who, capable of no articulate sound,
Mars all things with his imitative lisp,
How he would place his hand beside his ear,
His little hand, the small forefinger up,
And bid us listen! And I deem it wise
To make him Nature's play-mate. He knows well
The evening-star; and once, when he awoke
In most distressful mood (some inward pain
Had made up that strange thing, an infant's dream,—) 100
I hurried with him to our orchard-plot,
And he beheld the moon, and, hushed at once,
Suspends his sobs, and laughs most silently,
While his fair eyes, that swam with undropped tears,
Did glitter in the yellow moon-beam! Well!—
It is a father's tale: But if that Heaven
Should give me life, his childhood shall grow up
Familiar with these songs, that with the night
He may associate joy.—Once more, farewell,
Sweet Nightingale! once more, my friends! farewell. 110

Kubla Khan: Or, A Vision in a Dream

A Fragment

In the summer of the year 1797, the Author, then in ill health, had retired to a lonely farm house between Porlock and Linton, on the Exmoor confines of Somerset and Devonshire. In consequence of a slight indisposition, an anodyne had been prescribed, from the effects of which he fell asleep in his chair at the moment that he was reading the following sentence, or words of the same substance, in 'Purchas's Pilgrimage': 'Here the Khan Kubla commanded a palace to be built, and a stately garden thereunto. And thus ten miles of fertile ground were inclosed with a wall.' The author continued for about three hours in a profound sleep, at least of the external senses, during which time he has the most vivid confidence, that he could not have composed less than from two to three hundred lines; if that indeed can be called composition in which all the images rose up before him as things, with a parallel production of the correspondent expressions, without any sensation or consciousness of effort. On awaking he appeared to himself to have a distinct recollection of the whole, and taking his pen, ink, and paper, instantly and eagerly wrote down the lines that are here preserved. At this moment he was unfortunately called out by a person on business from Porlock, and detained by him above an hour, and on his return to his room, found, to his no small surprise and mortification, that though he still retained some vague and dim recollection of the general purport of the vision, yet, with the exception of some eight or ten scattered lines and images, all the rest had passed away like the images on the surface of a stream into which a stone has been cast, but, alas! without the after restoration of the latter! . . .°

KUBLA KHAN

In Xanadu did Kubla Khan
A stately pleasure-dome decree:
Where Alph, the sacred river, ran
Through caverns measureless to man
  Down to a sunless sea.
So twice five miles of fertile ground
With walls and towers were girdled round:
And there were gardens bright with sinuous rills
Where blossomed many an incense-bearing tree;
And here were forests ancient as the hills, 10
Enfolding sunny spots of greenery.

But oh! that deep romantic chasm which slanted
Down the green hill athwart a cedarn cover!
A savage place! as holy and enchanted
As e'er beneath a waning moon was haunted
By woman wailing for her demon-lover!
And from this chasm, with ceaseless turmoil seething,
As if this earth in fast thick pants were breathing,
A mighty fountain momently was forced:
Amid whose swift half-intermitted burst 20
Huge fragments vaulted like rebounding hail,
Or chaffy grain beneath the thresher's flail:
And 'mid these dancing rocks at once and ever
It flung up momently the sacred river.
Five miles meandering with a mazy motion
Through wood and dale the sacred river ran,
Then reached the caverns measureless to man,
And sank in tumult to a lifeless ocean:
And 'mid this tumult Kubla heard from far
Ancestral voices prophesying war! 30

The shadow of the dome of pleasure
Floated midway on the waves;
Where was heard the mingled measure
From the fountain and the caves.
It was a miracle of rare device,
A sunny pleasure-dome with caves of ice!

A damsel with a dulcimer
In a vision once I saw:
It was an Abyssinian maid,
And on her dulcimer she played, 40
Singing of Mount Abora.
Could I revive within me
Her symphony and song,
To such a deep delight 'twould win me,
That with music loud and long,
I would build that dome in air,
That sunny dome! those caves of ice!
And all who heard should see them there,
And all should cry, Beware! Beware!
His flashing eyes, his floating hair! 50

Weave a circle round him thrice,
And close your eyes with holy dread,
For he on honey-dew hath fed,
And drunk the milk of Paradise.

Recantation°

Illustrated in the Story of the Mad Ox

I

An Ox, long fed with musty hay,
  And work'd with yoke and chain,
Was turn'd out on an April day,
When fields are in their best array,
And growing grasses sparkle gay
  At once with Sun and rain.

II

The grass was fine, the Sun was bright—
  With truth I may aver it;
The ox was glad, as well he might,
Thought a green meadow no bad sight, 10
And frisk'd,—to shew his huge delight,
  Much like a beast of spirit.

III

Stop, neighbours, stop, why these alarms?
  The ox is only glad—
But still they pour from cots and farms—
Halloo! the parish is up in arms,
(A hoaxing-hunt has always charms)
  Halloo! the ox is mad.

IV

The frightened beast scamper'd about;
  Plunge! through the hedge he drove— 20
The mob pursue with hideous rout,
A bull-dog fastens on his snout;
He gores the dog! his tongue hangs out;
  He's mad, he's mad, by Jove!

V

'*Stop*, *Neighbours*, *stop!*' aloud did call
  A sage of sober hue.
But all at once, on him they fall,
And women squeak and children squall,
'What? would you have him toss us all?
  And damme! who are you?' 30

VI

Oh! hapless sage, his ears they stun,
  And curse him o'er and o'er!
'You bloody-minded dog! cries one,
To slit your windpipe were good fun,
'Od blast you for an *impious* son
  Of a presbyterian wh–re.'

VII

'You'd have him gore the parish-priest,
  And run against the altar—
You fiend!' The sage his warnings ceas'd,
And north and south, and west and east, 40
Halloo! they follow the poor beast,
  Mat, Dick, Tom, Bob and Walter.

VIII

Old Lewis, ('twas his evil day)
  Stood trembling in his shoes;
The ox was his—what cou'd he say?
His legs were stiffen'd with dismay,
The ox ran o'er him mid the fray,
  And give him his death's bruise.

IX

The frighted beast ran on—but here,
  (No tale, tho' in print, more true is) 50
My Muse stops short in mid career—
Nay, gentle reader, do not sneer!
I cannot chuse but drop a tear,
  A tear for good old Lewis!

X

The frighted beast ran through the town,
  All follow'd, boy and dad,
Bull-dog, Parson, Shopman, Clown:
The Publicans rush'd from the Crown,
'Halloo! hamstring him! cut him down!'
  THEY DROVE THE POOR OX MAD. 60

XI

Should you a Rat to madness teize,
  Why e'en a Rat may plague you:
There's no Philosopher but sees
That Rage and Fear are one disease—
Though that may burn, and this may freeze,
  They're both alike the Ague.

XII

And so this Ox, in frantic mood,
  Faced round like any Bull—
The mob turn'd tail, and he pursued,
Till they with heat and fright were stewed, 70
And not a chick of all this brood
  But had his belly full.

XIII

Old Nick's astride the beast, 'tis clear—
  Old Nicholas, to a tittle!
But all agree, he'd disappear,
Would but the Parson venture near,
And through his teeth, right o'er the steer,
  Squirt out some fasting-spittle.°

XIV

Achilles was a warrior fleet, 80
  The Trojans he could worry—
Our Parson too was swift of feet,
But shew'd it chiefly in retreat:
The victor Ox scour'd down the street,
  The mob fled hurry-scurry.

XV

Through gardens, lanes and fields new plough'd,
  Through his hedge, and through her hedge,
He plung'd and toss'd and bellow'd loud—
Till in his madness he grew proud
To see this helter-skelter crowd
  That had more wrath than courage. 90

XVI

Alas! to mend the breaches wide
  He made for these poor ninnies,
They all must work, whate'er betide,
Both days and months, and pay beside,
(Sad news for Avarice and for Pride)
  A sight of golden guineas!

XVII

But here once more to view did pop
  The man that kept his senses;
And now he cried—'Stop, neighbours, stop!
The Ox is mad! I would not swop, 100
No! not a school-boy's farthing-top
  For all the parish-fences.'

XVIII

'The Ox is mad! Ho! Dick, Bob, Mat!'
  What means this coward fuss?
'Ho! stretch this rope across the plat—
'Twill trip him up—or if not that,
Why, damme! we must lay him flat—
  See! here's my blunderbuss.'

XIX

'*A lying dog! just now he said*
  The Ox was only glad— 110
Let's break his presbyterian head!'
'Hush!' quoth the sage, 'you've been misled;
No quarrels now—let's all make head—
  YOU DROVE THE POOR OX MAD.'

XX

As thus I sat, in careless chat,
   With the morning's wet newspaper,
In eager haste, without his hat,
As blind and blund'ring as a bat,
In came that fierce Aristocrat,
   Our pursy Woollen-draper. 120

XXI

And so my Muse perforce drew bit;
   And in he rush'd and panted—
'Well, have you heard?' No, not a whit.
'What, *ha'nt* you heard?' Come, out with it!—
'That TIERNEY votes for Mister Pitt,
   And SHERIDAN's *recanted*!'

Lines

Written in the Album at Elbingerode,
in the Hartz Forest

I stood on Brocken's sovran height, and saw
Woods crowding upon woods, hills over hills,
A surging scene, and only limited
By the blue distance. Heavily my way
Downward I dragged through fir groves evermore,
Where bright green moss heaves in sepulchral forms
Speckled with sunshine; and, but seldom heard,
The sweet bird's song became a hollow sound;
And the breeze, murmuring indivisibly,
Preserved its solemn murmur most distinct 10
From many a note of many a waterfall,
And the brook's chatter; 'mid whose islet stones
The dingy kidling with its tinkling bell
Leaped frolicsome, or old romantic goat
Sat, his white beard slow waving. I moved on
In low and languid mood: for I had found
That outward forms, the loftiest, still receive
Their finer influence from the Life within;—

Fair cyphers else: fair, but of import vague
Or unconcerning, where the heart not finds 20
History or prophecy of friend, or child,
Or gentle maid, our first and early love,
Or father, or the venerable name
Of our adored country! O thou Queen,
Thou delegated Deity of Earth,
O dear, dear England! how my longing eye
Turned westward, shaping in the steady clouds
Thy sands and high white cliffs!

My native Land!
Filled with the thought of thee this heart was proud,
Yea, mine eye swam with tears: that all the view 30
From sovran Brocken, woods and woody hills,
Floated away, like a departing dream,
Feeble and dim! Stranger, these impulses
Blame thou not lightly; nor will I profane,
With hasty judgment or injurious doubt,
That man's sublimer spirit, who can feel
That God is everywhere! the God who framed
Mankind to be one mighty family,
Himself our Father, and the World our Home.

Love

All thoughts, all passions, all delights,
Whatever stirs this mortal frame,
All are but ministers of Love,
And feed his sacred flame.

Oft in my waking dreams do I
Live o'er again that happy hour,
When midway on the mount I lay,
Beside the ruined tower.

The moonshine, stealing o'er the scene
Had blended with the lights of eve; 10
And she was there, my hope, my joy,
My own dear Genevieve!

She leant against the armed man,
The statue of the armed knight;
She stood and listened to my lay,
    Amid the lingering light.

Few sorrows hath she of her own,
My hope! my joy! my Genevieve!
She loves me best, whene'er I sing
    The songs that make her grieve. 20

I played a soft and doleful air,
I sang an old and moving story—
An old rude song, that suited well
    That ruin wild and hoary.

She listened with a flitting blush,
With downcast eyes and modest grace;
For well she knew, I could not choose
    But gaze upon her face.

I told her of the Knight that wore
Upon his shield a burning brand; 30
And that for ten long years he wooed
    The Lady of the Land.

I told her how he pined: and ah!
The deep, the low, the pleading tone
With which I sang another's love,
    Interpreted my own.

She listened with a flitting blush,
With downcast eyes, and modest grace;
And she forgave me, that I gazed
    Too fondly on her face! 40

But when I told the cruel scorn
That crazed that bold and lovely Knight,
And that he crossed the mountain-woods,
    Nor rested day nor night;

That sometimes from the savage den,
And sometimes from the darksome shade,
And sometimes starting up at once
    In green and sunny glade,—

There came and looked him in the face
An angel beautiful and bright; 50
And that he knew it was a Fiend,
    This miserable Knight!

And that unknowing what he did,
He leaped amid a murderous band,
And saved from outrage worse than death
    The Lady of the Land!

And how she wept, and clasped his knees;
And how she tended him in vain—
And ever strove to expiate
    The scorn that crazed his brain;— 60

And that she nursed him in a cave;
And how his madness went away,
When on the yellow forest-leaves
    A dying man he lay;—

His dying words—but when I reached
That tenderest strain of all the ditty,
My faltering voice and pausing harp
    Disturbed her soul with pity!

All impulses of soul and sense
Had thrilled my guileless Genevieve; 70
The music and the doleful tale,
    The rich and balmy eve;

And hopes, and fears that kindle hope,
An undistinguishable throng,
And gentle wishes long subdued,
    Subdued and cherished long!

She wept with pity and delight,
She blushed with love, and virgin-shame;
And like the murmur of a dream,
I heard her breathe my name. 80

Her bosom heaved—she stepped aside,
As conscious of my look she stept—
Then suddenly, with timorous eye
She fled to me and wept.

She half inclosed me with her arms,
She pressed me with a meek embrace;
And bending back her head, looked up,
And gazed upon my face.

'Twas partly love, and partly fear,
And partly 'twas a bashful art, 90
That I might rather feel, than see,
The swelling of her heart.

I calmed her fears, and she was calm,
And told her love with virgin pride;
And so I won my Genevieve,
My bright and beauteous Bride.

Apologia Pro Vita Sua [*Defence of his own life*]

The poet in his lone yet genial hour
Gives to his eyes a magnifying power:
Or rather he emancipates his eyes
From the black shapeless accidents of size—
In unctuous cones of kindling coal,
Or smoke upwreathing from the pipe's trim bole,
His gifted ken can see
Phantoms of sublimity.

Dejection: An Ode°

Late, late yestreen I saw the new Moon,
With the old Moon in her arms;
And I fear, I fear, my Master dear!
We shall have a deadly storm.
Ballad of Sir Patrick Spence.

I

Well! If the Bard was weather-wise, who made
The grand old ballad of Sir Patrick Spence,
This night, so tranquil now, will not go hence
Unroused by winds, that ply a busier trade
Than those which mould yon cloud in lazy flakes,
Or the dull sobbing draft, that moans and rakes
Upon the strings of this Eolian lute,°
Which better far were mute.
For lo! the New-moon winter-bright!
And overspread with phantom light, 10
(With swimming phantom light o'erspread
But rimmed and circled by a silver thread)
I see the old Moon in her lap, foretelling
The coming on of rain and squally blast.
And oh! that even now the gust were swelling,
And the slant night-shower driving loud and fast!
Those sounds which oft have raised me, whilst they awed,
And sent my soul abroad,
Might now perhaps their wonted impulse give,
Might startle this dull pain, and make it move and live! 20

II

A grief without a pang, void, dark, and drear,
A stifled, drowsy, unimpassioned grief,
Which finds no natural outlet, no relief,
In word, or sigh, or tear—
O Lady! in this wan and heartless mood,
To other thoughts by yonder throstle woo'd,
All this long eve, so balmy and serene,
Have I been gazing on the western sky,
And its peculiar tint of yellow green:
And still I gaze—and with how blank an eye! 30

And those thin clouds above, in flakes and bars,
That give away their motion to the stars;
Those stars, that glide behind them or between,
Now sparkling, now bedimmed, but always seen:
Yon crescent Moon as fixed as if it grew
In its own cloudless, starless lake of blue;
I see them all so excellently fair,
I see, not feel, how beautiful they are!

III

My genial spirits fail;
And what can these avail 40
To lift the smothering weight from off my breast?
It were a vain endeavour,
Though I should gaze for ever
On that green light that lingers in the west:
I may not hope from outward forms to win
The passion and the life, whose fountains are within.

IV

O Lady! we receive but what we give,
And in our life alone does nature live:
Ours is her wedding-garment, ours her shroud!
And would we aught behold, of higher worth, 50
Than that inanimate cold world allowed
To the poor loveless ever-anxious crowd,
Ah! from the soul itself must issue forth
A light, a glory, a fair luminous cloud
Enveloping the Earth—
And from the soul itself must there be sent
A sweet and potent voice, of its own birth,
Of all sweet sounds the life and element!

V

O pure of heart! thou need'st not ask of me
What this strong music in the soul may be! 60
What, and wherein it doth exist,
This light, this glory, this fair luminous mist,
This beautiful and beauty-making power.

Joy, virtuous Lady! Joy that ne'er was given,
Save to the pure, and in their purest hour,
Life, and Life's effluence, cloud at once and shower,
Joy, Lady! is the spirit and the power,
Which wedding Nature to us gives in dower
A new Earth and new Heaven,
Undreamt of by the sensual and the proud— 70
Joy is the sweet voice, Joy the luminous cloud—
We in ourselves rejoice!
And thence flows all that charms or ear or sight,
All melodies the echoes of that voice,
All colours a suffusion from that light.

VI

There was a time when, though my path was rough,
This joy within me dallied with distress,
And all misfortunes were but as the stuff
Whence Fancy made me dreams of happiness:
For hope grew round me, like the twining vine, 80
And fruits, and foliage, not my own, seemed mine.
But now afflictions bow me down to earth:
Nor care I that they rob me of my mirth,
But oh! each visitation
Suspends what nature gave me at my birth,
My shaping spirit of Imagination.
For not to think of what I needs must feel,
But to be still and patient, all I can;
And haply by abstruse research to steal
From my own nature all the natural man— 90
This was my sole resource, my only plan:
Till that which suits a part infects the whole,
And now is almost grown the habit of my soul.

VII

Hence, viper thoughts, that coil around my mind,
Reality's dark dream!
I turn from you, and listen to the wind,
Which long has raved unnoticed. What a scream
Of agony by torture lengthened out

That lute sent forth! Thou Wind, that rav'st without,
 Bare crag, or mountain-tairn,* or blasted tree, 100
Or pine-grove whither woodman never clomb,
Or lonely house, long held the witches' home,
 Methinks were fitter instruments for thee,
Mad Lutanist! who in this month of showers,
Of dark brown gardens, and of peeping flowers,
Mak'st Devils' yule, with worse than wintry song,
The blossoms, buds, and timorous leaves among.
 Thou Actor, perfect in all tragic sounds!
Thou mighty Poet, e'en to frenzy bold!
  What tell'st thou now about? 110
  'Tis of the rushing of an host in rout,
 With groans, of trampled men, with smarting wounds—
At once they groan with pain, and shudder with the cold!
But hush! there is a pause of deepest silence!
 And all that noise, as of a rushing crowd,
With groans, and tremulous shudderings—all is over—
 It tells another tale, with sounds less deep and loud!
  A tale of less affright,
  And tempered with delight,
As Otway's self had framed the tender lay, 120
  'Tis of a little child
  Upon a lonesome wild,
Not far from home, but she hath lost her way:
And now moans low in bitter grief and fear,
And now screams loud, and hopes to make her mother hear.

VIII

'Tis midnight, but small thoughts have I of sleep:
Full seldom may my friend such vigils keep!
Visit her, gentle Sleep! with wings of healing,
 And may this storm be but a mountain-birth,
May all the stars hang bright above her dwelling, 130
 Silent as though they watched the sleeping Earth!

* Tairn is a small lake, generally if not always applied to the lakes up in the mountains and which are the feeders of those in the valleys. This address to the Storm-wind, will not appear extravagant to those who have heard it at night and in a mountainous country.

With light heart may she rise,
Gay fancy, cheerful eyes,
Joy lift her spirit, joy attune her voice;
To her may all things live, from pole to pole,
Their life the eddying of her living soul!
O simple spirit, guided from above,
Dear Lady! friend devoutest of my choice,
Thus mayest thou ever, evermore rejoice.

Hymn

BEFORE SUN-RISE, IN THE VALE OF CHAMOUNI

Besides the Rivers, Arve and Arveiron, which have their sources in the foot of Mont Blanc, five conspicuous torrents rush down its sides; and within a few paces of the Glaciers, the Gentiana Major grows in immense numbers, with its 'flowers of loveliest blue.'

Hast thou a charm to stay the morning-star
In his steep course? So long he seems to pause
On thy bald awful head, O sovran Blanc,
The Arve and Arveiron at thy base
Rave ceaselessly; but thou, most awful Form!
Risest from forth thy silent sea of pines,
How silently! Around thee and above
Deep is the air and dark, substantial, black,
An ebon mass: methinks thou piercest it,
As with a wedge! But when I look again, 10
It is thine own calm home, thy crystal shrine,
Thy habitation from eternity!
O dread and silent Mount! I gazed upon thee,
Till thou, still present to the bodily sense,
Didst vanish from my thought: entranced in prayer
I worshipped the Invisible alone.

Yet, like some sweet beguiling melody,
So sweet, we know not we are listening to it,
Thou, the meanwhile, wast blending with my thought,
Yea, with my life and life's own secret joy: 20
Till the dilating Soul, enrapt, transfused,
Into the mighty vision passing—there
As in her natural form, swelled vast to Heaven!

Awake, my soul! not only passive praise
Thou owest! not alone these swelling tears,
Mute thanks and secret ecstasy! Awake,
Voice of sweet song! Awake, my heart, awake!
Green vales and icy cliffs, all join my Hymn.

Thou first and chief, sole sovereign of the Vale!
O struggling with the darkness all the night, 30
And visited all night by troops of stars,
Or when they climb the sky or when they sink:
Companion of the morning-star at dawn,
Thyself Earth's rosy star, and of the dawn
Co-herald: wake, O wake, and utter praise!
Who sank thy sunless pillars deep in Earth?
Who filled thy countenance with rosy light?
Who made thee parent of perpetual streams?

And you, ye five wild torrents fiercely glad!
Who called you forth from night and utter death, 40
From dark and icy caverns called you forth,
Down those precipitous, black, jagged Rocks,
For ever shattered and the same for ever?
Who gave you your invulnerable life,
Your strength, your speed, your fury, and your joy,
Unceasing thunder and eternal foam?
And who commanded (and the silence came),
Here let the billows stiffen, and have rest?

Ye ice-falls! ye that from the mountain's brow
Adown enormous ravines slope amain— 50
Torrents, methinks, that heard a mighty voice,
And stopped at once amid their maddest plunge!
Motionless torrents! silent cataracts!
Who made you glorious as the gates of Heaven
Beneath the keen full moon? Who bade the sun
Clothe you with rainbows? Who, with living flowers
Of loveliest blue, spread garlands at your feet?—
God! let the torrents, like a shout of nations,
Answer! and let the ice-plains echo, God!
God! sing ye meadow-streams with gladsome voice! 60

Ye pine-groves, with your soft and soul-like sounds!
And they too have a voice, yon piles of snow,
And in their perilous fall shall thunder, God!

Ye living flowers that skirt the eternal frost!
Ye wild goats sporting round the eagle's nest!
Ye eagles, play-mates of the mountain-storm!
Ye lightnings, the dread arrows of the clouds!
Ye signs and wonders of the element!
Utter forth God, and fill the hills with praise!

Thou too, hoar Mount! with thy sky-pointing peaks, 70
Oft from whose feet the avalanche, unheard,
Shoots downward, glittering through the pure serene
Into the depth of clouds, that veil thy breast—
Thou too again, stupendous Mountain! thou
That as I raise my head, awhile bowed low
In adoration, upward from thy base
Slow travelling with dim eyes suffused with tears,
Solemnly seemest, like a vapoury cloud,
To rise before me—Rise, O ever rise,
Rise like a cloud of incense, from the Earth! 80
Thou kingly Spirit throned among the hills,
Thou dread ambassador from Earth to Heaven,
Great hierarch! tell thou the silent sky,
And tell the stars, and tell yon rising sun
Earth, with her thousand voices, praises God.

Inscription

FOR A FOUNTAIN ON A HEATH

This Sycamore, oft musical with bees,—
Such tents the Patriarchs loved! O long unharmed
May all its aged boughs o'er-canopy
The small round basin, which this jutting stone
Keeps pure from falling leaves! Long may the Spring,
Quietly as a sleeping infant's breath,
Send up cold waters to the traveller
With soft and even pulse! Nor ever cease

Yon tiny cone of sand its soundless dance,
Which at the bottom, like a Fairy's page, 10
As merry and no taller, dances still,
Nor wrinkles the smooth surface of the Fount.
Here twilight is and coolness: here is moss,
A soft seat, and a deep and ample shade.
Thou may'st toil far and find no second tree.
Drink, Pilgrim, here; Here rest! and if thy heart
Be innocent, here too shalt thou refresh
Thy Spirit, listening to some gentle sound,
Or passing gale or hum of murmuring bees!

Answer to a Child's Question

Do you ask what the birds say? The sparrow, the dove,
The linnet and thrush say, 'I love and I love!'
In the winter they're silent—the wind is so strong;
What it says, I don't know, but it sings a loud song.
But green leaves, and blossoms, and sunny warm weather,
And singing, and loving—all come back together.
But the lark is so brimful of gladness and love,
The green fields below him, the blue sky above,
That he sings, and he sings; and for ever sings he—
'I love my Love, and my Love loves me!' 10

The Knight's Tomb

Where is the grave of Sir Arthur O'Kellyn?
Where may the grave of that good man be?—
By the side of a spring, on the breast of Helvellyn,
Under the twigs of a young birch tree!
The oak that in summer was sweet to hear,
And rustled its leaves in the fall of the year,
And whistled and roared in the winter alone,
Is gone,—and the birch in its stead is grown.—
The Knight's bones are dust,
And his good sword rust;— 10
His soul is with the saints, I trust.

The Pains of Sleep

Ere on my bed my limbs I lay,
It hath not been my use to pray
With moving lips or bended knees;
But silently, by slow degrees,
My spirit I to Love compose,
In humble trust mine eye-lids close,
With reverential resignation,
No wish conceived, no thought exprest,
Only a sense of supplication;
A sense o'er all my soul imprest 10
That I am weak, yet not unblest,
Since in me, round me, every where
Eternal strength and wisdom are.

But yester-night I prayed aloud
In anguish and in agony,
Up-starting from the fiendish crowd
Of shapes and thoughts that tortured me:
A lurid light, a trampling throng,
Sense of intolerable wrong,
And whom I scorned, those only strong! 20
Thirst of revenge, the powerless will
Still baffled, and yet burning still!
Desire with loathing strangely mixed
On wild or hateful objects fixed.
Fantastic passions! maddening brawl!
And shame and terror over all!
Deeds to be hid which were not hid,
Which all confused I could not know
Whether I suffered, or I did:
For all seemed guilt, remorse or woe, 30
My own or others still the same
Life-stifling fear, soul-stifling shame.

So two nights passed: the night's dismay
Saddened and stunned the coming day.
Sleep, the wide blessing, seemed to me
Distemper's worst calamity.

The third night, when my own loud scream
Had waked me from the fiendish dream,
O'ercome with sufferings strange and wild,
I wept as I had been a child; 40
And having thus by tears subdued
My anguish to a milder mood,
Such punishments, I said, were due
To natures deepliest stained with sin,—
For aye entempesting anew
The unfathomable hell within,
The horror of their deeds to view,
To know and loathe, yet wish and do!
Such griefs with such men well agree,
But wherefore, wherefore fall on me? 50
To be beloved is all I need,
And whom I love, I love indeed.

What Is Life?

Resembles life what once was deem'd of light,
Too ample in itself for human sight?
An absolute self—an element ungrounded—
All that we see, all colours of all shade
By encroach of darkness made?—
Is very life by consciousness unbounded?
And all the thoughts, pains, joys of mortal breath,
A war-embrace of wrestling life and death?

Constancy to an Ideal Object

Since all that beat about in Nature's range,
Or veer or vanish; why shouldst thou remain
The only constant in a world of change,
O yearning thought! that liv'st but in the brain?
Call to the hours, that in the distance play,
The faery people of the future day—
Fond thought! not one of all that shining swarm

Will breathe on thee with life-enkindling breath,
Till when, like strangers shelt'ring from a storm,
Hope and Despair meet in the porch of Death! 10
Yet still thou haunt'st me; and though well I see,
She is not thou, and only thou art she,
Still, still as though some dear embodied good,
Some living love before my eyes there stood
With answering look a ready ear to lend,
I mourn to thee and say—'Ah! loveliest friend!
That this the meed of all my toils might be,
To have a home, an English home, and thee!'
Vain repetition! Home and Thou are one.
The peacefull'st cot, the moon shall shine upon, 20
Lulled by the thrush and wakened by the lark,
Without thee were but a becalmed bark,
Whose helmsman on an ocean waste and wide
Sits mute and pale his mouldering helm beside.
And art thou nothing? Such thou art, as when
The woodman winding westward up the glen
At wintry dawn, where o'er the sheep-track's maze
The viewless snow-mist weaves a glist'ning haze,
Sees full before him, gliding without tread,
An image with a glory round its head;° 30
The enamoured rustic worships its fair hues,
Nor knows he makes the shadow he pursues!

Metrical Feet

Lesson for a Boy

Trōchĕe trīps frŏm lōng tŏ shōrt;
From long to long in solemn sort
Slōw Spōndĕe stālks; strōng fo͞ot! yet ill able
Ēvĕr tŏ cōme ŭp wĭth Dāctȳ̆l trĭsȳllăblĕ.
Ĭāmbĭcs mārch frŏm shōrt tŏ lōng;—
Wĭth ă lēap ănd ă bo͞und thĕ swĭft Ānăpaĕsts thrōng;
One syllable long, with one short at each side,
Ămphībrăchy̆s hāstes wĭth ă stātely̆ stride;—
Fīrst ănd lāst bēĭng lōng, mīddlĕ shōrt, Am̄phĭmācer
Strīkes hĭs thūndērĭng ho͞ofs līke ă pro͞ud hīgh-brĕd Rācer. 10

If Derwent be innocent, steady, and wise,
And delight in the things of earth, water, and skies;
Tender warmth at his heart, with these metres to show it,
With sound sense in his brains, may make Derwent a poet,—
May crown him with fame, and must win him the love
Of his father on earth and his Father above.
My dear, dear child!
Could you stand upon Skiddaw, you would not from its whole ridge
See a man who so loves you as your fond S. T. Coleridge.

Time, Real and Imaginary

An Allegory

On the wide level of a mountain's head,
(I knew not where, but 'twas some faery place)
Their pinions, ostrich-like, for sails out-spread,
Two lovely children run an endless race,
A sister and a brother!
This far outstripp'd the other;
Yet ever runs she with reverted face,
And looks and listens for the boy behind:
For he, alas! is blind!
O'er rough and smooth with even step he passed, 10
And knows not whether he be first or last.

To William Wordsworth

Composed on the Night after his Recitation of a Poem on the Growth of an Individual Mind

Friend of the wise! and teacher of the good!
Into my heart have I received that lay
More than historic, that prophetic lay
Wherein (high theme by thee first sung aright)
Of the foundations and the building up
Of a Human Spirit thou hast dared to tell

What may be told, to the understanding mind
Revealable; and what within the mind
By vital breathings secret as the soul
Of vernal growth, oft quickens in the heart 10
Thoughts all too deep for words!—

Theme hard as high!
Of smiles spontaneous, and mysterious fears
(The first-born they of Reason and twin-birth),
Of tides obedient to external force,
And currents self-determined, as might seem,
Or by some inner power; of moments awful,
Now in thy inner life, and now abroad,
When power streamed from thee, and thy soul received
The light reflected, as a light bestowed—
Of fancies fair, and milder hours of youth, 20
Hyblean murmurs of poetic thought
Industrious in its joy, in vales and glens
Native or outland, lakes and famous hills!
Or on the lonely high-road, when the stars
Were rising; or by secret mountain-streams,
The guides and the companions of thy way!

Of more than Fancy, of the Social Sense
Distending wide, and man beloved as man,
Where France in all her towns lay vibrating
Like some becalmed bark beneath the burst 30
Of Heaven's immediate thunder, when no cloud
Is visible, or shadow on the main.
For thou wert there, thine own brows garlanded,
Amid the tremor of a realm aglow,
Amid a mighty nation jubilant,
When from the general heart of human kind
Hope sprang forth like a full-born Deity!
—Of that dear Hope afflicted and struck down,
So summoned homeward, thenceforth calm and sure
From the dread watch-tower of man's absolute self, 40
With light unwaning on her eyes, to look
Far on—herself a glory to behold,
The Angel of the vision! Then (last strain)
Of Duty, chosen laws controlling choice,

Action and joy!—An Orphic song indeed,
A song divine of high and passionate thoughts
To their own music chanted!

O great Bard!
Ere yet that last strain dying awed the air,
With steadfast eye I viewed thee in the choir
Of ever-enduring men. The truly great 50
Have all one age, and from one visible space
Shed influence! They, both in power and act,
Are permanent, and Time is not with them,
Save as it worketh for them, they in it.
Nor less a sacred roll, than those of old,
And to be placed, as they, with gradual fame
Among the archives of mankind, thy work
Makes audible a linkéd lay of Truth,
Of Truth profound a sweet continuous lay,
Not learnt, but native, her own natural notes! 60
Ah! as I listened with a heart forlorn,
The pulses of my being beat anew:
And even as life returns upon the drowned,
Life's joy rekindling roused a throng of pains—
Keen pangs of Love, awakening as a babe
Turbulent, with an outcry in the heart;
And fears self-willed, that shunned the eye of hope;
And hope that scarce would know itself from fear;
Sense of past youth, and manhood come in vain,
And genius given, and knowledge won in vain; 70
And all which I had culled in wood-walks wild,
And all which patient toil had reared, and all,
Commune with thee had opened out—but flowers
Strewed on my corse, and borne upon my bier,
In the same coffin, for the self-same grave!

That way no more! and ill beseems it me,
Who came a welcomer in herald's guise,
Singing of glory, and futurity,
To wander back on such unhealthful road,
Plucking the poisons of self-harm! And ill 80
Such intertwine beseems triumphal wreaths
Strew'd before thy advancing!

Nor do thou,
Sage Bard! impair the memory of that hour
Of thy communion with my nobler mind
By pity or grief, already felt too long!
Nor let my words import more blame than needs.
The tumult rose and ceased: for peace is nigh
Where wisdom's voice has found a listening heart.
Amid the howl of more than wintry storms,
The halcyon hears the voice of vernal hours 90
Already on the wing.

Eve following eve,
Dear tranquil time, when the sweet sense of Home
Is sweetest! moments for their own sake hailed
And more desired, more precious, for thy song,
In silence listening, like a devout child,
My soul lay passive, by thy various strain
Driven as in surges now beneath the stars,
With momentary stars of my own birth,
Fair constellated foam, still darting off
Into the darkness; now a tranquil sea, 100
Outspread and bright, yet swelling to the moon.

And when—O Friend! my comforter and guide!
Strong in thyself, and powerful to give strength!—
Thy long sustained Song finally closed,
And thy deep voice had ceased—yet thou thyself
Wert still before my eyes, and round us both
That happy vision of beloved faces—
Scarce conscious, and yet conscious of its close
I sate, my being blended in one thought
(Thought was it? or aspiration? or resolve?) 110
Absorbed, yet hanging still upon the sound—
And when I rose, I found myself in prayer.

The Pang More Sharp Than All

An Allegory

I

He too has flitted from his secret nest,
Hope's last and dearest Child without a name!—
Has flitted from me, like the warmthless flame,
That makes false promise of a place of rest
To the tir'd Pilgrim's still believing mind;—
Or like some Elfin Knight in kingly court,
Who having won all guerdons in his sport,
Glides out of view, and whither none can find!

II

Yes! He hath flitted from me—with what aim,
Or why, I know not! 'Twas a home of bliss, 10
And he was innocent, as the pretty shame
Of babe, that tempts and shuns the menaced kiss,
From its twy-cluster'd hiding place of snow!
Pure as the babe, I ween, and all aglow
As the dear hopes, that swell the mother's breast—
Her eyes down gazing o'er her clasped charge;—
Yet gay as that twice happy father's kiss,
That well might glance aside, yet never miss,
Where the sweet mark emboss'd so sweet a targe—
Twice wretched he who hath been doubly blest! 20

III

Like a loose blossom on a gusty night
He flitted from me—and has left behind
(As if to them his faith he ne'er did plight)
Of either sex and answerable mind
Two playmates, twin-births of his foster-dame:—
The one a steady lad (Esteem he hight)
And Kindness is the gentler sister's name.
Dim likeness now, though fair she be and good,
Of that bright Boy who hath us all forsook;—
But in his full-eyed aspect when she stood, 30
And while her face reflected every look,
And in reflection kindled—she became
So like Him, that almost she seem'd the same!

IV

Ah! He is gone, and yet will not depart!—
Is with me still, yet I from Him exiled!
For still there lives within my secret heart
The magic image of the magic Child,
Which there He made up-grow by his strong art,
As in that crystal orb—wise Merlin's feat,—
The wondrous 'World of Glass,' wherein inisl'd 40
All long'd-for things their beings did repeat;—
And there He left it, like a Sylph beguiled,
To live and yearn and languish incomplete!

V

Can wit of man a heavier grief reveal?
Can sharper pang from hate or scorn arise?—
Yes! one more sharp there is that deeper lies,
Which fond Esteem but mocks when he would heal.
Yet neither scorn nor hate did it devise,
But sad compassion and atoning zeal!
One pang more blighting-keen than hope betray'd! 50
And this it is my woeful hap to feel,
When, at her Brother's hest, the twin-born Maid
With face averted and unsteady eyes
Her truant playmate's faded robe puts on;
And inly shrinking from her own disguise
Enacts the faery Boy that's lost and gone.
O worse than all! O pang all pangs above
Is Kindness counterfeiting absent Love!

A Tombless Epitaph

'Tis true, Idoloclastes Satyrane!°
(So call him, for so mingling blame with praise,
And smiles with anxious looks, his earliest friends,
Masking his birth-name, wont to character
His wild-wood fancy and impetuous zeal,)
'Tis true that, passionate for ancient truths,
And honouring with religious love the great
Of elder times, he hated to excess,

With an unquiet and intolerant scorn,
The hollow puppets of a hollow age, 10
Ever idolatrous, and changing ever
Its worthless idols! learning, power, and time,
(Too much of all) thus wasting in vain war
Of fervid colloquy. Sickness, 'tis true,
Whole years of weary days, besieged him close,
Even to the gates and inlets of his life!
But it is true, no less, that strenuous, firm,
And with a natural gladness, he maintained
The citadel unconquered, and in joy
Was strong to follow the delightful Muse. 20
For not a hidden path, that to the shades
Of the beloved Parnassian forest leads,°
Lurked undiscovered by him; not a rill
There issues from the fount of Hippocrene,
But he had traced it upward to its source,
Through open glade, dark glen, and secret dell,
Knew the gay wild flowers on its banks, and culled
Its med'cinable herbs. Yea, oft alone,
Piercing the long-neglected holy cave,
The haunt obscure of old Philosophy, 30
He bade with lifted torch its starry walls
Sparkle, as erst they sparkled to the flame
Of odorous lamps tended by Saint and Sage.
O framed for calmer times and nobler hearts!
O studious Poet, eloquent for truth!
Philosopher! contemning wealth and death,
Yet docile, childlike, full of Life and Love!
Here, rather than on monumental stone,
This record of thy worth thy Friend inscribes,
Thoughtful, with quiet tears upon his cheek. 40

The Visionary Hope

Sad lot, to have no hope! Though lowly kneeling
He fain would frame a prayer within his breast,
Would fain entreat for some sweet breath of healing,
That his sick body might have ease and rest;
He strove in vain! the dull sighs from his chest

Against his will the stifling load revealing,
Though Nature forced; though like some captive guest,
Some royal prisoner at his conqueror's feast,
An alien's restless mood but half concealing,
The sternness on his gentle brow confessed, 10
Sickness within and miserable feeling:
Though obscure pangs made curses of his dreams,
And dreaded sleep, each night repelled in vain,
Each night was scattered by its own loud screams:
Yet never could his heart command, though fain,
One deep full wish to be no more in pain.

That Hope, which was his inward bliss and boast,
Which waned and died, yet ever near him stood,
Though changed in nature, wander where he would—
For Love's despair is but Hope's pining ghost! 20
For this one hope he makes his hourly moan,
He wishes and can wish for this alone!
Pierced, as with light from Heaven, before its gleams
(So the love-stricken visionary deems)
Disease would vanish, like a summer shower,
Whose dews fling sunshine from the noon-tide bower!
Or let it stay! yet this one Hope should give
Such strength that he would bless his pains and live.

Limbo°

'Tis a strange place, this Limbo!—not a Place,
Yet name it so;—where Time and weary Space
Fettered from flight, with night-mare sense of fleeing,
Strive for their last crepuscular half-being;—
Lank Space, and scytheless Time with branny hands
Barren and soundless as the measuring sands,
Not mark'd by flit of Shades,—unmeaning they
As moonlight on the dial of the day!
But that is lovely—looks like human Time,—
An old man with a steady look sublime, 10
That stops his earthly task to watch the skies;
But he is blind—a Statue hath such eyes;—

Yet having moonward turn'd his face by chance,
Gazes the orb with moon-like countenance,
With scant white hairs, with foretop bald and high,
He gazes still,—his eyeless face all eye;—
As 'twere an organ full of silent sight,
His whole face seemeth to rejoice in light!
Lip touching lip, all moveless, bust and limb—
He seems to gaze at that which seems to gaze on him! 20
  No such sweet sights doth Limbo den immure,
Wall'd round, and made a spirit-jail secure,
By the mere horror of blank Naught-at-all,
Whose circumambience doth these ghosts enthrall.
A lurid thought is growthless, dull Privation,
Yet that is but a Purgatory curse;
Hell knows a fear far worse,
A fear—a future state;—'tis positive Negation!

Ne Plus Ultra°

Sole Positive of Night!
Antipathist of Light!
Fate's only essence! primal scorpion rod—
The one permitted opposite of God!—
Condensed blackness and abysmal storm
Compacted to one sceptre
Arms the Grasp enorm—
The Intercepter—
The Substance that still casts the shadow Death!—
The Dragon foul and fell— 10
The unrevealable,
And hidden one, whose breath
Gives wind and fuel to the fires of Hell!
Ah! sole despair
Of both th'eternities in Heaven!
Sole interdict of all-bedewing prayer,
The all-compassionate!
Save to the Lampads Seven°
Revealed to none of all th'Angelic State,
Save to the Lampads Seven, 20
That watch the throne of Heaven!

On Donne's Poetry°

With Donne, whose muse on dromedary trots,
Wreathe iron pokers into true-love knots;
Rhyme's sturdy cripple, fancy's maze and clue,
Wit's forge and fire-blast, meaning's press and screw.

Song

From *Zapolya*

A sunny shaft did I behold,
  From sky to earth it slanted:
And poised therein a bird so bold—
  Sweet bird, thou wert enchanted!
He sank, he rose, he twinkled, he trolled
  Within that shaft of sunny mist;
His eyes of fire, his beak of gold,
  All else of amethyst!

And thus he sang: 'Adieu! adieu!
Love's dreams prove seldom true. 10
The blossoms they make no delay:
The sparkling dew-drops will not stay.
  Sweet month of May,
    We must away;
      Far, far away!
        To-day! to-day!'

Hunting Song

From *Zapolya*

Up, up! ye dames, ye lasses gay!
To the meadows trip away.
'Tis you must tend the flocks this morn,
And scare the small birds from the corn.
Not a soul at home may stay:
  For the shepherds must go
  With lance and bow
To hunt the wolf in the woods to-day.

Leave the hearth and leave the house
To the cricket and the mouse: 10
Find grannam out a sunny seat,
With babe and lambkin at her feet.
Not a soul at home may stay:
For the shepherds must go
With lance and bow
To hunt the wolf in the woods to-day.

Fancy in Nubibus [*in the clouds*]

Or The Poet in the Clouds

O! it is pleasant, with a heart at ease,
Just after sunset, or by moonlight skies,
To make the shifting clouds be what you please,
Or let the easily persuaded eyes
Own each quaint likeness issuing from the mould
Of a friend's fancy; or with head bent low
And cheek aslant see rivers flow of gold
'Twixt crimson banks; and then, a traveller, go
From mount to mount through Cloudland, gorgeous land!
Or list'ning to the tide, with closed sight, 10
Be that blind bard, who on the Chian strand°
By those deep sounds possessed with inward light,
Beheld the Iliad and the Odyssee
Rise to the swelling of the voiceful sea.

A Character°

A bird, who for his other sins
Had liv'd amongst the Jacobins;
Though like a kitten amid rats,
Or callow tit in nest of bats,
He much abhorr'd all democrats;
Yet nathless stood in ill report
Of wishing ill to Church and Court,

Tho' he'd nor claw, nor tooth, nor sting,
And learnt to pipe God save the King;
Tho' each day did new feathers bring, 10
All swore he had a leathern wing;
Nor polish'd wing, nor feather'd tail,
Nor down-clad thigh would aught avail;
And tho'—his tongue devoid of gall—
He civilly assur'd them all:—
'A bird am I of Phoebus' breed,°
And on the sunflower cling and feed;
My name, good Sirs, is Thomas Tit!'
The bats would hail him brother cit,
Or, at the furthest, cousin-german. 20
At length the matter to determine,
He publicly denounced the vermin;
He spared the mouse, he prais'd the owl;
But bats were neither flesh nor fowl.
Blood-sucker, vampire, harpy, goul,
Came in full clatter from his throat,
Till his old nest-mates chang'd their note
To hireling, traitor, and turncoat,—
A base apostate who had sold
His very teeth and claws for gold;— 30
And then his feathers!—sharp the jest—
No doubt he feather'd well his nest!
'A Tit indeed! aye, tit for tat—
With place and title, brother Bat,
We soon shall see how well he'll play
Count Goldfinch, or Sir Joseph Jay!'
  Alas, poor Bird! and ill-bestarr'd—
Or rather let us say, poor Bard!
And henceforth quit the allegoric,
  With metaphor and simile, 40
For simple facts and style historic:—
Alas, poor Bard! no gold had he;
Behind another's team he stept,
And plough'd and sow'd, while others reapt;
The work was his, but theirs the glory,
Sic vos non vobis, his whole story.°
Besides, whate'er he wrote or said
Came from his heart as well as head;

And though he never left in lurch
His king, his country, or his church, 50
'Twas but to humour his own cynical
Contempt of doctrines Jacobinical;
To his own conscience only hearty,
'Twas but by chance he serv'd the party;—
The self-same things had said and writ,
Had Pitt been Fox, and Fox been Pitt;
Content his own applause to win,
Would never dash thro' thick and thin,
And he can make, so say the wise,
No claim who makes no sacrifice;— 60
And bard still less:—what claim had he
Who swore it vex'd his soul to see
So grand a cause, so proud a realm,
With Goose and Goody at the helm;°
Who long ago had fall'n asunder
But for their rivals' baser blunder,
The coward whine and Frenchified
Slaver and slang of the other side?—
 Thus, his own whim his only bribe,
Our Bard pursued his old A. B. C. 70
Contented if he could subscribe
In fullest sense his name Ἔστησε [*Esteesee*];
('Tis Punic Greek for 'he hath stood!')°
Whate'er the men, the cause was good;
And therefore with a right good will,
Poor fool, he fights their battles still.
Tush! squeak'd the Bats;—a mere bravado
To whitewash that base renegado;
'Tis plain unless you're blind or mad,
His conscience for the bays he barters;— 80
And true it is—as true as sad—
These circlets of green baize he had—
But then, alas! they were his garters!
 Ah! silly Bard, unfed, untended,
His lamp but glimmer'd in its socket;
He lived unhonour'd and unfriended
With scarce a penny in his pocket;—
Nay–tho' he hid it from the many—
With scarce a pocket for his penny!

Youth and Age

Verse, a breeze mid blossoms straying,
Where Hope clung feeding, like a bee—
Both were mine! Life went a-maying
With Nature, Hope, and Poesy,
When I was young!
When I was young?—Ah, woful when!
Ah! for the change 'twixt Now and Then!
This breathing house not built with hands,
This body that does me grievous wrong,
O'er aery cliffs and glittering sands, 10
How lightly then it flashed along:—
Like those trim skiffs, unknown of yore,
On winding lakes and rivers wide,
That ask no aid of sail or oar,
That fear no spite of wind or tide!
Nought cared this body for wind or weather
When Youth and I lived in't together.

Flowers are lovely; Love is flower-like;
Friendship is a sheltering tree;
O! the joys, that came down shower-like, 20
Of Friendship, Love, and Liberty,
Ere I was old!
Ere I was old? Ah woful Ere,
Which tells me, Youth's no longer here!
O Youth! for years so many and sweet,
'Tis known, that Thou and I were one,
I'll think it but a fond conceit—
It cannot be that Thou art gone!
Thy vesper-bell hath not yet toll'd:—
And thou wert aye a masker bold! 30
What strange disguise hast now put on,
To make believe, that thou art gone?
I see these locks in silvery slips,
This drooping gait, this altered size:
But springtide blossoms on thy lips,
And tears take sunshine from thine eyes!
Life is but thought: so think I will
That Youth and I are house-mates still.

Dew-drops are the gems of morning,
But the tears of mournful eve! 40
Where no hope is, life's a warning
That only serves to make us grieve,
When we are old:
That only serves to make us grieve
With oft and tedious taking-leave,
Like some poor nigh-related guest,
That may not rudely be dismist;
Yet hath outstay'd his welcome while,
And tells the jest without the smile.

Lines

Suggested by the Last Words of Berengarius°

ob. Anno Dom. 1088

No more 'twixt conscience staggering and the Pope
Soon shall I now before my God appear,
By him to be acquitted, as I hope;
By him to be condemned, as I fear.—

REFLECTION ON THE ABOVE

Lynx amid moles! had I stood by thy bed,
Be of good cheer, meek soul! I would have said:
I see a hope spring from that humble fear.
All are not strong alike through storms to steer
Right onward. What? though dread of threatened death
And dungeon torture made thy hand and breath 10
Inconstant to the truth within thy heart?
That truth, from which, through fear, thou twice didst start,
Fear haply told thee, was a learned strife,
Or not so vital as to claim thy life:
And myriads had reached Heaven, who never knew
Where lay the difference 'twixt the false and true!

Ye, who secure 'mid trophies not your own,
Judge him who won them when he stood alone,
And proudly talk of recreant Berengare—

O first the age, and then the man compare! 20
That age how dark! congenial minds how rare!
No host of friends with kindred zeal did burn!
No throbbing hearts awaited his return!
Prostrate alike when prince and peasant fell,
He only disenchanted from the spell,
Like the weak worm that gems the starless night,
Moved in the scanty circlet of his light:
And was it strange if he withdrew the ray
That did but guide the night-birds to their prey?

The ascending day-star with a bolder eye 30
Hath lit each dew-drop on our trimmer lawn!
Yet not for this, if wise, shall we decry
The spots and struggles of the timid dawn;
Lest so we tempt th' approaching noon to scorn
The mists and painted vapours of our morn.

Work Without Hope°

Lines Composed 21st February, 1825

All Nature seems at work. Slugs leave their lair—
The bees are stirring—birds are on the wing—
And Winter slumbering in the open air,
Wears on his smiling face a dream of Spring!
And I the while, the sole unbusy thing,
Nor honey make, nor pair, nor build, nor sing.

Yet well I ken the banks where amaranths blow,
Have traced the fount whence streams of nectar flow.
Bloom, O ye amaranths! bloom for whom ye may,
For me ye bloom not! Glide, rich streams, away! 10
With lips unbrightened, wreathless brow, I stroll:
And would you learn the spells that drowse my soul?
Work without hope draws nectar in a sieve,
And hope without an object cannot live.

Duty Surviving Self-Love

The only sure Friend of declining Life

A SOLILOQUY

Unchanged within to see all changed without
Is a blank lot and hard to bear, no doubt.
Yet why at others' wanings should'st thou fret?
Then only might'st thou feel a just regret,
Hadst thou withheld thy love or hid thy light
In selfish forethought of neglect and slight.
O wiselier then, from feeble yearnings freed,
While, and on whom, thou may'st—shine on! nor heed
Whether the object by reflected light
Return thy radiance or absorb it quite: 10
And though thou notest from thy safe recess
Old friends burn dim, like lamps in noisome air,
Love them for what they are; nor love them less,
Because to thee they are not what they were.

The Improvisatore

Or 'John Anderson, my Jo, John'°

Scene—A spacious drawing-room, with music-room adjoining.

Katharine. What are the words?

Eliza. Ask our friend, the Improvisatore; here he comes. Kate has a favour to ask of you, Sir; it is that you will repeat the ballad that Mr. —— sang so sweetly.

Friend. It is in Moore's Irish Melodies; but I do not recollect the words distinctly. The moral of them, however, I take to be this:—

> Love would remain the same if true,
> When we were neither young nor new;
> Yea, and in all within the will that came,
> By the same proofs would show itself the same.

Eliz. What are the lines you repeated from Beaumont and Fletcher, which my mother admired so much? It begins with something about two vines so close that their tendrils intermingle.

Fri. You mean Charles' speech to Angelina, in *The Elder Brother.*

> We'll live together, like two neighbour vines,
> Circling our souls and loves in one another!
> We'll spring together, and we'll bear one fruit;
> One joy shall make us smile, and one grief mourn;
> One age go with us, and one hour of death
> Shall close our eyes, and one grave make us happy.

Kath. A precious boon, that would go far to reconcile one to old age—this love—*if* true! But is there any such true love?

Fri. I hope so.

Kath. But do you believe it?

Eliz. (*eagerly*). I am sure he does.

Fri. From a man turned of fifty, Katharine, I imagine, expects a less confident answer.

Kath. A more sincere one, perhaps.

Fri. Even though he should have obtained the nick-name of Improvisatore, by perpetrating charades and extempore verses at Christmas times?

Eliz. Nay, but be serious.

Fri. Serious! Doubtless. A grave personage of my years giving a love-lecture to two young ladies, cannot well be otherwise. The difficulty, I suspect, would be for them to remain so. It will be asked whether I am not the 'elderly gentleman' who sate 'despairing beside a clear stream,' with a willow for his wig-block.

Eliz. Say another word, and we will call it downright affectation.

Kath. No! we will be affronted, drop a courtesy, and ask pardon for our presumption in expecting that Mr. —— would waste his sense on two insignificant girls.

Fri. Well, well, I will be serious. Hem! Now then commences the discourse; Mr. Moore's song being the text. Love, as distinguished from Friendship, on the one hand, and from the passion that too often usurps its name, on the other—

Lucius (*Eliza's brother, who had just joined the trio, in a whisper to the Friend*). But is not Love the union of both?

Fri. (*aside to Lucius*). He never loved who thinks so.

Eliz. Brother, we don't want you. There! Mrs. H. cannot arrange the flower vase without you. Thank you, Mrs. Hartman.

Luc. I'll have my revenge! I know what I will say!

Eliz. Off! Off! Now, dear sir,—Love, you were saying—

Fri. Hush! Preaching, you mean, Eliza.

Eliz. (*impatiently*). Pshaw!

Fri. Well then, I was saying that love, truly such, is itself not the most common thing in the world: and mutual love still less so. But that enduring personal attachment, so beautifully delineated by Erin's sweet melodist, and still more touchingly, perhaps, in the well-known ballad, 'John Anderson, my Jo, John,' in addition to a depth and constancy of character of no every-day occurrence, supposes a peculiar sensibility and tenderness of nature; a constitutional communicativeness and utterancy of heart and soul; a delight in the detail of sympathy, in the outward and visible signs of the sacrament within—to count, as it were, the pulses of the life of love. But above all, it supposes a soul which, even in the pride and summer-tide of life—even in the lustihood of health and strength, had felt oftenest and prized highest that which age cannot take away and which, in all our loving, is *the* Love;—

Eliz. There is something *here* (*pointing to her heart*) that seems to understand you, but wants the word that would make it understand itself.

Kath. I, too, seem to feel what you mean. Interpret the feeling for us.

Fri. —I mean that willing sense of the insufficingness of the self for itself, which predisposes a generous nature to see, in the total being of another, the supplement and completion of its own;—that quiet perpetual seeking which the presence of the beloved object modulates, not suspends, where the heart momently finds, and, finding, again seeks on;—lastly, when 'life's changeful orb has pass'd the full,' a confirmed faith in the nobleness of humanity, thus brought home and pressed, as it were, to the very bosom of hourly experience; it supposes, I say, a heartfelt reverence for worth, not the less deep because divested of its solemnity by habit, by familiarity, by mutual infirmities, and even by a feeling of modesty which will arise in delicate minds, when they are conscious of possessing the same or the correspondent excellence in their own characters. In short, there must be a mind, which, while it feels the beautiful and the excellent in the beloved as its own, and by right of love appropriates it, can call Goodness its playfellow; and dares make sport of time and infirmity, while, in the person of a thousand-foldly endeared partner, we feel for aged virtue the caressing fondness that belongs to the innocence of childhood, and repeat the same attentions and tender courtesies which had been dictated by the same affection to the same object when attired in feminine loveliness or in manly beauty.

Eliz. What a soothing—what an elevating idea!

Kath. If it be not only a mere fancy.

Fri. At all events, these qualities which I have enumerated, are rarely found united in a single individual. How much more rare must it be, that two such individuals should meet together in this wide world under circumstances that admit of their union as Husband and Wife. A person may be highly estimable on the whole, nay, amiable as neighbour, friend, housemate—in short, in all the concentric circles of attachment save only the last and inmost; and yet from how many causes be estranged from the highest perfection in this! Pride, coldness, or fastidiousness of nature, worldly cares, an anxious or ambitious disposition, passion for display, a sullen temper,—one or the other—too often proves 'the dead fly in the compost of spices,' and any one is enough to unfit it for the precious balm of unction. For some mighty good sort of people, too, there is not seldom a sort of solemn saturnine, or, if you will, ursine vanity, that keeps itself alive by sucking the paws of its own self-importance. And as this high sense, or rather sensation of their own value is, for the most part, grounded on negative qualities, so they have no better means of preserving the same but by negatives—that is, by not doing or saying any thing, that might be put down for fond, silly, or nonsensical;—or (to use their own phrase) by never forgetting themselves, which some of their acquaintance are uncharitable enough to think the most worthless object they could be employed in remembering.

Eliz. (*in answer to a whisper from Katharine*). To a hair! He must have sate for it himself. Save me from such folks! But they are out of the question.

Fri. True! but the same effect is produced in thousands by the too general insensibility to a very important truth; this, namely, that the misery of human life is made up of large masses, each separated from the other by certain intervals. One year, the death of a child; years after, a failure in trade; after another longer or shorter interval, a daughter may have married unhappily;—in all but the singularly unfortunate, the integral parts that compose the sum total of the unhappiness of a man's life, are easily counted, and distinctly remembered. The happiness of life, on the contrary, is made up of minute fractions—the little, soon-forgotten charities of a kiss, a smile, a kind look, a heartfelt compliment in the disguise of playful raillery, and the countless other infinitesimals of pleasurable thought and genial feeling.

Kath. Well, Sir; you have said quite enough to make me despair of finding a 'John Anderson, my Jo, John', with whom to totter down the hill of life.

Fri. Not so! Good men are not, I trust, so much scarcer than good women, but that what another would find in you, you may hope to find in another. But well, however, may that boon be rare, the possession of which would be more than an adequate reward for the rarest virtue.

Eliz. Surely, he, who has described it so well, must have possessed it?

Fri. If he were worthy to have possessed it, and had believingly anticipated and not found it, how bitter the disappointment!

(*Then, after a pause of a few minutes*),

ANSWER, *ex improviso* [improvising]

Yes, yes! that boon, life's richest treat,
He had, or fancied that he had;
Say, 'twas but in his own conceit—
  The fancy made him glad!
Crown of his cup, and garnish of his dish,
The boon, prefigured in his earliest wish,
The fair fulfilment of his poesy,
When his young heart first yearn'd for sympathy!

But e'en the meteor offspring of the brain
            Unnourished wane; 10
Faith asks her daily bread,
And Fancy must be fed.
Now so it chanced—from wet or dry,
It boots not how—I know not why—
She missed her wonted food; and quickly
Poor Fancy stagger'd and grew sickly.
Then came a restless state, 'twixt yea and nay,
His faith was fix'd, his heart all ebb and flow;
Or like a bark, in some half-shelter'd bay,
Above its anchor driving to and fro. 20

That boon, which but to have possess'd
In a belief, gave life a zest—
Uncertain both what it had been,
And if by error lost, or luck;
And what it was;—an evergreen
Which some insidious blight had struck,
Or annual flower, which, past its blow,
No vernal spell shall e'er revive;

Uncertain, and afraid to know,
  Doubts toss'd him to and fro: 30
Hope keeping Love, Love Hope alive,
Like babes bewildered in the snow,
That cling and huddle from the cold
In hollow tree or ruin'd fold.

Those sparkling colours, once his boast
    Fading, one by one away,
Thin and hueless as a ghost,
    Poor Fancy on her sick bed lay;
Ill at distance, worse when near,
Telling her dreams to jealous Fear! 40
Where was it then, the sociable sprite
That crown'd the Poet's cup and deck'd his dish!
Poor shadow cast from an unsteady wish,
Itself a substance by no other right
But that it intercepted Reason's light;
It dimm'd his eye, it darken'd on his brow,
A peevish mood, a tedious time, I trow!
    Thank Heaven! 'tis not so now.

O bliss of blissful hours!
The boon of Heaven's decreeing, 50
While yet in Eden's bowers
Dwelt the first husband and his sinless mate!
The one sweet plant, which, piteous Heaven agreeing,
They bore with them thro' Eden's closing gate!
Of life's gay summer tide the sovran Rose!
Late autumn's amaranth, that more fragrant blows
When passion's flowers all fall or fade;
If this were ever his, in outward being,
Or but his own true love's projected shade,
Now that at length by certain proof he knows, 60
That whether real or a magic show,
Whate'er it was, it is no longer so;
Though heart be lonesome, hope laid low,
Yet, Lady! deem him not unblest:
The certainty that struck hope dead,
Hath left contentment in her stead:
    And that is next to best!

Alice Du Clos

Or The Forked Tongue

A BALLAD

'One word with two meanings is the traitor's shield and shaft: and a slit tongue be his blazon!'—*Caucasian Proverb.*

'The Sun is not yet risen,
But the dawn lies red on the dew:
Lord Julian has stolen from the hunters away,
Is seeking, Lady! for you.
Put on your dress of green,
  Your buskins and your quiver;
Lord Julian is a hasty man,
  Long waiting brook'd he never.
I dare not doubt him, that he means
  To wed you on a day, 10
Your lord and master for to be,
  And you his lady gay.
O Lady! throw your book aside!
I would not that my Lord should chide.'

Thus spake Sir Hugh the vassal knight
  To Alice, child of old Du Clos,
As spotless fair, as airy light
  As that moon-shiny doe,
The gold star on its brow, her sire's ancestral crest!
For ere the lark had left his nest, 20
  She in the garden bower below
Sate loosely wrapt in maiden white,
Her face half drooping from the sight,
  A snow-drop on a tuft of snow!
O close your eyes, and strive to see
The studious maid, with book on knee,—
  Ah! earliest-open'd flower;
While yet with keen unblunted light
The morning star shone opposite
  The lattice of her bower— 30
Alone of all the starry host,
  As if in prideful scorn
Of flight and fear he stay'd behind,
  To brave th'advancing morn.

O! Alice could read passing well,
  And she was conning then
Dan Ovid's mazy tale of loves,
  And gods, and beasts, and men.

The vassal's speech, his taunting vein,
It thrill'd like venom thro' her brain; 40
  Yet never from the book
She rais'd her head, nor did she deign
  The knight a single look.

'Off, traitor friend! how dar'st thou fix
  Thy wanton gaze on me?
And why, against my earnest suit,
  Does Julian send by thee?

Go, tell thy Lord, that slow is sure:
  Fair speed his shafts to-day!
I follow here a stronger lure, 50
  And chase a gentler prey.'

She said: and with a baleful smile
  The vassal knight reel'd off—
Like a huge billow from a bark
  Toil'd in the deep sea-trough,
That shouldering sideways in mid plunge,
  Is travers'd by a flash.
And staggering onward, leaves the ear
  With dull and distant crash.

And Alice sate with troubled mien 60
A moment; for the scoff was keen,
  And thro' her veins did shiver!
Then rose and donn'd her dress of green,
  Her buskins and her quiver.

There stands the flow'ring may-thorn tree!
From thro' the veiling mist you see
  The black and shadowy stem;—
Smit by the sun the mist in glee
Dissolves to lightsome jewelry—
  Each blossom hath its gem! 70

With tear-drop glittering to a smile,
The gay maid on the garden-stile
  Mimics the hunter's shout.
'Hip! Florian, hip! To horse, to horse!
  Go, bring the palfrey out.

My Julian's out with all his clan,
  And, bonny boy, you wis,
Lord Julian is a hasty man,
  Who comes late, comes amiss.'

Now Florian was a stripling squire, 80
  A gallant boy of Spain,
That toss'd his head in joy and pride,
Behind his Lady fair to ride,
  But blush'd to hold her train.

The huntress is in her dress of green,—
And forth they go; she with her bow,
  Her buskins and her quiver!—
The squire—no younger e'er was seen—
With restless arm and laughing een,
  He makes his javelin quiver. 90

And had not Ellen stay'd the race,°
And stopp'd to see, a moment's space,
  The whole great globe of light
Give the last parting kiss-like touch
To the eastern ridge, it lack'd not much,
  They had o'erta'en the knight.

It chanced that up the covert lane,
  Where Julian waiting stood,
A neighbour knight prick'd on to join
  The huntsmen in the wood. 100

And with him must Lord Julian go,
  Tho' with an anger'd mind:
Betroth'd not wedded to his bride,
  In vain he sought, 'twixt shame and pride,
  Excuse to stay behind.

He bit his lip, he wrung his glove,
He look'd around, he look'd above,
  But pretext none could find or frame.
Alas! alas! and well-a-day!
It grieves me sore to think, to say, 110
That names so seldom meet with Love,
  Yet Love wants courage without a name!

Straight from the forest's skirt the trees
  O'er-branching, made an aisle,
Where hermit old might pace and chaunt
  As in a minster's pile.

From underneath its leafy screen,
  And from the twilight shade,
You pass at once into a green,
  A green and lightsome glade. 120

And there Lord Julian sate on steed;
  Behind him, in a round,
Stood knight and squire, and menial train;
Against the leash the greyhounds strain;
  The horses paw'd the ground.

When up the alley green, Sir Hugh
  Spurr'd in upon the sward,
And mute, without a word, did he
  Fall in behind his lord.

Lord Julian turn'd his steed half round.— 130
  'What! doth not Alice deign
To accept your loving convoy, knight?
Or doth she fear our woodland sleight,
  And join us on the plain?'

With stifled tones the knight replied,
And look'd askance on either side,—
  'Nay, let the hunt proceed!—
The Lady's message that I bear,
I guess would scantly please your ear,
  And less deserves your heed. 140

You sent betimes. Not yet unbarr'd
  I found the middle door;—
Two stirrers only met my eyes,
  Fair Alice, and one more.

I came unlook'd for; and, it seem'd,
  In an unwelcome hour;
And found the daughter of Du Clos
  Within the lattic'd bower.

But hush! the rest may wait. If lost,
  No great loss, I divine; 150
And idle words will better suit
  A fair maid's lips than mine.'

'God's wrath! speak out, man,' Julian cried,
  O'ermaster'd by the sudden smart;—
And feigning wrath, sharp, blunt, and rude,
The knight his subtle shift pursued.—
'Scowl not at me; command my skill,
To lure your hawk back, if you will,
  But not a woman's heart.

  "Go! (said she) tell him,—slow is sure; 160
  Fair speed his shafts to-day!
I follow here a stronger lure,
  And chase a gentler prey."

  The game, pardie, was full in sight,
That then did, if I saw aright,
  The fair dame's eyes engage;
For turning, as I took my ways,
I saw them fix'd with steadfast gaze
  Full on her wanton page.'

The last word of the traitor knight 170
  It had but entered Julian's ear,—
From two o'erarching oaks between,
With glist'ning helm-like cap is seen,
  Borne on in giddy cheer,

A youth, that ill his steed can guide;
Yet with reverted face doth ride,
As answering to a voice,
That seems at once to laugh and chide—
'Not mine, dear mistress,' still he cried,
' 'Tis this mad filly's choice.' 180

With sudden bound, beyond the boy,
See! see! that face of hope and joy,
That regal front! those cheeks aglow!
Thou needed'st but the crescent sheen,
A quiver'd Dian to have been,
Thou lovely child of old Du Clos!

Dark as a dream Lord Julian stood,
Swift as a dream, from forth the wood,
Sprang on the plighted Maid!
With fatal aim, and frantic force, 190
The shaft was hurl'd!—a lifeless corse,
Fair Alice from her vaulting horse,
Lies bleeding on the glade.

Self-Knowledge

—E coelo descendit *γνῶθι σεαυτόν* [From heaven descended the 'Know thyself' (*gnōthi seauton*)]—Juvenal, xi. 27.

Γνῶθι σεαυτόν!—and is this the prime
And heaven-sprung adage of the olden time!—
Say, canst thou make thyself?—Learn first that trade;—
Haply thou mayst know what thyself had made.
What hast thou, Man, that thou dar'st call thine own?—
What is there in thee, Man, that can be known?—
Dark fluxion, all unfixable by thought,
A phantom dim of past and future wrought,
Vain sister of the worm,—life, death, soul, clod—
Ignore thyself, and strive to know thy God! 10

Love's Apparition and Evanishment

An Allegoric Romance

Like a lone Arab, old and blind,
Some caravan had left behind,
Who sits beside a ruin'd well,
Where the shy sand-asps bask and swell;
And now he hangs his aged head aslant,
And listens for a human sound—in vain!
And now the aid, which Heaven alone can grant,
Upturns his eyeless face from Heaven to gain;—
Even thus, in vacant mood, one sultry hour,
Resting my eye upon a drooping plant, 10
With brow low bent, within my garden bower,
I sate upon the couch of camomile;
And—whether 'twas a transient sleep, perchance,
Flitted across the idle brain, the while
I watch'd the sickly calm with aimless scope,
In my own heart; or that, indeed a trance,
Turn'd my eye inward—thee, O genial Hope,
Love's elder sister! thee did I behold,
Drest as a bridesmaid, but all pale and cold,
With roseless cheek, all pale and cold and dim, 
Lie lifeless at my feet!
And then came Love, a sylph in bridal trim,
And stood beside my seat;
She bent, and kiss'd her sister's lips,
As she was wont to do;—
Alas! 'twas but a chilling breath
Woke just enough of life in death
To make Hope die anew.

Epitaph

Stop, Christian Passer-by!—Stop, child of God,
And read with gentle breast. Beneath this sod
A poet lies, or that which once seem'd he.
O, lift one thought in prayer for S. T. C.;
That he who many a year with toil of breath
Found death in life, may here find life in death!
Mercy for praise—to be forgiven for fame°
He ask'd, and hoped, through Christ. Do thou the same!

BIOGRAPHIA LITERARIA;

OR

Biographical Sketches

OF

MY LITERARY LIFE

AND

OPINIONS

So wenig er auch bestimmt seyn mag andere zu belehren, so wünscht er doch sich denen mitzutheilen, die er sich gleichgesinnt weiss oder hofft, deren Anzahl aber in der Breite der Welt zerstreut ist: er wünscht sein Verhältniss zu den ältesten Freunden wieder anzuknüpfen, mit neuen es fortzusetzen, und in der letzen Generation sich wieder andere für sein übrige Lebenszeit zu gewinnen. Er wünscht der Jugend die Umwege zu ersparen, auf denen er sich selbst verirrte.

GOETHE

TRANSLATION. Little call as he may have to instruct others, he wishes nevertheless to open out his heart to such as he either knows or hopes to be of like mind with himself, but who are widely scattered in the world: he wishes to knit anew his connections with his oldest friends, to continue those recently formed, and to win other friends among the rising generation for the remaining course of his life. He wishes to spare the young those circuitous paths, on which he himself had lost his way.

Biographia Literaria°

CHAPTER I

The motives of the present work—Reception of the author's first publication—The discipline of his taste at school—The effect of contemporary writers on youthful minds—Bowles's sonnets—Comparison between the poets before and since Mr Pope.

IT has been my lot to have had my name introduced both in conversation, and in print, more frequently than I find it easy to explain, whether I consider the fewness, unimportance, and limited circulation of my writings, or the retirement and distance, in which I have lived, both from the literary and political world. Most often it has been connected with some charge, which I could not acknowledge, or some principle which I had never entertained. Nevertheless, had I had no other motive, or incitement, the reader would not have been troubled with this exculpation. What my additional purposes were, will be seen in the following pages. It will be found, that the least of what I have written concerns myself personally. I have used the narration chiefly for the purpose of giving a continuity to the work, in part for the sake of the miscellaneous reflections suggested to me by particular events, but still more as introductory to the statement of my principles in politics, religion, and philosophy, and the application of the rules, deduced from philosophical principles, to poetry and criticism. But of the objects, which I proposed to myself, it was not the least important to effect, as far as possible, a settlement of the long continued controversy concerning the true nature of poetic diction: and at the same time to define with the utmost impartiality the real *poetic* character of the poet, by whose writings this controversy was first kindled, and has been since fuelled and fanned.

In 1794, when I had barely passed the verge of manhood, I published a small volume of juvenile poems. They were received with a degree of favour, which, young as I was, I well knew, was bestowed on them not so much for any positive merit, as because they were considered buds of hope, and promises of better works to come. The critics of that day, the most flattering, equally with the severest, concurred in objecting to them, obscurity, a general turgidness of diction, and a profusion of new coined double epithets.* The first is the fault which a writer is the least able to detect in his own compositions: and my mind was not then sufficiently disciplined to receive the authority of others, as a substitute for my own conviction. Satisfied that the thoughts, such as they were, could not have been expressed otherwise, or at least more perspicuously, I forgot to enquire, whether the thoughts themselves did not demand a degree of attention unsuitable to the nature and objects of poetry. This remark however applies chiefly, though not exclusively to the *Religious Musings*. The remainder of the charge I admitted to its full extent, and not without sincere acknowledgements to both my private and public censors for their friendly admonitions. In the after editions, I pruned the double epithets with no sparing hand, and used my best efforts to tame the swell and glitter both of thought and diction; though in truth, these parasite plants of youthful poetry had insinuated themselves into my longer poems with such intricacy of union, that I was often obliged to omit disentangling the weed, from the fear of snapping the flower. From that period to the date of the present work I have published nothing, with my name, which could by any possibility have come before the board of anonymous criticism. Even the three or four poems, printed with the works of a friend, as far as they were censured at all, were charged with the same or similar defects, though I am persuaded not with equal justice: with an excess of ornament, in addition to strained and elaborate diction. (*Vide* [see] the criticisms on the 'Ancient Mariner,' in the Monthly and Critical Reviews of the first

* The authority of Milton and Shakespeare may be usefully pointed out to young authors. In the Comus, and earlier poems of Milton there is a superfluity of double epithets; while in the Paradise Lost we find very few, in the Paradise Regained scarce any. The same remark holds almost equally true, of the Love's Labour Lost, Romeo and Juliet, Venus and Adonis, and Lucrece compared with the Lear, Macbeth, Othello, and Hamlet of our great dramatist. The rule for the admission of double epithets seems to be this: either that they should be already denizens of our language, such as blood-stained, terror-stricken, self-applauding: or when a new epithet, or one found in books only, is hazarded, that it, at least, be one word, not two words made one by mere virtue of the printer's hyphen. A language which, like the English, is almost without cases, is indeed in its very genius unfitted for compounds. If a writer, every time a compounded word suggests itself to him, would seek for some other mode of expressing the same sense, the chances are always greatly in favour of his finding a better word. 'Tanquam scopulum sic vites insolens verbum [shun the unfamiliar word as you would a rock],' is the wise advice of Caesar to the Roman orators, and the precept applies with double force to the writers in our own language. But it must not be forgotten, that the same Caesar wrote a grammatical treatise for the purpose of reforming the ordinary language by bringing it to a greater accordance with the principles of logic or universal grammar.

volume of the Lyrical Ballads.) May I be permitted to add, that, even at the early period of my juvenile poems, I saw and admitted the superiority of an austerer, and more natural style, with an insight not less clear, than I at present possess. My judgement was stronger, than were my powers of realizing its dictates; and the faults of my language, though indeed partly owing to a wrong choice of subjects, and the desire of giving a poetic colouring to abstract and metaphysical truths, in which a new world then seemed to open upon me, did yet, in part likewise, originate in unfeigned diffidence of my own comparative talent.—During several years of my youth and early manhood, I reverenced those, who had reintroduced the manly simplicity of the Grecian, and of our own elder poets, with such enthusiasm, as made the hope seem presumptuous of writing successfully in the same style. Perhaps a similar process has happened to others; but my earliest poems were marked by an ease and simplicity, which I have studied, perhaps with inferior success, to impress on my later compositions.

At school I enjoyed the inestimable advantage of a very sensible, though at the same time, a very severe master. He* early moulded my taste to the preference of Demosthenes to Cicero, of Homer and Theocritus to Virgil, and again of Virgil to Ovid. He habituated me to compare Lucretius, (in such extracts as I then read), Terence, and above all the chaster poems of Catullus, not only with the Roman poets of the, so-called, Silver and Brazen Ages; but with even those of the Augustan era: and on grounds of plain sense and universal logic to see and assert the superiority of the former, in the truth and nativeness, both of their thoughts and diction. At the same time that we were studying the Greek tragic poets, he made us read Shakespeare and Milton as lessons: and they were the lessons too, which required most time and trouble to *bring up*, so as to escape his censure. I learnt from him, that poetry, even that of the loftiest, and, seemingly, that of the wildest odes, had a logic of its own, as severe as that of science; and more difficult, because more subtle, more complex, and dependent on more, and more fugitive causes. In the truly great poets, he would say, there is a reason assignable, not only for every word, but for the position of every word; and I well remember, that availing himself of the synonyms to the Homer of Didymus, he made us attempt to show, with regard to each, why it would not have answered the same purpose; and wherein consisted the peculiar fitness of the word in the original text.

* The Revd James Boyer, many years Head Master of the Grammar School, Christ's Hospital.

In our own English compositions (at least for the last three years of our school education) he showed no mercy to phrase, metaphor, or image, unsupported by a sound sense, or where the same sense might have been conveyed with equal force and dignity in plainer words. Lute, harp, and lyre, muse, muses, and inspirations, Pegasus, Parnassus, and Hipocrene, were all an abomination to him. In fancy I can almost hear him now, exclaiming 'Harp? Harp? Lyre? Pen and ink, boy, you mean! Muse, boy, Muse? your Nurse's daughter, you mean! Pierian spring? Oh 'aye! the cloister-pump, I suppose!' Nay certain introductions, similes, and examples, were placed by name on a list of interdiction. Among the similes, there was, I remember, that of the Manchineel fruit, as suiting equally well with too many subjects; in which however it yielded the palm at once to the example of Alexander and Clytus, which was equally good and apt, whatever might be the theme. Was it ambition? Alexander and Clytus!—Flattery? Alexander and Clytus!—Anger? Drunkenness? Pride? Friendship? Ingratitude? Late repentance? Still, still Alexander and Clytus! At length, the praises of agriculture having been exemplified in the sagacious observation, that had Alexander been holding the plough, he would not have run his friend Clytus through with a spear, this tried, and serviceable old friend was banished by public edict *in secula seculorum* [for ever and ever]. I have sometimes ventured to think, that a list of this kind, or an *index expurgatorius*° of certain well-known and ever returning phrases, both introductory, and transitional, including the large assortment of modest egotisms, and flattering illeisms, etc. etc. might be hung up in our lawcourts, and both Houses of Parliament, with great advantage to the public, as an important saving of national time, an incalculable relief to His Majesty's ministers, but above all, as ensuring the thanks of country attorneys, and their clients, who have private bills to carry through the House.

Be this as it may, there was one custom of our master's, which I cannot pass over in silence, because I think it imitable and worthy of imitation. He would often permit our theme exercises, under some pretext of want of time, to accumulate, till each lad had four or five to be looked over. Then placing the whole number abreast on his desk, he would ask the writer, why this or that sentence might not have found as appropriate a place under this or that other thesis: and if no satisfying answer could be returned, and two faults of the same kind were found in one exercise, the irrevocable verdict followed, the exercise was torn up, and another on the same subject to be produced, in addition to the tasks of the day. The reader will, I trust, excuse this tribute of recollection to a man, whose severities, even now, not seldom furnish the dreams, by which the blind fancy would fain interpret to the mind the painful sensations of distempered sleep; but neither lessen nor dim the deep sense of my moral and intellectual obligations. He sent us to the university excellent Latin and Greek scholars, and tolerable Hebraists. Yet our classical knowledge was the least of the good gifts, which we derived from his zealous and conscientious tutorage. He is now gone to his final reward, full of years, and full of honours, even of those honours, which were dearest to his heart, as gratefully bestowed by that school, and still binding him to the interests of that school, in which he had been himself educated, and to which during his whole life he was a dedicated thing.

From causes, which this is not the place to investigate, no models of past times, however perfect, can have the same vivid effect on the youthful mind, as the productions of contemporary genius. The discipline, my mind had undergone, 'Ne falleretur rotundo sono et versuum cursu, cincinnis et floribus; sed ut inspiceret quidnam subesset, quae sedes, quod firmamentum, quis fundus verbis; an figurae essent mera ornatura et orationis fucus: vel sanguinis e materiae ipsius corde effluentis rubor quidam nativus et incalescentia genuina [So that it was not misled by the smooth sound and flow of the verses, their ringlets and flowers, but examined what lay beneath them, what was their ground, their firmament, their foundation; whether the figures were mere ornamentation and the paint of rhetoric, or a natural flush and genuine warmth of the blood flowing from the heart of the matter itself (*CC*)°]'; removed all obstacles to the appreciation of excellence in style without diminishing my delight. That I was thus prepared for the perusal of Mr Bowles's° sonnets and earlier poems, at once increased their influence, and my enthusiasm. The great works of past ages seem to a young man things of another race, in respect to which his faculties must remain passive and submiss, even as to the stars and mountains. But the writings of a contemporary, perhaps not many years elder than himself, surrounded by the same circumstances, and disciplined by the same manners, possess a reality for him, and inspire an actual friendship as of a man for a man. His very admiration is the wind which fans and feeds his hope. The poems themselves assume the properties of flesh and blood. To recite, to extol, to contend for them is but the payment of a debt due to one, who exists to receive it.

There are indeed modes of teaching which have produced, and are producing, youths of a very different stamp; modes of teaching, in comparison with which we have been called on to despise our great public schools, and universities

> In whose halls are hung
> Armoury of the invincible knights of old—

modes, by which children are to be metamorphosed into prodigies. And prodigies with a vengeance have I known thus produced! Prodigies of self-conceit, shallowness, arrogance, and infidelity! Instead of storing the memory, during the period when the memory is the predominant faculty, with facts for the after exercise of the judgement; and instead of awakening by the noblest models the fond and unmixed love and admiration, which is the natural and graceful temper of early youth; these nurselings of improved pedagogy are taught to dispute and decide; to suspect all, but their own and their lecturer's wisdom; and to hold nothing sacred from their contempt, but their own contemptible arrogance: boy-graduates in all the technicals, and in all the dirty passions and impudence, of anonymous criticism. To such dispositions alone can the admonition of Pliny be requisite, 'Neque enim debet operibus ejus obesse, quod vivit. An si inter eos, quos nunquam vidimus, floruisset, non solum libros ejus, verum etiam imagines conquireremus, ejusdem nunc honor praesentis, et gratia quasi satietate languescet? At hoc pravum, malignumque est, non admirari hominem admiratione dignissimum, quia videre, complecti, nec laudare tantum, verum etiam amare contingit [Let it not be any prejudice to his merit that he is a contemporary writer. Had he flourished in some distant age, not only his works, but the very pictures and statues of him would have been passionately inquired after; and shall we then, from a sort of satiety, and merely because he is present among us, suffer his talents to languish and fade away unhonoured and unadmired? It is surely a very perverse and envious disposition, to look with indifference upon a man worthy of the highest approbation, for no other reason but because we have it in our power to see him and to converse familiarly with him, and not only to give him our applause, but to receive him into our friendship (LCL)].' Plin. *Epist. Lib. I.*

I had just entered on my seventeenth year, when the sonnets of Mr Bowles, twenty in number, and just then published in a quarto pamphlet, were first made known and presented to me, by a schoolfellow who had quitted us for the university, and who, during the whole time that he was in our first form (or in our school language a Grecian) had been my patron and protector. I refer to Dr Middleton, the truly learned, and every way excellent Bishop of Calcutta:

> Qui laudibus amplis
> Ingenium celebrare meum, calamumque solebat,
> Calcar agens animo validum. Non omnia terrae
> Obruta! Vivit amor, vivit dolor! Ora negatur
> Dulcia conspicere; at flere et meminisse* relictum est.

[Who, with lavish praises, was wont to celebrate my genius and my pen, setting a sharp spur to my spirit. Not everything is buried in the earth. Love lives, grief lives on! We are denied the sight of those sweet features; but it is left for us to weep and to remember. (*CC*)]

Petr. *Ep. Lib. I. Ep. I.*

It was a double pleasure to me, and still remains a tender recollection, that I should have received from a friend so revered the first knowledge of a poet, by whose works, year after year, I was so enthusiastically delighted and inspired. My earliest acquaintances will not have forgotten the undisciplined eagerness and impetuous zeal, with which I laboured to make proselytes, not only of my companions, but of all with whom I conversed, of whatever rank, and in whatever place. As my school finances did not permit me to purchase copies, I made, within less than a year and an half, more than forty transcriptions, as the best presents I could offer to those, who had in any way won my regard. And with almost equal delight did I receive the three or four following publications of the same author.

Though I have seen and known enough of mankind to be well aware, that I shall perhaps stand alone in my creed, and that it will be well, if I subject myself to no worse charge than that of singularity; I am not therefore deterred from avowing, that I regard, and ever have regarded the obligations of intellect among the most sacred of the claims of gratitude. A valuable thought, or a particular train of thoughts, gives me additional pleasure, when I can safely refer and attribute it to the conversation or correspondence of another. My obligations to Mr Bowles were indeed important, and for radical good. At a very premature age, even before my fifteenth year, I had bewildered myself in metaphysics, and in theological controversy.

* I am most happy to have the necessity of informing the reader, that since this passage was written, the report of Dr Middleton's death on his voyage to India has been proved erroneous. He lives and long may he live; for I dare prophesy, that with his life only will his exertions for the temporal and spiritual welfare of his fellow men be limited.

Nothing else pleased me. History, and particular facts, lost all interest in my mind. Poetry (though for a schoolboy of that age, I was above par in English versification, and had already produced two or three compositions which, I may venture to say, without reference to my age, were somewhat above mediocrity, and which had gained me more credit, than the sound, good sense of my old master was at all pleased with) poetry itself, yea novels and romances, became insipid to me. In my friendless wanderings on our *leave-days*,* (for I was an orphan, and had scarce any connections in London) highly was I delighted, if any passenger, especially if he were dressed in black, would enter into conversation with me. For I soon found the means of directing it to my favourite subjects

> Of providence, fore-knowledge, will, and fate,
> Fix'd fate, free will, fore-knowledge absolute,
> And found no end in wandering mazes lost.

This preposterous pursuit was, beyond doubt, injurious, both to my natural powers, and to the progress of my education. It would perhaps have been destructive, had it been continued; but from this I was auspiciously withdrawn, partly indeed by an accidental introduction to an amiable family, chiefly however, by the genial influence of a style of poetry, so tender, and yet so manly, so natural and real, and yet so dignified, and harmonious, as the sonnets, etc. of Mr Bowles! Well were it for me perhaps, had I never relapsed into the same mental disease; if I had continued to pluck the flower and reap the harvest from the cultivated surface, instead of delving in the unwholesome quicksilver mines of metaphysic depths. But if in after time I have sought a refuge from bodily pain and mismanaged sensibility in abstruse researches, which exercised the strength and subtlety of the understanding without awakening the feelings of the heart; still there was a long and blessed interval, during which my natural faculties were allowed to expand, and my original tendencies to develop themselves: my fancy, and the love of nature, and the sense of beauty in forms and sounds.

The second advantage, which I owe to my early perusal, and admiration of these poems (to which let me add, though known to me at a somewhat later period, the Lewesdon Hill of Mr Crowe) bears more immediately on my present subject. Among those with whom I conversed, there were, of course, very many who had formed their

* The Christ's Hospital phrase, not for holidays altogether, but for those on which the boys are permitted to go beyond the precincts of the school.

taste, and their notions of poetry, from the writings of Mr Pope and his followers: or to speak more generally, in that school of French poetry, condensed and invigorated by English understanding, which had predominated from the last century. I was not blind to the merits of this school, yet as from inexperience of the world, and consequent want of sympathy with the general subjects of these poems, they gave me little pleasure, I doubtless undervalued the kind, and with the presumption of youth withheld from its masters the legitimate name of poets. I saw, that the excellence of this kind consisted in just and acute observations on men and manners in an artificial state of society, as its matter and substance: and in the logic of wit, conveyed in smooth and strong epigrammatic couplets, as its form. Even when the subject was addressed to the fancy, or the intellect, as in the Rape of the Lock, or the Essay on Man; nay, when it was a consecutive narration, as in that astonishing product of matchless talent and ingenuity, Pope's Translation of the Iliad; still a point was looked for at the end of each second line, and the whole was as it were a sorites, or, if I may exchange a logical for a grammatical metaphor, a *conjunction disjunctive*, of epigrams. Meantime the matter and diction seemed to me characterized not so much by poetic thoughts, as by thoughts translated into the language of poetry. On this last point, I had occasion to render my own thoughts gradually more and more plain to myself, by frequent amicable disputes concerning Darwin's Botanic Garden,° which, for some years, was greatly extolled, not only by the reading public in general, but even by those, whose genius and natural robustness of understanding enabled them afterwards to act foremost in dissipating these 'painted mists' that occasionally rise from the marshes at the foot of Parnassus. During my first Cambridge vacation, I assisted a friend in a contribution for a literary society in Devonshire: and in this I remember to have compared Darwin's work to the Russian palace of ice, glittering, cold and transitory. In the same essay too, I assigned sundry reasons, chiefly drawn from a comparison of passages in the Latin poets with the original Greek, from which they were borrowed, for the preference of Collins's odes to those of Gray; and of the simile in Shakespeare

> How like a younker or a prodigal,
> The skarfed bark puts from her native bay
> Hugg'd and embraced by the strumpet wind!
> How like a prodigal doth she return,
> With over-weather'd ribs and ragged sails,
> Lean, rent, and beggar'd by the strumpet wind!

to the imitation in The Bard;

> Fair laughs the morn, and soft the zephyr blows
> While proudly riding o'er the azure realm
> In gallant trim the gilded vessel goes,
> YOUTH at the prow and PLEASURE at the helm,
> Regardless of the sweeping whirlwinds sway,
> That hush'd in grim repose, expects its evening prey.

(In which, by the by, the words 'realm' and 'sway' are rhymes dearly purchased.) I preferred the original on the ground, that in the imitation it depended wholly in the compositor's putting, or not putting a small capital, both in this, and in many other passages of the same poet, whether the words should be personifications, or mere abstracts. I mention this, because in referring various lines in Gray to their original in Shakespeare and Milton; and in the clear perception how completely all the propriety was lost in the transfer; I was, at that early period, led to a conjecture, which, many years afterwards was recalled to me from the same thought having been started in conversation, but far more ably, and developed more fully, by Mr Wordsworth; namely, that this style of poetry, which I have characterized above, as translations of prose thoughts into poetic language, had been kept up by, if it did not wholly arise from, the custom of writing Latin verses, and the great importance attached to these exercises, in our public schools. Whatever might have been the case in the fifteenth century, when the use of the Latin tongue was so general among learned men, that Erasmus is said to have forgotten his native language; yet in the present day it is not to be supposed, that a youth can *think* in Latin, or that he can have any other reliance on the force or fitness of his phrases, but the authority of the author from whence he has adopted them. Consequently he must first prepare his thoughts, and then pick out, from Virgil, Horace, Ovid, or perhaps more compendiously from his Gradus,* halves and quarters of lines, in which to embody them.

* In the Nutricia of Politian there occurs this line:

> 'Pura coloratos interstrepit unda lapillos.'

Casting my eye on a university prize poem, I met this line,

> 'Lactea purpureos interstrepit unda lapillos.'°

Now look out in the Gradus for *purus*, and you find as the first synonym, *lacteus*; for *coloratus*, and the first synonym is *purpureus*. I mention this by way of elucidating one of the most ordinary processes in the *ferrumination* of these centos.

I never object to a certain degree of disputatiousness in a young man from the age of seventeen to that of four or five and twenty, provided I find him always arguing on one side of the question. The controversies, occasioned by my unfeigned zeal for the honour of a favourite contemporary, then known to me only by his works, were of great advantage in the formation and establishment of my taste and critical opinions. In my defence of the lines running into each other, instead of closing at each couplet; and of natural language, neither bookish, nor vulgar, neither redolent of the lamp, or of the kennel, such as *I will remember thee*; instead of the same thought tricked up in the rag-fair finery of,

> . . . Thy image on her wing
> Before my FANCY's eye shall MEMORY bring,

I had continually to adduce the metre and diction of the Greek poets from Homer to Theocritus inclusive; and still more of our elder English poets from Chaucer to Milton. Nor was this all. But as it was my constant reply to authorities brought against me from later poets of great name, that no authority could avail in opposition to truth, nature, logic, and the laws of universal grammar; actuated too by my former passion for metaphysical investigations; I laboured at a solid foundation, on which permanently to ground my opinions, in the component faculties of the human mind itself, and their comparative dignity and importance. According to the faculty or source, from which the pleasure given by any poem or passage was derived, I estimated the merit of such poem or passage. As the result of all my reading and meditation, I abstracted two critical aphorisms, deeming them to comprise the conditions and criteria of poetic style; first, that not the poem which we have *read*, but that to which we *return*, with the greatest pleasure, possesses the genuine power, and claims the name of *essential poetry*. Second, that whatever lines can be translated into other words of the same language, without diminution of their significance, either in sense, or association, or in any worthy feeling, are so far vicious in their diction. Be it however observed, that I excluded from the list of worthy feelings, the pleasure derived from mere novelty, in the reader, and the desire of exciting wonderment at his powers in the author. Oftentimes since then, in perusing French tragedies, I have fancied two marks of admiration at the end of each line, as hieroglyphics of the author's own admiration at his own cleverness. Our genuine admiration of a great poet is a continuous undercurrent of feeling; it is everywhere present, but seldom anywhere as a separate excitement. I was wont boldly to affirm, that it would be scarcely more difficult to push a stone out from the pyramids with the bare hand, than to alter a word, or the position of a word, in Milton or Shakespeare, (in their most important works at least) without making the author say something else, or something worse, than he does say. One great distinction, I appeared to myself to see plainly, between, even the characteristic faults of our elder poets, and the false beauty of the moderns. In the former, from Donne to Cowley, we find the most fantastic out-of-the-way thoughts, but in the most pure and genuine mother English; in the latter, the most obvious thoughts, in language the most fantastic and arbitrary. Our faulty elder poets sacrificed the passion, and passionate flow of poetry, to the subtleties of intellect, and to the starts of wit; the moderns to the glare and glitter of a perpetual, yet broken and heterogeneous imagery, or rather to an amphibious something, made up, half of image, and half of abstract meaning.* The one sacrificed the heart to the head; the other both heart and head to point and drapery.

The reader must make himself acquainted with the general style of composition that was at that time deemed poetry, in order to understand and account for the effect produced on me by the Sonnets, the Monody at Matlock, and the Hope, of Mr Bowles; for it is peculiar to original genius to become less and less striking, in proportion to its success in improving the taste and judgement of its contemporaries. The poems of West indeed had the merit of chaste and manly diction, but they were cold, and, if I may so express it, only *dead-coloured*; while in the best of Warton's there is a stiffness, which too often gives them the appearance of imitations from the Greek. Whatever relation therefore of cause or impulse Percy's collection of ballads may bear to the most popular poems of the present day; yet in the more sustained and elevated style, of the then living poets Bowles and Cowper† were,

* I remember a ludicrous instance in the poem of a young tradesman:

> No more will I endure love's pleasing pain,
> Or round my *heart's leg* tie his galling chain.

† Cowper's Task was published some time before the sonnets of Mr Bowles; but I was not familiar with it till many years afterwards. The vein of satire which runs through that excellent poem, together with the sombre hue of its religious opinions, would probably, *at that time*, have prevented its laying any strong hold on my affections. The love of nature seems to have led Thomson to a cheerful religion; and a gloomy religion to have led Cowper to a love of nature. The one would carry his fellow men along with him into nature; the other flies to nature from his fellow men. In chastity of diction however, and the harmony of blank verse, Cowper leaves Thomson unmeasurably below him; yet still I feel the latter to have been the *born poet*.

to the best of my knowledge, the first who combined natural thoughts with natural diction; the first who reconciled the heart with the head.

It is true, as I have before mentioned, that from diffidence in my own powers, I for a short time adopted a laborious and florid diction, which I myself deemed, if not absolutely vicious, yet of very inferior worth. Gradually, however, my practice conformed to my better judgement; and the compositions of my twenty-fourth and twenty-fifth year (*ex. gr.* [*e*(*xempli*) *g*(*ratia*), for example] the shorter blank-verse poems, the lines which are now adopted in the introductory part of the Vision in the present collection in Mr Southey's Joan of Arc, 2nd book, 1st edition, and the tragedy of Remorse) are not more below my present ideal in respect of the general tissue of the style, than those of the latest date. Their faults were at least a remnant of the former leaven, and among the many who have done me the honour of putting my poems in the same class with those of my betters, the one or two, who have pretended to bring examples of affected simplicity from my volume, have been able to adduce but one instance, and that out of a copy of verses half ludicrous, half splenetic, which I intended, and had myself characterized, as *sermoni propriora* [more suitable for prose].

Every reform, however necessary, will by weak minds be carried to an excess, that itself will need reforming. The reader will excuse me for noticing, that I myself was the first to expose *risu honesto* [with honest laughter] the three sins of poetry, one or the other of which is the most likely to beset a young writer. So long ago as the publication of the second number of the Monthly Magazine, under the name of Nehemiah Higgenbottom I contributed three sonnets, the first of which had for its object to excite a good-natured laugh at the spirit of doleful egotism, and at the recurrence of favourite phrases, with the double defect of being at once trite, and licentious. The second, on low, creeping language and thoughts, under the pretence of simplicity. And the third, the phrases of which were borrowed entirely from my own poems, on the indiscriminate use of elaborate and swelling language and imagery. The reader will find them in the note below,*

* SONNET I

Pensive at eve, on the *hard* world I mused,
And *my poor* heart was sad; so at the MOON
I gazed, and sighed, and sighed; for ah how soon
Eve saddens into night! mine eyes perused
With tearful vacancy the *dampy* grass
That wept and glitter'd in the *paly* ray:
And I *did pause me*, on my lonely way

[*cont.*]

and will I trust regard them as reprinted for biographical purposes, and not for their poetic merits. So general at that time, and so decided was the opinion concerning the characteristic vices of my style, that a celebrated physician (now, alas! no more) speaking of me in other

And *mused me*, on the *wretched ones* that pass
O'er the bleak heath of sorrow. But alas!
Most of *myself* I thought! when it befel,
That the *soothe* spirit of the *breezy* wood
Breath'd in mine ear: 'All this is very well,
But much of ONE thing, is for NO thing good.'
Oh *my poor heart's* INEXPLICABLE SWELL!

SONNET II

Oh I do love thee, meek SIMPLICITY!
For of thy lays the lulling simpleness
Goes to my heart, and soothes each small distress,
Distress tho' small, yet haply great to me,
'Tis true on Lady Fortune's gentlest pad
I amble on; and yet I know not why
So sad I am! but should a friend and I
Frown, pout and part, then I am *very* sad.
And then with sonnets and with sympathy
My dreamy bosom's mystic woes I pall;
Now of my false friend plaining plaintively,
Now raving at mankind in general;
But whether sad or fierce, 'tis simple all,
All very simple, meek SIMPLICITY!

SONNET III

And this reft house is that, the which he built,
Lamented Jack! and here his malt he pil'd,
Cautious in vain! these rats, that squeak so wild,
Squeak not unconscious of their father's guilt.
Did he not see her gleaming thro' the glade!
Belike 'twas she, the maiden all forlorn.
What tho' she milk no cow with crumpled horn,
Yet, *aye* she haunts the dale where *erst* she stray'd;
And *aye*, beside her stalks her amorous knight!
Still on his thighs their wonted brogues are worn,
And thro' those brogues, still tatter'd and betorn,
His hindward charms gleam an unearthly white.
Ah! thus thro' broken clouds at night's high Noon
Peeps in fair fragments forth the full-orb'd harvest-moon!

The following anecdote will not be wholly out of place here, and may perhaps amuse the reader. An amateur performer in verse expressed to a common friend, a strong desire to be introduced to me, but hesitated in accepting my friend's immediate offer, on

respects with his usual kindness to a gentleman, who was about to meet me at a dinner party, could not however resist giving him a hint not to mention the 'House that Jack built' in my presence, for 'that I was as sore as a boil about that sonnet'; he not knowing, that I was myself the author of it.

CHAPTER II

Supposed irritability of men of genius—Brought to the test of facts—Causes and occasions of the charge—Its injustice.

I HAVE often thought, that it would be neither uninstructive nor unamusing to analyse, and bring forward into distinct consciousness, that complex feeling, with which readers in general take part against the author, in favour of the critic; and the readiness with which they apply to all poets the old sarcasm of Horace upon the scribblers of his time: 'Genus irritabile vatum [the irritable race of poets].' A debility and dimness of the imaginative power, and a consequent necessity of reliance on the immediate impressions of the senses, do, we well know, render the mind liable to superstition and fanaticism. Having a deficient portion of internal and proper warmth, minds of this class seek in the crowd *circum fana* [around the temples] for a warmth in common, which they do not possess singly. Cold and phlegmatic in their own nature, like damp hay, they heat and inflame by co-acervation; or like bees they become restless and irritable through the increased temperature of collected multitudes. Hence the German word for fanaticism (such at least was its original import) is derived

the score that he was, he must acknowledge, the author of a confounded severe epigram on my Ancient Mariner, which had given me great pain. I assured my friend that if the epigram was a good one, it would only increase my desire to become acquainted with the author, and begged to hear it recited: when, to my no less surprise than amusement, it proved to be one which I had myself some time before written and inserted in the Morning Post.

To the author of the Ancient Mariner.

Your poem must eternal be,
Dear sir! it cannot fail,
For 'tis incomprehensible
And without head or tail.

from the swarming of bees, namely, *Schwärmen*, *Schwärmerey*. The passion being in an inverse proportion to the insight, *that* the more vivid, as *this* the less distinct; anger is the inevitable consequence. The absence of all foundation within their own minds for that, which they yet believe both true and indispensable for their safety and happiness, cannot but produce an uneasy state of feeling, an involuntary sense of fear from which nature has no means of rescuing herself but by anger. Experience informs us that the first defence of weak minds is to recriminate.

> There's no Philosopher but sees,
> That rage and fear are one disease,
> Tho' that may burn, and this may freeze,
> They're both alike the ague.
>
> Mad Ox.

But where the ideas are vivid, and there exists an endless power of combining and modifying them, the feelings and affections blend more easily and intimately with these ideal creations, than with the objects of the senses; the mind is affected by thoughts, rather than by things; and only then feels the requisite interest even for the most important events, and accidents, when by means of meditation they have passed into thoughts. The sanity of the mind is between superstition with fanaticism on the one hand; and enthusiasm with indifference and a diseased slowness to action on the other. For the conceptions of the mind may be so vivid and adequate, as to preclude that impulse to the realizing of them, which is strongest and most restless in those, who possess more than mere talent (or the faculty of appropriating and applying the knowledge of others) yet still want something of the creative, and self-sufficing power of absolute genius. For this reason therefore, they are men of commanding genius. While the former rest content between thought and reality, as it were in an intermundium of which their own living spirit supplies the substance, and their imagination the ever-varying form; the latter must impress their preconceptions on the world without, in order to present them back to their own view with the satisfying degree of clearness, distinctness, and individuality. These in tranquil times are formed to exhibit a perfect poem in palace or temple or landscape garden; or a tale of romance in canals that join sea with sea, or in walls of rock, which shouldering back the billows imitate the power, and supply the benevolence of nature to sheltered navies; or in aqueducts that arching the wide vale from mountain to mountain give a Palmyra to the desert.

But alas! in times of tumult they are the men destined to come forth as the shaping spirit of ruin, to destroy the wisdom of ages in order to substitute the fancies of a day, and to change kings and kingdoms, as the wind shifts and shapes the clouds.* The records of biography seem to confirm this theory. The men of the greatest genius, as far as we can judge from their own works or from the accounts of their contemporaries, appear to have been of calm and tranquil temper, in all that related to themselves. In the inward assurance of permanent fame, they seem to have been either indifferent or resigned, with regard to immediate reputation. Through all the works of Chaucer there reigns a cheerfulness, a manly hilarity, which makes it almost impossible to doubt a correspondent habit of feeling in the author himself. Shakespeare's evenness and sweetness of temper were almost proverbial in his own age. That this did not arise from ignorance of his own comparative greatness, we have abundant proof in his sonnets, which could scarcely have been known to Mr Pope,† when he asserted, that

* Of old things all are over old,
Of good things none are good enough:—
We'll show that we can help to frame
A world of other stuff.

I too will have my kings, that take
From me the sign of life and death;
Kingdoms shall shift about, like clouds,
Obedient to my breath.

Wordsworth's ROB ROY

† Mr Pope was under the common error of his age, an error, far from being sufficiently exploded even at the present day. It consists (as I explained at large, and proved in detail in my public lectures) in mistaking for the *essentials* of the Greek stage certain rules, which the wise poets imposed upon themselves, in order to render all the remaining parts of the drama consistent with those, that had been forced upon them by circumstances independent of their will; out of which circumstances the drama itself arose. The circumstances in the time of Shakespeare, which it was equally out of his power to alter, were different, and such as, in my opinion, allowed a far wider sphere, and a deeper and more human interest. Critics are too apt to forget, that *rules* are but means to an end; consequently where the ends are different, the rules must be likewise so. We must have ascertained what the end *is*, before we can determine what the rules *ought* to be. Judging under this impression, I did not hesitate to declare my full conviction, that the consummate judgement of Shakespeare, not only in the general construction, but in all the *detail*, of his dramas impressed me with greater wonder, than even the might of his genius, or the depth of his philosophy. The substance of these lectures I hope soon to publish; and it is but a debt of justice to myself and my friends to notice, that the first course of lectures, which differed from the following courses only, by occasionally varying the illustrations of the same thoughts, was addressed to very numerous, and I need not add, respectable audiences at the royal institution, before Mr Schlegel gave his lectures on the same subjects at Vienna.

our great bard 'grew immortal in his own despite.' Speaking of one whom he had celebrated, and contrasting the duration of his works with that of his personal existence, Shakespeare adds:

> Your name from hence immortal life shall have,
> Tho' I once gone to all the world must die;
> The earth can yield me but a common grave,
> When you entombed in men's eyes shall lie.
> Your monument shall be my gentle verse,
> Which eyes not yet created shall o'er-read;
> And *tongues to be* your being shall rehearse,
> When all the breathers of this world are dead:
> You still shall live, such virtue hath my pen,
> Where breath most breathes, e'en in the mouth of men.
>
> Sonnet 81st

I have taken the first that occurred; but Shakespeare's readiness to praise his rivals, *ore pleno* [with full voice], and the confidence of his own equality with those whom he deemed most worthy of his praise, are alike manifested in the 86th sonnet.

> Was it the proud full sail of his great verse
> Bound for the praise of all-too-precious you,
> That did my ripe thoughts in my brain inhearse,
> Making their tomb, the womb wherein they grew?
> Was it his spirit, by spirits taught to write
> Above a mortal pitch that struck me dead?
> No, neither he, nor his compeers by night
> Giving him aid, my verse astonished.
> He, nor that affable familiar ghost,
> Which nightly gulls him with intelligence,
> As victors of my silence cannot boast;
> I was not sick of any fear from thence!
> But when your countenance fill'd up his line,
> Then lack'd I matter, that enfeebled mine.

In Spenser indeed, we trace a mind constitutionally tender, delicate, and, in comparison with his three great compeers, I had almost said, *effeminate*; and this additionally saddened by the unjust persecution of Burleigh, and the severe calamities, which overwhelmed his latter days. These causes have diffused over all his compositions 'a melancholy grace,' and have drawn forth occasional strains, the more pathetic from their gentleness. But nowhere do we find the least trace of irritability, and still less of quarrelsome or affected contempt of his censurers.

The same calmness, and even greater self-possession, may be affirmed of Milton, as far as his poems, and poetic character are concerned. He reserved his anger, for the enemies of religion, freedom, and his country. My mind is not capable of forming a more august conception, than arises from the contemplation of this great man in his latter days: poor, sick, old, blind, slandered, persecuted,

> Darkness before, and danger's voice behind,

in an age in which he was as little understood by the party, for whom, as by that, against whom he had contended; and among men before whom he strode so far as to dwarf himself by the distance; yet still listening to the music of his own thoughts, or if additionally cheered, yet cheered only by the prophetic faith of two or three solitary individuals, he did nevertheless

> . . . Argue not
> Against Heaven's hand or will, nor bate a jot
> Of heart or hope; but still bore up and steer'd
> Right onward.

From others only do we derive our knowledge that Milton, in his latter day, had his scorners and detractors; and even in his day of youth and hope, that he had enemies would have been unknown to us, had they not been likewise the enemies of his country.

I am well aware, that in advanced stages of literature, when there exist many and excellent models, a high degree of talent, combined with taste and judgement, and employed in works of imagination, will acquire for a man the name of a great genius; though even that analogon of genius, which, in certain states of society, may even render his writings more popular than the absolute reality could have done, would be sought for in vain in the mind and temper of the author himself. Yet even in instances of this kind, a close examination will often detect, that the irritability, which has been attributed to the author's genius as its cause, did really originate in an ill conformation of body, obtuse pain, or constitutional defect of pleasurable sensation. What is charged to the author, belongs to the man, who would probably have been still more impatient, but for the humanizing influences of the very pursuit, which yet bears the blame of his irritability.

How then are we to explain the easy credence generally given to this charge, if the charge itself be not, as we have endeavoured to show, supported by experience? This seems to me of no very difficult solution. In whatever country literature is widely diffused, there will be many who mistake an intense desire to possess the reputation of poetic genius, for the actual powers, and original tendencies which constitute it. But men, whose dearest wishes are fixed on objects wholly out of their own power, become in all cases more or less impatient and prone to anger. Besides, though it may be paradoxical to assert, that a man can know one thing, and believe the opposite, yet assuredly, a vain person may have so habitually indulged the wish, and persevered in the attempt to appear, what he is not, as to become himself one of his own proselytes. Still, as this counterfeit and artificial persuasion must differ, even in the person's own feelings, from a real sense of inward power, what can be more natural, than that this difference should betray itself in suspicious and jealous irritability? Even as the flowery sod, which covers a hollow, may be often detected by its shaking and trembling.

But, alas! the multitude of books, and the general diffusion of literature, have produced other, and more lamentable effects in the world of letters, and such as are abundant to explain, though by no means to justify, the contempt with which the best-grounded complaints of injured genius are rejected as frivolous, or entertained as matter of merriment. In the days of Chaucer and Gower, our language might (with due allowance for the imperfections of a simile) be compared to a wilderness of vocal reeds, from which the favourites only of Pan or Apollo could construct even the rude Syrinx;° and from this the constructors alone could elicit strains of music. But now, partly by the labours of successive poets, and in part by the more artificial state of society and social intercourse, language, mechanized as it were into a barrel-organ, supplies at once both instrument and tune. Thus even the deaf may play, so as to delight the many. Sometimes (for it is with similes, as it is with jests at a wine table, one is sure to suggest another) I have attempted to illustrate the present state of our language, in its relation to literature, by a press-room of larger and smaller stereotype pieces, which, in the present Anglo-Gallican fashion of unconnected, epigrammatic periods, it requires but an ordinary portion of ingenuity to vary indefinitely, and yet still produce something which, if not sense, will be so like it, as to do as well. Perhaps better; for it spares the reader the trouble of thinking; prevents vacancy, while it indulges indolence; and secures the memory from all danger of an intellectual plethora. Hence of all trades, literature at present demands the least talent or information; and, of all modes of literature, the manufacturing of poems. The difference indeed between these and the works of genius, is not less than between an egg, and an egg-shell; yet at a distance they both look alike. Now it is no less remarkable than true, with how little examination works of polite literature are commonly perused, not only by the mass of readers, but by men of first-rate ability, till some accident or chance discussion* have roused their attention, and put them on their guard. And hence individuals below mediocrity not less in natural power than in acquired knowledge; nay, bunglers that had failed in the lowest mechanic crafts, and whose presumption is in due proportion to their want of sense and sensibility; men, who being first scribblers from idleness and ignorance next become libellers from envy and malevolence; have been able to drive a successful trade in the

* In the course of my lectures, I had occasion to point out the almost faultless position and choice of words, in Mr Pope's *original* compositions, particularly in his satires and moral essays, for the purpose of comparing them with his translation of Homer, which I do not stand alone in regarding as the main source of our pseudo-poetic diction. And this, by the by, is an additional confirmation of a remark made, I believe, by Sir Joshua Reynolds, that next to the man who formed and elevated the taste of the public, he that corrupted it, is commonly the greatest genius. Among other passages, I analysed sentence by sentence, and almost word by word, the popular lines

As when the moon, resplendent lamp of light, etc.

much in the same way as has been since done, in an excellent article on Chalmers's British Poets in the Quarterly Review. The impression on the audience in general was sudden and evident: and a number of enlightened and highly educated individuals, who at different times afterwards addressed me on the subject, expressed their wonder, that truth so obvious should not have struck them before; but at the same time acknowledged (so much had they been accustomed, in reading poetry, to receive pleasure from the separate images and phrases successively, without asking themselves whether the collective meaning was sense or nonsense) that they might in all probability have read the same passage again twenty times with undiminished admiration, and without once reflecting, that '*αστρα φαεινην αμφι σεληνην φαινετ' αριπρεπεα*' (i.e. the stars around, or near the full moon, shine pre-eminently bright) conveys a just and happy image of a moonlight sky: while it is difficult to determine whether in the lines,

Around *her throne* the vivid planets *roll*,
And stars *unnumber'd gild* the *glowing pole*,

the sense, or the diction be the more absurd. My answer was; that though I had derived peculiar advantages from my school discipline, and though my *general* theory of poetry was the same then as now, I had yet experienced the same sensations myself, and felt almost as if I had been newly couched,° when by Mr Wordsworth's conversation, I had been induced to re-examine with impartial strictness Gray's celebrated elegy. I had long before detected the defects in the Bard but the Elegy I had considered as proof against all fair attacks; and to this day I cannot read either, without delight, and a portion of enthusiasm. At all events, whatever pleasure I may have lost by the clearer perception of the faults in certain passages, has been more than repaid to me, by the additional delight with which I read the remainder.

employment of the booksellers, nay have raised themselves into temporary name and reputation with the public at large, by that most powerful of all adulation, the appeal to the bad and malignant passions of mankind.* But as it is the nature of scorn, envy, and all malignant propensities to require a quick change of objects, such writers are sure, sooner or later to awake from their dream of vanity to disappointment and neglect with embittered and envenomed feelings. Even during their short-lived success, sensible in spite of themselves on what a shifting foundation it rested, they resent the mere refusal of praise, as a robbery, and at the justest censures kindle at once into violent and undisciplined abuse; till the acute disease changing into chronical, the more deadly as the less violent, they become the fit instruments of literary detraction, and moral slander. They are then no longer to be questioned without exposing the complainant to ridicule, because, forsooth, they are anonymous critics, and authorized as 'synodical individuals'† to speak of themselves *plurali majestatico* [with the royal plural]! As if literature formed a caste, like that of the pariahs in Hindostan, who, however maltreated, must not dare to deem themselves wronged! As if that, which in all other cases adds a deeper dye to slander, the circumstance of its being anonymous, here acted only to make the slanderer inviolable! Thus, in part, from the

* Especially 'in this age of personality, this age of literary and political gossiping, when the meanest insects are worshipped with a sort of Egyptian superstition, if only the brainless head be atoned for by the sting of personal malignity in the tail! When the most vapid satires have become the objects of a keen public interest, purely from the number of contemporary characters named in the patchwork notes (which possess, however, the comparative merit of being more poetical than the text) and because, to increase the stimulus, the author has sagaciously left his own name for whispers and conjectures! In an age, when even sermons are published with a double appendix stuffed with names—in a generation so transformed from the characteristic reserve of Britons, that from the ephemeral sheet of a London newspaper, to the everlasting Scotch professorial quarto, almost every publication exhibits or flatters the epidemic distemper; that the very "last year's rebuses" in the Ladies Diary, are answered in a serious elegy "on my father's death" with the name and habitat of the elegiac Oedipus subscribed; and "other ingenious solutions were likewise given" to the said rebuses—not as heretofore by Crito, Philander, A, B, Y, etc. but by fifty or sixty plain English surnames at full length with their several places of abode! In an age, when a bashful Philalethes, or Phileleutheros is as rare on the title-pages, and among the signatures of our magazines, as a real name used to be in the days of our shy and notice-shunning grandfathers! When (more exquisite than all) I see an epic poem (spirits of Maro and Maeonides make ready to welcome your new compeer!) advertised with the special recommendation, that the said epic poem contains more than an hundred names of *living* persons.'

Friend No. 10°

† A phrase of Andrew Marvell's.

accidental tempers of individuals (men of undoubted talent, but not men of genius) tempers rendered yet more irritable by their desire to appear men of genius; but still more effectively by the excesses of the mere counterfeits both of talent and genius; the number too being so incomparably greater of those who are thought to be, than of those who really are men of real genius; and in part from the natural, but not therefore the less partial and unjust distinction, made by the public itself between literary, and all other property; I believe the prejudice to have arisen, which considers an unusual irascibility concerning the reception of its products as characteristic of genius. It might correct the moral feelings of a numerous class of readers, to suppose a review set on foot, the object of which was to criticize all the chief works presented to the public by our ribbon-weavers, calico-printers, cabinet-makers, and china manufacturers; a review conducted in the same spirit, and which should take the same freedom with personal character, as our literary journals. They would scarcely, I think, deny their belief, not only that the 'genus irritabile [irritable race]' would be found to include many other species besides that of bards; but that the irritability of trade would soon reduce the resentments of poets into mere shadow-fights (σκιομαχιας) in the comparison. Or is wealth the only rational object of human interest? Or even if this were admitted, has the poet no property in his works? Or is it a rare, or culpable case, that he who serves at the altar of the Muses, should be compelled to derive his maintenance from the altar, when too he has perhaps deliberately abandoned the fairest prospects of rank and opulence in order to devote himself, an entire and undistracted man, to the instruction or refinement of his fellow citizens? Or should we pass by all higher objects and motives, all disinterested benevolence, and even that ambition of lasting praise which is at once the crutch and ornament, which at once supports and betrays, the infirmity of human virtue; is the character and property of the individual, who labours for our intellectual pleasures, less entitled to a share of our fellow feeling, than that of the wine-merchant or milliner? Sensibility indeed, both quick and deep, is not only a characteristic feature, but may be deemed a component part, of genius. But it is no less an essential mark of true genius, that its sensibility is excited by any other cause more powerfully, than by its own personal interests; for this plain reason, that the man of genius lives most in the ideal world, in which the present is still constituted by the future or the past; and because his feelings have been habitually associated with thoughts and images, to the number, clearness, and vivacity of which the sensation of self is always in an inverse proportion. And yet, should he perchance have occasion to repel some false charge, or to rectify some erroneous censure, nothing is more common, than for the many to mistake the general liveliness of his manner and language whatever is the subject, for the effects of peculiar irritation from its accidental relation to himself.*

For myself, if from my own feelings, or from the less suspicious test of the observations of others, I had been made aware of any literary testiness or jealousy; I trust, that I should have been, however, neither silly or arrogant enough, to have burthened the imperfection on genius. But an experience (and I should not need documents in abundance to prove my words, if I added) a tried experience of twenty years, has taught me, that the original sin of my character consists in a careless indifference to public opinion, and to the attacks of those who influence it; that praise and admiration have become yearly, less and less desirable, except as marks of sympathy; nay that it is difficult and distressing to me, to think with any interest even about the sale and profit of my works, important, as in my present circumstances, such considerations must needs be. Yet it never occurred to me to believe or fancy, that the *quantum* [amount] of intellectual power bestowed on me by nature or education was in any way connected with this habit of my feelings; or that it needed any other parents or fosterers, than constitutional indolence, aggravated into languor by ill health; the accumulating embarrassments of procrastination; the mental cowardice, which is the inseparable companion of procrastination, and which makes us anxious to think and converse on anything rather than on what concerns ourselves; in fine, all those close vexations, whether chargeable on my faults or my fortunes which leave me but little grief to spare for evils comparatively distant and alien.

Indignation at literary wrongs, I leave to men born under happier

* This is one instance among many of deception, by the telling the half of a fact, and omitting the other half, when it is from their mutual counteraction and neutralization, that the whole truth arises, as a *tertium aliquid* [third thing] different from either. Thus in Dryden's famous line 'Great wit' (which here means genius) 'to madness sure is near allied.' Now as far as the profound sensibility, which is doubtless *one* of the components of genius, were alone considered, single and unbalanced, it might be fairly described as exposing the individual to a greater chance of mental derangement; but then a more than usual rapidity of association, a more than usual power of passing from thought to thought, and image to image, is a component equally essential; and in the due modification of each by the other the genius itself consists; so that it would be as just as fair to describe the earth, as in imminent danger of exorbitating, or of falling into the sun, according as the assertor of the absurdity confined his attention either to the projectile or to the attractive force exclusively.

stars. I cannot afford it. But so far from condemning those who can, I deem it a writer's duty, and think it creditable to his heart, to feel and express a resentment proportioned to the grossness of the provocation, and the importance of the object. There is no profession on earth, which requires an attention so early, so long, or so unintermitting as that of poetry; and indeed as that of literary composition in general, if it be such, as at all satisfies the demands both of taste and of sound logic. How difficult and delicate a task even the mere mechanism of verse is, may be conjectured from the failure of those, who have attempted poetry late in life. Where then a man has, from his earliest youth, devoted his whole being to an object, which by the admission of all civilized nations in all ages is honourable as a pursuit, and glorious as an attainment; what of all that relates to himself and his family, if only we except his moral character, can have fairer claims to his protection, or more authorize acts of self-defence, than the elaborate products of his intellect, and intellectual industry? Prudence itself would command us to show, even if defect or diversion of natural sensibility had prevented us from feeling, a due interest and qualified anxiety for the offspring and representatives of our nobler being. I know it, alas! by woeful experience! I have laid too many eggs in the hot sands of this wilderness the world, with ostrich carelessness and ostrich oblivion. The greater part indeed have been trod under foot, and are forgotten; but yet no small number have crept forth into life, some to furnish feathers for the caps of others, and still more to plume the shafts in the quivers of my enemies, of them that unprovoked have lain in wait against my soul.

> Sic vos, non vobis mellificatis, apes!
>
> [So you make honey, bees, but not for yourselves!]

An instance in confirmation of the Note, p. [177], occurs to me as I am correcting this sheet, with the Faithful Shepherdess open before me. Mr Seward first traces Fletcher's lines:

> More soul diseases than e'er yet the hot
> Sun bred thro' his burnings, while the dog
> Pursues the raging lion, throwing the fog
> And deadly vapor from his angry breath,
> Filling the lower world with plague and death.—

to Spenser's Shepherd's Calendar,

> The rampant lion hunts he fast
> With dogs of noisome breath;
> Whose baleful barking brings, in haste,
> Pyne, plagues, and dreary death!

[*cont.*]

CHAPTER III

The author's obligations to critics, and the probable occasion—Principles of modern criticism—Mr Southey's works and character.

To anonymous critics in reviews, magazines, and news-journals of various name and rank, and to satirists with or without a name, in verse or prose, or in verse text aided by prose comment, I do seriously believe and profess, that I owe full two-thirds of whatever reputation and publicity I happen to possess. For when the name of an individual has occurred so frequently, in so many works, for so great a length of time, the readers of these works (which with a shelf or two of Beauties, Elegant Extracts and Anas, form nine-tenths of the reading of the reading public*) cannot but be familiar with the name, without

He then takes occasion to introduce Homer's simile of the sight of Achilles' shield to Priam compared with the Dog Star, literally thus—'For this indeed is most splendid, but it was made an evil sign, and brings many a consuming disease to wretched mortals.' Nothing can be more simple as a description, or more accurate as a simile; which (says Mr S.) is thus finely translated by Mr Pope:

> Terrific Glory! for his burning breath
> Taints the *red* air with fevers, plagues, and death!

Now here (not to mention the tremendous bombast) the *Dog Star*, so called, is turned into a real Dog, a very odd Dog, a Fire, Fever, Plague, and death-breathing, *red*-air-tainting Dog: and the whole visual likeness is lost, while the likeness in the effects is rendered absurd by the exaggeration. In Spenser and Fletcher the thought is justifiable; for the images are at least consistent, and it was the intention of the writers to mark the seasons by this allegory of visualized puns.

* For as to the devotees of the circulating libraries, I dare not compliment their pass-time, or rather kill-time, with the name of reading. Call it rather a sort of beggarly day-dreaming, during which the mind of the dreamer furnishes for itself nothing but laziness and a little mawkish sensibility; while the whole *matériel* and imagery of the doze is supplied *ab extra* [from without] by a sort of mental camera obscura° manufactured at the printing office, which *pro tempore* [temporarily] fixes, reflects and transmits the moving phantasms of one man's delirium, so as to people the barrenness of an hundred other brains afflicted with the same trance or suspension of all common sense and all definite purpose. We should therefore transfer this species of amusement, (if indeed those can be said to retire *a musis* [away from the Muses], who were never in their company, or relaxation be attributable to those, whose bows are never bent) from the genus, reading, to that comprehensive class characterized by the power of reconciling the two contrary yet coexisting propensities of human nature, namely;

distinctly remembering whether it was introduced for an eulogy or for censure. And this becomes the more likely, if (as I believe) the habit of perusing periodical works may be properly added to Averroes'* catalogue of anti-mnemonics, or weakeners of the memory. But where this has not been the case, yet the reader will be apt to suspect, that there must be something more than usually strong and extensive in a reputation, that could either require or stand so merciless and long-continued a cannonading. Without any feeling of anger therefore (for which indeed, on my own account, I have no pretext) I may yet be allowed to express some degree of surprise, that after having run the critical gauntlet for a certain class of faults which I *had*, nothing having come before the judgement seat in the interim, I should, year after year, quarter after quarter, month after month (not to mention sundry petty periodicals of still quicker revolution, 'or weekly or diurnal') have been for at least 17 years consecutively dragged forth by them into the foremost ranks of the proscribed, and forced to abide the brunt of abuse, for faults directly opposite, and which I certainly had not. How shall I explain this?

Whatever may have been the case with others, I certainly cannot attribute this persecution to personal dislike, or to envy, or to feelings of vindictive animosity. Not to the former, for, with the exception of a very few who are my intimate friends, and were so before they were known as authors, I have had little other acquaintance with literary characters, than what may be implied in an accidental introduction, or casual meeting in a mixed company. And, as far as words and looks can be trusted, I must believe that, even in these instances, I had

indulgence of sloth, and hatred of vacancy. In addition to novels and tales of chivalry in prose or rhyme, (by which last I mean neither rhythm nor metre) this genus comprises as its species, gaming, swinging, or swaying on a chair or gate; spitting over a bridge; smoking; snuff-taking; tête-à-tête quarrels after dinner between husband and wife; conning word by word all the advertisements of the Daily Advertizer in a public house on a rainy day, etc. etc. etc.

* Ex. gr. 'Pediculos e capillis excerptos in arenam jacere incontusos [throwing to the ground lice picked from the hair, without crushing them]'; eating of unripe fruit; gazing on the clouds, and (*in genere* [the same in kind]) on movable things suspended in the air; riding among a multitude of camels; frequent laughter; listening to a series of jests and humorous anecdotes, as when (so to modernize the learned Saracen's meaning) one man's droll story of an Irishman inevitably occasions another's droll story of a Scotchman, which again by the same sort of conjunction disjunctive leads to some *étourderie* [blunder] of a Welshman, and that again to some sly hit of a Yorkshireman; the habit of reading tombstones in church-yards, etc. By the by, this catalogue, strange as it may appear, is not insusceptible of a sound psychological commentary.

excited no unfriendly disposition.* Neither by letter, or in conversation, have I ever had dispute or controversy beyond the common social interchange of opinions. Nay, where I had reason to suppose my convictions fundamentally different, it has been my habit, and I may

* Some years ago, a gentleman, the chief writer and conductor of a celebrated review, distinguished by its hostility to Mr Southey,° spent a day or two at Keswick. That he was, without diminution on this account, treated with every hospitable attention by Mr Southey and myself, I trust I need not say. But one thing I may venture to notice; that at no period of my life do I remember to have received so many, and such high-coloured compliments in so short a space of time. He was likewise circumstantially informed by what series of accidents it had happened, that Mr Wordsworth, Mr Southey, and I had become neighbours; and how utterly unfounded was the supposition, that we considered ourselves, as belonging to any common school, but that of good sense confirmed by the long-established models of the best times of Greece, Rome, Italy, and England; and still more groundless the notion, that Mr Southey (for as to myself I have published so little, and that little, of so little importance, as to make it almost ludicrous to mention my name at all) could have been concerned in the formation of a poetic sect with Mr Wordsworth, when so many of his works had been published not only previously to any acquaintance between them; but before Mr Wordsworth himself had written anything but in a diction ornate, and uniformly sustained; when too the slightest examination will make it evident, that between those and the after writings of Mr Southey, there exists no other difference than that of a progressive degree of excellence from progressive development of power, and progressive facility from habit and increase of experience. Yet among the first articles which this man wrote after his return from Keswick, we were characterized as 'the School of whining and hypochondriacal poets that haunt the Lakes.' In reply to a letter from the same gentleman, in which he had asked me, whether I was in earnest in preferring the style of Hooker to that of Dr Johnson; and Jeremy Taylor to Burke; I stated, somewhat at large, the comparative excellences and defects which characterized our best prose writers, from the reformation, to the first half of Charles II; and that of those who had flourished during the present reign, and the preceding one. About twelve months afterwards, a review appeared on the same subject, in the concluding paragraph of which the reviewer asserts, that his chief motive for entering into the discussion was to separate a rational and qualified admiration of our elder writers, from the indiscriminate enthusiasm of a recent school, who praised what they did not understand, and caricatured what they were unable to imitate. And, that no doubt might be left concerning the persons alluded to, the writer annexes the names of Miss Baillie, R. Southey, Wordsworth and Coleridge. For that which follows, I have only ear-say evidence; but yet such as demands my belief; viz. that on being questioned concerning this apparently wanton attack, more especially with reference to Miss Baillie, the writer had stated as his motives, that this lady when at Edinburgh had declined a proposal of introducing him to her; that Mr Southey had written against him; and Mr Wordsworth had talked contemptuously of him; but that as to Coleridge he had noticed him merely because the names of Southey and Wordsworth and Coleridge always went together. But if it were worth while to mix together, as ingredients, half the anecdotes which I either myself know to be true, or which I have received from men incapable of intentional falsehood, concerning the characters, qualifications, and motives of our anonymous critics, whose decisions are oracles for our reading public; I might safely borrow the words of the

add, the impulse of my nature, to assign the grounds of my belief, rather than the belief itself; and not to express dissent, till I could establish some points of complete sympathy, some grounds common to both sides, from which to commence its explanation.

Still less can I place these attacks to the charge of envy. The few pages, which I have published, are of too distant a date; and the extent of their sale a proof too conclusive against their having been popular at any time; to render probable, I had almost said possible, the excitement of envy on their account; and the man who should envy me on any other, verily he must be envy-mad!

Lastly, with as little semblance of reason, could I suspect any animosity towards me from vindictive feelings as the cause. I have before said, that my acquaintance with literary men has been limited and distant; and that I have had neither dispute nor controversy. From my first entrance into life, I have, with few and short intervals, lived either abroad or in retirement. My different essays on subjects of national interest, published at different times, first in the Morning Post and then in the Courier, with my courses of lectures on the principles of criticism as applied to Shakespeare and Milton, constitute my whole publicity; the only occasions on which I *could* offend any member of the republic of letters. With one solitary exception in which my words were first misstated and then wantonly applied to an individual, I could never learn, that I had excited the displeasure of any among my literary contemporaries. Having announced my intention to give a course of lectures on the characteristic merits and defects of English poetry in its different eras; first, from Chaucer to Milton; second, from Dryden inclusive to Thomson; and third, from Cowper to the present day; I changed my plan, and confined my disquisition to the two former eras, that I might furnish no possible pretext for the unthinking to misconstrue, or the malignant to misapply my words, and having stamped their own meaning on them, to pass them as current coin in the marts of garrulity or detraction.

Praises of the unworthy are felt by ardent minds as robberies of the deserving; and it is too true, and too frequent, that Bacon, Harrington,

apocryphal Daniel; 'Give me leave, O Sovereign Public, and I shall slay this dragon without sword or staff.' For the compound would be as the 'Pitch, and fat, and hair, which Daniel took, and did seethe them together, and made lumps thereof, and put into the dragon's mouth, and so the dragon burst in sunder; and Daniel said "Lo; these are the Gods ye worship".'

Machiavel, and Spinoza, are *not* read, because Hume, Condillac, and Voltaire *are*. But in promiscuous company no prudent man will oppugn the merits of a contemporary in his own supposed department; contenting himself with praising in his turn those whom *he* deems excellent. If I should ever deem it my duty at all to oppose the pretensions of individuals, I would oppose them in books which could be weighed and answered, in which I could evolve the whole of my reasons and feelings, with their requisite limits and modifications; not in irrecoverable conversation, where however strong the reasons might be, the feelings that prompted them would assuredly be attributed by some one or other to envy and discontent. Besides I well know, and I trust, have acted on that knowledge, that it must be the ignorant and injudicious who extol the unworthy; and the eulogies of critics without taste or judgement are the natural reward of authors without feeling or genius. 'Sint unicuique sua premia [Let each have his own reward].'

How then, dismissing, as I do, these three causes, am I to account for attacks, the long continuance and inveteracy of which it would require all three to explain? The solution may seem to have been given, or at least suggested, in a note to a preceding page. I was in habits of intimacy with Mr Wordsworth and Mr Southey! This, however, transfers, rather than removes, the difficulty. Be it, that by an unconscionable extension of the old adage, 'noscitur a socio [he is known by the company he keeps]' my literary friends are never under the waterfall of criticism, but I must be wet through with the spray; yet how came the torrent to descend upon them?

First then, with regard to Mr Southey. I well remember the general reception of his earlier publications: viz. the poems published with Mr Lovell under the names of Moschus and Bion; the two volumes of poems under his own name, and the Joan of Arc. The censures of the critics by profession are extant, and may be easily referred to:—careless lines, inequality in the merit of the different poems, and (in the lighter works) a predilection for the strange and whimsical; in short, such faults as might have been anticipated in a young and rapid writer, were indeed sufficiently enforced. Nor was there at that time wanting a party spirit to aggravate the defects of a poet, who with all the courage of uncorrupted youth had avowed his zeal for a cause, which he deemed that of liberty, and his abhorrence of oppression by whatever name consecrated. But it was as little objected by others, as dreamt of by the poet himself, that he *preferred* careless and prosaic lines on rule and of forethought, or indeed that he pretended to any other art or theory of poetic diction, besides that which we may all learn from Horace, Quintilian, the admirable dialogue de Causis Corruptae Eloquentiae, or Strada's Prolusions; if indeed natural good sense and the early study of the best models in his own language had not infused the same maxims more securely, and, if I may venture the expression, more vitally. All that could have been fairly deduced was, that in his taste and estimation of writers Mr Southey agreed far more with Warton, than with Johnson. Nor do I mean to deny, that at all times Mr Southey was of the same mind with Sir Philip Sidney in preferring an excellent ballad in the humblest style of poetry to twenty indifferent poems that strutted in the highest. And by what have his works, published since then, been characterized, each more strikingly than the preceding, but by greater splendour, a deeper pathos, profounder reflections, and a more sustained dignity of language and of metre? Distant may the period be, but whenever the time shall come, when all his works shall be collected by some editor worthy to be his biographer, I trust that an *excerpta* [selections] of all the passages, in which his writings, name, and character have been attacked, from the pamphlets and periodical works of the last twenty years, may be an accompaniment. Yet that it would prove medicinal in after times, I dare not hope; for as long as there are readers to be delighted with calumny, there will be found reviewers to calumniate. And such readers will become in all probability more numerous, in proportion as a still greater diffusion of literature shall produce an increase of sciolists; and sciolism bring with it petulance and presumption. In times of old, books were as religious oracles; as literature advanced, they next became venerable preceptors; they then descended to the rank of instructive friends; and as their numbers increased, they sunk still lower to that of entertaining companions; and at present they seem degraded into culprits to hold up their hands at the bar of every self-elected, yet not the less peremptory, judge, who chooses to write from humour or interest, from enmity or arrogance, and to abide the decision (in the words of Jeremy Taylor) 'of him that reads in malice, or him that reads after dinner.'

The same gradual retrograde movement may be traced, in the relation which the authors themselves have assumed towards their readers. From the lofty address of Bacon: 'these are the meditations of Francis of Verulam, which that posterity should be possessed of, he deemed *their* interest': or from dedication to monarch or pontiff, in which the honour given was asserted in equipoise, to the patronage acknowledged from Pindar's

. . . ἐπ' ἄλλοι-
σι δ' ἄλλοι μεγάλοι. τό δ' ἔσχατον κορυ-
φοῦται βασιλεῦσι. μηκέτι
Πάπταινε πόρσιον.
Εἴη σέ τε τοῦτον
'Υψοῦ χρόνον πατεῖν, ἐμέ
Τε τοσσάδε νικαφόροις
'Ομιλεῖν, πρόφαντον σοφίαν καθ' Ἕλ-
λανας ἐόντα παντᾶ.

[Some men are great in one thing; others in another: but the crowning summit is for kings. Refrain from peering too far. Heaven grant that thou mayest plant thy feet on high, so long as thou livest, and that I may consort with victors for all my days, and be foremost in the lore of song among Greeks in every land. (LCL)]

Olymp. Od. I

Poets and philosophers, rendered diffident by their very number, addressed themselves to '*learned* readers'; then, aimed to conciliate the graces of 'the *candid* reader'; till, the critic still rising as the author sunk, the amateurs of literature collectively were erected into a municipality of judges, and addressed as 'the town'! And now finally, all men being supposed able to read, and all readers able to judge, the multitudinous public, shaped into personal unity by the magic of abstraction, sits nominal despot on the throne of criticism. But, alas! as in other despotisms, it but echoes the decisions of its invisible ministers, whose intellectual claims to the guardianship of the muses seem, for the greater part, analogous to the physical qualifications which adapt their oriental brethren for the superintendance of the harem. Thus it is said, that St Nepomuc was installed the guardian of bridges because he had fallen over one, and sunk out of sight; thus too St Cecilia is said to have been first propitiated by musicians, because having failed in her own attempts, she had taken a dislike to the art, and all its successful professors. But I shall probably have occasion hereafter to deliver my convictions more at large concerning this state of things, and its influences on taste, genius and morality.

In the 'Thalaba', the 'Madoc' and still more evidently in the unique* 'Cid,' the 'Kehama,' and as last, so best, the 'Don Roderick',

* I have ventured to call it 'unique'; not only because I know no work of the kind in our language (if we except a few chapters of the old translation of Froissart) none, which uniting the charms of romance and history, keeps the imagination so constantly on the wing, and yet leaves so much for after reflection; but likewise, and chiefly, because it is a

Southey has given abundant proof, 'se cogitâsse quám sit magnum dare aliquid in manus hominum: nec persuadere sibi posse, non saepe tractandum quod placere et semper et omnibus cupiat [that he has thought what a serious thing it is to place a work in the hands of the public; and he cannot help but be persuaded that he should constantly revise a work that he wants to please everyone for all time (LCL)].' Plin. Ep. Lib. 7. Ep. 17. But on the other hand I guess, that Mr Southey was quite unable to comprehend, wherein could consist the crime or mischief of printing half a dozen or more playful poems; or to speak more generally, compositions which would be enjoyed or passed over, according as the taste and humour of the reader might chance to be; provided they contained nothing immoral. In the present age 'periturae parcere chartae [to spare paper that would be wasted anyway]' is emphatically an unreasonable demand. The merest trifle, he ever sent abroad, had tenfold better claims to its ink and paper, than all the silly criticisms, which prove no more, than that the critic was not one of those, for whom the trifle was written; and than all the grave exhortations to a greater reverence for the public. As if the passive page of a book, by having an epigram or doggrel tale impressed on it, instantly assumed at once locomotive power and a sort of ubiquity, so as to flutter and buzz in the ear of the public to the sore annoyance of the said mysterious personage. But what gives an additional and more ludicrous absurdity to these lamentations is the curious fact, that if in a volume of poetry the critic should find poem or passage which he deems more especially worthless, he is sure to select and reprint it in the review; by which, on his own grounds, he wastes as much more paper than the author, as the copies of a fashionable review are more numerous than those of the original book; in some, and those the most prominent instances, as ten thousand to five hundred. I know nothing that surpasses the vileness of deciding on the merits of a poet or painter (not by characteristic defects; for where there is genius, these always point to his characteristic beauties; but) by accidental failures or faulty passages; except the impudence of defending it, as the proper duty, and most instructive part, of criticism. Omit or pass slightly over, the expression, grace, and grouping of Raphael's figures; but ridicule in detail the knitting-needles and broom-twigs, that are to represent trees in his backgrounds; and never let him hear the last of

compilation, which in the various excellencies of translation, selection, and arrangement, required and proves greater genius in the compiler, as living in the present state of society, than in the original composers.

his gallipots! Admit, that the Allegro and Penseroso of Milton are not without merit; but repay yourself for this concession, by reprinting at length the two poems on the University Carrier! As a fair specimen of his sonnets, quote 'a Book was writ of late called Tetrachordon'; and as characteristic of his rhythm and metre cite his literal translation of the first and second psalm! In order to justify yourself, you need only assert, that had you dwelt chiefly on the beauties and excellencies of the poet, the admiration of these might seduce the attention of future writers from the objects of their love and wonder, to an imitation of the few poems and passages in which the poet was most unlike himself.

But till reviews are conducted on far other principles, and with far other motives; till in the place of arbitrary dictation and petulant sneers, the reviewers support their decisions by reference to fixed canons of criticism, previously established and deduced from the nature of man; reflecting minds will pronounce it arrogance in them thus to announce themselves to men of letters, as the guides of their taste and judgement. To the purchaser and mere reader it is, at all events, an injustice. He who tells me that there are defects in a new work, tells me nothing which I should not have taken for granted without his information. But he, who points out and elucidates the beauties of an original work, does indeed give me interesting information, such as experience would not have authorized me in anticipating. And as to compositions which the authors themselves announce with 'Haec ipsi novimus esse nihil [we ourselves know these things to be nothing],' why should we judge by a different rule two printed works, only because the one author was alive, and the other in his grave? What literary man has not regretted the prudery of Sprat in refusing to let his friend Cowley appear in his slippers and dressing-gown? I am not perhaps the only one who has derived an innocent amusement from the riddles, conundrums, trisyllable lines, etc. etc. of Swift and his correspondents, in hours of languor when to have read his more finished works would have been useless to myself, and, in some sort, an act of injustice to the author. But I am at a loss to conceive by what perversity of judgement, these relaxations of his genius could be employed to diminish his fame as the writer of 'Gulliver's Travels,' and the 'Tale of a Tub.' Had Mr Southey written twice as many poems of inferior merit, or partial interest, as have enlivened the journals of the day, they would have added to his honour with good and wise men, not merely or principally as proving the versatility of his talents, but as evidences of the purity of that mind, which even in its levities never wrote a line, which it need regret on any moral account.

I have in imagination transferred to the future biographer the duty of contrasting Southey's fixed and well-earned fame, with the abuse and indefatigable hostility of his anonymous critics from his early youth to his ripest manhood. But I cannot think so ill of human nature as not to believe, that these critics have already taken shame to themselves, whether they consider the object of their abuse in his moral or his literary character. For reflect but on the variety and extent of his acquirements! He stands second to no man, either as an historian or as a bibliographer; and when I regard him, as a popular essayist (for the articles of his compositions in the reviews are for the greater part essays on subjects of deep or curious interest rather than criticisms on particular works*), I look in vain for any writer, who has conveyed so much information, from so many and such recondite sources, with so many just and original reflections, in a style so lively and poignant, yet so uniformly classical and perspicuous; no one in short who has combined so much wisdom with so much wit; so much truth and knowledge with so much life and fancy. His prose is always intelligible and always entertaining. In poetry he has attempted almost every species of composition known before, and he has added new ones; and if we except the highest lyric (in which how few, how very few even of the greatest minds have been fortunate), he has attempted every species successfully: from the political song of the day, thrown off in the playful overflow of honest joy and patriotic exultation, to the wild ballad;† from epistolary ease and graceful narrative, to the austere and impetuous moral declamation; from the pastoral claims and wild streaming lights of the 'Thalaba,' in which sentiment and imagery have given permanence even to the excitement of curiosity; and from the full blaze of the 'Kehama,' (a gallery of finished pictures in one splendid fancy piece, in which, notwithstanding, the moral grandeur rises gradually above the brilliance of the colouring and the boldness and novelty of the machinery) to the more sober beauties of the 'Madoc'; and lastly, from the 'Madoc' to his 'Roderick,' in which, retaining all his former excellencies of a poet eminently inventive and picturesque, he has surpassed himself in language and metre, in the construction of the whole, and in the splendour of particular passages.

Here then shall I conclude? No! The characters of the deceased, like

* See the articles on Methodism, in the Quarterly Review; the small volume on the New System of Education, etc.

† See the incomparable 'Return to Moscow,' and the 'Old Woman of Berkeley.'

the encòmia on tombstones, as they are described with religious tenderness, so are they read, with allowing sympathy indeed, but yet with rational deduction. There are men, who deserve a higher record; men with whose characters it is the interest of their contemporaries, no less than that of posterity, to be made acquainted; while it is yet possible for impartial censure, and even for quick-sighted envy, to cross-examine the tale without offence to the courtesies of humanity; and while the eulogist detected in exaggeration or falsehood must pay the full penalty of his baseness in the contempt which brands the convicted flatterer. Publicly has Mr Southey been reviled by men, who (I would feign hope for the honour of human nature) hurled firebrands against a figure of their own imagination, publicly have his talents been depreciated, his principles denounced; as publicly do I therefore, who have known him intimately, deem it my duty to leave recorded, that it is Southey's almost unexampled felicity, to possess the best gifts of talent and genius free from all their characteristic defects. To those who remember the state of our public schools and universities some twenty years past, it will appear no ordinary praise in any man to have passed from innocence into virtue, not only free from all vicious habit, but unstained by one act of intemperance, or the degradations akin to intemperance. That scheme of head, heart, and habitual demeanour, which in his early manhood, and first controversial writings, Milton, claiming the privilege of self-defence, asserts of himself, and challenges his calumniators to disprove; this will his schoolmates, his fellow collegians, and his maturer friends, with a confidence proportioned to the intimacy of their knowledge, bear witness to, as again realized in the life of Robert Southey. But still more striking to those, who by biography or by their own experience are familiar with the general habits of genius, will appear the poet's matchless industry and perseverance in his pursuits; the worthiness and dignity of those pursuits; his generous submission to tasks of transitory interest, or such as *his* genius alone could make otherwise; and that having thus more than satisfied the claims of affection or prudence, he should yet have made for himself time and power, to achieve more, and in more various departments than almost any other writer has done, though employed wholly on subjects of his own choice and ambition. But as Southey possesses, and is not possessed by, his genius, even so is he the master even of his virtues. The regular and methodical tenor of his daily labours, which would be deemed rare in the most mechanical pursuits, and might be envied by the mere man of business, loses all semblance of formality in the dignified simplicity of his manners, in the spring and healthful cheerfulness of his spirits. Always employed, his friends find him always at leisure. No less punctual in trifles, than steadfast in the performance of highest duties, he inflicts none of those small pains and discomforts which irregular men scatter about them, and which in the aggregate so often become formidable obstacles both to happiness and utility; while on the contrary he bestows all the pleasures, and inspires all that ease of mind on those around him or connected with him, which perfect consistency, and (if such a word might be framed) absolute *reliability*, equally in small as in great concerns, cannot but inspire and bestow: when this too is softened without being weakened by kindness and gentleness, I know few men who so well deserve the character which an ancient attributes to Marcus Cato, namely, that he was likest virtue, inasmuch as he seemed to act aright, not in obedience to any law or outward motive, but by the necessity of a happy nature, which could not act otherwise. As son, brother, husband, father, master, friend, he moves with firm yet light steps, alike unostentatious, and alike exemplary. As a writer, he has uniformly made his talents subservient to the best interests of humanity, of public virtue, and domestic piety; his cause has ever been the cause of pure religion and of liberty, of national independence and of national illumination. When future critics shall weigh out his guerdon of praise and censure, it will be Southey the poet only, that will supply them with the scanty materials for the latter. They will likewise not fail to record, that as no man was ever a more constant friend, never had poet more friends and honourers among the good of all parties; and that quacks in education, quacks in politics, and quacks in criticism were his only enemies.*

* It is not easy to estimate the effects which the example of a young man as highly distinguished for strict purity of disposition and conduct, as for intellectual power and literary acquirements, may produce on those of the same age with himself, especially on those of similar pursuits and congenial minds. For many years, my opportunities of intercourse with Mr Southey have been rare, and at long intervals; but I dwell with unabated pleasure on the strong and sudden, yet I trust not fleeting influence, which my moral being underwent on my acquaintance with him at Oxford, whither I had gone at the commencement of our Cambridge vacation on a visit to an old schoolfellow. Not indeed on my moral or religious principles, for they had never been contaminated; but in awakening the sense of the duty and dignity of making my actions accord with those principles, both in word and deed. The irregularities only not universal among the young men of my standing, which I always knew to be wrong, I then learnt to feel as degrading; learnt to know that an opposite conduct, which was at that time considered by us as the easy virtue of cold and selfish prudence, might originate in the noblest emotions, in views the most disinterested and imaginative. It is not however from

CHAPTER IV

The Lyrical Ballads with the preface—Mr Wordsworth's earlier poems—On fancy and imagination—The investigation of the distinction important to the fine arts.

I HAVE wandered far from the object in view, but as I fancied to myself readers who would respect the feelings that had tempted me from the main road; so I dare calculate on not a few, who will warmly sympathize with them. At present it will be sufficient for my purpose, if I have proved, that Mr Southey's writings no more than my own, furnished the original occasion to this fiction of a new school of poetry, and of clamours against its supposed founders and proselytes.

As little do I believe that 'Mr Wordsworth's Lyrical Ballads' were in themselves the cause. I speak exclusively of the two volumes so entitled. A careful and repeated examination of these confirms me in the belief, that the omission of less than an hundred lines would have precluded nine-tenths of the criticism on this work. I hazard this declaration, however, on the supposition, that the reader had taken it up, as he would have done any other collection of poems purporting to derive their subjects or interests from the incidents of domestic or ordinary life, intermingled with higher strains of meditation which the poet utters in his own person and character; with the proviso, that they

grateful recollections only, that I have been impelled thus to leave these, my deliberate sentiments on record; but in some sense as a debt of justice to the man, whose name has been so often connected with mine, for evil to which he is a stranger. As a specimen I subjoin part of a note, from 'the Beauties of the Anti-jacobin,' in which, having previously informed the public that I had been dishonoured at Cambridge for preaching deism, at a time when for my youthful ardour in defence of Christianity, I was decried as a bigot by the proselytes of French Phi- (or to speak more truly, Psi)losophy,° the writer concludes with these words; 'since this time he has left his native country, commenced citizen of the world, left his poor children fatherless, and his wife destitute. *Ex his disce* [Learn from these], his friends, Lamb and Southey.' With severest truth it may be asserted, that it would not be easy to select two men more exemplary in their domestic affections, than those whose names were thus printed at full length as in the same rank of morals with a denounced infidel and fugitive, who had left his children fatherless and his wife destitute! Is it surprising, that many good men remained longer than perhaps they otherwise would have done, adverse to a party, which encouraged and openly rewarded the authors of such atrocious calumnies! 'Qualis es, nescio; sed per quales agis, scio et doleo [What sort of man you are, I do not know; but I know and deplore the sort of men through whom you act]'.

were perused without knowledge of, or reference to, the author's peculiar opinions, and that the reader had not had his attention previously directed to those peculiarities. In these, as was actually the case with Mr Southey's earlier works, the lines and passages which might have offended the general taste, would have been considered as mere inequalities, and attributed to inattention, not to perversity of judgement. The men of business who had passed their lives chiefly in cities, and who might therefore be expected to derive the highest pleasure from acute notices of men and manners conveyed in easy, yet correct and pointed language; and all those who, reading but little poetry, are most stimulated with that species of it, which seems most distant from prose, would probably have passed by the volume altogether. Others more catholic in their taste, and yet habituated to be most pleased when most excited, would have contented themselves with deciding, that the author had been successful in proportion to the elevation of his style and subject. Not a few perhaps, might by their admiration of 'the lines written near Tintern Abbey,' those 'left upon a Seat under a Yew Tree,' the 'old Cumberland beggar,' and 'Ruth,' have been gradually led to peruse with kindred feeling the 'Brothers,' the 'Hart-leap well,' and whatever other poems in that collection may be described as holding a middle place between those written in the highest and those in the humblest style; as for instance between the 'Tintern Abbey,' and 'the Thorn,' or the 'Simon Lee.' Should their taste submit to no further change, and still remain unreconciled to the colloquial phrases, or the imitations of them, that are, more or less, scattered through the class last mentioned; yet even from the small number of the latter, they would have deemed them but an inconsiderable subtraction from the merit of the whole work; or, what is sometimes not unpleasing in the publication of a new writer, as serving to ascertain the natural tendency, and consequently the proper direction of the author's genius.

In the critical remarks therefore, prefixed and annexed to the 'Lyrical Ballads,' I believe, that we may safely rest, as the true origin of the unexampled opposition which Mr Wordsworth's writings have been since doomed to encounter. The humbler passages in the poems themselves were dwelt on and cited to justify the rejection of the theory. What in and for themselves would have been either forgotten or forgiven as imperfections, or at least comparative failures, provoked direct hostility when announced as intentional, as the result of choice after full deliberation. Thus the poems, admitted by all as excellent, joined with those which had pleased the far greater number, though they formed two-thirds of the whole work, instead of being deemed (as in all right they should have been, even if we take for granted that the reader judged aright) an atonement for the few exceptions, gave wind and fuel to the animosity against both the poems and the poet. In all perplexity there is a portion of fear, which predisposes the mind to anger. Not able to deny that the author possessed both genius and a powerful intellect, they felt very positive, but were not quite certain, that he might not be in the right, and they themselves in the wrong; an unquiet state of mind, which seeks alleviation by quarrelling with the occasion of it, and by wondering at the perverseness of the man, who had written a long and argumentative essay to persuade them, that

Fair is foul, and foul is fair;

in other words, that they had been all their lives admiring without judgement, and were now about to censure without reason.*

* In opinions of long continuance, and in which we had never before been molested by a single doubt, to be suddenly convinced of an error, is almost like being convicted of a fault. There is a state of mind, which is the direct antithesis of that, which takes place when we make a bull.° The bull namely consists in the bringing together two incompatible thoughts, with the sensation, but without the sense, of their connection. The psychological condition, or that which constitutes the possibility of this state, being such disproportionate vividness of two distant thoughts, as extinguishes or obscures the consciousness of the intermediate images or conceptions, or wholly abstracts the attention from them. Thus in the well-known bull, 'I was a fine child, but they changed me'; the first conception expressed in the word 'I,' is that of personal identity—*Ego contemplans* [I contemplating]: the second expressed in the word 'me,' is the visual image or object by which the mind represents to itself its past condition, or rather, its personal identity under the form in which it imagined itself previously to have existed,—*Ego contemplatus* [I contemplated]. Now the change of one visual image for another involves in itself no absurdity, and becomes absurd only by its immediate juxtaposition with the first thought, which is rendered possible by the whole attention being successively absorbed in each singly, so as not to notice the interjacent notion, 'changed' which by its incongruity with the first thought, 'I,' constitutes the bull. Add only, that this process is facilitated by the circumstance of the words 'I,' and 'me,' being sometimes equivalent, and sometimes having a distinct meaning; sometimes, namely, signifying the act of self-consciousness, sometimes the external image in and by which the mind represents that act to itself, the result and symbol of its individuality. Now suppose the direct contrary state, and you will have a distinct sense of the connection between two conceptions, without that sensation of such connection which is supplied by habit. The man feels, as if he were standing on his head, though he cannot but see, that he is truly standing on his feet. This, as a painful sensation, will of course have a tendency to associate itself with the person who occasions it; even as persons, who have been by painful means restored from derangement, are known to feel an involuntary dislike towards their physician.

That this conjecture is not wide from the mark, I am induced to believe from the noticeable fact, which I can state on my own knowledge, that the same general censure should have been grounded almost by each different person on some different poem. Among those, whose candour and judgement I estimate highly, I distinctly remember six who expressed their objections to the 'Lyrical Ballads' almost in the same words, and altogether to the same purport, at the same time admitting, that several of the poems had given them great pleasure; and, strange as it might seem, the composition which one had cited as execrable, another had quoted as his favourite. I am indeed convinced in my own mind, that could the same experiment have been tried with these volumes, as was made in the well-known story of the picture, the result would have been the same; the parts which had been covered by the number of the black spots on the one day, would be found equally albo *lapide notatae* [marked with a *white* stone] on the succeeding.

However this may be, it is assuredly hard and unjust to fix the attention on a few separate and insulated poems with as much aversion, as if they had been so many plague-spots on the whole work, instead of passing them over in silence, as so much blank paper, or leaves of bookseller's catalogue; especially, as no one pretends to have found immorality or indelicacy; and the poems therefore, at the worst, could only be regarded as so many light or inferior coins in a roleau of gold, not as so much alloy in a weight of bullion. A friend whose talents I hold in the highest respect, but whose judgement and strong sound sense I have had almost continued occasion to revere, making the usual complaints to me concerning both the style and subjects of Mr Wordsworth's minor poems; I admitted that there were some few of the tales and incidents, in which I could not myself find a sufficient cause for their having been recorded in metre. I mentioned the 'Alice Fell' as an instance; 'nay,' replied my friend with more than usual quickness of manner, 'I cannot agree with you *there!* that I own *does* seem to me a remarkably pleasing poem.' In the 'Lyrical Ballads' (for my experience does not enable me to extend the remark equally unqualified to the two subsequent volumes) I have heard at different times, and from different individuals every single poem extolled and reprobated, with the exception of those of loftier kind, which as was before observed, seem to have won universal praise. This fact of itself would have made me diffident in my censures, had not a still stronger ground been furnished by the strange contrast of the heat and long continuance of the opposition, with the nature of the faults stated as justifying it. The seductive faults, the *dulcia vitia* of Cowley, Marini, or Darwin might reasonably be thought capable of corrupting the public judgement for half a century, and require a twenty years' war, campaign after campaign, in order to dethrone the usurper and re-establish the legitimate taste. But that a downright simpleness, under the affectation of simplicity, prosaic words in feeble metre, silly thoughts in childish phrases, and a preference of mean, degrading, or at best trivial associations and characters, should succeed in forming a school of imitators, a company of almost *religious* admirers, and this too among young men of ardent minds, liberal education, and not

with academic laurels unbestowed;

and that this bare and bald counterfeit of poetry, which is characterized as below criticism, should for nearly twenty years have well-nigh engrossed criticism, as the main, if not the only, butt of review, magazine, pamphlets, poem, and paragraph;—this is indeed matter of wonder! Of yet greater is it, that the contest should still continue as undecided* as that between Bacchus and the frogs in Aristophanes; when the former descended to the realms of the departed to bring back the spirit of old and genuine poesy.

* Without however the apprehensions attributed to the pagan reformer of the poetic republic. If we may judge from the preface to the recent collection of his poems, Mr W. would have answered with Xanthias—

Συ δ'ουκ εδεισας τον ψοφον των ρηματων,
Και τας απειλας; ΞΑΝ. ου μα Δι', ουδ' εφροντισα.

[But weren't *you* frightened at those dreadful threats and shoutings? XANTHIAS. Frightened? Not a bit. I cared not.]

And here let me dare hint to the authors of the numerous parodies, and pretended imitations of Mr Wordsworth's style, that at once to conceal and convey wit and wisdom in the semblance of folly and dullness, as is done in the clowns and fools, nay even in the Dogberry, of our Shakespeare, is doubtless a proof of genius, or at all events, of satiric talent; but that the attempt to ridicule a silly and childish poem, by writing another still sillier and still more childish, can only prove (if it prove anything at all) that the parodist is a still greater blockhead than the original writer, and what is far worse, a malignant coxcomb to boot. The talent for mimicry seems strongest where the human race are most degraded. The poor, naked, half-human savages of New Holland were found excellent mimics: and in civilized society, minds of the very lowest stamp alone satirize by copying. At least the difference, which must blend with and balance the likeness, in order to constitute a just imitation, existing here merely in caricature, detracts from the libeller's heart, without adding an iota to the credit of his understanding.

Χορος Βατραχων; Διονυσος,

Χ.	*βρεκεκεκεξ, κοαξ, κοαξ!*
Δ.	*αλλ' εξολοισθ' αυτω κοαξ.*
	ουδεν γαρ εστι, ἤ κόαξ.
	οιμωζετ'· ου μοι μελει.
Χ.	*αλλα μην κεκραξομεσθα*
	γ'οποσον η φαρυγξ αν ημων
	χανδανῃ δὶ ημερας
	βρεκεκεκεξ, κοαξ, κοαξ!
Δ.	*τουτῳ γαρ ου νικησετε.*
Χ.	*ουδε μεν ημας συ παντως.*
Δ.	*ουδε μεν υμεις γε δη με*
	ουδεποτε· κεκραξομαι γαρ
	κἄν με δ'ει δι' ημερας,
	εως ἄν ὑμήων επικρατησω τῷ κοαξ!
Χ.	*βρεκεκεκεξ, ΚΟΑΞ, ΚΟΑΞ!*

[Chorus of Frogs; Dionysus

FROGS:	Brekekekex, ko-ax, ko-ax.
DIONYSUS:	Hang you, and your ko-axing too!
	There's nothing but ko-ax with you.
	Go hang yourselves, for what care I?
FROGS:	All the same we'll shout and cry,
	Stretching all our throats with song,
	Shouting, crying, all day long.
FR. and DI.:	Brekekekex, ko-ax, ko-ax.
DIONYSUS:	In this you'll never, never win.
FROGS:	This you shall not beat us in.
DIONYSUS:	No, nor ye prevail o'er me.
	Never! never! I'll my song
	Shout, if need be, all day long,
	Until I've learned to master your ko-ax.
FROGS:	Brekekekex, ko-ax, ko-ax. (LCL)]

During the last year of my residence at Cambridge, I became acquainted with Mr Wordsworth's first publication entitled 'Descriptive Sketches'; and seldom, if ever, was the emergence of an original poetic genius above the literary horizon more evidently announced. In the form, style, and manner of the whole poem, and in the structure of the particular lines and periods, there is an harshness and acerbity connected and combined with words and images all a-glow, which might recall those products of the vegetable world, where gorgeous blossoms rise out of the hard and thorny rind and shell, within which the rich fruit was elaborating. The language was not only peculiar and strong, but at times knotty and contorted, as by its own impatient strength; while the novelty and struggling crowd of images acting in conjunction with the difficulties of the style, demanded always a greater closeness of attention, than poetry, (at all events, than descriptive poetry) has a right to claim. It not seldom therefore justified the complaint of obscurity. In the following extract I have sometimes fancied, that I saw an emblem of the poem itself, and of the author's genius as it was then displayed.

'Tis storm; and hid in mist from hour to hour,
All day the floods a deepening murmur pour;
The sky is veiled, and every cheerful sight:
Dark is the region as with coming night;
And yet what frequent bursts of overpowering light!
Triumphant on the bosom of the storm,
Glances the fire-clad eagle's wheeling form;
Eastward, in long perspective glittering, shine
The wood-crowned cliffs that o'er the lake recline;
Wide o'er the Alps a hundred streams unfold,
At once to pillars turn'd that flame with gold;
Behind his sail the peasant strives to shun
The West, that burns like one dilated sun,
Where in a mighty crucible expire
The mountains, glowing hot, like coals of fire.

The poetic Psyche, in its process to full development, undergoes as many changes as its Greek name-sake, the butterfly.* And it is remarkable how soon genius clears and purifies itself from the faults and errors of its earliest products; faults which, in its earliest compositions, are the more obtrusive and confluent, because as heterogeneous elements, which had only a temporary use, they constitute the very ferment, by which themselves are carried off. Or

* The fact, that in Greek Psyche is the common name for the soul, and the butterfly, is thus alluded to in the following stanza from an unpublished poem of the author:

The butterfly the ancient Grecians made
The soul's fair emblem, and its only name—
But of the soul, escaped the slavish trade
Of mortal life! For in this earthly frame
Ours is the reptile's lot, much toil, much blame,
Manifold motions making little speed,
And to deform and kill the things, whereon we feed.
S.T.C.

we may compare them to some diseases, which must work on the humours, and be thrown out on the surface, in order to secure the patient from their future recurrence. I was in my twenty-fourth year, when I had the happiness of knowing Mr Wordsworth personally, and while memory lasts, I shall hardly forget the suddèn effect produced on my mind, by his recitation of a manuscript poem, which still remains unpublished, but of which the stanza, and tone of style, were the same as those of the 'Female Vagrant' as originally printed in the first volume of the 'Lyrical Ballads.' There was here, no mark of strained thought, or forced diction, no crowd or turbulence of imagery, and, as the poet hath himself well described in his lines 'on revisiting the Wye,' manly reflection, and human associations had given both variety, and an additional interest to natural objects, which in the passion and appetite of the first love they had seemed to him neither to need or permit. The occasional obscurities, which had risen from an imperfect control over the resources of his native language, had almost wholly disappeared, together with that worse defect of arbitrary and illogical phrases, at once hackneyed, and fantastic, which hold so distinguished a place in the technique of ordinary poetry, and will, more or less, alloy the earlier poems of the truest genius, unless the attention has been specifically directed to their worthlessness and incongruity.* I did not perceive anything particular in the mere style of the poem alluded to during its recitation, except indeed such difference as was not separable from the thought and manner; and the Spenserian stanza, which always, more or less, recalls to the reader's mind Spenser's own style, would doubtless have authorized in my

* Mr Wordsworth, even in his two earliest, 'the Evening Walk' and the 'Descriptive Sketches,' is more free from this latter defect than most of the young poets his contemporaries. It may however be exemplified, together with the harsh and obscure construction, in which he more often offended, in the following lines:—

> 'Mid stormy vapours ever driving by,
> Where ospreys, cormorants, and herons cry;
> Where hardly given the hopeless waste to cheer,
> Denied the bread of life the foodful ear,
> Dwindles the pear on autumn's latest spray,
> And *apple sickens* pale in summer's ray;
> *Ev'n here content has fixed her smiling reign*
> *With independence, child of high disdain.*

I hope, I need not say, that I have quoted these lines for no other purpose than to make my meaning fully understood. It is to be regretted that Mr Wordsworth has not republished these two poems entire.

then opinion a more frequent descent to the phrases of ordinary life, than could without an ill effect have been hazarded in the heroic couplet. It was not however the freedom from false taste, whether as to common defects, or to those more properly his own, which made so unusual an impression on my feelings immediately, and subsequently on my judgement. It was the union of deep feeling with profound thought; the fine balance of truth in observing with the imaginative faculty in modifying the objects observed; and above all the original gift of spreading the tone, the atmosphere, and with it the depth and height of the ideal world around forms, incidents, and situations, of which, for the common view, custom had bedimmed all the lustre, had dried up the sparkle and the dewdrops. 'To find no contradiction in the union of old and new; to contemplate the Ancient of days and all his works with feelings as fresh, as if all had then sprang forth at the first creative fiat; characterizes the mind that feels the riddle of the world, and may help to unravel it. To carry on the feelings of childhood into the powers of manhood; to combine the child's sense of wonder and novelty with the appearances, which every day for perhaps forty years had rendered familiar;

> With sun and moon and stars throughout the year,
> And man and woman;

this is the character and privilege of genius, and one of the marks which distinguish genius from talents. And therefore is it the prime merit of genius and its most unequivocal mode of manifestation, so to represent familiar objects as to awaken in the minds of others a kindred feeling concerning them and that freshness of sensation which is the constant accompaniment of mental, no less than of bodily, convalescence. Who has not a thousand times seen snow fall on water? Who has not watched it with a new feeling, from the time that he has read Burns's comparison of sensual pleasure

> To snow that falls upon a river
> A moment white—then gone for ever!

In poems, equally as in philosophic disquisitions, genius produces the strongest impressions of novelty, while it rescues the most admitted truths from the impotence caused by the very circumstance of their universal admission. Truths of all others the most awful and mysterious, yet being at the same time of universal interest, are too often considered as *so* true, that they lose all the life and efficiency of truth, and lie bedridden in the dormitory of the soul, side by side, with the most despised and exploded errors.' The Friend,* page 76, No. 5.

This excellence, which in all Mr Wordsworth's writings is more or less predominant, and which constitutes the character of his mind, I no sooner felt, than I sought to understand. Repeated meditations led me first to suspect, (and a more intimate analysis of the human faculties, their appropriate marks, functions, and effects matured my conjecture into full conviction) that fancy and imagination were two distinct and widely different faculties, instead of being, according to the general belief, either two names with one meaning, or at furthest, the lower and higher degree of one and the same power. It is not, I own, easy to conceive a more apposite translation of the Greek *phantasia*, than the Latin *imaginatio*; but it is equally true that in all societies there exists an instinct of growth, a certain collective, unconscious good sense working progressively to desynonymize† those words originally of the same meaning, which the conflux of dialects had supplied to the more homogeneous languages, as the Greek and German: and which the same cause, joined with accidents of translation from original works of different countries, occasion in mixed languages like our own. The first and most important point to

* As 'the Friend' was printed on stamped sheets, and sent only by the post to a very limited number of subscribers, the author has felt less objection to quote from it, though a work of his own. To the public at large indeed it is the same as a volume in manuscript.

† This is effected either by giving to the one word a general, and to the other an exclusive use; as 'to put on the back' and 'to indorse'; or by an actual distinction of meanings as 'naturalist,' and 'physician'; or by difference of relation as 'I' and 'Me'; (each of which the rustics of our different provinces still use in all the cases singular of the first personal pronoun). Even the mere difference, or corruption, in the *pronunciation* of the same word, if it have become general, will produce a new word with a distinct signification; thus 'property' and 'propriety,' the latter of which, even to the time of Charles II was the *written* word for all the senses of both. Thus too 'mister' and 'master,' both hasty pronunciations of the same word 'magister'; 'mistress,' and 'miss,' 'if,' and 'give,' etc. etc. There is a sort of *minim immortal* among the *animalcula infusoria*° which has not naturally either birth, or death, absolute beginning, or absolute end: for at a certain period a small point appears on its back, which deepens and lengthens till the creature divides into two, and the same process recommences in each of the halves now become integral. This may be a fanciful, but it is by no means a bad emblem of the formation of words, and may facilitate the conception, how immense a nomenclature may be organized from a few simple sounds by rational beings in a social state. For each new application, or excitement of the same sound, will call forth a different sensation, which cannot but affect the pronunciation. The after recollection of the sound, without the same vivid sensation, will modify it still further; till at length all trace of the original likeness is worn away.

be proved is, that two conceptions perfectly distinct are confused under one and the same word, and (this done) to appropriate that word exclusively to one meaning, and the synonym (should there be one) to the other. But if (as will be often the case in the arts and sciences) no synonym exists, we must either invent or borrow a word. In the present instance the appropriation had already begun, and been legitimated in the derivative adjective: Milton had a highly *imaginative*; Cowley a very *fanciful* mind. If therefore I should succeed in establishing the actual existences of two faculties generally different, the nomenclature would be at once determined. To the faculty by which I had characterized Milton, we should confine the term *imagination*; while the other would be contra-distinguished as *fancy*. Now were it once fully ascertained, that this division is no less grounded in nature, than that of delirium from mania, or Otway's

> Lutes, lobsters, seas of milk, and ships of amber,

from Shakespeare's

> What! have his daughters brought him to this pass?

or from the preceding apostrophe to the elements; the theory of the fine arts, and of poetry in particular, could not, I thought, but derive some additional and important light. It would in its immediate effects furnish a torch of guidance to the philosophical critic; and ultimately to the poet himself. In energetic minds, truth soon changes by domestication into power; and from directing in the discrimination and appraisal of the product, becomes influencive in the production. To admire on principle, is the only way to imitate without loss of originality.

It has been already hinted, that metaphysics and psychology have long been my hobby-horse. But to have a hobby-horse, and to be vain of it, are so commonly found together, that they pass almost for the same. I trust therefore, that there will be more good humour than contempt, in the smile with which the reader chastises my self-complacency, if I confess myself uncertain, whether the satisfaction from the perception of a truth new to myself may not have been rendered more poignant by the conceit, that it would be equally so to the public. There was a time, certainly, in which I took some little credit to myself, in the belief that I had been the first of my countrymen, who had pointed out the diverse meaning of which the two terms were capable, and analysed the faculties to which they should be appropriated. Mr W. Taylor's recent volume of synonyms I have not yet seen;* but his specification of the terms in question has been clearly shown to be both insufficient and erroneous by Mr Wordsworth in the preface added to the late collection of his 'Lyrical Ballads and other poems.' The explanation which Mr Wordsworth has himself given, will be found to differ from mine, chiefly perhaps, as our objects are different. It could scarcely indeed happen otherwise, from the advantage I have enjoyed of frequent conversation with him on a subject to which a poem of his own first directed my attention and my conclusions concerning which, he had made more lucid to myself by many happy instances drawn from the operation of natural objects on the mind. But it was Mr Wordsworth's purpose to consider the influences of fancy and imagination as they are manifested in poetry, and from the different effects to conclude their diversity in kind; while it is my object to investigate the seminal principle, and then from the kind to deduce the degree. My friend has drawn a masterly sketch of the branches with their poetic fruitage. I wish to

* I ought to have added, with the exception of a single sheet which I accidentally met with at the printers. Even from this scanty specimen, I found it impossible to doubt the talent, or not to admire the ingenuity of the author. That his distinctions were for the greater part unsatisfactory to *my* mind, proves nothing against their accuracy; but it may possibly be serviceable to him in case of a second edition, if I take this opportunity of suggesting the query; whether he may not have been occasionally misled, by having assumed, as to me he appeared to have done, the non-existence of *any* absolute synonyms in our language? Now I cannot but think, that there are many which remain for our posterity to distinguish and appropriate, and which I regard as so much reversionary wealth in our mother tongue. When two distinct meanings are confounded under one or more words, (and such must be the case, as sure as our knowledge is progressive and of course imperfect) erroneous consequences will be drawn, and what is true in one sense of the word, will be affirmed as true *in toto* [altogether]. Men of research startled by the consequences, seek in the things themselves (whether in or out of the mind) for a knowledge of the fact, and having discovered the difference, remove the equivocation either by the substitution of a new word, or by the appropriation of one of the two or more words, that had before been used promiscuously. When this distinction has been so naturalized and of such general currency, that the language itself does as it were think for us (like the sliding rule which is the mechanic's safe substitute for arithmetical knowledge) we then say, that it is evident to common sense. Common sense, therefore, differs in different ages. What was born and christened in the schools passes by degrees into the world at large, and becomes the property of the market and the tea-table. At least I can discover no other meaning of the term, *common sense*, if it is to convey any specific difference from sense and judgement *in genere* [in kind], and where it is not used scholastically for the universal reason. Thus in the reign of Charles II the philosophic world was called to arms by the moral sophisms of Hobbes, and the ablest writers exerted themselves in the detection of an error, which a schoolboy would now be able to confute by the mere recollection, that *compulsion* and *obligation* conveyed two ideas perfectly disparate, and that what appertained to the one, had been falsely transferred to the other by a mere confusion of terms.

add the trunk, and even the roots as far as they lift themselves above ground, and are visible to the naked eye of our common consciousness.

Yet even in this attempt I am aware, that I shall be obliged to draw more largely on the reader's attention, than so immethodical a miscellany can authorize; when in such a work (the *Ecclesiastical Policy*) of such a mind as Hooker's, the judicious author, though no less admirable for the perspicuity than for the port and dignity of his language; and though he wrote for men of learning in a learned age; saw nevertheless occasion to anticipate and guard against 'complaints of obscurity,' as often as he was to trace his subject 'to the highest well-spring and fountain.' Which, (continues he) 'because men are not accustomed to, the pains we take are more needful a great deal, than acceptable; and the matters we handle, seem by reason of newness (till the mind grow better acquainted with them) dark and intricate.' I would gladly therefore spare both myself and others this labour, if I knew how without it to present an intelligible statement of my poetic creed; not as my opinions, which weigh for nothing, but as deductions from established premisses conveyed in such a form, as is calculated either to effect a fundamental conviction, or to receive a fundamental confutation. If I may dare once more adopt the words of Hooker, 'they, unto whom we shall seem tedious, are in no wise injured by us, because it is in their own hands to spare that labour, which they are not willing to endure.' Those at least, let me be permitted to add, who have taken so much pains to render me ridiculous for a perversion of taste, and have supported the charge by attributing strange notions to me on no other authority than their own conjectures, owe it to themselves as well as to me not to refuse their attention to my own statement of the theory, which I do acknowledge; or shrink from the trouble of examining the grounds on which I rest it, or the arguments which I offer in its justification.

CHAPTER V

On the law of association—Its history traced from Aristotle to Hartley.°

THERE have been men in all ages, who have been impelled as by an instinct to propose their own nature as a problem, and who devote their attempts to its solution. The first step was to construct a table of distinctions, which they seem to have formed on the principle of the absence or presence of the will. Our various sensations, perceptions, and movements were classed as active or passive, or as media partaking of both. A still finer distinction was soon established between the voluntary and the spontaneous. In our perceptions we seem to ourselves merely passive to an external power, whether as a mirror reflecting the landscape, or as a blank canvas on which some unknown hand paints it. For it is worthy of notice, that the latter, or the system of idealism, may be traced to sources equally remote with the former, or materialism; and Berkeley can boast an ancestry at least as venerable as Gassendi or Hobbes. These conjectures, however, concerning the mode in which our perceptions originated, could not alter the natural difference of things and thoughts. In the former, the cause appeared wholly external, while in the latter, sometimes our will interfered as the producing or determining cause, and sometimes our nature seemed to act by a mechanism of its own, without any conscious effort of the will, or even against it. Our inward experiences were thus arranged in three separate classes, the passive sense, or what the school-men call the merely receptive quality of the mind; the voluntary, and the spontaneous, which holds the middle place between both. But it is not in human nature to meditate on any mode of action, without enquiring after the law that governs it; and in the explanation of the spontaneous movements of our being, the metaphysician took the lead of the anatomist and natural philosopher. In Egypt, Palestine, Greece, and India the analysis of the mind had reached its noon and manhood, while experimental research was still in its dawn and infancy. For many, very many centuries, it has been difficult to advance a new truth, or even a new error, in the philosophy of the intellect or morals. With regard, however, to the laws that direct the spontaneous movements of thought and the principle of their intellectual mechanism there exists, it has been asserted, an important exception most honourable to the moderns, and in the merit of which our own country claims the largest share. Sir James Mackintosh (who amid the variety of his talents and attainments is not of less repute for the depth and accuracy of his philosophical enquiries, than for the eloquence with which he is said to render their most difficult results perspicuous, and the driest attractive) affirmed in the lectures, delivered by him at Lincoln's Inn Hall, that the law of association as established in the contemporaneity of the original impressions, formed the basis of all true psychology; and any ontological or metaphysical science not contained in such (i.e. empirical) psychology was but a web of abstractions and generalizations. Of this prolific truth, of this great fundamental law, he declared Hobbes to have been the original discoverer, while its full application to the whole intellectual system we owe to David Hartley; who stood in the same relation to Hobbes as Newton to Kepler; the law of association being that to the mind, which gravitation is to matter.

Of the former clause in this assertion, as it respects the comparative merits of the ancient metaphysicians, including their commentators, the schoolmen, and of the modern French and British philosophers from Hobbes to Hume, Hartley and Condillac, this is not the place to speak. So wide indeed is the chasm between this gentleman's philosophical creed and mine, that so far from being able to join hands, we could scarce make our voices intelligible to each other: and to bridge it over, would require more time, skill and power than I believe myself to possess. But the latter clause involves for the greater part a mere question of fact and history, and the accuracy of the statement is to be tried by documents rather than reasoning.

First then, I deny Hobbes's claim *in toto* [altogether]: for he had been anticipated by Descartes whose work 'De Methodo' preceded Hobbes's 'De Natura Humana,' by more than a year. But what is of much more importance, Hobbes builds nothing on the principle which he had announced. He does not even announce it, as differing in any respect from the general laws of material motion and impact: nor was it, indeed, possible for him so to do, compatibly with his system, which was exclusively material and mechanical. Far otherwise is it with Descartes; greatly as he too in his after writings (and still more egregiously his followers De la Forge, and others) obscured the truth by their attempts to explain it on the theory of nervous fluids, and material configurations. But in his interesting work 'De Methodo,' Descartes relates the circumstance which first led him to meditate on this subject, and which since then has been often noticed and employed as an instance and illustration of the law. A child who with its eyes bandaged had lost several of his fingers by amputation, continued to complain for many days successively of pains, now in this joint and now in that of the very fingers which had been cut off. Descartes was led by this incident to reflect on the uncertainty with which we attribute any particular place to any inward pain or uneasiness, and proceeded after long consideration to establish it as a general law; that contemporaneous impressions, whether images or sensations, recall each other mechanically. On this principle, as a groundwork, he built up the whole system of human language, as one continued process of association. He showed, in what sense not only general terms, but generic images (under the name of abstract ideas) actually existed, and in what consists their nature and power. As one word may become the general exponent of many, so by association a simple image may represent a whole class. But in truth Hobbes himself makes no claims to any discovery, and introduces this law of association, or (in his own language) *discursus mentalis* [mental discourse], as an admitted fact, in the solution alone of which, this by causes purely physiological, he arrogates any originality. His system is briefly this; whenever the senses are impinged on by external objects, whether by the rays of light reflected from them, or by effluxes of their finer particles, there results a correspondent motion of the innermost and subtlest organs. This motion constitutes a representation, and there remains an impression of the same, or a certain disposition to repeat the same motion. Whenever we feel several objects at the same time, the impressions that are left (or in the language of Mr Hume, the ideas) are linked together. Whenever therefore any one of the movements, which constitute a complex impression, are renewed through the senses, the others succeed mechanically. It follows of necessity therefore that Hobbes, as well as Hartley and all others who derive association from the connection and interdependence of the supposed matter, the movements of which constitute our thoughts, must have reduced all its forms to the one law of time. But even the merit of announcing this law with philosophic precision cannot be fairly conceded to him. For the objects of any two ideas* need

* I here use the word 'idea' in Mr Hume's sense on account of its general currency among the English metaphysicians; though against my own judgement, for I believe that the vague use of this word has been the cause of much error and more confusion. The word, *Ιδεα* [Idea], in its original sense as used by Pindar, Aristophanes, and in the gospel of Matthew, represented the visual abstraction of a distant object, when we see the whole without distinguishing its parts. Plato adopted it as a technical term, and as the antithesis to *Ειδωλα*, or sensuous images; the transient and perishable emblems, or mental words, of ideas. The ideas themselves he considered as mysterious powers, living, seminal, formative, and exempt from time. In this sense the word became the property of the Platonic school; and it seldom occurs in Aristotle, without some such phrase annexed to it, as according to Plato, or as Plato says. Our English writers to the end of Charles II's reign, or somewhat later, employed it either in the original sense, or Platonically, or in a sense nearly correspondent to our present use of the substantive, Ideal, always however opposing it, more or less, to image, whether of present or absent objects. The reader will not be displeased with the following interesting exemplification from Bishop Jeremy Taylor. 'St Lewis the King sent Ivo Bishop of Chartres on an embassy, and he told, that he met a grave and stately matron on the way with a censer of fire in one hand, and a vessel of water in the other; and observing her to have a

not have coexisted in the same sensation in order to become mutually associable. The same result will follow when one only of the two ideas has been represented by the senses, and the other by the memory.

Long however before either Hobbes or Descartes the law of association had been defined, and its important functions set forth by Melanchthon, Ammerbach, and Ludovicus Vives; more especially by the last. *Phantasia*, it is to be noticed, is employed by Vives to express the mental power of comprehension, or the active function of the mind; and *imaginatio* for the receptivity (*vis receptiva* [receptive power]) of impressions, or for the passive perception. The power of combination he appropriates to the former: 'quae singula et simpliciter acceperat imaginatio, ea conjungit et disjungit phantasia [things the imagination had received simply and one at a time the fancy joins and unjoins].' And the law by which the thoughts are spontaneously presented follows thus; 'quae simul sunt a phantasia comprehensa si alterutrum occurrat, solet secum alterum representare [if one of two things that fancy has grasped at the same time offers itself, it usually presents the other thing along with itself].' To time therefore he subordinates all the other exciting causes of association. The soul proceeds 'a causa ad effectum, ab hoc ad instrumentum, a parte ad totum [from cause to effect, from this to the instrument, from a part to the whole]'; thence to the place, from place to person, and from this to whatever preceded or followed, all as being parts of a total impression, each of which may recall the other. The apparent springs 'Saltus vel transitus etiam longissimos [the very longest leaps and transitions]', he explains by the same thought having been a component part of two or more total impressions. Thus 'ex Scipione venio in cogitationem potentiae Turcicae propter victorias ejus in eâ parte Asiae in qua regnabat Antiochus [from Scipio I pass to the thought of the Turkish

melancholy, religious, and fantastic deportment and look, he asked her what those symbols meant, and what she meant to do with her fire and water; she answered, my purpose is with the fire to burn paradise, and with my water to quench the flames of hell, that men may serve God purely for the love of God. But we rarely meet with such spirits which love virtue so metaphysically as to *abstract her from all sensible compositions, and love the purity of idea*.' Descartes having introduced into his philosophy the fanciful hypothesis of *material ideas*, or certain configurations of the brain, which were as so many moulds to the influxes of the external world; Mr Locke adopted the term, but extended its signification to whatever is the immediate object of the mind's attention or consciousness. Mr Hume distinguishing those representations which are accompanied with a sense of a present object, from those reproduced by the mind itself, designated the former by *impressions*, and confined the word *idea* to the latter.

power, because his victories were in the region of Asia in which Antiochus ruled].

But from Vives I pass at once to the source of his doctrines, and (as far as we can judge from the remains yet extant of Greek philosophy) as to the first, so to the fullest and most perfect enunciation of the associative principle, viz. to the writings of Aristotle; and of these principally to the books 'De Anima,' 'De Memoria,' and that which is entitled in the old translations 'Parva Naturalia.' Inasmuch as later writers have either deviated from, or added to his doctrines, they appear to me to have introduced either error or groundless supposition.

In the first place it is to be observed, that Aristotle's positions on this subject are unmixed with fiction. The wise Stagyrite speaks of no successive particles propagating motion like billiard balls (as Hobbes); nor of nervous or animal spirits, where inanimate and irrational solids are thawed down, and distilled, or filtrated by ascension, into living and intelligent fluids, that etch and re-etch engravings on the brain, (as the followers of Descartes, and the humoral pathologists in general); nor of an oscillating ether which was to effect the same service for the nerves of the brain considered as solid fibres, as the animal spirits perform for them under the notion of hollow tubes (as Hartley teaches)—nor finally, (with yet more recent dreamers) of chemical compositions by elective affinity, or of an electric light at once the immediate object and the ultimate organ of inward vision, which rises to the brain like an Aurora Borealis, and there disporting in various shapes (as the balance of plus and minus, or negative and positive, is destroyed or re-established) images out both past and present. Aristotle delivers a just theory, without pretending to an hypothesis; or in other words a comprehensive survey of the different facts, and of their relations to each other without *supposition*, i.e. a fact *placed under* a number of facts, as their common support and explanation; though in the majority of instances these hypotheses or suppositions better deserve the name of *Υποποιησεῖς*, or *suffictions.*°
He uses indeed the word *Κινησεῖς* [motions], to express what we call representations or ideas, but he carefully distinguishes them from material motion, designating the latter always by annexing the words *Εν τοπῳ*, or *κατα τοπον* [in space . . . as regards space]. On the contrary in his treatise 'De Anima,' he excludes place and motion from all the operations of thought, whether representations or volitions, as attributes utterly and absurdly heterogeneous.

The general law of association, or more accurately, the common condition under which all exciting causes act, and in which they may be generalized, according to Aristotle is this. Ideas by having been together acquire a power of recalling each other; or every partial representation awakes the total representation of which it had been a part. In the practical determination of this common principle to particular recollections, he admits five agents or occasioning causes: 1st, connection in time, whether simultaneous, preceding or successive; 2nd, vicinity or connection in space; 3rd, interdependence or necessary connection, as cause and effect; 4th, likeness; and 5th, contrast. As an additional solution of the occasional seeming chasms in the continuity of reproduction he proves, that movements or ideas possessing one or the other of these five characters had passed through the mind as intermediate links, sufficiently clear to recall other parts of the same total impressions with which they had coexisted, though not vivid enough to excite that degree of attention which is requisite for distinct recollection, or as we may aptly express it, *after-consciousness*. In association then consists the whole mechanism of the reproduction of impressions, in the Aristotelian psychology. It is the universal law of the passive fancy and mechanical memory; that which supplies to all other faculties their objects, to all thought the elements of its materials.

In consulting the excellent commentary of St Thomas Aquinas on the Parva Naturalia of Aristotle, I was struck at once with its close resemblance to Hume's essay on association. The main thoughts were the same in both, the order of the thoughts was the same, and even the illustrations differed only by Hume's occasional substitution of more modern examples. I mentioned the circumstance to several of my literary acquaintances, who admitted the closeness of the resemblance, and that it seemed too great to be explained by mere coincidence; but they thought it improbable that Hume should have held the pages of the angelic Doctor worth turning over. But some time after Mr Payne, of the King's mews, showed Sir James Mackintosh some odd volumes of St Thomas Aquinas, partly perhaps from having heard that Sir James (then Mr) Mackintosh had in his lectures passed a high encomium on this canonized philosopher, but chiefly from the fact, that the volumes had belonged to Mr Hume, and had here and there marginal marks and notes of reference in his own handwriting. Among these volumes was that which contains the *Parva Naturalia*, in the old Latin version, swathed and swaddled in the commentary aforementioned!

It remains then for me, first to state wherein Hartley differs from

Aristotle; then, to exhibit the grounds of my conviction, that he differed only to err; and next as the result, to show, by what influences of the choice and judgement the associative power becomes either memory or fancy; and, in conclusion, to appropriate the remaining offices of the mind to the reason, and the imagination. With my best efforts to be as perspicuous as the nature of language will permit on such a subject, I earnestly solicit the good wishes and friendly patience of my readers, while I thus go 'sounding on my dim and perilous way.'

CHAPTER VI

That Hartley's system, as far as it differs from that of Aristotle, is neither tenable in theory, nor founded in facts.

OF Hartley's hypothetical vibrations in his hypothetical oscillating ether of the nerves, which is the first and most obvious distinction between his system and that of Aristotle, I shall say little. This, with all other similar attempts to render *that* an object of the sight which has no relation to sight, has been already sufficiently exposed by the younger Reimarus, Maass, etc. as outraging the very axioms of mechanics in a scheme, the merit of which consists in its being mechanical. Whether any other philosophy be possible, but the mechanical; and again, whether the mechanical system can have any claim to be called philosophy; are questions for another place. It is, however, certain, that as long as we deny the former, and affirm the latter, we must bewilder ourselves, whenever we would pierce into the *adyta* [innermost sanctuaries] of causation; and all that laborious conjecture can do, is to fill up the gaps of fancy. Under that despotism of the eye (the emancipation from which Pythagoras by his numeral, and Plato by his musical, symbols, and both by geometric discipline, aimed at, as the first προπαιδευτικον [propaedeutic, preliminary education] of the mind)—under this strong sensuous influence, we are restless because invisible things are not the objects of vision; and metaphysical systems, for the most part, become popular, not for their truth, but in proportion as they attribute to causes a susceptibility of being seen, if only our visual organs were sufficiently powerful.

From an hundred possible confutations let one suffice. According to this system the idea or vibration *a* from the external object A becomes associable with the idea or vibration *m* from the external object M, because the oscillation *a* propagated itself so as to reproduce the oscillation *m*. But the original impression from M was essentially different from the impression A: unless therefore different causes may produce the same effect, the vibration *a* could never produce the vibration *m*: and this therefore could never be the means, by which *a* and *m* are associated. To understand this, the attentive reader need only be reminded, that the ideas are themselves, in Hartley's system, nothing more than their appropriate configurative vibrations. It is a mere delusion of the fancy to conceive the pre-existence of the ideas, in any chain of association, as so many differently coloured billiard-balls in contact, so that when an object, the billiard-stick, strikes the first or white ball, the same motion propagates itself through the red, green, blue, black, etc. and sets the whole in motion. No! we must suppose the very same force, which constitutes the white ball, to constitute the red or black; or the idea of a circle to constitute the idea of a triangle; which is impossible.

But it may be said, that, by the sensations from the objects A and M, the nerves have acquired a disposition to the vibrations *a* and *m*, and therefore *a* need only be repeated in order to reproduce *m*. Now we will grant, for a moment, the possibility of such a disposition in a material nerve, which yet seems scarcely less absurd than to say, that a weather-cock had acquired a habit of turning to the east, from the wind having been so long in that quarter: for if it be replied, that we must take in the circumstance of life, what then becomes of the mechanical philosophy? And what is the nerve, but the flint which the wag placed in the pot as the first ingredient of his stone-broth, requiring only salt, turnips and mutton, for the remainder! But if we waive this, and presuppose the actual existence of such a disposition; two cases are possible. Either, every idea has its own nerve and correspondent oscillation, or this is not the case. If the latter be the truth, we should gain nothing by these dispositions; for then, every nerve having several dispositions, when the motion of any other nerve is propagated into it, there will be no ground or cause present, why exactly the oscillation *m* should arise, rather than any other to which it was equally predisposed. But if we take the former, and let every idea have a nerve of its own, then every nerve must be capable of propagating its motion into many other nerves; and again, there is no reason assignable, why the vibration *m* should arise, rather than any other *ad libitum* [at one's pleasure].

It is fashionable to smile at Hartley's vibrations and vibratiuncles;

and his work has been re-edited by Priestley, with the omission of the material hypothesis. But Hartley was too great a man, too coherent a thinker, for this to have been done, either consistently or to any wise purpose. For all other parts of his system, as far as they are peculiar to that system, once removed from their mechanical basis, not only lose their main support, but the very motive which led to their adoption. Thus the principle of contemporaneity, which Aristotle had made the common condition of all the laws of association, Hartley was constrained to represent as being itself the sole law. For to what law can the action of material atoms be subject, but that of proximity in place? And to what law can their motions be subjected, but that of time? Again, from this results inevitably, that the will, the reason, the judgement, and the understanding, instead of being the determining causes of association, must needs be represented as its creatures, and among its mechanical effects. Conceive, for instance, a broad stream, winding through a mountainous country with an indefinite number of currents, varying and running into each other according as the gusts chance to blow from the opening of the mountains. The temporary union of several currents in one, so as to form the main current of the moment, would present an accurate image of Hartley's theory of the will.

Had this been really the case, the consequence would have been, that our whole life would be divided between the despotism of outward impressions, and that of senseless and passive memory. Take his law in its highest abstraction and most philosophical form, viz. that every partial representation recalls the total representation of which it was a part; and the law becomes nugatory, were it only from its universality. In practice it would indeed be mere lawlessness. Consider, how immense must be the sphere of a total impression from the top of St Paul's church; and how rapid and continuous the series of such total impressions. If therefore we suppose the absence of all interference of the will, reason, and judgement, one or other of two consequences must result. Either the ideas (or relicts of such impression) will exactly imitate the order of the impression itself, which would be absolute delirium: or any one part of that impression might recall any other part, and (as from the law of continuity, there must exist in every total impression some one or more parts, which are components of some other following total impression, and so on *ad infinitum* [to infinity]) any part of any impression might recall any part of any other, without a cause present to determine what it should be. For to bring in the will, or reason, as causes of their own cause, that is, as at once causes and effects, can satisfy those only who in their pretended evidences of a God having first demanded organization, as the sole cause and ground of intellect, will then coolly demand the pre-existence of intellect, as the cause and groundwork of organization. There is in truth but one state to which this theory applies at all, namely, that of complete light-headedness; and even to this it applies but partially, because the will, and reason are perhaps never wholly suspended.

A case of this kind occurred in a Catholic town in Germany a year or two before my arrival at Göttingen, and had not then ceased to be a frequent subject of conversation. A young woman of four or five and twenty, who could neither read, nor write, was seized with a nervous fever; during which, according to the asseverations of all the priests and monks of the neighbourhood, she became possessed, and, as it appeared, by a very learned devil. She continued incessantly talking Latin, Greek, and Hebrew, in very pompous tones and with most distinct enunciation. This possession was rendered more probable by the known fact, that she was or had been an heretic. Voltaire humorously advises the devil to decline all acquaintance with medical men; and it would have been more to his reputation, if he had taken this advice in the present instance. The case had attracted the particular attention of a young physician, and by his statement many eminent physiologists and psychologists visited the town, and cross-examined the case on the spot. Sheets full of her ravings were taken down from her own mouth, and were found to consist of sentences, coherent and intelligible each for itself, but with little or no connection with each other. Of the Hebrew, a small portion only could be traced to the Bible; the remainder seemed to be in the rabbinical dialect. All trick or conspiracy was out of the question. Not only had the young woman ever been an harmless, simple creature; but she was evidently labouring under a nervous fever. In the town, in which she had been resident for many years as a servant in different families, no solution presented itself. The young physician, however, determined to trace her past life step by step; for the patient herself was incapable of returning a rational answer. He at length succeeded in discovering the place, where her parents had lived: travelled thither, found them dead, but an uncle surviving; and from him learnt, that the patient had been charitably taken by an old Protestant pastor at nine years old, and had remained with him some years, even till the old man's death. Of this pastor the uncle knew nothing, but that he was a very good man. With great difficulty, and after much search, our young medical philosopher discovered a niece of the pastor's, who had lived with him as his housekeeper, and had inherited his effects. She remembered the girl; related, that her venerable uncle had been too indulgent, and could not bear to hear the girl scolded; that she was willing to have kept her, but that after her patron's death, the girl herself refused to stay. Anxious enquiries were then, of course, made concerning the pastor's habits; and the solution of the phenomenon was soon obtained. For it appeared, that it had been the old man's custom, for years, to walk up and down a passage of his house into which the kitchen door opened, and to read to himself with a loud voice, out of his favourite books. A considerable number of these were still in the niece's possession. She added, that he was a very learned man and a great Hebraist. Among the books were found a collection of rabbinical writings, together with several of the Greek and Latin fathers; and the physician succeeded in identifying so many passages with those taken down at the young woman's bedside, that no doubt could remain in any rational mind concerning the true origin of the impressions made on her nervous system.

This authenticated case furnishes both proof and instance, that relics of sensation may exist for an indefinite time in a latent state, in the very same order in which they were originally impressed; and as we cannot rationally suppose the feverish state of the brain to act in any other way than as a stimulus, this fact (and it would not be difficult to adduce several of the same kind) contributes to make it even probable, that all thoughts are in themselves imperishable; and, that if the intelligent faculty should be rendered more comprehensive, it would require only a different and apportioned organization, the body celestial instead of the body terrestrial, to bring before every human soul the collective experience of its whole past existence. And this, this, perchance, is the dread book of judgement, in whose mysterious hieroglyphics every idle word is recorded! Yea, in the very nature of a living spirit, it may be more possible that heaven and earth should pass away, than that a single act, a single thought, should be loosened or lost from that living chain of causes, to all whose links, conscious or unconscious, the free will, our only absolute self, is coextensive and co-present. But not now dare I longer discourse of this, waiting for a loftier mood, and a nobler subject, warned from within and from without, that it is profanation to speak of these mysteries* τοῖς μηδέποτε

* 'To those to whose imagination it has never been presented, how beautiful is the countenance of justice and wisdom; and that neither the morning nor the evening star

φαντασθεῖσιν, ὡς καλὸν τὸ τῆς δικαιοσύνης καὶ σωφροσύνης πρόσωπον, καὶ ὡς οὔτε ἕσπερος οὔτε ἕωος οὕτω καλά. Τὸν γάρ ὁρῶντα πρὸς τὸ ὁρώμενον συγγενὲς καὶ ὁμοῖον ποιησαμενον δεῖ ἐπιβάλλειν τῇ θεᾳ· οὐ γὰρ ἂν πώποτε εἶδεν Ὀφθαλμος Ἥλιον ἡλιοειδὴς μὴ γεγενήμενος, οὔδε το Καλον ἂν ἴδῃ Ψύχη μὴ κάλη γενομένη. Plotinus.

CHAPTER VII

Of the necessary consequences of the Hartleian theory—Of the original mistake or equivocation which procured admission for the theory—Memoria technica [Artificial Memory].

WE will pass by the utter incompatibility of such a law (if law it may be called, which would itself be the slave of chances) with even that appearance of rationality forced upon us by the outward phenomena of human conduct, abstracted from our own consciousness. We will agree to forget this for the moment, in order to fix our attention on that subordination of final to efficient causes in the human being, which flows of necessity from the assumption, that the will, and with the will all acts of thought and attention, are parts and products of this blind mechanism, instead of being distinct powers, whose function it is to control, determine, and modify the phantasmal chaos of association. The soul becomes a mere *ens logicum* [logical entity]; for as a real separable being, it would be more worthless and ludicrous, than the grimalkins in the cat-harpsichord, described in the Spectator.° For these did form a part of the process; but in Hartley's scheme the soul is present only to be pinched or stroked, while the very squeals or purring are produced by an agency wholly independent and alien. It involves all the difficulties, all the incomprehensibility (if it be not indeed, ὡς ἔμοιγε δοκεῖ [as it seems to me], the absurdity) of intercommunion between substances that have no one property in common, without any of the convenient consequences that bribed the judgement to the admission of the dualistic hypothesis. Accordingly, this *caput mortuum* ['dead head', a technical term in chemistry

are so fair. For in order to direct the view aright, it behoves that the beholder should have made himself congenerous and similar to the object beheld. Never could the eye have beheld the sun, had not its own essence been soliform' (i.e. preconfigured to light by a similarity of essence with that of light) 'neither can a soul not beautiful attain to an intuition of beauty.'

referring to the residue left after distillation] of the Hartleian process has been rejected by his followers, and the consciousness considered as a result, as a tune, the common product of the breeze and the harp: though this again is the mere remotion of one absurdity to make way for another, equally preposterous. For what is harmony but a mode of relation, the very *esse* [being] of which is *percipi* [to be perceived]? An *ens rationale* [rational entity], which pre-supposes the power, that by perceiving creates it? The razor's edge becomes a saw to the armed vision;° and the delicious melodies of Purcell or Cimarosa might be disjointed stammerings to a hearer, whose partition of time should be a thousand times subtler than ours. But this obstacle too let us imagine ourselves to have surmounted, and 'at one bound high overleap all bound!' Yet according to this hypothesis the disquisition, to which I am at present soliciting the reader's attention, may be as truly said to be written by Saint Paul's church, as by me: for it is the mere motion of my muscles and nerves; and these again are set in motion from external causes equally passive, which external causes stand themselves in interdependent connection with everything that exists or has existed. Thus the whole universe co-operates to produce the minutest stroke of every letter, save only that I myself, and I alone, have nothing to do with it, but merely the causeless and effectless beholding of it when it is done. Yet scarcely can it be called a beholding; for it is neither an act nor an effect; but an impossible creation of a something-nothing out of its very contrary! It is the mere quick-silver plating behind a looking-glass; and in this alone consists the poor worthless I! The sum total of my moral and intellectual intercourse dissolved into its elements are reduced to extension, motion, degrees of velocity, and those diminished copies of configurative motion, which form what we call notions, and notions of notions. Of such philosophy well might Butler say—

> The metaphysics but a puppet motion
> That goes with screws, the notion of a notion;
> The copy of a copy and lame draught
> Unnaturally taken from a thought:
> That counterfeits all pantomimic tricks,
> And turns the eyes, like an old crucifix;
> That counterchanges whatsoe'er it calls
> B' another name, and makes it true or false;
> Turns truth to falsehood, falsehood into truth,
> By virtue of the Babylonian's tooth.
>
> Miscellaneous Thoughts

The inventor of the watch did not in reality invent it; he only looked on, while the blind causes, the only true artists, were unfolding themselves. So must it have been too with my friend Allston, when he sketched his picture of the dead man revived by the bones of the prophet Elijah. So must it have been with Mr Southey and Lord Byron, when the one fancied himself composing his 'Roderick,' and the other his 'Childe Harold.' The same must hold good of all systems of philosophy; of all arts, governments, wars by sea and by land; in short, of all things that ever have been or that ever will be produced. For according to this system it is not the affections and passions that are at work, in as far as they are sensations or thoughts. We only fancy, that we act from rational resolves, or prudent motives, or from impulses of anger, love, or generosity. In all these cases the real agent is a something-nothing-everything, which does all of which we know, and knows nothing of all that itself does.

The existence of an infinite spirit, of an intelligent and holy will, must on this system be mere articulated motions of the air. For as the function of the human understanding is no other than merely (to appear to itself) to combine and to apply the phenomena of the association; and as these derive all their reality from the primary sensations; and the sensations again all their reality from the impressions *ab extra* [from outside]; a God not visible, audible, or tangible, can exist only in the sounds and letters that form his name and attributes. If in ourselves there be no such faculties as those of the will and the scientific reason, we must either have an innate idea of them, which would overthrow the whole system; or we can have no idea at all. The process, by which Hume degraded the notion of cause and effect into a blind product of delusion and habit, into the mere sensation of proceeding life (*nisus vitalis* [vital urge]) associated with the images of the memory; this same process must be repeated to the equal degradation of every fundamental idea in ethics or theology.

Far, very far am I from burthening with the odium of these consequences the moral characters of those who first formed, or have since adopted the system! It is most noticeable of the excellent and pious Hartley, that in the proofs of the existence and attributes of God, with which his second volume commences, he makes no reference to the principles or results of the first. Nay, he assumes, as his foundations, ideas which, if we embrace the doctrines of his first volume, can exist nowhere but in the vibrations of the ethereal medium common to the nerves and to the atmosphere. Indeed the whole of the second volume is, with the fewest possible exceptions, independent of his peculiar system. So true is it, that the faith, which saves and sanctifies, is a collective energy, a total act of the whole moral being; that its living sensorium is in the heart; and that no errors of the understanding can be morally arraigned unless they have proceeded from the heart.—But whether they be such, no man can be certain in the case of another, scarcely perhaps even in his own. Hence it follows by inevitable consequence, that man may perchance determine, what is an heresy; but God only can know, who is an heretic. It does not, however, by any means follow, that opinions fundamentally false are harmless. An hundred causes may coexist to form one complex antidote. Yet the sting of the adder remains venomous, though there are many who have taken up the evil thing; and it hurted them not! Some indeed there seem to have been, in an unfortunate neighbour nation at least, who have embraced this system with a full view of all its moral and religious consequences; some—

> . . . who deem themselves most free,
> When they within this gross and visible sphere
> Chain down the winged thought, scoffing assent,
> Proud in their meanness; and themselves they cheat
> With noisy emptiness of learned phrase,
> Their subtle fluids, impacts, essences,
> Self-working tools, uncaus'd effects, and all
> Those blind omniscients, those Almighty slaves,
> Untenanting Creation of its God!

Such men need discipline, not argument; they must be made better men, before they can become wiser.

The attention will be more profitably employed in attempting to discover and expose the paralogisms, by the magic of which such a faith could find admission into minds framed for a nobler creed. These, it appears to me, may be all reduced to one sophism as their common genus; the mistaking the conditions of a thing for its causes and essence; and the process by which we arrive at the knowledge of a faculty, for the faculty itself. The air I breathe, is the condition of my life, not its cause. We could never have learnt that we had eyes but by the process of seeing; yet having seen we know that the eyes must have pre-existed in order to render the process of sight possible. Let us cross-examine Hartley's scheme under the guidance of this distinction; and we shall discover, that contemporaneity (Leibniz's *Lex Continui* [Law of Continuity]) is the limit and condition of the laws of mind, itself being rather a law of matter, at least of phenomena considered as material. At the utmost, it is to thought the same, as the law of gravitation is to locomotion. In every voluntary movement we first counteract gravitation, in order to avail ourselves of it. It must exist, that there may be a something to be counteracted, and which by its reaction, aids the force that is exerted to resist it. Let us consider, what we do when we leap. We first resist the gravitating power by an act purely voluntary, and then by another act, voluntary in part, we yield to it in order to light on the spot, which we had previously proposed to ourselves. Now let a man watch his mind while he is composing; or, to take a still more common case, while he is trying to recollect a name; and he will find the process completely analogous. Most of my readers will have observed a small water-insect on the surface of rivulets, which throws a cinque-spotted shadow fringed with prismatic colours on the sunny bottom of the brook; and will have noticed, how the little animal *wins* its way up against the stream, by alternate pulses of active and passive motion, now resisting the current, and now yielding to it in order to gather strength and a momentary fulcrum for a further propulsion. This is no unapt emblem of the mind's self-experience in the act of thinking. There are evidently two powers at work, which relatively to each other are active and passive; and this is not possible without an intermediate faculty, which is at once both active and passive. (In philosophical language, we must denominate this intermediate faculty in all its degrees and determinations, the imagination. But in common language, and especially on the subject of poetry, we appropriate the name to a superior degree of the faculty, joined to a superior voluntary control over it.)

Contemporaneity then, being the common condition of all the laws of association, and a component element in all the *materia subjecta* [subject-matter], the parts of which are to be associated, must needs be co-present with all. Nothing, therefore, can be more easy than to pass off on an incautious mind this constant companion of each, for the essential substance of all. But if we appeal to our own consciousness, we shall find that even time itself, as the cause of a particular act of association, is distinct from contemporaneity, as the condition of all association. Seeing a mackerel it may happen, that I immediately think of gooseberries, because I at the same time ate mackerel with gooseberries as the sauce. The first syllable of the latter word, being that which had coexisted with the image of the bird so called, I may then think of a goose. In the next moment the image of a swan may arise before me, though I had never seen the two birds together. In the two former instances, I am conscious that their coexistence in time was the circumstance, that enabled me to recollect them; and equally conscious am I, that the latter was recalled to me by the joint operation of likeness and contrast. So it is with cause and effect; so too with order. So am I able to distinguish whether it was proximity in time, or continuity in space, that occasioned me to recall B on the mention of A. They cannot be indeed separated from contemporaneity; for that would be to separate them from the mind itself. The act of consciousness is indeed identical with time considered in its essence. (I mean time *per se* [in itself], as contradistinguished from our notion of time; for this is always blended with the idea of space, which as the contrary of time, is therefore its measure.) Nevertheless the accident of seeing two objects at the same moment acts, as a distinguishable cause from that of having seen them in the same place: and the true practical general law of association is this; that whatever makes certain parts of a total impression more vivid or distinct than the rest, will determine the mind to recall these in preference to others equally linked together by the common condition of contemporaneity, or (what I deem a more appropriate and philosophical term) of *continuity*. But the will itself by confining and intensifying* the attention may arbitrarily give vividness or distinctness to any object whatsoever; and from hence we may deduce the uselessness if not the absurdity of certain recent schemes which promise an artificial memory, but which in reality can only produce a confusion and debasement of the fancy. Sound logic, as the habitual subordination of the individual to the species, and of the species to the genus; philosophical knowledge of facts under the relation of cause and effect; a cheerful and communicative temper that disposes us to notice the similarities and contrasts of things, that we may be able to illustrate the one by the other; a quiet conscience; a condition free from anxieties; sound health, and above all (as far as relates to passive remembrance) a healthy digestion; these are the best, these are the only arts of memory.

* I am aware, that this word occurs neither in Johnson's Dictionary or in any classical writer. But the word, '*to intend*,' which Newton and others before him employ in this sense, is now so completely appropriated to another meaning, that I could not use it without ambiguity: while to paraphrase the sense, as by *render intense*, would often break up the sentence and destroy that harmony of the position of the words with the logical position of the thoughts, which is a beauty in all composition, and more especially desirable in a close philosophical investigation. I have therefore hazarded the word, *intensify*; though, I confess, it sounds uncouth to my own ear.

CHAPTER VIII

The system of dualism introduced by Descartes—Refined first by Spinoza and afterwards by Leibniz into the doctrine of Harmonia praestabilita [pre-established harmony]—*Hylozoism—Materialism—Neither of these systems on any possible theory of association supplies or supersedes a theory of perception, or explains the formation of the associable.*

To the best of my knowledge Descartes was the first philosopher, who introduced the absolute and essential heterogeneity of the soul as intelligence, and the body as matter. The assumption, and the form of speaking, have remained, though the denial of all other properties to matter but that of extension, on which denial the whole system of dualism is grounded, has been long exploded. For since impenetrability is intelligible only as a mode of resistance; its admission places the essence of matter in an act or power, which it possesses in common with spirit; and body and spirit are therefore no longer absolutely heterogeneous, but may without any absurdity be supposed to be different modes, or degrees in perfection, of a common substratum. To this possibility, however, it was not the fashion to advert. The soul was a thinking substance; and body a space-filling substance. Yet the apparent action of each on the other pressed heavy on the philosopher on the one hand; and no less heavily on the other hand pressed the evident truth, that the law of causality holds only between homogeneous things, i.e. things having some common property; and cannot extend from one world into another, its opposite. A close analysis evinced it to be no less absurd, than the question whether a man's affection for his wife, lay north-east, or south-west of the love he bore towards his child? Leibniz's doctrine of a pre-established harmony, which he certainly borrowed from Spinoza, who had himself taken the hint from Descartes's animal machines, was in its common interpretation too strange to survive the inventor—too repugnant to our common sense (which is not indeed entitled to a judicial voice in the courts of scientific philosophy; but whose whispers still exert a strong secret influence). Even Wolff the admirer, and illustrious systematizer of the Leibnizian doctrine, contents himself with defending the possibility of the idea, but does not adopt it as a part of the edifice.

The hypothesis of hylozoism on the other side, is the death of all rational physiology, and indeed of all physical science; for that requires a limitation of terms, and cannot consist with the arbitrary power of multiplying attributes by occult qualities. Besides, it answers no purpose; unless indeed a difficulty can be solved by multiplying it, or that we can acquire a clearer notion of our soul, by being told that we have a million souls, and that every atom of our bodies has a soul of its own. Far more prudent is it to admit the difficulty once for all, and then let it lie at rest. There is a sediment indeed at the bottom of the vessel, but all the water above it is clear and transparent. The hylozoist only shakes it up, and renders the whole turbid.

But it is not either the nature of man, or the duty of the philosopher to despair concerning any important problem until, as in the squaring of the circle, the impossibility of a solution has been demonstrated. How the *esse* [being] assumed as originally distinct from the *scire* [knowing], can ever unite itself with it; how being can transform itself into a knowing, becomes conceivable on one only condition; namely, if it can be shown that the *vis representativa* [representing power], or the sentient, is itself a species of being; i.e. either as a property or attribute, or as an hypostasis or self-subsistence. The former is indeed the assumption of materialism; a system which could not but be patronized by the philosopher, if only it actually performed what it promises. But how any affection from without can metamorphose itself into perception or will; the materialist has hitherto left, not only as incomprehensible as he found it, but has aggravated it into a comprehensible absurdity. For, grant that an object from without could act upon the conscious self, as on a consubstantial object; yet such an affection could only engender something homogeneous with itself. Motion could only propagate motion. Matter has no inward. We remove one surface, but to meet with another. We can but divide a particle into particles; and each atom comprehends in itself the properties of the material universe. Let any reflecting mind make the experiment of explaining to itself the evidence of our sensuous intuitions, from the hypothesis that in any given perception there is a something which has been communicated to it by an impact or an impression *ab extra* [from outside]. In the first place, by the impact on the percipient or *ens representans* [representing entity] not the object itself, but only its action or effect, will pass into the same. Not the iron tongue, but its vibrations, pass into the metal of the bell. Now in our immediate perception, it is not the mere power or act of the object, but the object itself, which is immediately present. We might indeed attempt to explain this result by a chain of deductions and conclusions; but that, first, the very faculty of deducing and concluding would equally demand an explanation; and secondly, that there exists in fact no such intermediation by logical notions, such as those of cause and effect. It is the object itself, not the product of a syllogism, which is present to our consciousness. Or would we explain this supervention of the object to the sensation, by a productive faculty set in motion by an impulse; still the transition, into the percipient, of the object itself, from which the impulse proceeded, assumes a power that can permeate and wholly possess the soul,

> And like a God by spiritual art,
> Be all in all, and all in every part.
>
> Cowley

And how came the percipient here? And what is become of the wonder-promising matter, that was to perform all these marvels by force of mere figure, weight, and motion? The most consistent proceeding of the dogmatic materialist is to fall back into the common rank of soul-and-bodyists; to affect the mysterious, and declare the whole process a revelation given, and not to be understood, which it would be profane to examine too closely. *Datur non intelligitur* [It is given, not understood]. But a revelation unconfirmed by miracles, and a faith not commanded by the conscience, a philosopher may venture to pass by, without suspecting himself of any irreligious tendency.

Thus as materialism has been generally taught, it is utterly unintelligible, and owes all its proselytes to the propensity so common among men, to mistake distinct images for clear conceptions; and vice versa, to reject as inconceivable whatever from its own nature is unimaginable. But as soon as it becomes intelligible, it ceases to be materialism. In order to explain thinking, as a material phenomenon, it is necessary to refine matter into a mere modification of intelligence, with the twofold function of appearing and perceiving. Even so did Priestley in his controversy with Price!° He stripped matter of all its material properties; substituted spiritual powers; and when we expected to find a body, behold! we had nothing but its ghost! the apparition of a defunct substance!

I shall not dilate further on this subject; because it will (if God grant health and permission) be treated of at large and systematically in a work, which I have many years been preparing, on the Productive Logos human and divine; with, and as the introduction to, a full commentary on the Gospel of St John. To make myself intelligible as far as my present subject requires, it will be sufficient briefly to observe—1. That all association demands and presupposes the existence of the thoughts and images to be associated.—2. The hypothesis of an external world exactly correspondent to those images or modifications of our own being, which alone (according to this system) we actually behold, is as thorough idealism as Berkeley's, inasmuch as it equally (perhaps, in a more perfect degree) removes all reality and immediateness of perception, and places us in a dream-world of phantoms and spectres, the inexplicable swarm and equivocal generation of motions in our own brains.—3. That this hypothesis neither involves the explanation, nor precludes the necessity, of a mechanism and co-adequate forces in the percipient, which at the more than magic touch of the impulse from without is to create anew for itself the correspondent object. The formation of a copy is not solved by the mere pre-existence of an original; the copyist of Raphael's Transfiguration must repeat more or less perfectly the process of Raphael. It would be easy to explain a thought from the image on the retina, and that from the geometry of light, if this very light did not present the very same difficulty. We might as rationally chant the Brahmin creed of the tortoise that supported the bear, that supported the elephant, that supported the world, to the tune of 'This is the house that Jack built.' The *sic Deo placitum est* [thus has it pleased God] we all admit as the sufficient cause, and the divine goodness as the sufficient reason; but an answer to the whence? and why? is no answer to the how? which alone is the physiologist's concern. It is a mere *sophisma pigrum* [lazy sophism], and (as Bacon hath said) the arrogance of pusillanimity, which lifts up the idol of a mortal's fancy and commands us to fall down and worship it, as a work of divine wisdom, an ancile or palladium fallen from heaven. By the very same argument the supporters of the Ptolemaic system might have rebuffed the Newtonian, and pointing to the sky with self-complacent grin* have appealed to common sense, whether the sun did not move and the earth stand still.

CHAPTER IX

Is philosophy possible as a science, and what are its conditions?—Giordano Bruno—Literary aristocracy, or the existence of a tacit compact among the learned as a privileged order—The author's obligations to the mystics;—to Immanuel Kant—The difference

* 'And Coxcombs vanquish Berkeley with a grin.' *Pope.*

between the letter and the spirit of Kant's writings, and a vindication of prudence in the teaching of philosophy—Fichte's attempt to complete the critical system—Its partial success and ultimate failure—Obligations to Schelling; and among English writers to Saumarez.

AFTER I had successively studied in the schools of Locke, Berkeley, Leibniz, and Hartley, and could find in neither of them an abiding place for my reason, I began to ask myself; is a system of philosophy, as different from mere history and historic classification, possible? If possible, what are its necessary conditions? I was for a while disposed to answer the first question in the negative, and to admit that the sole practicable employment for the human mind was to observe, to collect, and to classify. But I soon felt, that human nature itself fought up against this wilful resignation of intellect; and as soon did I find, that the scheme taken with all its consequences and cleared of all inconsistencies was not less impracticable, than contra-natural. Assume in its full extent the position, *nihil in intellectu quod non prius in sensu* [there is nothing in the mind that was not in the senses first], without Leibniz's qualifying *praeter ipsum intellectum* [except mind itself], and in the same sense, in which it was understood by Hartley and Condillac: and what Hume had demonstratively deduced from this concession concerning cause and effect, will apply with equal and crushing force to all the other* eleven categorical forms, and the logical functions corresponding to them. How can we make bricks without straw? Or build without cement? We learn all things indeed by occasion of experience; but the very facts so learnt force us inward on the antecedents, that must be presupposed in order to render experience itself possible. The first book of Locke's Essay° (if the supposed error, which it labours to subvert, be not a mere thing of straw, an absurdity which no man ever did, or indeed ever could believe) is formed on a *Σόφισμα Ετεροζητησέως* [Sophism of Looking for something else], and involves the old mistake of *cum hoc: ergo, propter hoc* [with this, therefore because of this].

The term, philosophy, defines itself as an affectionate seeking after the truth; but truth is the correlative of being. This again is no way conceivable, but by assuming as a postulate, that both are *ab initio* [from the beginning], identical and co-inherent; that intelligence and

* *Videlicet* [= Viz., 'namely']; quantity, quality, relation, and mode, each consisting of three subdivisions. *Vide* [See] Kritik der reinen Vernunft, p. 95, and 106. See too the judicious remarks in° Locke and Hume.

being are reciprocally each other's substrate. I presumed that this was a possible conception (i.e. that it involved no logical inconsonance) from the length of time during which the scholastic definition of the Supreme Being, as *actus purissimus sine ullâ potentialitate* [the purest act without any potentiality], was received in the schools of theology, both by the pontifician and the reformed divines. The early study of Plato and Plotinus, with the commentaries and the Theologia Platonica, of the illustrious Florentine;° of Proclus, and Gemistus Pletho; and at a later period of the 'De Immenso et Innumerabili,' and the 'De la causa, principio et uno,' of the philosopher of Nola,° who could boast of a Sir Philip Sidney, and Fulke Greville among his patrons, and whom the idolaters of Rome burnt as an atheist in the year 1600; had all contributed to prepare my mind for the reception and welcoming of the *Cogito quia sum, et sum quia Cogito* [I think because I am, and I am because I think]; a philosophy of seeming hardihood, but certainly the most ancient, and therefore presumptively the most natural.

Why need I be afraid? Say rather how dare I be ashamed of the Teutonic theosophist, Jacob Behmen?° Many indeed, and gross were his delusions; and such as furnish frequent and ample occasion for the triumph of the learned over the poor ignorant shoemaker, who had dared think for himself. But while we remember that these delusions were such, as might be anticipated from his utter want of all intellectual discipline, and from his ignorance of rational psychology, let it not be forgotten that the latter defect he had in common with the most learned theologians of his age. Neither with books, nor with book-learned men was he conversant. A meek and shy quietist, his intellectual powers were never stimulated into feverous energy by crowds of proselytes, or by the ambition of proselyting. Jacob Behmen was an enthusiast, in the strictest sense, as not merely distinguished, but as contradistinguished, from a fanatic. While I in part translate the following observations from a contemporary writer of the Continent,° let me be permitted to premise, that I might have transcribed the substance from memoranda of my own, which were written many years before his pamphlet was given to the world; and that I prefer another's words to my own, partly as a tribute due to priority of publication; but still more from the pleasure of sympathy in a case where coincidence only was possible.

Whoever is acquainted with the history of philosophy, during the two or three last centuries, cannot but admit, that there appears to have existed a sort of secret and tacit compact among the learned, not to pass beyond a certain limit in speculative science. The privilege of free thought, so highly extolled, has at no time been held valid in actual practice, except within this limit; and not a single stride beyond it has ever been ventured without bringing obloquy on the transgressor. The few men of genius among the learned class, who actually did overstep this boundary, anxiously avoided the appearance of having so done. Therefore the true depth of science, and the penetration to the inmost centre, from which all the lines of knowledge diverge to their ever distant circumference, was abandoned to the illiterate and the simple, whom unstilled yearning, and an original ebulliency of spirit, had urged to the investigation of the indwelling and living ground of all things. These then, because their names had never been enrolled in the guilds of the learned, were persecuted by the registered liverymen as interlopers on their rights and privileges. All without distinction were branded as fanatics and fantasts; not only those, whose wild and exorbitant imaginations had actually engendered only extravagant and grotesque phantasms, and whose productions were, for the most part, poor copies and gross caricatures of genuine inspiration; but the truly inspired likewise, the originals themselves! And this for no other reason, but because they were the *unlearned*, men of humble and obscure occupations. When, and from whom among the literati by profession, have we ever heard the divine doxology repeated, 'I thank thee O father! Lord of Heaven and Earth! because thou hast hid these things from the wise and prudent, and hast revealed them unto babes.' No! the haughty priests of learning, not only banished from the schools and marts of science all, who had dared draw living waters from the fountain, but drove them out of the very temple, which meantime 'the buyers, and sellers, and money-changers' were suffered to make 'a den of thieves.'

And yet it would not be easy to discover any substantial ground for this contemptuous pride in those literati, who have most distinguished themselves by their scorn of Behmen, De Thoyras, George Fox, etc.; unless it be, that *they* could write orthographically, make smooth periods, and had the fashions of authorship almost literally at their fingers' ends, while the latter, in simplicity of soul, made their words immediate echoes of their feelings. Hence the frequency of those phrases among them, which have been mistaken for pretences to immediate inspiration; as for instance, 'it was delivered unto me,' 'I strove not to speak,' 'I said, I will be silent,' 'but the word was in [my] heart as a burning fire,' 'and I could not forbear.' Hence too the unwillingness to give offence; hence the foresight, and the dread of the clamours, which would be raised against them, so frequently avowed in the writings of these men, and expressed, as was natural, in the words of the only book, with which they were familiar. 'Woe is me that I am become a man of strife, and a man of contention,—I love peace: the souls of men are dear unto me: yet because I seek for Light every one of them doth curse me!' O! it requires deeper feeling, and a stronger imagination, than belong to most of those, to whom reasoning and fluent expression have been as a trade learnt in boyhood, to conceive with what might, with what inward strivings and commotion, the perception of a new and vital truth takes possession of an uneducated man of genius. His meditations are almost inevitably employed on the eternal, or the everlasting; for 'the world is not his friend, nor the world's law.' Need we then be surprised, that under an excitement at once so strong and so unusual, the man's body should sympathize with the struggles of his mind; or that he should at times be so far deluded, as to mistake the tumultuous sensations of his nerves, and the coexisting spectres of his fancy, as parts or symbols of the truths which were opening on him? It has indeed been plausibly observed, that in order to derive any advantage, or to collect any intelligible meaning, from the writings of these ignorant mystics, the reader must bring with him a spirit and judgement superior to that of the writers themselves:

> And what he brings, what needs he elsewhere seek?
>
> Paradise Regained

—A sophism, which I fully agree with Warburton, is unworthy of Milton; how much more so of the awful person, in whose mouth he has placed it? One assertion I will venture to make, as suggested by my own experience, that there exist folios on the human understanding, and the nature of man, which would have a far juster claim to their high rank and celebrity, if in the whole huge volume there could be found as much fullness of heart and intellect, as burst forth in many a simple page of George Fox, Jacob Behmen, and even of Behmen's commentator, the pious and fervid William Law.

The feeling of gratitude, which I cherish towards these men, has caused me to digress further than I had foreseen or proposed; but to have passed them over in an historical sketch of my literary life and opinions, would have seemed to me like the denial of a debt, the concealment of a boon. For the writings of these mystics acted in no slight degree to prevent my mind from being imprisoned within the outline of any single dogmatic system. They contributed to keep alive the *heart* in the *head*; gave me an indistinct, yet stirring and working presentment, that all the products of the mere reflective faculty partook of death, and were as the rattling twigs and sprays in winter, into which a sap was yet to be propelled, from some root to which I had not penetrated, if they were to afford my soul either food or shelter. If they were too often a moving cloud of smoke to me by day, yet they were always a pillar of fire throughout the night, during my wanderings through the wilderness of doubt, and enabled me to skirt, without crossing, the sandy deserts of utter unbelief. That the system is capable of being converted into an irreligious pantheism, I well know. The Ethics of Spinoza, may, or may not, be an instance. But at no time could I believe, that in itself and essentially it is incompatible with religion, natural, or revealed: and now I am most thoroughly persuaded of the contrary. The writings of the illustrious sage of Königsberg, the founder of the Critical Philosophy, more than any other work, at once invigorated and disciplined my understanding. The originality, the depth, and the compression of the thoughts; the novelty and subtlety, yet solidity and importance, of the distinctions; the adamantine chain of the logic; and I will venture to add (paradox as it will appear to those who have taken their notion of Immanuel Kant from reviewers and Frenchmen) the clearness and evidence, of the 'Critique of the Pure Reason'; of the Judgement; of the 'Metaphysical Elements of Natural Philosophy,' and of his 'Religion within the Bounds of Pure Reason,' took possession of me as with a giant's hand. After fifteen years' familiarity with them, I still read these and all his other productions with undiminished delight and increasing admiration. The few passages that remained obscure to me, after due efforts of thought, (as the chapter on original apperception), and the apparent contradictions which occur, I soon found were hints and insinuations referring to ideas, which Kant either did not think it prudent to avow, or which he considered as consistently *left behind* in a pure analysis, not of human nature *in toto* [altogether], but of the speculative intellect alone. Here therefore he was constrained to commence at the point of reflection, or natural consciousness: while in his moral system he was permitted to assume a higher ground (the autonomy of the will) as a postulate deducible from the unconditional command, or (in the technical language of his school) the categorical imperative, of the conscience. He had been in imminent danger of persecution during the reign of the late king of Prussia, that strange compound of lawless debauchery, and priest-ridden superstition: and it is probable that he had little inclination, in his old age, to act over again the fortunes, and hair-breadth escapes of Wolff. The expulsion of the first among Kant's disciples, who attempted to complete his system, from the University of Jena, with the confiscation and prohibition of the obnoxious work by the joint efforts of the courts of Saxony and Hanover, supplied experimental proof, that the venerable old man's caution was not groundless. In spite therefore of his own declarations, I could never believe, it was possible for him to have meant no more by his *Noumenon*, or thing in itself, than his mere words express; or that in his own conception he confined the whole plastic power to the forms of the intellect, leaving for the external cause, for the *materiale* [stuff] of our sensations, a matter without form, which is doubtless inconceivable. I entertained doubts likewise, whether in his own mind, he even laid all the stress, which he appears to do on the moral postulates.

An idea, in the highest sense of that word, cannot be conveyed but by a symbol; and, except in geometry, all symbols of necessity involve an apparent contradiction. *Φώνησε Συνέτοισιν* [He spoke to the wise]: and for those who could not pierce through this symbolic husk, his writings were not intended. Questions which cannot be fully answered without exposing the respondent to personal danger, are not entitled to a fair answer; and yet to say this openly, would in many cases furnish the very advantage, which the adversary is insidiously seeking after. Veracity does not consist in saying, but in the intention of communicating truth; and the philosopher who cannot utter the whole truth without conveying falsehood, and at the same time, perhaps, exciting the most malignant passions, is constrained to express himself either mythically or equivocally. When Kant therefore was importuned to settle the disputes of his commentators himself, by declaring what he meant, how could he decline the honours of martyrdom with less offence, than by simply replying, 'I meant what I said, and at the age of near four score, I have something else, and more important to do, than to write a commentary on my own works.'

Fichte's Wissenschaftslehre, or *Lore* of Ultimate Science, was to add the keystone of the arch: and by commencing with an act, instead of a thing or substance, Fichte assuredly gave the first mortal blow to Spinozism, as taught by Spinoza himself; and supplied the idea of a system truly metaphysical, and of a *metaphysique* [metaphysics] truly systematic: (i.e. having its spring and principle within itself). But this fundamental idea he overbuilt with a heavy mass of mere notions, and psychological acts of arbitrary reflection. Thus his theory degenerated into a crude egoismus,* a boastful and hyperstoic hostility to Nature, as lifeless, godless, and altogether unholy: while his religion consisted in the assumption of a mere *ordo ordinans* [ordering order], which we were permitted *exotericé* [popularly] to call God; and his ethics in an ascetic, and almost monkish, mortification of the natural passions and desires.

* The following burlesque on the Fichtean Egoismus may, perhaps, be amusing to the few who have studied the system, and to those who are unacquainted with it, may convey as tolerable a likeness of Fichte's idealism as can be expected from an avowed caricature.

The categorical imperative, or the annunciation of the new Teutonic God, *ΕΓΩΕΝΚΑΙΠΑΝ* [*Egoenkaipan*, I the one and all]: a dithyrambic Ode, by Querkopf von Klubstick, Grammarian, and Subrector in Gymnasio [in the grammar school or high school]****.

Eu! Dei vices gerens, ipse Divus [Hurrah! God's vice-regent, myself God],
(*Speak English*, *Friend!*) the God Imperativus [God the Imperative (in a grammatical sense)],
Here on this market-cross aloud I cry:
I, I, I! I itself I!
The form and the substance, the what and the why,
The when and the where, and the low and the high,
The inside and outside, the earth and the sky,
I, you, and he, and he, you and I,
All souls and all bodies are I itself I!
All I itself I!
(Fools! a truce with this starting!)
All my I! all my I!
He's a heretic dog who but adds Betty Martin!
Thus cried the God with high imperial tone:
In robe of stiffest state, that scoff'd at beauty,
A pronoun-verb imperative he shone—
Then substantive and plural-singular grown
He thus spake on! Behold in I alone
(For ethics boast a syntax of their own)
Or if in ye, yet as I doth depute ye,
In O! I, you, the vocative of duty!
I of the world's whole Lexicon the root!
Of the whole universe of touch, sound, sight
The genitive and ablative to boot:
The accusative of wrong, the nom'native of right,
And in all cases the case absolute!
Self-construed, I all other moods decline:
Imperative, from nothing we derive us;
Yet as a super-postulate of mine,
Unconstrued antecedence I assign
To X, Y, Z, the God infinitivus [God the infinitive]!

In Schelling's 'Natur-Philosophie,' and the 'System des transcendentalen Idealismus,' I first found a genial coincidence with much that I had toiled out for myself, and a powerful assistance in what I had yet to do.

I have introduced this statement, as appropriate to the narrative nature of this sketch; yet rather in reference to the work which I have announced in a preceding page, than to my present subject. It would be but a mere act of justice to myself, were I to warn my future readers, that an identity of thought, or even similarity of phrase will not be at all times a certain proof that the passage has been borrowed from Schelling, or that the conceptions were originally learnt from him. In this instance, as in the dramatic lectures of Schlegel to which I have before alluded, from the same motive of self-defence against the charge of plagiarism, many of the most striking resemblances, indeed all the main and fundamental ideas, were born and matured in my mind before I had ever seen a single page of the German philosopher; and I might indeed affirm with truth, before the more important works of Schelling had been written, or at least made public. Nor is this coincidence at all to be wondered at. We had studied in the same school; been disciplined by the same preparatory philosophy, namely, the writings of Kant; we had both equal obligations to the polar logic and dynamic philosophy of Giordano Bruno; and Schelling has lately, and, as of recent acquisition, avowed that same affectionate reverence for the labours of Behmen, and other mystics, which I had formed at a much earlier period. The coincidence of Schelling's system with certain general ideas of Behmen, he declares to have been mere coincidence; while my obligations have been more direct. He needs give to Behmen only feelings of sympathy; while I owe him a debt of gratitude. God forbid! that I should be suspected of a wish to enter into a rivalry with Schelling for the honours so unequivocally his right, not only as a great and original genius, but as the founder of the Philosophy of Nature, and as the most successful improver of the dynamic system* which, begun by Bruno, was reintroduced (in a more

* It would be an act of high and almost criminal injustice to pass over in silence the name of Mr Richard Saumarez, a gentleman equally well known as a medical man and as a philanthropist, but who demands notice on the present occasion as the author of 'a new System of Physiology' in two volumes octavo, published 1797; and in 1812 of 'an Examination of the natural and artificial Systems of Philosophy which now prevail' in one volume octavo, entitled, 'The Principles of physiological and physical Science.' The latter work is not quite equal to the former in style or arrangement; and there is a greater necessity of distinguishing the principles of the author's philosophy from his

philosophical form, and freed from all its impurities and visionary accompaniments) by Kant; in whom it was the native and necessary growth of his own system. Kant's followers, however, on whom (for the greater part) their master's cloak had fallen without, or with a very scanty portion of, his spirit, had adopted his dynamic ideas, only as a more refined species of mechanics. With exception of one or two fundamental ideas, which cannot be withheld from Fichte, to Schelling we owe the completion, and the most important victories, of this revolution in philosophy. To me it will be happiness and honour enough, should I succeed in rendering the system itself intelligible to my countrymen, and in the application of it to the most awful of subjects for the most important of purposes. Whether a work is the offspring of a man's own spirit, and the product of original thinking, will be discovered by those who are its sole legitimate judges, by better tests than the mere reference to dates. For readers in general, let whatever shall be found in this or any future work of mine, that resembles, or coincides with, the doctrines of my German predecessor, though contemporary, be wholly attributed to him: provided, that the absence of distinct references to his books, which I could not at all times make with truth as designating citations or thoughts

conjectures concerning colour, the atmospheric matter, comets, etc. which whether just or erroneous are by no means necessary consequences of that philosophy. Yet even in this department of this volume, which I regard as comparatively the inferior work, the reasonings by which Mr Saumarez invalidates the immanence of an infinite power in any finite substance are the offspring of no common mind; and the experiment on the expansibility of the air is at least plausible and highly ingenious. But the merit, which will secure both to the book and to the writer a high and honourable name with posterity, consists in the masterly force of reasoning, and the copiousness of induction, with which he has assailed, and (in my opinion) subverted the tyranny of the mechanic system in physiology; established not only the existence of final causes, but their necessity and efficiency in every system that merits the name of philosophical; and substituting life and progressive power, for the contradictory *inert force*, has a right to be known and remembered as the first instaurator of the dynamic philosophy in England. The author's views, as far as concerns himself, are unborrowed and completely his own, as he neither possessed nor do his writings discover, the least acquaintance with the works of Kant, in which the germs of the philosophy exist; and his volumes were published many years before the full development of these germs by Schelling. Mr Saumarez's detection of the Braunonian system° was no light or ordinary service at the time; and I scarcely remember in any work on any subject a confutation so thoroughly satisfactory. It is sufficient at this time to have stated the fact; as in the preface to the work, which I have already announced on the Logos, I have exhibited in detail the merits of this writer, and genuine philosopher, who needed only have taken his foundations somewhat deeper and wider to have superseded a considerable part of my labours.

actually derived from him; and which, I trust, would, after this general acknowledgement be superfluous; be not charged on me as an ungenerous concealment or intentional plagiarism. I have not indeed (*eheu! res angusta domi!* [alas! the narrow circumstances at home!]) been hitherto able to procure more than two of his books, viz. the 1st volume of his collected Tracts, and his System of Transcendental Idealism; to which, however, I must add a small pamphlet against Fichte, the spirit of which was to *my* feelings painfully incongruous with the principles, and which (with the usual allowance afforded to an antithesis) displayed the love of wisdom rather than the wisdom of love. I regard truth as a divine ventriloquist: I care not from whose mouth the sounds are supposed to proceed, if only the words are audible and intelligible. 'Albeit, I must confess to be half in doubt, whether I should bring it forth or no, it being so contrary to the eye of the world, and the world so potent in most men's hearts, that I shall endanger either not to be regarded or not to be understood.' Milton: *Reason of Church Government.*

And to conclude the subject of citation, with a cluster of citations, which as taken from books, not in common use, may contribute to the reader's amusement, as a voluntary before a sermon. 'Dolet mihi quidem deliciis literarum inescatos subito jam homines adeo esse, praesertim qui Christianos se profitentur, ut legere nisi quod ad delectationem facit, sustineant nihil: unde et disciplinae severiores et philosophia ipsa jam fere prorsus etiam a doctis negliguntur. Quod quidem propositum studiorum, nisi mature corrigitur, tam magnum rebus incommodum dabit, quám dedit Barbaries olim. Pertinax res Barbaries est, fateor: sed minus potest tamen, quám illa mollities et *persuasa prudentia* literarum, quae si *ratione* caret, sapientiae virtutisque *specie* mortales miserè circumducit. Succedet igitur, ut arbitror, haud ita multo post, pro rusticanâ seculi nostri ruditate captatrix illa *communiloquentia* robur animi virilis omne, omnem virtutem masculam profligatura, nisi cavetur. [In very truth, it grieveth me that men, those especially who profess themselves to be Christians, should be so taken with the sweet baits of literature that they can endure to read nothing but what gives them immediate gratification, no matter how low or sensual it may be. Consequently, the more austere and disciplinary branches of philosophy itself are almost wholly neglected, even by the learned.—A course of study (if such reading, with such a purpose in view, could deserve that name) which, if not corrected in time, will occasion worse consequences than even barbarism did in the times of our forefathers. Barbarism is, I own, a wilful headstrong thing; but with all its blind obstinacy it has less power of doing harm than this self-sufficient, self-satisfied *plain good common-sense* sort of writing, this prudent saleable popular style of composition, if it be deserted by reason and scientific insight; pitiably decoying the minds of men by an imposing show of amiableness, and practical wisdom, so that the delighted reader knowing nothing knows all about almost everything. There will succeed therefore in my opinion, and that too within no long time, to the rudeness and rusticity of our age, that ensnaring meretricious popularness in literature, with all the tricksy humilities of the ambitious candidates for the favourable suffrages of the judicious public, which if we do not take good care will break up and scatter before it all robustness and manly vigour of intellect, all masculine fortitude of virtue (Coleridge's own translation, from the *Friend*)].

Simon Grynaeus, candido lectori [to the candid reader], prefixed to the Latin translation of Plato, by Marsilius Ficinus. Lugduni, 1557. A too prophectic remark, which has been in fulfilment from the year 1680, to the present 1815. NB. By 'persuasa prudentia,' Grynaeus means self-complacent common sense as opposed to science and philosophic reason.

'Est medius ordo et velut equestris Ingeniorum quidem sagacium et rebus humanis commodorum, non tamen in primam magnitudinem patentium. Eorum hominum, ut ita dicam, major annona est. Sedulum esse, nihil temerè loqui, assuescere labori, et imagine prudentiae et modestiae tegere angustiores partes captûs dum exercitationem et usum, quo isti in civilibus rebus pollent, pro natura et magnitudine ingenii plerique accipiunt [There is a middle rank (like that of knights among the nobility) of wise men, and useful in affairs, which yet reach not to the first height and greatness. Of them there is a more plentiful store. To be diligent, to speak nothing rashly, to accustom oneself to labour, and [by] the show of wisdom [and modesty] to hide the weaker part of the wits while custom and use, which in affairs of State make them able, is by the most part taken for nature, and greatness of mind (*CC*)].' Barclaii Argenis, p. 71.

'As therefore, physicians are many times forced to leave such methods of curing as themselves know to be fittest, and being overruled by the sick man's impatience, are fain to try the best they can: in like sort, considering how the case doth stand with the present age, full of tongue and weak of brain, behold we would (*if our subject permitted it*) yield to the stream thereof. That way we would be contented to prove our thesis, which being the worse in itself, notwithstanding is now by reason of common imbecility the fitter and likelier to be brooked.'—Hooker.

If this fear could be rationally entertained in the controversial age of Hooker, under the then robust discipline of the scholastic logic, pardonably may a writer of the present times anticipate a scanty audience for abstrusest themes, and truths that can neither be communicated or received without effort of thought, as well as patience of attention.

> Che s'io non erro al calcular de' punti,
> Par ch'*Asinina* Stella a noi predomini,
> E 'l Somaro e 'l castron si sian congiunti.
> Il tempo d'Apuleio piu non si nomini:
> Che se allora un sol Huom sembrava un Asino,
> Mille Asini á miei dì rassembran Huomini!
>
> [For if I err not in calculating the points,
> as asinine star seems to rule us, and the
> Donkey and the Mule are in conjunction.
> Let the Time of Apuleius be named no more!
> For if then one Man alone seemed to be an Ass,
> a thousand Asses in my days resemble Men. (*CN*)]
>
> Di Salvator Rosa, *Satir*. I. l. 10

CHAPTER X

A chapter of digression and anecdotes, as an interlude preceding that on the nature and genesis of the imagination or plastic power—On pedantry and pedantic expressions—Advice to young authors respecting publication—Various anecdotes of the author's literary life, and the progress of his opinions in religion and politics.

'*Esemplastic*. The word is not in Johnson, nor have I met with it elsewhere.' Neither have I! I constructed it myself from the Greek words, εις εν πλαττειν [*eis en plattein*] i.e. to shape into one; because, having to convey a new sense, I thought that a new term would both aid the recollection of my meaning, and prevent its being confounded with the usual import of the word, imagination. 'But this is pedantry!' Not necessarily so, I hope. If I am not misinformed, pedantry consists in the use of words unsuitable to the time, place, and company. The language of the market would be in the schools as pedantic, though it might not be reprobated by that name, as the language of the schools in the market. The mere man of the world, who insists that no other terms but such as occur in common conversation should be employed in a scientific disquisition, and with no greater precision, is as truly a pedant as the man of letters, who either overrating the acquirements of his auditors, or misled by his own familiarity with technical or scholastic terms, converses at the wine-table with his mind fixed on his museum or laboratory; even though the latter pedant instead of desiring his wife to make the tea, should bid her add to the quant. suff. of thea sinensis the oxyd of hydrogen saturated with caloric.° To use the colloquial (and in truth somewhat vulgar) metaphor, if the pedant of the cloister, and the pedant of the lobby, both smell equally of the shop, yet the odour from the Russian binding of good old authentic-looking folios and quartos is less annoying than the steams from the tavern or bagnio. Nay, though the pedantry of the scholar should betray a little ostentation, yet a well-conditioned mind would more easily, methinks, tolerate the fox-brush of learned vanity, than the *sans culotterie* of a contemptuous ignorance, that assumes a merit from mutilation in the self-consoling sneer at the pompous incumbrance of tails.°

The first lesson of philosophic discipline is to wean the student's attention from the degrees of things, which alone form the vocabulary of common life, and to direct it to the kind abstracted from degree. Thus the chemical student is taught not to be startled at disquisitions on the heat in ice, or on latent and fixible light. In such discourse the instructor has no other alternative than either to use old words with new meanings (the plan adopted by Darwin in his Zoonomia); or to introduce new terms, after the example of Linnaeus, and the framers of the present chemical nomenclature. The latter mode is evidently preferable, were it only that the former demands a twofold exertion of thought in one and the same act. For the reader (or hearer) is required not only to learn and bear in mind the new definition; but to unlearn, and keep out of his view, the old and habitual meaning; a far more difficult and perplexing task, and for which the mere semblance of eschewing pedantry seems to me an inadequate compensation. Where, indeed, it is in our power to recall an appropriate term that had without sufficient reason become obsolete, it is doubtless a less evil to restore than to coin anew. Thus to express in one word, all that appertains to the perception considered as passive, and merely recipient, I have adopted from our elder classics the word *sensuous*;

because *sensual* is not at present used, except in a bad sense, or at least as a moral distinction, while *sensitive* and *sensible* would each convey a different meaning. Thus too I have followed Hooker, Sanderson, Milton, etc. in designating the immediateness of any act or object of knowledge by the word *intuition*, used sometimes subjectively, sometimes objectively, even as we use the word, thought; now as *the* thought, or act of thinking, and now as *a* thought, or the object of our reflection; and we do this without confusion or obscurity. The very words, *objective* and *subjective*, of such constant recurrence in the schools of yore, I have ventured to reintroduce, because I could not so briefly, or conveniently by any more familiar terms distinguish the *percipere* [perceiving] from the *percipi* [being perceived]. Lastly, I have cautiously discriminated the terms, the *reason*, and the *understanding*, encouraged and confirmed by the authority of our genuine divines, and philosophers, before the revolution.

> . . . both life, and sense,
> Fancy, and *understanding*: whence the soul
> *Reason* receives, and REASON is her *being*,
> DISCURSIVE OR INTUITIVE. Discourse*
> Is oftest yours, the latter most is ours,
> Differing but in *degree*, in *kind* the same.
>
> Paradise Lost, *Book V*

I say, that I was *confirmed* by authority so venerable: for I had previous and higher motives in my own conviction of the importance, nay, of the necessity of the distinction, as both an indispensable condition and a vital part of all sound speculation in metaphysics, ethical or theological. To establish this distinction was one main object of The Friend; if even in a biography of my own literary life I can with propriety refer to a work, which was printed rather than published, or so published that it had been well for the unfortunate author, if it had remained in manuscript! I have even at this time bitter cause for remembering that, which a number of my subscribers have but a trifling motive for forgetting. This effusion might have been spared; but I would fain flatter myself, that the reader will be less austere than an oriental professor of the bastinado, who during an attempt to

* But for sundry notes on Shakespeare, etc. which have fallen in my way, I should have deemed it unnecessary to observe, that *discourse* here, or elsewhere does not mean what we now call discoursing; but the discursion of the mind, the processes of generalization and subsumption, of deduction and conclusion. Thus, philosophy has hitherto been discursive: while geometry is always and essentially intuitive.

extort *per argumentum baculinum* [by the argument of the stick] a full confession from a culprit, interrupted his outcry of pain by reminding him, that it was 'a mere digression!' All this noise, Sir! is nothing to the point, and no sort of answer to my questions! Ah! but (replied the sufferer) it is the most pertinent reply in nature to your blows.

An imprudent man of common goodness of heart, cannot but wish to turn even his imprudences to the benefit of others, as far as this is possible. If therefore any one of the readers of this semi-narrative should be preparing or intending a periodical work, I warn him, in the first place, against trusting in the number of names on his subscription list. For he cannot be certain that the names were put down by sufficient authority; or (should that be ascertained) it still remains to be known, whether they were not extorted by some over-zealous friend's importunity; whether the subscriber had not yielded his name, merely from want of courage to answer, no! and with the intention of dropping the work as soon as possible. One gentleman procured me nearly a hundred names for The Friend, and not only took frequent opportunity to remind me of his success in his canvass, but laboured to impress my mind with the sense of the obligation, I was under to the subscribers; for (as he very pertinently admonished me) 'fifty-two shillings a year was a large sum to be bestowed on one individual, where there were so many objects of charity with strong claims to the assistance of the benevolent.' Of these hundred patrons ninety threw up the publication before the fourth number, without any notice; though it was well known to them, that in consequence of the distance, and the slowness and irregularity of the conveyance, I was compelled to lay in a stock of stamped paper for at least eight weeks beforehand; each sheet of which stood me in five pence previous to its arrival at my printer's; though the subscription money was not to be received till the twenty-first week after the commencement of the work; and lastly, though it was in nine cases out of ten impracticable for me to receive the money for two or three numbers without paying an equal sum for the postage.

In confirmation of my first caveat, I will select one fact among many. On my list of subscribers, among a considerable number of names equally flattering, was that of an Earl of Cork, with his address. He might as well have been an Earl of Bottle, for aught *I* knew of him, who had been content to reverence the peerage *in abstracto* [in the abstract], rather than *in concretis* [in the concrete]. Of course The Friend was regularly sent as far, if I remember right, as the eighteenth number: i.e. till a fortnight before the subscription was to be paid.

And lo! just at this time I received a letter from his Lordship, reproving me in language far more lordly than courteous for my impudence in directing my pamphlets to him, who knew nothing of me or my work! Seventeen or eighteen numbers of which, however, his Lordship was pleased to retain, probably for the culinary or post-culinary conveniences of his servants.

Secondly, I warn all others from the attempt to deviate from the ordinary mode of publishing a work by *the trade*. I thought indeed, that to the purchaser it was indifferent, whether thirty per cent of the purchase money went to the booksellers or to the Government; and that the convenience of receiving the work by the post at his own door would give the preference to the latter. It is hard, I own, to have been labouring for years, in collecting and arranging the materials; to have spent every shilling that could be spared after the necessaries of life had been furnished, in buying books, or in journeys for the purpose of consulting them or of acquiring facts at the fountain-head; then to buy the paper, pay for the printing, etc. all at least fifteen per cent beyond what the trade would have paid; and then after all to give thirty per cent not of the net profits, but of the gross results of the sale, to a man who has merely to give the books shelf or warehouse room, and permit his apprentice to hand them over the counter to those who may ask for them; and this too copy by copy, although if the work be on any philosophical or scientific subject, it may be years before the edition is sold off. All this, I confess, must seem an hardship, and one, to which the products of industry in no other mode of exertion are subject. Yet even this is better, far better, than to attempt in any way to unite the functions of author and publisher. But the most prudent mode is to sell the copyright, at least of one or more editions, for the most that the trade will offer. By few only can a large remuneration be expected; but fifty pounds and ease of mind are of more real advantage to a literary man, than the chance of five hundred with the certainty of insult and degrading anxieties. I shall have been grievously misunderstood, if this statement should be interpreted as written with the desire of detracting from the character of booksellers or publishers. The individuals did not make the laws and customs of their trade, but as in every other trade take them as they find them. Till the evil can be proved to be removable and without the substitution of an equal or greater inconvenience, it were neither wise or manly even to complain of it. But to use it as a pretext for speaking, or even for thinking, or feeling, unkindly or opprobriously of the tradesmen, as individuals, would be something worse than unwise or even than unmanly; it would be immoral and calumnious! My motives point in a far different direction and to far other objects, as will be seen in the conclusion of the chapter.

A learned and examplary old clergyman, who many years ago went to his reward followed by the regrets and blessings of his flock, published at his own expense two volumes octavo, entitled, a new Theory of Redemption. The work was most severely handled in the Monthly or Critical Review, I forget which, and this unprovoked hostility became the good old man's favourite topic of conversation among his friends. Well! (he used to exclaim) in the second edition, I shall have an opportunity of exposing both the ignorance and the malignity of the anonymous critic. Two or three years however passed by without any tidings from the bookseller, who had undertaken the printing and publication of the work, and who was perfectly at his ease, as the author was known to be a man of large property. At length the accounts were written for; and in the course of a few weeks they were presented by the rider for the house, in person. My old friend put on his spectacles, and holding the scroll with no very firm hand, began—*Paper, so much*: O moderate enough—not at all beyond my expectation! *Printing, so much*: well! moderate enough! *Stitching, covers, advertisements, carriage, etc. so much.*—Still nothing amiss. *Selleridge* (for orthography is no necessary part of a bookseller's literary acquirements) *£3. 3s.* Bless me! only three guineas for the what d'ye call it? the *selleridge*? No more, Sir! replied the rider. Nay, but that is *too* moderate! rejoined my old friend. Only three guineas for *selling* a thousand copies of a work in two volumes? O Sir! (cries the young traveller) you have mistaken the word. There have been none of them *sold*; they have been sent back from London long ago; and this £3. 3*s.* is for the *cellarage*, or warehouse room in our book *cellar*. The work was in consequence preferred from the ominous cellar of the publisher's, to the author's garret; and on presenting a copy to an acquaintance the old gentleman used to tell the anecdote with great humour and still greater good nature.

With equal lack of worldly knowledge, I was a far more than equal sufferer for it, at the very outset of my authorship. Toward the close of the first year from the time, that in an inauspicious hour I left the friendly cloisters, and the happy grove of quiet, ever honoured Jesus College, Cambridge, I was persuaded by sundry philanthropists and anti-polemists to set on foot a periodical work, entitled The Watchman, that (according to the general motto of the work) *all might know the truth, and that the truth might make us free*! In order to exempt it from the stamp tax, and likewise to contribute as little as possible to the supposed guilt of a war against freedom, it was to be published on every eighth day, thirty-two pages, large octavo, closely printed, and price only fourpence. Accordingly with a flaming prospectus, '*Knowledge is Power*,' *etc. to cry the state of the political atmosphere*, and so forth, I set off on a tour to the North, from Bristol to Sheffield, for the purpose of procuring customers, preaching by the way in most of the great towns, as an hireless volunteer, in a blue coat and white waistcoat, that not a rag of the woman of Babylon might be seen on me. For I was at that time and long after, though a Trinitarian (i.e. *ad normam Platonis* [after the model of Plato]) in philosophy, yet a zealous Unitarian in religion; more accurately, I was a *psilanthropist*, one of those who believe our Lord to have been the real son of Joseph, and who lay the main stress on the resurrection rather than on the crucifixion. O! never can I remember those days with either shame or regret. For I was most sincere, most disinterested! My opinions were indeed in many and most important points erroneous; but my heart was single. Wealth, rank, life itself then seemed cheap to me, compared with the interests of (what I believed to be) the truth, and the will of my maker. I cannot even accuse myself of having been actuated by vanity; for in the expansion of my enthusiasm I did not think of myself at all.

My campaign commenced at Birmingham; and my first attack was on a rigid Calvinist, a tallow chandler by trade. He was a tall dingy man, in whom length was so predominant over breadth, that he might almost have been borrowed for a foundry poker. O that face! a face κατ' εμφασιν [emphatically]! I have it before me at this moment. The lank, black, twine-like hair, *pingui-nitescent*, cut in a straight line along the black stubble of his thin gunpowder eyebrows, that looked like a scorched *aftermath* from a last week's shaving. His coat collar behind in perfect unison, both of colour and lustre with the coarse yet glib cordage, that I suppose he called his hair, and which with a *bend* inward at the nape of the neck (the only approach to flexure in his whole figure) slunk in behind his waistcoast; while the countenance lank, dark, very hard, and with strong perpendicular furrows, gave me a dim notion of someone looking at me through a *used* gridiron, all soot, grease, and iron! But he was one of the thoroughbred, a true lover of liberty, and (I was informed) had proved to the satisfaction of many, that Mr Pitt was one of the horns of the second beast in the Revelations, *that spoke like a dragon*. A person, to whom one of my letters of recommendation had been addressed, was my introducer. It was a new event in my life, my first stroke in the new business I had undertaken of an author, yea, and of an author trading on his own account. My companion after some imperfect sentences and a multitude of hums and haws abandoned the cause to his client; and I commenced an harangue of half an hour to Phileleutheros [Lover of Liberty], the tallow chandler, varying my notes through the whole gamut of eloquence from the ratiocinative to the declamatory, and in the latter from the pathetic to the indignant. I argued, I described, I promised, I prophesied; and beginning with the captivity of nations I ended with the near approach of the millennium, finishing the whole with some of my own verses describing that glorious state out of the *Religious Musings*:

> . . . Such delights,
> As float to earth, permitted visitants!
> When in some hour of solemn jubilee
> The massive gates of Paradise are thrown
> Wide open: and forth come in fragments wild
> Sweet echoes of unearthly melodies,
> And odours snatch'd from beds of Amaranth,
> And they that from the crystal river of life
> Spring up on freshen'd wings, ambrosial gales!
>
> *Religious Musings*, l. 356

My taper man of lights listened with perseverant and praiseworthy patience, though (as I was afterwards told on complaining of certain gales that were not altogether ambrosial) it was a *melting* day with him. And what, Sir! (he said after a short pause) might the cost be? *Only fourpence* (O! how I felt the anti-climax, the abysmal bathos of that *fourpence*!) *only fourpence*, *Sir*, *each number*, *to be published on every eighth day*. That comes to a deal of money at the end of a year. And how much did you say there was to be for the money? *Thirty-two pages*, *Sir! large octavo*, *closely printed*. Thirty and two pages? Bless me, why except what I does in a family way on the Sabbath, that's more than I ever reads, Sir! all the year round. I am as great a one, as any man in Brummagem, Sir! for liberty and truth and all them sort of things, but as to this (no offence, I hope, Sir!) I must beg to be excused.

So ended my first canvass: from causes that I shall presently mention, I made but one other application in person. This took place at Manchester, to a stately and opulent wholesale dealer in cottons. He took my letter of introduction, and having perused it, measured me from head to foot and again from foot to head, and then asked if I had any bill or invoice of the thing; I presented my prospectus to him; he rapidly skimmed and hummed over the first side, and still more rapidly the second and concluding page; crushed it within his fingers and the palm of his hand; then most deliberately and significantly, rubbed and smoothed one part against the other; and lastly putting it into his pocket turned his back on me with an '*overrun* with these articles!' and so without another syllable retired into his counting-house. And I can truly say, to my unspeakable amusement.

This I have said, was my second and last attempt. On returning baffled from the first, in which I had vainly essayed to repeat the miracle of Orpheus with the Brummagem patriot, I dined with the tradesman who had introduced me to him. After dinner he importuned me to smoke a pipe with him, and two or three other illuminati of the same rank. I objected, both because I was engaged to spend the evening with a minister and his friends, and because I had never smoked except once or twice in my lifetime, and then it was herb tobacco mixed with Oronooko. On the assurance however that the tobacco was equally mild, and seeing too that it was of a yellow colour; (not forgetting the lamentable difficulty, I have always experienced, in saying, No! and in abstaining from what the people about me were doing) I took half a pipe, filling the lower half of the bowl with salt. I was soon however compelled to resign it, in consequence of a giddiness and distressful feeling in my eyes, which as I had drank but a single glass of ale, must, I knew, have been the effect of the tobacco. Soon after, deeming myself recovered, I sallied forth to my engagement, but the walk and the fresh air brought on all the symptoms again, and I had scarcely entered the minister's drawing-room, and opened a small packet of letters, which he had received from Bristol for me; ere I sunk back on the sofa in a sort of swoon rather than sleep. Fortunately I had found just time enough to inform him of the confused state of my feelings, and of the occasion. For here and thus I lay, my face like a wall that is whitewashing, *deathy* pale and with the cold drops of perspiration running down it from my forehead, while one after another there dropped in the different gentlemen, who had been invited to meet, and spend the evening with me, to the number of from fifteen to twenty. As the poison of tobacco acts but for a short time, I at length awoke from insensibility, and looked round on the party, my eyes dazzled by the candles which had been lighted in the interim. By way of relieving my embarrassment one of the gentlemen began the conversation, with 'Have you seen a paper today, Mr Coleridge?' 'Sir! (I replied, rubbing my eyes) I am far from convinced, that a Christian is permitted to read either newspapers or any other works of merely political and temporary interest.' This remark so ludicrously inapposite to, or rather, incongruous with, the purpose, for which I was known to have visited Birmingham, and to assist me in which they were all then met, produced an involuntary and general burst of laughter; and seldom indeed have I passed so many delightful hours, as I enjoyed in that room from the moment of that laugh to an early hour the next morning. Never, perhaps, in so mixed and numerous a party have I since heard conversation sustained with such animation, enriched with such variety of information and enlivened with such a flow of anecdote. Both then and afterwards they all joined in dissuading me from proceeding with my scheme; assured me in the most friendly and yet most flattering expressions, that the employment was neither fit for me, nor I fit for the employment. Yet if I had determined on persevering in it, they promised to exert themselves to the utmost to procure subscribers, and insisted that I should make no more applications in person, but carry on the canvass by proxy. The same hospitable reception, the same dissuasion, and (that failing) the same kind exertions in my behalf, I met with at Manchester, Derby, Nottingham, Sheffield, indeed, at every place in which I took up my sojourn. I often recall with affectionate pleasure the many respectable men who interested themselves for me, a perfect stranger to them, not a few of whom I can still name among my friends. They will bear witness for me, how opposite even then my principles were to those of Jacobinism or even of democracy, and can attest the strict accuracy of the statement which I have left on record in the 10th and 11th numbers of The Friend.

From this rememberable tour I returned with nearly a thousand names on the subscription list of the Watchman; yet more than half-convinced, that prudence dictated the abandonment of the scheme. But for this very reason I persevered in it; for I was at that period of my life so completely hag-ridden by the fear of being influenced by selfish motives that to know a mode of conduct to be the dictate of prudence was a sort of presumptive proof to my feelings, that the contrary was the dictate of duty. Accordingly, I commenced the work, which was announced in London by long bills in letters larger than had ever been seen before, and which (I have been informed, for I did not see them myself) eclipsed the glories even of the lottery puffs. But, alas! the publication of the very first number was delayed beyond the day announced for its appearance. In the second number an essay against fast-days, with a most censurable application of a text from Isaiah for its motto, lost me near five hundred of my subscribers at one blow. In the two following numbers I made enemies of all my Jacobin and Democratic patrons; for disgusted by their infidelity, and their adoption of French morals with French *psilosophy* [shallow wisdom; false philosophy]; and perhaps thinking, that charity ought to begin nearest home; instead of abusing the Government and the aristocrats chiefly or entirely, as had been expected of me, I levelled my attacks at 'modern patriotism,' and even ventured to declare my belief that whatever the motives of Ministers might have been for the sedition (or as it was then the fashion to call them, the *gagging*) bills,° yet the bills themselves would produce an effect to be desired by all the true friends of freedom, as far as they should contribute to deter men from openly declaiming on subjects, the principles of which they had never bottomed, and from 'pleading to the poor and ignorant, instead of pleading *for* them.' At the same time I avowed my conviction, that national education and a concurring spread of the Gospel were the indispensable condition of any true political amelioration. Thus by the time the seventh number was published, I had the mortification (but why should I say this, when in truth I cared too little for anything that concerned my worldly interests to be at all mortified about it?) of seeing the preceding numbers exposed in sundry old iron shops for a penny a piece. At the ninth number I dropped the work. But from the London publisher I could not obtain a shilling; he was a —— and set me at defiance. From other places I procured but little, and after such delays as rendered that little worth nothing: and I should have been inevitably thrown into jail by my Bristol printer, who refused to wait even for a month, for a sum between eighty and ninety pounds, if the money had not been paid for me by a man by no means affluent, a dear friend who attached himself to me from my first arrival at Bristol, who has continued my friend with a fidelity unconquered by time or even by my own apparent neglect; a friend from whom I never received an advice that was not wise, or a remonstrance that was not gentle and affectionate.

Conscientiously an opponent of the first revolutionary war, yet with my eyes thoroughly opened to the true character and impotence of the favourers of revolutionary principles in England, principles which I held in abhorrence (for it was part of my political creed, that whoever ceased to act as an individual by making himself a member of any society not sanctioned by his Government, forfeited the rights of a citizen)—a vehement anti-ministerialist, but after the invasion of Switzerland a more vehement anti-Gallican, and still more intensely an anti-Jacobin, I retired to a cottage at Stowey, and provided for my scanty maintenance by writing verses for a London morning paper. I saw plainly, that literature was not a profession, by which I could expect to live; for I could not disguise from myself, that whatever my talents might or might not be in other respects, yet they were not of the sort that could enable me to become a popular writer; and that whatever my opinions might be in themselves, they were almost equidistant from all the three prominent parties, the Pittites, the Foxites, and the Democrats. Of the unsaleable nature of my writings I had an amusing memento one morning from our own servant girl. For happening to rise at an earlier hour than usual, I observed her putting an extravagant quantity of paper into the grate in order to light the fire, and mildly checked her for her wastefulness; la, Sir! (replied poor Nanny) why, it is only 'Watchmen.'

I now devoted myself to poetry and to the study of ethics and psychology; and so profound was my admiration at this time of Hartley's Essay on Man, that I gave his name to my first-born. In addition to the gentleman, my neighbour, whose garden joined on to my little orchard, and the cultivation of whose friendship had been my sole motive in choosing Stowey for my residence, I was so fortunate as to acquire, shortly after my settlement there, an invaluable blessing in the society and neighbourhood of one, to whom I could look up with equal reverence, whether I regarded him as a poet, a philosopher, or a man. His conversation extended to almost all subjects, except physics and politics; with the latter he never troubled himself. Yet neither my retirement nor my utter abstraction from all the disputes of the day could secure me in those jealous times from suspicion and obloquy, which did not stop at me, but extended to my excellent friend, whose perfect innocence was even adduced as a proof of his guilt. One of the many busy *sycophants** of that day (I here use the word sycophant, in its original sense, as a wretch who flatters the prevailing party by informing against his neighbours, under pretence that they are exporters of prohibited figs or fancies! for the moral application of the term it matters not which)—one of these sycophantic law-mongrels, discoursing on the politics of the neighbourhood, uttered the following deep remark: 'As to Coleridge, there is not so much harm in

* *Συκους φαινειν* [*sykous phainein*], to show or detect figs, the exportation of which from Attica was forbidden by the laws.

him, for he is a whirl-brain that talks whatever comes uppermost; but that ——! he is the dark traitor. You never hear *him* say a syllable on the subject.'

Now that the hand of providence has disciplined all Europe into sobriety, as men tame wild elephants, by alternate blows and caresses; now that Englishmen of all classes are restored to their old English notions and feelings; it will with difficulty be credited, how great an influence was at that time possessed and exerted by the spirit of secret defamation (the too constant attendant on party zeal!) during the restless interim from 1793 to the commencement of the Addington administration, or the year before the truce of Amiens. For by the latter period the minds of the partisans, exhausted by excess of stimulation and humbled by mutual disappointment, had become languid. The same causes, that inclined the nation to peace, disposed the individuals to reconciliation. Both parties had found themselves in the wrong. The one had confessedly mistaken the moral character of the revolution, and the other had miscalculated both its moral and its physical resources. The experiment was made at the price of great, almost we may say, of humiliating sacrifices; and wise men foresaw that it would fail, at least in its direct and ostensible object. Yet it was purchased cheaply, and realized an object of equal value, and, if possible, of still more vital importance. For it brought about a national unanimity unexampled in our history since the reign of Elizabeth; and providence, never wanting to a good work when men have done their parts, soon provided a common focus in the cause of Spain, which made us all once more Englishmen by at once gratifying and correcting the predilections of both parties. The sincere reverers of the throne felt the cause of loyalty ennobled by its alliance with that of freedom; while the *honest* zealots of the people could not but admit, that freedom itself assumed a more winning form, humanized by loyalty and consecrated by religious principle. The youthful enthusiasts who, flattered by the morning rainbow of the French Revolution, had made a boast of expatriating their hopes and fears, now disciplined by the succeeding storms and sobered by increase of years, had been taught to prize and honour the spirit of nationality as the best safeguard of national independence, and this again as the absolute prerequisite and necessary basis of popular rights.

If in Spain too disappointment has nipped our too forward expectations, yet all is not destroyed that is checked. The crop was perhaps springing up too rank in the stalk, to *kern* well; and there were, doubtless, symptoms of the Gallican blight on it. If superstition and despotism have been suffered to let in their wolvish sheep to trample and eat it down even to the surface, yet the roots remain alive, and the second growth may prove all the stronger and healthier for the temporary interruption. At all events, to us heaven has been just and gracious. The people of England did their best, and have received their rewards. Long may we continue to deserve it! Causes, which it had been too generally the habit of former statesmen to regard as belonging to another world, are now admitted by all ranks to have been the main agents of our success. 'We fought from heaven; the stars in their courses fought against Sisera.' If then unanimity grounded on moral feelings has been among the least equivocal sources of our national glory, that man deserves the esteem of his countrymen, even as patriots, who devotes his life and the utmost efforts of his intellect to the preservation and continuance of that unanimity by the disclosure and establishment of principles. For by these all opinions must be ultimately tried; and (as the feelings of men are worthy of regard only as far as they are the representatives of their fixed opinions) on the knowledge of these all unanimity, not accidental and fleeting, must be grounded. Let the scholar, who doubts this assertion, refer only to the speeches and writings of Edmund Burke at the commencement of the American War, and compare them with his speeches and writings at the commencement of the French Revolution. He will find the principles exactly the same and the deductions the same; but the practical inferences almost opposite, in the one case from those drawn in the other; yet in both equally legitimate and in both equally confirmed by the results. Whence gained he this superiority of foresight? Whence arose the striking difference, and in most instances even the discrepancy between the grounds assigned by him, and by those who voted with him, on the same questions? How are we to explain the notorious fact, that the speeches and writings of Edmund Burke are more interesting at the present day, than they were found at the time of their first publication; while those of his illustrious confederates are either forgotten, or exist only to furnish proofs, that the same conclusion, which one man had deduced scientifically, may be brought out by another in consequence of errors that luckily chanced to neutralize each other. It would be unhandsome as a conjecture, even were it not, as it actually is, false in point of fact, to attribute this difference to deficiency of talent on the part of Burke's friends, or of experience, or of historical knowledge. The satisfactory solution is, that Edmund Burke possessed and had sedulously sharpened that eye, which sees all things, actions, and events, in relation to the laws that determine their existence and circumscribe their possibility. He referred habitually to principles. He was a scientific statesman; and therefore a seer. For every principle contains in itself the germs of a prophecy; and as the prophetic power is the essential privilege of science, so the fulfilment of its oracles supplies the outward and (to men in general) the only test of its claim to the title. Wearisome as Burke's refinements appeared to his Parliamentary auditors, yet the cultivated classes throughout Europe have reason to be thankful, that

> . . . he went on refining,
> And thought of convincing, while they thought of dining.

Our very sign boards (said an illustrious friend to me) give evidence, that there has been a Titian in the world. In like manner, not only the debates in Parliament, not only our proclamations and state papers, but the essays and leading paragraphs of our journals are so many remembrancers of Edmund Burke. Of this the reader may easily convince himself, if either by recollection or reference he will compare the Opposition newspapers at the commencement and during the five or six following years of the French Revolution with the sentiments, and grounds of argument assumed in the same class of journals at present, and for some years past.

Whether the spirit of Jacobinism, which the writings of Burke exorcised from the higher and from the literary classes, may not like the ghost in Hamlet, be heard moving and mining in the underground chambers with an activity the more dangerous because less noisy, may admit of a question. I have given my opinions on this point, and the grounds of them, in my letters to Judge Fletcher occasioned by his charge to the Wexford grand jury, and published in the *Courier*. Be this as it may, the evil spirit of jealousy, and with it the Cerberean whelps of feud and slander, no longer walk their rounds, in cultivated society.

Far different were the days to which these anecdotes have carried me back. The dark guesses of some zealous quidnunc met with so congenial a soil in the grave alarm of a titled Dogberry of our neighbourhood, that a spy was actually sent down from the Government *pour surveillance* [to keep watch over] of myself and friend. There must have been not only abundance, but variety of these 'honourable men' at the disposal of Ministers: for this proved a very honest fellow. After three weeks' truly Indian perseverance in tracking us (for we were commonly together) during all which time seldom were we out of doors, but he contrived to be within hearing (and all the while utterly unsuspected; how indeed could such a suspicion enter our fancies?) he not only rejected Sir Dogberry's request that he would try yet a little longer, but declared to him his belief, that both my friend and myself were as good subjects, for aught he could discover to the contrary, as any in His Majesty's dominions. He had repeatedly hid himself, he said, for hours together behind a bank at the sea-side (our favourite seat) and overheard our conversation. At first he fancied, that we were aware of our danger; for he often heard me talk of one *Spy Nozy* [Spinoza], which he was inclined to interpret of himself, and of a remarkable feature belonging to him; but he was speedily convinced that it was the name of a man who had made a book and lived long ago. Our talk ran most upon books, and we were perpetually desiring each other to look at this, and to listen to that; but he could not catch a word about politics. Once he had joined me on the road (this occurred, as I was returning home alone from my friend's house, which was about three miles from my own cottage); and passing himself off as a traveller, he had entered into conversation with me, and talked of purpose in a *democrat* way in order to draw me out. The result, it appears, not only convinced him that I was no friend of Jacobinism; but (he added) I had 'plainly made it out to be such a silly as well as wicked thing, that he felt ashamed, though he had only *put it on*.' I distinctly remembered the occurrence, and had mentioned it immediately on my return, repeating what the traveller with his Bardolph nose had said, with my own answer; and so little did I suspect the true object of my 'tempter ere accuser,' that I expressed with no small pleasure my hope and belief, that the conversation had been of some service to the poor misled malcontent. This incident therefore prevented all doubt as to the truth of the report, which through a friendly medium came to me from the master of the village inn, who had been ordered to entertain the Government gentleman in his best manner, but above all to be silent concerning such a person being in his house. At length, he received Sir Dogberry's commands to accompany his guest at the final interview; and after the absolving suffrage of the gentleman honoured with the confidence of Ministers answered, as follows, to the following queries. *D*. Well, landlord! and what do you know of the person in question? *L*. I see him often pass by with maister ——, my landlord (i.e. the owner of the house) and sometimes with the newcomers at Holford; but I never said a word to him or he to me. *D*. But do you not know, that he has distributed papers and hand-bills of a seditious nature among the common people! *L*. No, your honour! I never heard of such a thing. *D*. Have you not seen this Mr Coleridge, or heard of, his haranguing and talking to knots and clusters of the inhabitants?—What are you grinning at, Sir! *L.* Beg your honour's pardon! but I was only thinking, how they'd have stared at him. If what I have heard be true, your honour! they would not have understood a word, he said. When our vicar was here, Dr L. the master of the great school and canon of Windsor, there was a great dinner party at maister ——'s; and one of the farmers, that was there, told us that he and the Doctor talked real Hebrew Greek at each other for an hour together after dinner. *D.* Answer the question, Sir! Does he ever harangue the people? *L.* I hope, your honour an't angry with me. I can say no more than I know. I never saw him talking with anyone, but my landlord, and our curate, and the strange gentleman. *D.* Has he not been seen wandering on the hills towards the Channel, and along the shore, with books and papers in his hand, taking charts and maps of the country? *L.* Why, as to that, your honour! I own, I have heard; I am sure, I would not wish to say ill of anybody; but it is certain, that I have heard—*D.* Speak out man! don't be afraid, you are doing your duty to your King and Government. What have you heard? *L.* Why, folks do say, your honour! as how that he is a poet, and that he is going to put Quantock and all about here in print; and as they be so much together, I suppose that the strange gentleman has some *consarn* in the business.—So ended this formidable inquisition, the latter part of which alone requires explanation, and at the same time entitles the anecdote to a place in my literary life. I had considered it as a defect in the admirable poem of the Task, that the subject, which gives the title to the work, was not, and indeed could not be, carried on beyond the three or four first pages, and that throughout the poem the connections are frequently awkward, and the transitions abrupt and arbitrary. I sought for a subject, that should give equal room and freedom for description, incident, and impassioned reflections on men, nature, and society, yet supply in itself a natural connection to the parts, and unity to the whole. Such a subject I conceived myself to have found in a stream, traced from its source in the hills among the yellow-red moss and conical glass-shaped tufts of bent, to the first break or fall, where its drops became audible, and it begins to form a channel; thence to the peat and turf barn, itself built of the same dark squares as it sheltered; to the sheep-fold; to the first cultivated plot of ground; to the lonely cottage and its bleak garden won from the heath; to the hamlet, the villages, the market town, the manufactories, and the seaport. My walks therefore were almost daily on the top of Quantock, and among its sloping coombs. With my pencil and memorandum book in my hand, I was *making studies*, as the artists call them, and often moulding my thoughts into verse, with the objects and imagery immediately before my senses. Many circumstances, evil and good, intervened to prevent the completion of the poem, which was to have been entitled 'The Brook.' Had I finished the work, it was my purpose in the heat of the moment to have dedicated it to our then committee of public safety as containing the charts and maps, with which I was to have supplied the French Government in aid of their plans of invasion. And these too for a tract of coast that from Clevedon to Minehead scarcely permits the approach of a fishing boat!

All my experience from my first entrance into life to the present hour is in favour of the warning maxim, that the man, who opposes *in toto* [altogether] the political or religious zealots of his age, is safer from their obloquy than he who differs from them in one or two points or perhaps only in degree. By that transfer of the feelings of private life into the discussion of public questions, which is the queen bee in the hive of party fanaticism, the partisan has more sympathy with an intemperate opposite than with a moderate friend. We now enjoy an intermission, and long may it continue! In addition to far higher and more important merits, our present Bible societies and other numerous associations for national or charitable objects, may serve perhaps to carry off the superfluous activity and fervour of stirring minds in innocent hyperboles and the bustle of management. But the poison-tree is not dead, though the sap may for a season have subsided to its roots. At least let us not be lulled into such a notion of our entire security, as not to keep watch and ward, even on our best feelings. I have seen gross intolerance shown in support of toleration; sectarian antipathy most obtrusively displayed in the promotion of an undistinguishing comprehension of sects; and acts of cruelty (I had almost said) of treachery, committed in furtherance of an object vitally important to the cause of humanity; and all this by men too of naturally kind dispositions and exemplary conduct.

The magic rod of fanaticism is preserved in the very *adyta* [innermost sanctuaries] of human nature; and needs only the re-exciting warmth of a master hand to bud forth afresh and produce the old fruits. The horror of the Peasants' War in Germany, and the direful effects of the Anabaptists' tenets (which differed only from those of Jacobinism by the substitution of theological for philosophical jargon) struck all Europe for a time with affright. Yet little more than a century was sufficient to obliterate all effective memory of these events. The same principles with similar though less dreadful consequences were again at work from the imprisonment of the first Charles to the restoration of his son. The fanatic maxim of extirpating fanaticism by persecution produced a civil war. The war ended in the victory of the insurgents; but the temper survived, and Milton had abundant grounds for asserting, that 'Presbyter was but old priest writ large!' One good result, thank heaven! of this zealotry was the re-establishment of the Church. And now it might have been hoped, that the mischievous spirit would have been bound for a season, 'and a seal set upon him that he might deceive the nation no more.' But no! The ball of persecution was taken up with undiminished vigour by the persecuted. The same fanatic principle, that under the solemn oath and covenant had turned cathedrals into stables, destroyed the rarest trophies of art and ancestral piety, and hunted the brightest ornaments of learning and religion into holes and corners, now marched under episcopal banners, and having first crowded the prisons of England emptied its whole vial of wrath on the miserable Covenanters of Scotland. (Laing's History of Scotland.—Walter Scott's bards, ballads, etc.) A merciful providence at length constrained both parties to join against a common enemy. A wise Government followed; and the established Church became, and now is, not only the brightest example, but our best and only sure bulwark, of toleration! The true and indispensable bank against a new inundation of persecuting zeal—*Esto perpetua* [May it last forever]!

A long interval of quiet succeeded; or rather, the exhaustion had produced a cold fit of the agüe which was symptomatized by indifference among the many, and a tendency to infidelity or scepticism in the educated classes. At length those feelings of disgust and hatred, which for a brief while the multitude had attached to the crimes and absurdities of sectarian and democratic fanaticism, were transferred to the oppressive privileges of the *noblesse*, and the luxury, intrigues and favouritism of the continental courts. The same principles dressed in the ostentatious garb of a fashionable philosophy once more rose triumphant and effected the French Revolution. And have we not within the last three or four years had reason to apprehend, that the detestable maxims and correspondent measures of the late French despotism had already bedimmed the public recollections of democratic frenzy; had drawn off to other objects the electric force of the feelings which had massed and upheld those recollections; and that a favourable concurrence of occasions was alone wanting to awaken the thunder and precipitate the lightning from the opposite quarter of the political heaven? (See The Friend, p. 110.)

In part from constitutional indolence, which in the very heyday of hope had kept my enthusiasm in check, but still more from the habits and influences of a classical education and academic pursuits, scarcely had a year elapsed from the commencement of my literary and political adventures before my mind sunk into a state of thorough disgust and despondency, both with regard to the disputes and the parties disputant. With more than poetic feeling I exclaimed:

The sensual and the dark rebel in vain,
Slaves by their own compulsion! In mad game
They break their manacles, to wear the *name*
Of freedom, graven on an heavier chain.
O liberty! with profitless endeavour
Have I pursued thee many a weary hour;
But thou nor swell'st the victor's pomp, nor ever
Didst breathe thy soul in forms of human power!
Alike from all, howe'er they praise thee
(Nor prayer nor boastful name delays thee)
From superstition's harpy minions
And factious blasphemy's obscener slaves,
Thou speedest on thy cherub pinions,
The guide of homeless winds and playmate of the waves!

France, *a Palinodia*

I retired to a cottage in Somersetshire at the foot of Quantock, and devoted my thoughts and studies to the foundations of religion and morals. Here I found myself all afloat. Doubts rushed in; broke upon me 'from the fountains of the great deep,' and fell 'from the windows of heaven.' The fontal truths of natural religion and the books of Revelation alike contributed to the flood; and it was long ere my ark touched on an Ararat, and rested. The idea of the Supreme Being appeared to me to be as necessarily implied in all particular modes of being as the idea of infinite space in all the geometrical figures by which space is limited. I was pleased with the Cartesian opinion, that the idea of God is distinguished from all other ideas by involving its reality; but I was not wholly satisfied. I began then to ask myself, what proof I had of the outward existence of anything? Of this sheet of paper for instance, as a thing in itself, separate from the phenomenon or image in my perception. I saw, that in the nature of things such proof is impossible; and that of all modes of being, that are not objects of the senses, the existence is assumed by a logical necessity arising from the constitution of the mind itself, by the absence of all motive to doubt it, not from any absolute contradiction in the supposition of the contrary. Still the existence of a being, the ground of all existence, was not yet the existence of a moral creator, and governor. 'In the position, that all reality is either contained in the necessary being as an attribute, or exists through him, as its ground, it remains undecided whether the properties of intelligence and will are to be referred to the Supreme Being in the former or only in the latter sense; as inherent attributes, or only as consequences that have existence in other things through him. Thus organization, and motion, are regarded as *from* God not *in* God. Were the latter the truth, then notwithstanding all the pre-eminence which must be assigned to the Eternal First from the sufficiency, unity, and independence of his being, as the dread ground of the universe, his nature would yet fall far short of that, which we are bound to comprehend in the idea of God. For without any knowledge or determining resolve of its own it would only be a blind necessary ground of other things and other spirits; and thus would be distinguished from the fate of certain ancient philosophers in no respect, but that of being more definitely and intelligibly described.' Kant's *einzig möglicher Beweisgrund: vermischte Schriften*, Zweiter Band, § 102, and 103.

For a very long time indeed I could not reconcile personality with infinity; and my head was with Spinoza, though my whole heart remained with Paul and John. Yet there had dawned upon me, even before I had met with the Critique of the Pure Reason, a certain guiding light. If the mere intellect could make no certain discovery of a holy and intelligent first cause, it might yet supply a demonstration, that no legitimate argument could be drawn from the intellect *against* its truth. And what is this more than St Paul's assertion, that by wisdom (more properly translated by the powers of reasoning) no man ever arrived at the knowledge of God? What more than the sublimest, and probably the oldest, book on earth has taught us,

> Silver and gold man searcheth out:
> Bringeth the ore out of the earth, and darkness into light.
>
> But where findeth he wisdom?
> Where is the place of understanding?
>
> The abyss crieth; it is not in me!
> Ocean echoeth back; not in me!
>
> Whence then cometh wisdom?
> Where dwelleth understanding?

Hidden from the eyes of the living:
Kept secret from the fowls of heaven!

Hell and death answer;
We have heard the rumour thereof from afar!

God marketh out the road to it;
God knoweth its abiding place!

He beholdeth the ends of the earth;
He surveyeth what is beneath the heavens!

And as he weighed out the winds, and measured the sea,
And appointed laws to the rain,
And a path to the thunder,
A path to the flashes of the lightning!

Then did he see it,
And he counted it;
He searched into the depth thereof,
And with a line did he compass it round!

But to man he said,
The fear of the Lord is wisdom for THEE!
And to avoid evil,
That is *thy* understanding.

Job, Chap. 28th

I became convinced, that religion, as both the corner-stone and the keystone of morality, must have a moral origin; so far at least, that the evidence of its doctrines could not, like the truths of abstract science, be wholly independent of the will. It were therefore to be expected, that its fundamental truth would be such as might be denied; though only, by the fool, and even by the fool from the madness of the heart alone!

The question then concerning our faith in the existence of a God, not only as the ground of the universe by his essence, but as its maker and judge by his wisdom and holy will, appeared to stand thus. The sciential reason, whose objects are purely theoretical, remains neutral, as long as its name and semblance are not usurped by the opponents of the doctrine. But it then becomes an effective ally by exposing the false show of demonstration, or by evincing the equal demonstrability of the contrary from premisses equally logical. The understanding meantime suggests, the analogy of experience facilitates, the belief. Nature excites and recalls it, as by a perpetual revelation. Our feelings almost necessitate it; and the law of conscience peremptorily commands it. The arguments, that at all apply to it, are in its favour; and there is nothing against it, but its own sublimity. It could not be intellectually more evident without becoming morally less effective; without counteracting its own end by sacrificing the life of faith to the cold mechanism of a worthless because compulsory assent. The belief of a God and a future state (if a passive acquiescence may be flattered with the name of belief) does not indeed always beget a good heart; but a good heart so naturally begets the belief, that the very few exceptions must be regarded as strange anomalies from strange and unfortunate circumstances.

From these premisses I proceeded to draw the following conclusions. First, that having once fully admitted the existence of an infinite yet self-conscious Creator, we are not allowed to ground the irrationality of any other article of faith on arguments which would equally prove that to be irrational, which we had allowed to be real. Secondly, that whatever is deducible from the admission of a self-comprehending and creative spirit may be legitimately used in proof of the possibility of any further mystery concerning the divine nature. '*Possibilitatem* mysteriorum, (Trinitatis, etc.) contra insultus Infidelium et Hereticorum a contradictionibus vindico; haud quidem *veritatem*, quae revelatione solâ stabiliri possit [I am freeing the *possibility* of mysteries (of the Trinity etc.) from contradictions, against the attacks of Unbelievers and Heretics: not, indeed, the *truth*, which can be established only by revelation (*CC*)]'; says Leibniz in a letter to his Duke. He then adds the following just and important remark. 'In vain will tradition or texts of Scripture be adduced in support of a doctrine, donec clava impossibilitatis et contradictionis e manibus horum Herculum extorta fuerit [until the club of impossibility and contradiction has been wrested from the hands of these Herculeses (*CC*)]. For the heretic will still reply, that texts, the literal sense of which is not so much *above* as directly *against* all reason, must be understood figuratively, as Herod is a fox, etc.'

These principles I held, philosophically, while in respect of revealed religion I remained a zealous Unitarian. I considered the idea of the Trinity a fair scholastic inference from the being of God, as a creative intelligence; and that it was therefore entitled to the rank of an esoteric doctrine of natural religion. But seeing in the same no practical or moral bearing, I confined it to the schools of philosophy.

The admission of the logos, as *hypostasized* (i.e. neither a mere attribute or a personification) in no respect removed my doubts concerning the incarnation and the redemption by the cross; which I could neither reconcile in reason with the impassiveness of the Divine Being, nor in my moral feelings with the sacred distinction between things and persons, the vicarious payment of a debt and the vicarious expiation of guilt. A more thorough revolution in my philosophic principles, and a deeper insight into my own heart, were yet wanting. Nevertheless, I cannot doubt, that the difference of my metaphysical notions from those of Unitarians in general contributed to my final reconversion to the whole truth in Christ; even as according to his own confession the books of certain Platonic philosophers (*libri quorundam Platonicorum*) commenced the rescue of St Augustine's faith from the same error aggravated by the far darker accompaniment of the Manichaean heresy.

While my mind was thus perplexed, by a gracious providence for which I can never be sufficiently grateful, the generous and munificent patronage of Mr Josiah, and Mr Thomas Wedgwood enabled me to finish my education in Germany. Instead of troubling others with my own crude notions and juvenile compositions I was thenceforward better employed in attempting to store my own head with the wisdom of others. I made the best use of my time and means; and there is therefore no period of my life on which I can look back with such unmingled satisfaction. After acquiring a tolerable sufficiency in the German language* at Ratzeburg, which with my voyage and journey thither I have described in The Friend, I proceeded through Hanover to Göttingen.

* To those, who design to acquire the language of a country in the country itself, it may be useful, if I mention the incalculable advantage which I derived from learning all the words, that could possibly be so learnt, with the objects before me, and without the intermediation of the English terms. It was a regular part of my morning studies for the first six weeks of my residence at Ratzeburg, to accompany the good and kind old pastor, with whom I lived, from the cellar to the roof, through gardens, farmyard, etc. and to call every, the minutest, thing by its German name. Advertisements, farces, jest-books, and the conversation of children while I was at play with them, contributed their share to a more home-like acquaintance with the language, than I could have acquired from works of polite literature alone, or even from polite society. There is a passage of *hearty* sound sense in Luther's German letter on interpretation, to the translation of which I shall prefix, for the sake of those who read the German, yet are not likely to have dipped often in the massive folios of this heroic reformer, the simple, sinewy, idiomatic words of the original. 'Denn man muss nicht die Buchstaben in der Lateinischen Sprache fragen wie man soll Deutsch reden; sondern man muss die Mutter im Hause,

Here I regularly attended the lectures on physiology in the morning, and on natural history in the evening, under Blumenbach, a name as dear to every Englishman who has studied at that university, as it is venerable to men of science throughout Europe! Eichhorn's lectures on the New Testament were repeated to me from notes by a student from Ratzeburg, a young man of sound learning and indefatigable industry, who is now, I believe, a professor of the oriental languages at Heidelberg. But my chief efforts were directed towards a grounded knowledge of the German language and literature. From professor Tychsen I received as many lessons in the Gothic of Ulphilas as sufficed to make me acquainted with its grammar, and the radical words of most frequent occurrence; and with the occasional assistance of the same philosophical linguist, I read through* Ottfried's metrical

die Kinder auf den Gassen, den gemeinen Mann auf dem Markte, darum fragen: und denselbigen auf das Maul sehen wie sie reden, und darnach dolmetschen. So verstehen sie es denn, und merken dass man Deutsch mit ihnen redet.'

TRANSLATION

For one must not ask the letters in the Latin tongue, how one ought to speak German; but one must ask the mother in the house, the children in the lanes and alleys, the common man in the market, concerning this; yea, and look at the *moves* of their mouths while they are talking, and thereafter interpret. They understand you then, and mark that one talks German with them.

* This paraphrase, written about the time of Charlemagne, is by no means deficient in occasional passages of considerable poetic merit. There is a flow, and a tender enthusiasm in the following lines (at the conclusion of Chapter V) which even in the translation will not, I flatter myself, fail to interest the reader. Ottfried is describing the circumstances immediately following the birth of our Lord.

She gave with joy her virgin breast;
She hid it not, she bared the breast,
Which suckled that divinest babe!
Blessed, blessed were the breasts
Which the Saviour infant kiss'd;
And blessed, blessed was the mother
Who wrapp'd his limbs in swaddling clothes,
Singing placed him on her lap,
Hung o'er him with her looks of love,
And soothed him with a lulling motion.
Blessed! for she shelter'd him
From the damp and chilling air;
Blessed, blessed! for she lay
With such a babe in one blest bed,
Close as babes and mothers lie!
Blessed, blessed evermore,
With her virgin lips she kiss'd,

[*cont.*]

paraphrase of the Gospel, and the most important remains of the Theotiscan, or the transitional state of the Teutonic language from the Gothic to the old German of the Swabian period. Of this period (the polished dialect of which is analogous to that of our Chaucer, and which leaves the philosophic student in doubt, whether the language has not since then lost more in sweetness and flexibility, than it has gained in condensation and copiousness) I read with sedulous accuracy the Minnesinger (or singers of love, the Provençal poets of the Swabian court) and the metrical romances; and then laboured through sufficient specimens of the *master singers*, their degenerate successors; not however without occasional pleasure from the rude, yet interesting strains of Hans Sachs the cobbler of Nuremberg. Of this man's genius five folio volumes with double columns are extant in print, and nearly an equal number in manuscript; yet the indefatigable bard takes care to inform his readers, that he never made a shoe the less, but had virtuously reared a large family by the labour of his hands.

In Pindar, Chaucer, Dante, Milton, etc. etc. we have instances of the close connection of poetic genius with the love of liberty and of genuine reformation. The moral sense at least will not be outraged, if I add to the list the name of this honest shoemaker (a trade by the by remarkable for the production of philosophers and poets). His poem entitled the Morning Star, was the very first publication that appeared in praise and support of Luther; and an excellent hymn of Hans Sachs, which has been deservedly translated into almost all the European languages, was commonly sung in the Protestant churches, whenever the heroic reformer visited them.

In Luther's own German writings, and eminently in his translation of the Bible, the German language commenced. I mean the language as it is at present written; that which is called the High German, as

With her arms, and to her breast
She embraced the babe divine,
Her babe divine the virgin mother!
There lives not on this ring of earth
A mortal, that can sing her praise.
Mighty mother, virgin pure,
In the darkness and the night
For us she *bore* the heavenly Lord!

Most interesting is it to consider the effect, when the feelings are wrought above the natural pitch by the belief of something mysterious, while all the images are purely natural. Then it is, that religion and poetry strike deepest.

contradistinguished from the Platt-Teutsch, the dialect of the flat or northern countries, and from the Ober-Teutsch, the language of the middle and southern Germany. The High German is indeed a *lingua communis* [common tongue], not actually the native language of any province, but the choice and fragrancy of all the dialects. From this cause it is at once the most copious and the most grammatical of all the European tongues.

Within less than a century after Luther's death the German was inundated with pedantic barbarisms. A few volumes of this period I read through from motives of curiosity; for it is not easy to imagine anything more fantastic, than the very appearance of their pages. Almost every third word is a Latin word with a Germanized ending, the Latin portion being always printed in Roman letters, while in the last syllable the German character is retained.

At length, about the year 1620, Opitz arose, whose genius more nearly resembled that of Dryden than any other poet, who at present occurs to my recollection. In the opinion of Lessing, the most acute of critics, and of Adelung, the first of lexicographers, Opitz, and the Silesian poets, his followers, not only restored the language, but still remain the models of pure diction. A stranger has no vote on such a question; but after repeated perusal of the work my feelings justified the verdict, and I seemed to have acquired from them a sort of *tact* for what is genuine in the style of later writers.

Of the splendid era, which commenced with Gellert, Klopstock, Ramler, Lessing, and their compeers, I need not speak. With the opportunities which I enjoyed, it would have been disgraceful not to have been familiar with their writings; and I have already said as much, as the present biographical sketch requires, concerning the German philosophers, whose works, for the greater part, I became acquainted with at a far later period.

Soon after my return from Germany I was solicited to undertake the literary and political department in the Morning Post; and I acceded to the proposal on the condition, that the paper should thenceforwards be conducted on certain fixed and announced principles, and that I should be neither obliged or requested to deviate from them in favour of any party or any event. In consequence, that journal became and for many years continued anti-Ministerial indeed, yet with a very qualified approbation of the Opposition, and with far greater earnestness and zeal both anti-Jacobin and anti-Gallican. To this hour I cannot find reason to approve of the first war either in its commencement or its conduct. Nor can I understand, with what reason either Mr Percival (whom I am singular enough to regard as the best and wisest Minister of this reign) or the present administration, can be said to have pursued the plans of Mr Pitt. The love of their country, and perseverant hostility to French principles and French ambition are indeed honourable qualities common to them and to their predecessor. But it appears to me as clear as the evidence of facts can render any question of history, that the successes of the Percival and of the existing ministry have been owing to their having pursued measures the direct contrary to Mr Pitt's. Such for instance are the concentration of the national force to one object; the abandonment of the subsidising policy, so far at least as neither to goad or bribe the continental courts into war, till the convictions of their subjects had rendered it a war of their own seeking; and above all, in their manly and generous reliance on the good sense of the English people, and on that loyalty which is linked to the very heart* of the nation by the system of credit and the interdependence of property.

* Lord Grenville has lately reasserted (in the House of Lords) the imminent danger of a revolution in the earlier part of the war against France. I doubt not, that his Lordship is sincere; and it must be flattering to his feelings to believe it. But where are the evidences of the danger, to which a future historian can appeal? Or must he rest on an assertion? Let me be permitted to extract a passage on the subject from The Friend. 'I have said that to withstand the arguments of the lawless, the Anti-Jacobins proposed to suspend the law, and by the interposition of a particular statute to eclipse the blessed light of the universal sun, that spies and informers might tyrannize and escape in the ominous darkness. Oh! if these mistaken men intoxicated and bewildered with the panic of property, which they themselves were the chief agents in exciting, had ever lived in a country where there really existed a general disposition to change and rebellion! Had they ever travelled through Sicily; or through France at the first coming on of the revolution; or even alas! through too many of the provinces of a sister island; they could not but have shrunk from their own declarations concerning the state of feeling, and opinion at that time predominant throughout Great Britain. There was a time (heaven grant! that that time may have passed by) when by crossing a narrow strait, they might have learnt the true symptoms of approaching danger, and have secured themselves from mistaking the meetings and idle rant of such sedition, as shrunk appalled from the sight of a constable, for the dire murmuring and strange consternation which precedes the storm or earthquake of national discord. Not only in coffee-houses and public theatres, but even at the tables of the wealthy, they would have heard the advocates of existing Government defend their cause in the language and with the tone of men, who are conscious that they are in a minority. But in England, when the alarm was at its highest, there was not a city, no not a town or village, in which a man suspected of holding democratic principles could move abroad without receiving some unpleasant proof of the hatred, in which his supposed opinions were held by the great majority of the people; and the only instances of popular excess and indignation were in favour of the Government and the established Church. But why need I appeal to these invidious facts? Turn over the pages of history and seek for a single instance of a revolution having

Be this as it may, I am persuaded that the Morning Post proved a far more useful ally to the Government in its most important objects, in consequence of its being generally considered as moderately anti-Ministerial, than if it had been the avowed eulogist of Mr Pitt. (The few, whose curiosity or fancy should lead them to turn over the journals of that date, may find a small proof of this in the frequent charges made by the Morning Chronicle, that such and such essays or leading paragraphs had been sent from the Treasury.) The rapid and unusual increase in the sale of the Morning Post is a sufficient pledge, that genuine impartiality with a respectable portion of literary talent will secure the success of a newspaper without the aid of party or Ministerial patronage. But by impartiality I mean an honest and enlightened adherence to a code of intelligible principles previously announced, and faithfully referred to in support of every judgement on men and events; not indiscriminate abuse, not the indulgence of an editor's own malignant passions, and still less, if that be possible, a determination to make money by flattering the envy and cupidity, the vindictive restlessness and self-conceit of the half-witted vulgar; a determination almost fiendish, but which, I have been informed, has been boastfully avowed by one man, the most notorious of these mob-sycophants! From the commencement of the Addington administration to the present day, whatever I have written in the Morning Post, or (after that paper was transferred to other proprietors) in the

been effected without the concurrence of either the nobles, or the ecclesiastics, or the monied classes, in any country, in which the influences of property had ever been predominant, and where the interests of the proprietors were interlinked! Examine the revolution of the Belgic provinces under Philip II; the civil wars of France in the preceding generation; the history of the American Revolution, or the yet more recent events in Sweden and in Spain; and it will be scarcely possible not to perceive, that in England from 1791 to the Peace of Amiens there were neither tendencies to confederacy nor actual confederacies, against which the existing laws had not provided sufficient safeguards and an ample punishment. But alas! the panic of property had been struck in the first instance for party purposes; and when it became general, its propagators caught it themselves and ended in believing their own lie; even as our bulls in Borrowdale sometimes run mad with the echo of their own bellowing. The consequences were most injurious. Our attention was concentrated to a monster, which could not survive the convulsions, in which it had been brought forth: even the enlightened Burke himself too often talking and reasoning, as if a perpetual and organized anarchy had been a possible thing! Thus while we were warring against French doctrines, we took little heed, whether the means, by which we attempted to overthrow them, were not likely to aid and augment the far more formidable evil of French ambition. Like children we ran away from the yelping of a cur, and took shelter at the heels of a vicious war-horse.'

Courier, has been in defence or furtherance of the measures of Government.

> Things of this nature scarce survive the night
> That gives them birth; they perish in the sight,
> Cast by so far from *after-life*, that there
> Can scarcely aught be said, but that *they were!*
> Cartwright's *Prol. to the Royal Slave*

Yet in these labours I employed, and in the belief of partial friends wasted, the prime and manhood of my intellect. Most assuredly, they added nothing to my fortune or my reputation. The industry of the week supplied the necessities of the week. From Government or the friends of Government I not only never received remuneration, or ever expected it; but I was never honoured with a single acknowledgement, or expression of satisfaction. Yet the retrospect is far from painful or matter of regret. I am not indeed silly enough to take, as any thing more than a violent hyperbole of party debate, Mr Fox's assertion that the late war (I trust that the epithet is not prematurely applied) was a war produced by the Morning Post; or I should be proud to have the words inscribed on my tomb. As little do I regard the circumstance, that I was a specified object of Bonaparte's resentment during my residence in Italy in consequence of those essays in the Morning Post during the Peace of Amiens. (Of this I was warned, directly, by Baron von Humboldt, the Prussian plenipotentiary, who at that time was the Minister of the Prussian court at Rome; and indirectly, through his secretary, by Cardinal Fesch himself.) Nor do I lay any greater weight on the confirming fact, that an order for my arrest was sent from Paris, from which danger I was rescued by the kindness of a noble Benedictine, and the gracious connivance of that good old man, the present Pope. For the late tyrant's vindictive appetite was omnivorous, and preyed equally on a duc d'Enghien,*

* I seldom think of the murder of this illustrious Prince without recollecting the lines of Valerius Flaccus (Argonaut. Lib. I. 30.)

> . . . Super ipsius ingens
> Instat fama viri, virtusque haud laeta Tyranno;
> Ergo anteire metus, juvenemque exstinguere pergit.

[Moreover, above all, the great renown of the hero himself weighed upon his mind, and prowess never welcome to a tyrant. Wherefore he sought to forestall his fears and to destroy the son. (LCL)]

and the writer of a newspaper paragraph. Like a true vulture,* Napoleon with an eye not less telescopic, and with a taste equally coarse in his ravin, could descend from the most dazzling heights to pounce on the leveret in the brake, or even on the field-mouse amid the grass. But I do derive a gratification from the knowledge, that my essays contributed to introduce the practice of placing the questions and events of the day in a moral point of view; in giving a dignity to particular measures by tracing their policy or impolicy to permanent principles, and an interest to principles by the application of them to individual measures. In Mr Burke's writings indeed the germs of almost all political truths may be found. But I dare assume to myself the merit of having first explicitly defined and analysed the nature of Jacobinism; and that in distinguishing the Jacobin from the Republican, the Democrat, and the mere demagogue, I both rescued the word from remaining a mere term of abuse, and put on their guard many honest minds, who even in their heat of zeal against Jacobinism, admitted or supported principles from which the worst parts of that system may be legitimately deduced. That these are not necessary practical results of such principles, we owe to that fortunate inconsequence of our nature, which permits the heart to rectify the errors of the understanding. The detailed examination of the consular Government and its pretended constitution, and the proof given by me, that it was a consummate despotism in masquerade, extorted a recantation even from the Morning Chronicle, which had previously extolled this constitution as the perfection of a wise and regulated liberty. On every great occurrence I endeavoured to discover in past history the event, that most nearly resembled it. I procured, wherever it was possible, the contemporary historians, memorialists, and pamphleteers. Then fairly subtracting the points of difference from those of likeness, as the balance favoured the former or the latter, I conjectured that the result would be the same or different. In the series of essays† entitled 'a comparison of France under Napoleon

* *Θηρᾷ δὲ καὶ τὸν χῆνα καί τὴν Δορκάδα,*
Καὶ τὸν Λαγωὸν, καὶ τὸ τῶν Ταὺρων γένος.

[For he (the eagle) preys even upon the goose, and the antelope, and the hare, and the breed of bulls. (*CC*)]

Phile, *de animal. propriet.*

† A small selection from the numerous articles furnished by me to the Morning Post and Courier, chiefly as they regard the sources and effects of Jacobinism and the connection of certain systems of political economy with Jacobinical despotism, will form

with Rome under the first Cæsars,' and in those which followed 'on the probable final restoration of the Bourbons,' I feel myself authorized to affirm, by the effect produced on many intelligent men, that were the dates wanting, it might have been suspected that the essays had been written within the last twelve months. The same plan I pursued at the commencement of the Spanish revolution, and with the same success, taking the war of the United Provinces with Philip II, as the groundwork of the comparison. I have mentioned this from no motives of vanity, nor even from motives of self-defence, which would justify a certain degree of egotism, especially if it be considered, how often and grossly I have been attacked for sentiments, which I had exerted my best powers to confute and expose, and how grievously these charges acted to my disadvantage while I was in Malta. Or rather they would have done so, if my own feelings had not precluded the wish of a settled establishment in that island. But I have mentioned it from the full persuasion that, armed with the twofold knowledge of history and the human mind, a man will scarcely err in his judgement concerning the sum total of any future national event, if he have been able to procure the original documents of the past together with authentic accounts of the present, and if he have a philosophic tact for what is truly important in facts, and in most instances therefore for such facts as the dignity of history has excluded from the volumes of our modern compilers, by the courtesy of the age entitled historians.

To have lived in vain must be a painful thought to any man, and especially so to him who has made literature his profession. I should therefore rather condole than be angry with the mind, which could attribute to no worthier feelings, than those of vanity or self-love, the satisfaction which I acknowledge to have enjoyed from the republication of my political essays (either whole or as extracts) not only in many of our own provincial papers, but in the federal journals throughout America. I regarded it as some proof of my not having laboured altogether in vain, that from the articles written by me shortly before and at the commencement of the late unhappy war with

part of 'The Friend,' which I am now completing, and which will be shortly published, for I can scarcely say republished, with the numbers arranged in chapters according to their subjects.

> Accipe principium rursus, corpusque *coactum*
> Desere; mutata melior procede figura.

[Receive back thy life, quit the body that must die, and by a change of form come forth more beauteous than ever. (*CC*)]

America, not only the sentiments were adopted, but in some instances the very language, in several of the Massachusetts state papers.

But no one of these motives nor all conjointly would have impelled me to a statement so uncomfortable to my own feelings, had not my character been repeatedly attacked, by an unjustifiable intrusion on private life, as of a man incorrigibly idle, and who entrusted not only with ample talents, but favoured with unusual opportunities of improving them, had nevertheless suffered them to rust away without any efficient exertion either for his own good or that of his fellow creatures. Even if the compositions, which I have made public, and that too in a form the most certain of an extensive circulation, though the least flattering to an author's self-love, had been published in books, they would have filled a respectable number of volumes, though every passage of merely temporary interest were omitted. My prose writings have been charged with a disproportionate demand on the attention; with an excess of refinement in the mode of arriving at truths; with beating the ground for that which might have been run down by the eye; with the length and laborious construction of my periods; in short with obscurity and the love of paradox. But my severest critics have not pretended to have found in my compositions triviality, or traces of a mind that shrunk from the toil of thinking. No one has charged me with tricking out in other words the thoughts of others, or with hashing up anew the *crambe jam decies coctam* [ten-times-reheated cabbage] of English literature or philosophy. Seldom have I written that in a day, the acquisition or investigation of which had not cost me the previous labour of a month.

But are books the only channel through which the stream of intellectual usefulness can flow? Is the diffusion of truth to be estimated by publications; or publications by the truth, which they diffuse or at least contain? I speak it in the excusable warmth of a mind stung by an accusation, which has not only been advanced in reviews of the widest circulation, not only registered in the bulkiest works of periodical literature, but by frequency of repetition has become an admitted fact in private literary circles, and thoughtlessly repeated by too many who call themselves my friends, and whose own recollections ought to have suggested a contrary testimony. Would that the criterion of a scholar's utility were the number and moral value of the truths, which he has been the means of throwing into the general circulation; or the number and value of the minds, whom by his conversation or letters, he has excited into activity, and supplied with the germs of their aftergrowth! A distinguished rank might not indeed, even then, be awarded to my exertions, but I should dare look forward with confidence to an honourable acquittal. I should dare appeal to the numerous and respectable audiences, which at different times and in different places honoured my lecture-rooms with their attendance, whether the points of view from which the subjects treated of were surveyed, whether the grounds of my reasoning were such, as they had heard or read elsewhere, or have since found in previous publications. I can conscientiously declare, that the complete success of the *Remorse* on the first night of its representation did not give me as great or as heartfelt a pleasure, as the observation that the pit and boxes were crowded with faces familiar to me, though of individuals whose names I did not know, and of whom I knew nothing, but that they had attended one or other of my courses of lectures. It is an excellent though perhaps somewhat vulgar proverb, that there are cases where a man may be as well 'in for a pound as for a penny.' To those, who from ignorance of the serious injury I have received from this rumour of having dreamt away my life to no purpose, injuries which I unwillingly remember at all, much less am disposed to record in a sketch of my literary life; or to those, who from their own feelings, or the gratification they derive from thinking contemptuously of others, would like Job's comforters attribute these complaints, extorted from me by the sense of wrong, to self-conceit or presumptuous vanity, I have already furnished such ample materials, that I shall gain nothing by withholding the remainder. I will not therefore hesitate to ask the consciences of those, who from their long acquaintance with me and with the circumstances are best qualified to decide or be my judges, whether the restitution of the *suum cuique* [to each his own] would increase or detract from my literary reputation. In this exculpation I hope to be understood as speaking of myself comparatively, and in proportion to the claims, which others are entitled to make on my time or my talents. By what I *have* effected, am I to be judged by my fellow men; what I *could* have done, is a question for my own conscience. On my own account I may perhaps have had sufficient reason to lament my deficiency in self-control, and the neglect of concentring my powers to the realization of some permanent work. But to verse rather than to prose, if to either, belongs the 'voice of mourning' for

> Keen pangs of love awakening as a babe
> Turbulent, with an outcry in the heart,
> And fears self-will'd that shunn'd the eye of hope,

And hope that scarce would know itself from fear;
Sense of past youth, and manhood come in vain
And genius given and knowledge won in vain,
And all which I had cull'd in wood-walks wild
And all which patient toil had rear'd, and all
Commune with thee had open'd out—but flowers
Strew'd on my corpse, and borne upon my bier
In the same coffin, for the self-same grave!

S.T.C.

These will exist, for the future, I trust only in the poetic strains, which the feelings at the time called forth. In those only, gentle reader,

Affectus animi varios, bellumque sequacis
Perlegis invidiae; cursaque revolvis inanes;
Quas humilis tenero stylus olim effudit in aevo.
Perlegis et lacrymas, et quod pharetratus acutâ
Ille puer puero fecit mihi cuspide vulnus.
OMNIA PAULATIM CONSUMIT LONGIOR AETAS
VIVENDOQUE SIMUL MORIMUR, RAPIMURQUE MANENDO.
Ipse mihi collatus enim non ille videbor;
Frons alia est, moresque alii, nova mentis imago,
Vox aliudque sonat. Jamque observatio vitae
Multa dedit:—lugere nihil, ferre omnia; jamque
Paulatim lacrymas rerum experientia tersit.

[You read of various passions of the mind, of the warfare of persistent malice, you peruse the idle cares that once, in tender youth, my humble pen poured forth. You read too of tears, and of the wound given me, a boy, by that quivered boy with piercing barb. ADVANCING TIME DEVOURS ALL THINGS BY DEGREES, AND AS WE LIVE WE DIE, AND AS WE REST WE ARE HURRIED ONWARD. For, compared to myself, I shall not seem that self; my face is another, my ways are changed, I have a new sort of mind, my voice sounds otherwise. Already the study of life has given me much:—to grieve at nothing, to endure all things; and already experience has little by little wiped away my tears. (*CC*)]

CHAPTER XI

An affectionate exhortation to those who in early life feel themselves disposed to become authors.

IT was a favourite remark of the late Mr Whitbread's, that no man does anything from a single motive. The separate motives, or rather moods of mind, which produced the preceding reflections and anecdotes have been laid open to the reader in each separate instance. But an interest in the welfare of those, who at the present time may be in circumstances not dissimilar to my own at my first entrance into life, has been the constant accompaniment, and (as it were) the under-song of all my feelings. Whitehead exerting the prerogative of his laureateship addressed to youthful poets a poetic charge, which is perhaps the best, and certainly the most interesting, of his works. With no other privilege than that of sympathy and sincere good wishes, I would address an affectionate exhortation to the youthful literati, grounded on my own experience. It will be but short; for the beginning, middle, and end converge to one charge: never pursue literature as a trade. With the exception of one extraordinary man, I have never known an individual, least of all an individual of genius, healthy or happy without a profession, i.e. some regular employment, which does not depend on the will of the moment, and which can be carried on so far mechanically that an average quantum [amount] only of health, spirits, and intellectual exertion are requisite to its faithful discharge. Three hours of leisure, unannoyed by any alien anxiety, and looked forward to with delight as a change and recreation, will suffice to realize in literature a larger product of what is truly genial, than weeks of compulsion. Money and immediate reputation form only an arbitrary and accidental end of literary labour. The *hope* of increasing them by any given exertion will often prove a stimulant to industry; but the *necessity* of acquiring them will in all works of genius convert the stimulant into a narcotic. Motives by excess reverse their very nature, and instead of exciting, stun and stupefy the mind. For it is one contradistinction of genius from talent, that its predominant end is always comprised in the means; and this is one of the many points, which establish an analogy between genius and virtue. Now though talents may exist without genius, yet as genius cannot exist, certainly not manifest itself, without talents, I would advise every scholar, who feels the genial power working within him, so far to make a division between the two, as that he should devote his talents to the acquirement of competence in some known trade or profession, and his genius to objects of his tranquil and unbiased choice; while the consciousness of being actuated in both alike by the sincere desire to perform his duty, will alike ennoble both. My dear young friend (I would say) suppose yourself established in any honourable occupation. From the manufactory or counting-house, from the lawcourt, or from having visited your last patient, you return at evening,

> Dear tranquil time, when the sweet sense of home
> Is sweetest . . .

to your family, prepared for its social enjoyments, with the very countenances of your wife and children brightened, and their voice of welcome made doubly welcome, by the knowledge that, as far as they are concerned, you have satisfied the demands of the day by the labour of the day. Then, when you retire into your study, in the books on your shelves you revisit so many venerable friends with whom you can converse. Your own spirit scarcely less free from personal anxieties than the great minds, that in those books are still living for you! Even your writing-desk with its blank paper and all its other implements will appear as a chain of flowers, capable of linking your feelings as well as thoughts to events and characters past or to come; not a chain of iron which binds you down to think of the future and the remote by recalling the claims and feelings of the peremptory present. But why should I say *retire*? The habits of active life and daily intercourse with the stir of the world will tend to give you such self-command, that the presence of your family will be no interruption. Nay, the social silence, or undisturbing voices of a wife or sister will be like a restorative atmosphere, or soft music which moulds a dream without becoming its object. If facts are required to prove the possibility of combining weighty performances in literature with full and independent employment, the works of Cicero and Xenophon among the ancients; of Sir Thomas More, Bacon, Baxter, or to refer at once to later and contemporary instances, Darwin and Roscoe,° are at once decisive of the question.

But all men may not dare promise themselves a sufficiency of self-control for the imitation of those examples; though strict scrutiny should always be made, whether indolence, restlessness, or a vanity impatient for immediate gratification, have not tampered with the judgement and assumed the vizard of humility for the purposes of self-delusion. Still the Church presents to every man of learning and genius a profession, in which he may cherish a rational hope of being able to unite the widest schemes of literary utility with the strictest performance of professional duties. Among the numerous blessings of Christianity, the introduction of an established Church makes an especial claim on the gratitude of scholars and philosophers; in England, at least, where the principles of Protestantism have conspired with the freedom of the Government to double all its salutary powers by the removal of its abuses.

That not only the maxims, but the grounds of a pure morality, the mere fragments of which

> . . . the lofty grave tragedians taught
> In chorus or iambic, teachers best
> Of moral prudence, with delight received
> In brief sententious precepts;
>
> Paradise Regained

and that the sublime truths of the divine unity and attributes, which a Plato found most hard to learn and deemed it still more difficult to reveal; that these should have become the almost hereditary property of childhood and poverty, of the hovel and the workshop; that even to the unlettered they sound as commonplace, is a phenomenon, which must withhold all but minds of the most vulgar cast from undervaluing the services even of the pulpit and the reading-desk. Yet those, who confine the efficiency of an established Church to its public offices, can hardly be placed in a much higher rank of intellect. That to every parish throughout the kingdom there is transplanted a germ of civilization; that in the remotest villages there is a nucleus, round which the capabilities of the place may crystallize and brighten; a model sufficiently superior to excite, yet sufficiently near to encourage and facilitate, imitation; this, the inobtrusive, continuous agency of a Protestant Church establishment, this it is, which the patriot, and the philanthropist, who would fain unite the love of peace with the faith in the progressive amelioration of mankind, cannot estimate at too high a price. 'It cannot be valued with the gold of Ophir, with the precious onyx, or the sapphire. No mention shall be made of coral or of pearls; for the price of wisdom is above rubies.' The clergyman is with his parishioners and among them; he is neither in the cloistered cell, or in the wilderness, but a neighbour and a family man, whose education and rank admit him to the mansion of the rich landholder, while his duties make him the frequent visitor of the farmhouse and the cottage.

He is, or he may become, connected with the families of his parish or its vicinity by marriage. And among the instances of the blindness, or at best of the short-sightedness, which it is the nature of cupidity to inflict, I know few more striking, than the clamours of the farmers against Church property. Whatever was not paid to the clergyman would inevitably at the next lease be paid to the landholder, while, as the case at present stands, the revenues of the Church are in some sort the reversionary property of every family, that may have a member educated for the Church, or a daughter that may marry a clergyman. Instead of being foreclosed and immovable, it is in fact the only species of landed property, that is essentially moving and circulative. That there exist no inconveniences, who will pretend to assert? But I have yet to expect the proof, that the inconveniences are greater in this than in any other species; or that either the farmers or the clergy would be benefited by forcing the latter to become either Trullibers° or salaried placemen. Nay, I do not hesitate to declare my firm persuasion, that whatever reason of discontent the farmers may assign, the true cause is this; that they may cheat the parson, but cannot cheat the steward; and they are disappointed, if they should have been able to withhold only two pounds less than the legal claim, having expected to withhold five. At all events, considered relatively to the encouragement of learning and genius, the establishment presents a patronage at once so effective and unburthensome, that it would be impossible to afford the like or equal in any but a Christian and Protestant country. There is scarce a department of human knowledge without some bearing on the various critical, historical, philosophical, and moral truths, in which the scholar must be interested as a clergyman; no one pursuit worthy of a man of genius, which may not be followed without incongruity. To give the history of the Bible as a book, would be little less than to relate the origin or first excitement of all the literature and science, that we now possess. The very decorum, which the profession imposes, is favourable to the best purposes of genius, and tends to counteract its most frequent defects. Finally, that man must be deficient in sensibility, who would not find an incentive to emulation in the great and burning lights, which in a long series have illustrated the Church of England; who would not hear from within an echo to the voice from their sacred shrines,

> Et Pater Aeneas et avunculus excitat Hector.
>
> [Both his father, Aeneas, and his uncle, Hector, inspire him.]

But whatever be the profession or trade chosen, the advantages are many and important, compared with the state of a *mere* literary man, who in any degree depends on the sale of his works for the necessaries and comforts of life. In the former a man lives in sympathy with the world, in which he lives. At least he acquires a better and quicker tact for the knowledge of that, with which men in general can sympathize. He learns to manage his genius more prudently and efficaciously. His powers and acquirements gain him likewise more real admiration; for they surpass the legitimate expectations of others. He is something besides an author, and is not therefore considered merely as an author. The hearts of men are open to him, as to one of their own class; and whether he exerts himself or not in the conversational circles of his acquaintance, his silence is not attributed to pride, nor his communicativeness to vanity. To these advantages I will venture to add a superior chance of happiness in domestic life, were it only that it is as natural for the man to be out of the circle of his household during the day, as it is meritorious for the woman to remain for the most part within it. But this subject involves points of consideration so numerous and so delicate, and would not only permit, but require such ample documents from the biography of literary men, that I now merely allude to it *in transitu* [in passing]. When the same circumstance has occurred at very different times to very different persons, all of whom have some one thing in common; there is reason to suppose that such circumstance is not merely attributable to the persons concerned, but is in some measure occasioned by the one point in common to them all. Instead of the vehement and almost slanderous dehortation from marriage, which the misogyne, Boccaccio (*Vita e Costumi* di Dante, p. 12, 16) addresses to literary men, I would substitute the simple advice: be not *merely* a man of letters! Let literature be an honourable augmentation to your arms; but not constitute the coat, or fill the escutcheon!

To objections from conscience I can of course answer in no other way, than by requesting the youthful objector (as I have already done on a former occasion) to ascertain with strict self-examination, whether other influences may not be at work; whether spirits, 'not of health,' and with whispers 'not from heaven,' may not be walking in the twilight of his consciousness. Let him catalogue his scruples, and reduce them to a distinct intelligible form; let him be certain, that he has read with a docile mind and favourable dispositions the best and most fundamental works on the subject; that he has had both mind and heart opened to the great and illustrious qualities of the many renowned characters, who had doubted like himself, and whose researches had ended in the clear conviction, that their doubts had been groundless, or at least in no proportion to the counterweight. Happy will it be for such a man, if among his contemporaries elder than himself he should meet with one, who with similar powers, and feelings as acute as his own, had entertained the same scruples; had acted upon them; and who by after-research (when the step was, alas! irretrievable, but for that very reason his research undeniably disinterested) had discovered himself to have quarrelled with received opinions only to embrace errors, to have left the direction tracked out for him on the high road of honourable exertion, only to deviate into a labyrinth, where when he had wandered, till his head was giddy, his best good fortune was finally to have found his way out again, too late for prudence though not too late for conscience or for truth! Time spent in such delay is time won; for manhood in the meantime is advancing, and with it increase of knowledge, strength of judgement, and above all, temperance of feelings. And even if these should effect no change, yet the delay will at least prevent the final approval of the decision from being alloyed by the inward censure of the rashness and vanity, by which it had been precipitated. It would be a sort of irreligion, and scarcely less than a libel on human nature to believe, that there is any established and reputable profession or employment, in which a man may not continue to act with honesty and honour; and doubtless there is likewise none, which may not at times present temptations to the contrary. But woefully will that man find himself mistaken, who imagines that the profession of literature, or (to speak more plainly) the trade of authorship, besets its members with fewer or with less insidious temptations, than the Church, the law, or the different branches of commerce. But I have treated sufficiently on this unpleasant subject in an early chapter of this volume. I will conclude the present therefore with a short extract from Herder, whose name I might have added to the illustrious list of those, who have combined the successful pursuit of the Muses, not only with the faithful discharge, but with the highest honours and honourable emoluments, of an established profession. The translation the reader will find in a note below.* 'Am sorgfältigsten, meiden sie die Autorschaft. Zu früh

TRANSLATION

'With the greatest possible solicitude avoid authorship. Too early or immoderately employed, it makes the head *waste* and the heart empty; even were there no other worse consequences. A person, who reads only to print, in all probability reads amiss; and he,

oder unmässig gebraucht, macht sie den Kopf wüste und das Herz leer; wenn sie auch sonst keine üble Folgen gäbe. Ein Mensch, der nur lieset um zu drucken, lieset wahrscheinlich übel; und wer jeden Gedanken, der ihm aufstösst, durch Feder und Presse versendet, hat sie in kurzer Zeit alle versandt, und wird bald ein blosser Diener der Druckerey, ein Buchstabensetzer werden.' Herder.

CHAPTER XII

A chapter of requests and premonitions concerning the perusal or omission of the chapter that follows.

IN the perusal of philosophical works I have been greatly benefited by a resolve, which, in the antithetic form and with the allowed quaintness of an adage or maxim, I have been accustomed to word thus: 'until you understand a writer's ignorance, presume yourself ignorant of his understanding.' This golden rule of mine does, I own, resemble those of Pythagoras in its obscurity rather than in its depth. If however the reader will permit me to be my own Hierocles, I trust, that he will find its meaning fully explained by the following instances. I have now before me a treatise of a religious fanatic, full of dreams and supernatural experiences. I see clearly the writer's grounds, and their hollowness. I have a complete insight into the causes, which through the medium of his body had acted on his mind; and by application of received and ascertained laws I can satisfactorily explain to my own reason all the strange incidents, which the writer records of himself. And this I can do without suspecting him of any intentional falsehood. As when in broad daylight a man tracks the steps of a traveller, who had lost his way in a fog or by treacherous moonshine, even so, and with the same tranquil sense of certainty, can I follow the traces of this bewildered visionary. I understand his ignorance.

who sends away through the pen and the press every thought, the moment it occurs to him, will in a short time have sent all away, and will become a mere journeyman of the printing-office, a *compositor*.'

To which I may add from myself, that what medical physiologists affirm of certain secretions, applies equally to our thoughts; they too must be taken up again into the circulation, and be again and again resecreted in order to ensure a healthful vigour, both to the mind and to its intellectual offspring.

On the other hand, I have been reperusing with the best energies of my mind the Timaeus of Plato. Whatever I comprehend, impresses me with a reverential sense of the author's genius; but there is a considerable portion of the work, to which I can attach no consistent meaning. In other treatises of the same philosopher intended for the average comprehensions of men, I have been delighted with the masterly good sense, with the perspicuity of the language, and the aptness of the inductions. I recollect likewise, that numerous passages in this author, which I thoroughly comprehend, were formerly no less unintelligible to me, than the passages now in question. It would, I am aware, be quite fashionable to dismiss them at once as Platonic jargon. But this I cannot do with satisfaction to my own mind, because I have sought in vain for causes adequate to the solution of the assumed inconsistency. I have no insight into the possibility of a man so eminently wise, using words with such half-meanings to himself, as must perforce pass into no-meaning to his readers. When in addition to the motives thus suggested by my own reason, I bring into distinct remembrance the number and the series of great men, who after long and zealous study of these works had joined in honouring the name of Plato with epithets, that almost transcend humanity, I feel, that a contemptuous verdict on my part might argue want of modesty, but would hardly be received by the judicious, as evidence of superior penetration. Therefore, utterly baffled in all my attempts to understand the ignorance of Plato, I conclude myself ignorant of his understanding.

In lieu of the various requests which the anxiety of authorship addresses to the unknown reader, I advance but this one; that he will either pass over the following chapter altogether, or read the whole connectedly. The fairest part of the most beautiful body will appear deformed and monstrous, if dissevered from its place in the organic whole. Nay, on delicate subjects, where a seemingly trifling difference of more or less may constitute a difference in kind, even a faithful display of the main and supporting ideas, if yet they are separated from the forms by which they are at once clothed and modified, may perchance present a skeleton indeed; but a skeleton to alarm and deter. Though I might find numerous precedents, I shall not desire the reader to strip his mind of all prejudices, or to keep all prior systems out of view during his examination of the present. For in truth, such requests appear to me not much unlike the advice given to hypochondriacal patients in Dr Buchan's Domestic Medicine; *videlicet* [= viz., 'namely'], to preserve themselves uniformly tranquil and in good spirits. Till I had discovered the art of destroying the memory *a parte post* [from behind], without injury to its future operations, and without detriment to the judgement, I should suppress the request as premature; and therefore, however much I may wish to be read with an unprejudiced mind, I do not presume to state it as a necessary condition.

The extent of my daring is to suggest one criterion, by which it may be rationally conjectured beforehand, whether or no a reader would lose his time, and perhaps his temper, in the perusal of this, or any other treatise constructed on similar principles. But it would be cruelly misinterpreted, as implying the least disrespect either for the moral or intellectual qualities of the individuals thereby precluded. The criterion is this: if a man receives as fundamental facts, and therefore of course indemonstrable and incapable of further analysis, the general notions of matter, spirit, soul, body, action, passiveness, time, space, cause and effect, consciousness, perception, memory and habit; if he feels his mind completely at rest concerning all these, and is satisfied, if only he can analyse all other notions into some one or more of these supposed elements with plausible subordination and apt arrangement: to such a mind I would as courteously as possible convey the hint, that for him the chapter was not written.

> Vir bonus es, doctus, prudens; ast *haud tibi spiro.*
>
> [You are a good man, learned, prudent; but *I do not blow for you.*]

For these terms do in truth include all the difficulties, which the human mind can propose for solution. Taking them therefore in mass, and unexamined, it requires only a decent apprenticeship in logic, to draw forth their contents in all forms and colours, as the professors of legerdemain at our village fairs pull out ribbon after ribbon from their mouths. And not more difficult is it to reduce them back again to their different genera. But though this analysis is highly useful in rendering our knowledge more distinct, it does not really add to it. It does not increase, though it gives us a greater mastery over, the wealth which we before possessed. For forensic purposes, for all the established professions of society, this is sufficient. But for philosophy in its highest sense, as the science of ultimate truths, and therefore *scientia scientiarum* [the science of sciences], this mere analysis of terms is preparative only, though as a preparative discipline indispensable.

Still less dare a favourable perusal be anticipated from the proselytes of that compendious philosophy, which talking of mind but thinking of brick and mortar, or other images equally abstracted from body, contrives a theory of spirit by nicknaming matter, and in a few hours can qualify its dullest disciples to explain the *omne scibile* [everything knowable] by reducing all things to impressions, ideas, and sensations.

But it is time to tell the truth; though it requires some courage to avow it in an age and country, in which disquisitions on all subjects, not privileged to adopt technical terms or scientific symbols, must be addressed to the public. I say then, that it is neither possible or necessary for all men, or for many, to be philosophers. There is a philosophic (and inasmuch as it is actualized by an effort of freedom, an artificial) consciousness, which lies beneath or (as it were) behind the spontaneous consciousness natural to all reflecting beings. As the elder Romans distinguished their northern provinces into Cis-Alpine and Trans-Alpine, so may we divide all the objects of human knowledge into those on this side, and those on the other side of the spontaneous consciousness; *citra et trans conscientiam communem*. The latter is exclusively the domain of pure philosophy, which is therefore properly entitled *transcendental*, in order to discriminate it at once, both from mere reflection and re-presentation on the one hand, and on the other from those flights of lawless speculation which abandoned by all distinct consciousness, because transgressing the bounds and purposes of our intellectual faculties, are justly condemned, as *transcendent*.* The first range of hills, that encircles the scanty vale of

* This distinction between transcendental and transcendent is observed by our elder divines and philosophers, whenever they express themselves scholastically. Dr Johnson indeed has confounded the two words; but his own authorities do not bear him out. Of this celebrated dictionary I will venture to remark once for all, that I should suspect the man of a morose disposition who should speak of it without respect and gratitude as a most instructive and entertaining book, and hitherto, unfortunately, an indispensable book; but I confess, that I should be surprised at hearing from a philosophic and thorough scholar any but very qualified praises of it, as a dictionary. I am not now alluding to the number of genuine words omitted; for this is (and perhaps to a greater extent) true, as Mr Wakefield has noticed, of our best Greek lexicons, and this too after the successive labours of so many giants in learning. I refer at present both to omissions and commissions of a more important nature. What these are, *me saltem judice* [at least in my opinion], will be stated at full in The Friend, republished and completed.

I had never heard of the correspondence between Wakefield and Fox till I saw the account of it this morning (16th September 1815) in the Monthly Review. I was not a little gratified at finding, that Mr Wakefield had proposed to himself nearly the same plan for a Greek and English Dictionary, which I had formed, and began to execute, now ten years ago. But far, far more grieved am I, that he did not live to complete it. I

human life, is the horizon for the majority of its inhabitants. On its ridges the common sun is born and departs. From them the stars rise, and touching them they vanish. By the many, even this range, the natural limit and bulwark of the vale, is but imperfectly known. Its higher ascents are too often hidden by mists and clouds from uncultivated swamps, which few have courage or curiosity to penetrate. To the multitude below these vapours appear, now as the dark haunts of terrific agents, on which none may intrude with impunity; and now all aglow, with colours not their own, they are gazed at, as the splendid palaces of happiness and power. But in all ages there have been a few, who measuring and sounding the rivers of the vale at the feet of their furthest inaccessible falls have learnt, that the sources must be far higher and far inward; a few, who even in the level streams have detected elements, which neither the vale itself or the surrounding mountains contained or could supply. How and whence to these thoughts, these strong probabilities, the ascertaining vision, the intuitive knowledge, may finally supervene, can be learnt only by the fact. I might oppose to the question the words with which Plotinus* supposes Nature to answer a similar difficulty. 'Should any

cannot but think it a subject of most serious regret, that the same heavy expenditure, which is now employing in the republication of Stephanus augmented, had not been applied to a new lexicon on a more philosophical plan, with the English, German, and French synonyms as well as the Latin. In almost every instance the precise individual meaning might be given in an English or German word; whereas in Latin we must too often be contented with a mere general and inclusive term. How indeed can it be otherwise, when we attempt to render the most copious language of the world, the most admirable for the fineness of its distinctions, into one of the poorest and most vague languages? Especially, when we reflect on the comparative number of the works, still extant, written, while the Greek and Latin were living languages. Were I asked, what I deemed the greatest and most unmixed benefit, which a wealthy individual, or an association of wealthy individuals could bestow on their country and on mankind, I should not hesitate to answer, 'a philosophical English dictionary; with the Greek, Latin, German, French, Spanish and Italian synonyms, and with correspondent indexes.' That the learned languages might thereby be acquired, better, in half the time, is but a part, and not the most important part, of the advantages which would accrue from such a work. O! if it should be permitted by providence, that without detriment to freedom and independence our Government might be enabled to become more than a committee for war and revenue! There was a time, when everything was to be done by Government. Have we not flown off to the contrary extreme?

* Ennead, iii. l. 8. c. 3. The force of the Greek *συνιέναι* is imperfectly expressed by 'understand'; our own idiomatic phrase 'to go along with me' comes nearest to it. The passage, that follows, full of profound sense, appears to me evidently corrupt; and in fact no writer more wants, better deserves, or is less likely to obtain, a new and more correct edition.—*τὶ οὖν συνιέναὶ; ὅτι τὸ γενόμενον εστι θὲαμα ἐμὸν, σιωπησις* (*mallem* [I

one interrogate her, how she works, if graciously she vouchsafe to listen and speak, she will reply, it behoves thee not to disquiet me with interrogatories, but to understand in silence, even as I am silent, and work without words.'

Likewise in the fifth book of the fifth Ennead, speaking of the highest and intuitive knowledge as distinguished from the discursive, or in the language of Wordsworth,

> The vision and the faculty divine;

he says: 'it is not lawful to enquire from whence it sprang, as if it were a thing subject to place and motion, for it neither approached hither, nor again departs from hence to some other place; but it either appears to us or it does not appear. So that we ought not to pursue it with a view of detecting its secret source, but to watch in quiet till it suddenly shines upon us; preparing ourselves for the blessed spectacle as the eye waits patiently for the rising sun.' They and they only can acquire the philosophic imagination, the sacred power of self-intuition, who within themselves can interpret and understand the symbol, that the wings of the air-sylph are forming within the skin of the caterpillar; those only, who feel in their own spirits the same instinct, which impels the chrysalis of the horned fly to leave room in its involucrum for antennae yet to come. They know and feel, that the potential works in them, even as the actual works on them! In short, all the organs of sense are framed for a corresponding world of sense; and we have it. All the organs of spirit are framed for a correspondent world of spirit: though the latter organs are not developed in all alike. But they exist in all, and their first appearance discloses itself in the moral being. How else could it be, that even worldlings, not wholly debased, will contemplate the man of simple and disinterested goodness with contradictory feelings of pity and respect? 'Poor man! he is not made for *this* world.' Oh! herein they utter a prophecy of universal fulfilment; for man must either rise or sink.

should prefer], *θέαμα, ἐμοῦ σιωπωσῆς,) καὶ φύσει γενομενον θεώρημα και μοι γενομένη ἐκ θεωρίας τῆς ὡδί, την φύσιν ἔχειν φιλοθεάμονα ὑπαρχει* (*mallem*, *καὶ μοι ἧς γενομένη ἐκ θεωρίας αὐτῆς ὡδὶς*). 'What then are we to understand? That whatever is produced is an intuition, I silent; and that, which is thus generated, is by its nature a theorem, or form of contemplation; and the birth, which results to me from this contemplation, attains to have a contemplative nature.' So Synesius: *Ωδὶς ἱερα, Αρρητε γονά* [sacred travail, ineffable generation]. The after comparison of the process of the *natura naturans* [naturing nature (i.e. process, not product)] with that of the geometrician is drawn from the very heart of philosophy.

It is the essential mark of the true philosopher to rest satisfied with no imperfect light, as long as the impossibility of attaining a fuller knowledge has not been demonstrated. That the common consciousness itself will furnish proofs by its own direction, that it is connected with master currents below the surface, I shall merely assume as a postulate *pro tempore* [for the time being]. This having been granted, though but in expectation of the argument, I can safely deduce from it the equal truth of my former assertion, that philosophy cannot be intelligible to all, even of the most learned and cultivated classes. A system, the first principle of which it is to render the mind intuitive of the spiritual in man (i.e. of that which lies on the other side of our natural consciousness) must needs have a great obscurity for those, who have never disciplined and strengthened this ulterior consciousness. It must in truth be a land of darkness, a perfect Anti-Goshen, for men to whom the noblest treasures of their own being are reported only through the imperfect translation of lifeless and sightless notions. Perhaps, in great part, through words which are but the shadows of notions; even as the notional understanding itself is but the shadowy abstraction of living and actual truth. On the immediate, which dwells in every man, and on the original intuition, or absolute affirmation of it, (which is likewise in every man, but does not in every man rise into consciousness) all the certainty of our knowledge depends; and this becomes intelligible to no man by the ministry of mere words from without. The medium, by which spirits understand each other, is not the surrounding air; but the freedom which they possess in common, as the common ethereal element of their being, the tremulous reciprocations of which propagate themselves even to the inmost of the soul. Where the spirit of a man is not filled with the consciousness of freedom (were it only from its restlessness, as of one still struggling in bondage) all spiritual intercourse is interrupted, not only with others, but even with himself. No wonder then, that he remains incomprehensible to himself as well as to others. No wonder, that in the fearful desert of his consciousness, he wearies himself out with empty words, to which no friendly echo answers, either from his own heart, or the heart of a fellow being; or bewilders himself in the pursuit of notional phantoms, the mere refractions from unseen and distant truths through the distorting medium of his own unenlivened and stagnant understanding! To remain unintelligible to such a mind, exclaims Schelling on a like occasion, is honour and a good name before God and man.

The history of philosophy (the same writer observes) contains instances of systems, which for successive generations have remained enigmatic. Such he deems the system of Leibniz, whom another writer (rashly I think, and invidiously) extols as the only philosopher, who was himself deeply convinced of his own doctrines. As hitherto interpreted, however, they have not produced the effect, which Leibniz himself, in a most instructive passage, describes as the criterion of a true philosophy; namely, that it would at once explain and collect the fragments of truth scattered through systems apparently the most incongruous. The truth, says he, is diffused more widely than is commonly believed; but it is often painted, yet oftener masked, and is sometimes mutilated and sometimes, alas! in close alliance with mischievous errors. The deeper, however, we penetrate into the ground of things, the more truth we discover in the doctrines of the greater number of the philosophical sects. The want of substantial reality in the objects of the senses, according to the sceptics; the harmonies or numbers, the prototypes and ideas, to which the Pythagoreans and Platonists reduced all things; the one and all of Parmenides and Plotinus, without Spinozism;* the necessary

* This is happily effected in three lines by Synesius, in his Fourth Hymn:

'Εν καὶ Πάντα [one and all]—(taken by itself) is *Spinozism.*
'Εν δ' 'Απάντων [one of all]—a mere *anima Mundi* [World-spirit].
'Εν τε πρό παντων [one before all]—is mechanical Theism.

But unite all three, and the result is the Theism of Saint Paul and Christianity.

Synesius was censured for his doctrine of the Pre-existence of the Soul; but never, that I can find, arraigned or deemed heretical for his Pantheism, though neither Giordano Bruno, or Jacob Behmen ever avowed it more broadly.

Μύστας δὲ Νόος,
Τά τε καὶ τά λέγει,
Βύθον ἄῤῥητον
Αμφιχορεύων.
Σὺ τὸ τίκτον ἔφυς,
Σὺ τὸ τικτόμενον·
Σὺ τὸ φωτίζιον,
Σὺ τὸ λαμπόμενον·
Σύ τὸ φαινόμενον,
Σὺ τὸ κρυπτόμενον
Ιδίαῖς αὐγαῖς.
'Εν καὶ παντα,
'Εν καθ' ἑαυτο,
Καί διὰ πάντων·

[The mind initiated in the mysteries says such and such things, moving in harmony the while around Thy awful abysm. Thou art the Generator, Thou the Generated; Thou the Light that shineth, Thou the Illumined; Thou what is revealed, Thou that which is

connection of things according to the Stoics, reconcilable with the spontaneity of the other schools; the vital-philosophy of the Cabbalists and Hermetists, who assumed the universality of sensation; the substantial forms and entelechies of Aristotle and the schoolmen, together with the mechanical solution of all particular phenomena according to Democritus and the recent philosophers—all these we shall find united in one perspective central point, which shows regularity and a coincidence of all the parts in the very object, which from every other point of view must appear confused and distorted. The spirit of sectarianism has been hitherto our fault, and the cause of our failures. We have imprisoned our own conceptions by the lines, which we have drawn, in order to exclude the conceptions of others. J'ai trouvé que la plupart des sectes ont raison dans une bonne partie de ce qu'elles avancent, mais non pas tant en ce qu'elles nient [I have found that most sects are right in much that they affirm, but not so much so in what they deny].

A system, which aims to deduce the memory with all the other functions of intelligence, must of course place its first position from beyond the memory, and anterior to it, otherwise the principle of solution would be itself a part of the problem to be solved. Such a position therefore must, in the first instance be demanded, and the first question will be, by what right is it demanded? On this account I think it expedient to make some preliminary remarks on the introduction of postulates in philosophy. The word postulate is borrowed from the science of mathematics. (See Schell. abhandl. zur Erläuter. des id. der Wissenschaftslehre.) In geometry the primary construction is not demonstrated, but postulated. This first and most simple construction in space is the point in motion, or the line. Whether the point is moved in one and the same direction, or whether its direction is continually changed, remains as yet undetermined. But if the direction of the point have been determined, it is either by a point without it, and then there arises the straight line which encloses

hidden in thine own beams; The One and All, The One Self-contained and dispersed through all things. (*CC*)]

Pantheism is therefore not necessarily irreligious or heretical; though it may be taught atheistically. Thus Spinoza would agree with Synesius in calling God *Φυσις εν Νοεροις*, the Nature in Intelligences; but he could not subscribe to the preceding *Νοῦς καί Νοερος*, *i.e.* Himself Intelligence and Intelligent.

In this biographical sketch of my literary life I may be excused, if I mention here, that I had translated the eight Hymns of Synesius from the Greek into English Anacreontics before my 15th year.

no space; or the direction of the point is not determined by a point without it, and then it must flow back again on itself, that is, there arises a cyclical line, which does enclose a space. If the straight line be assumed as the positive, the cyclical is then the negation of the straight. It is a line, which at no point strikes out into the straight, but changes its direction continuously. But if the primary line be conceived as undetermined, and the straight line as determined throughout, then the cyclical is the third compounded of both. It is at once undetermined and determined; undetermined through any point without, and determined through itself. Geometry therefore supplies philosophy with the example of a primary intuition, from which every science that lays claim to evidence must take its commencement. The mathematician does not begin with a demonstrable proposition, but with an intuition, a practical idea.

But here an important distinction presents itself. Philosophy is employed on objects of the inner sense, and cannot, like geometry, appropriate to every construction a correspondent outward intuition. Nevertheless philosophy, if it is to arrive at evidence, must proceed from the most original construction, and the question then is, what is the most original construction or first productive act for the inner sense. The answer to this question depends on the direction which is given to the inner sense. But in philosophy the inner sense cannot have its direction determined by any outward object. To the original construction of the line, I can be compelled by a line drawn before me on the slate or on sand. The stroke thus drawn is indeed not the line itself, but only the image or picture of the line. It is not from it, that we first learn to know the line; but, on the contrary, we bring this stroke to the original line generated by the act of the imagination; otherwise we could not define it as without breadth or thickness. Still however this stroke is the sensuous image of the original or ideal line, and an efficient mean to excite every imagination to the intuition of it.

It is demanded then, whether there be found any means in philosophy to determine the direction of the inner sense, as in mathematics it is determinable by its specific image or outward picture. Now the inner sense has its direction determined for the greater part only by an act of freedom. One man's consciousness extends only to the pleasant or unpleasant sensations caused in him by external impressions; another enlarges his inner sense to a consciousness of forms and quantity; a third in addition to the image is conscious of the conception or notion of the thing; a fourth attains to a notion of his notions—he reflects on his own reflections; and thus we may say without impropriety, that the one possesses more or less inner sense, than the other. This more or less betrays already, that philosophy in its first principles must have a practical or moral, as well as a theoretical or speculative side. This difference in degree does not exist in the mathematics. Socrates in Plato shows, that an ignorant slave may be brought to understand and of himself to solve the most difficult geometrical problem. Socrates drew the figures for the slave in the sand. The disciples of the critical philosophy could likewise (as was indeed actually done by La Forge and some other followers of Descartes) represent the origin of our representations in copperplates; but no one has yet attempted it, and it would be utterly useless. To an Eskimo or New Zealander our most popular philosophy would be wholly unintelligible. The sense, the inward organ, for it is not yet born in him. So is there many a one among us, yes, and some who think themselves philosophers too, to whom the philosophic organ is entirely wanting. To such a man, philosophy is a mere play of words and notions, like a theory of music to the deaf, or like the geometry of light to the blind. The connection of the parts and their logical dependencies may be seen and remembered; but the whole is groundless and hollow, unsustained by living contact, unaccompanied with any realizing intuition which exists by and in the act that affirms its existence, which is known, because it is, and is, because it is known. The words of Plotinus, in the assumed person of nature, hold true of the philosophic energy. *Τὸ θεωροῦν μου θεώρημα ποιεῖ, ὥσπερ οἱ Γεωμετραι θεωρουντες γράφουσιν. ἀλλ' ἐμοῦ μὴ γραφούσης, θεωρούσης δὲ, ὑφίσηανται αἱ τῶν σωμάτων γράμμαι.* With me the act of contemplation makes the thing contemplated, as the geometricians contemplating describe lines correspondent; but I not describing lines, but simply contemplating, the representative forms of things rise up into existence.

The postulate of philosophy and at the same time the test of philosophic capacity, is no other than the heaven-descended Know thyself! (*E caelo descendit*, *Γνωθι σεαυτον* [It came down from heaven, *Know thyself*]). And this at once practically and speculatively. For as philosophy is neither a science of the reason or understanding only, nor merely a science of morals, but the science of being altogether, its primary ground can be neither merely speculative or merely practical, but both in one. All knowledge rests on the coincidence of an object with a subject. (My readers have been warned in a former chapter that for their convenience as well as the writer's, the term, subject is used by me in its scholastic sense as equivalent to mind or sentient being, and as the necessary correlative of object or *quicquid objicitur menti* [whatever is cast before the mind].) For we can *know* that only which is true: and the truth is universally placed in the coincidence of the thought with the thing, of the representation with the object represented.

Now the sum of all that is merely objective, we will henceforth call nature, confining the term to its passive and material sense, as comprising all the phenomena by which its existence is made known to us. On the other hand the sum of all that is subjective, we may comprehend in the name of the self or intelligence. Both conceptions are in necessary antithesis. Intelligence is conceived of as exclusively representative, nature as exclusively represented; the one as conscious, the other as without consciousness. Now in all acts of positive knowledge there is required a reciprocal concurrence of both, namely of the conscious being, and of that which is in itself unconscious. Our problem is to explain this concurrence, its possibility and its necessity.

During the act of knowledge itself, the objective and subjective are so instantly united, that we cannot determine to which of the two the priority belongs. There is here no first, and no second; both are coinstantaneous and one. While I am attempting to explain this intimate coalition, I must suppose it dissolved. I must necessarily set out from the one, to which therefore I give hypothetical antecedence, in order to arrive at the other. But as there are but two factors or elements in the problem, subject and object, and as it is left indeterminate from which of them I should commence, there are two cases equally possible.

1. EITHER THE OBJECTIVE IS TAKEN AS THE FIRST, AND THEN WE HAVE TO ACCOUNT FOR THE SUPERVENTION OF THE SUBJECTIVE, WHICH COALESCES WITH IT.

The notion of the subjective is not contained in the notion of the objective. On the contrary they mutually exclude each other. The subjective therefore must supervene to the objective. The conception of nature does not apparently involve the co-presence of an intelligence making an ideal duplicate of it, i.e. representing it. This desk for instance would (according to our natural notions) be, though there should exist no sentient being to look at it. This then is the problem of natural philosophy. It assumes the objective or unconscious nature as the first, and has therefore to explain how intelligence can supervene to it, or how itself can grow into intelligence. If it should appear, that all enlightened naturalists without having distinctly proposed the problem to themselves have yet constantly moved in the line of its solution, it must afford a strong presumption that the problem itself is founded in nature. For if all knowledge has as it were two poles reciprocally required and presupposed, all sciences must proceed from the one or the other, and must tend toward the opposite as far as the equatorial point in which both are reconciled and become identical. The necessary tendence therefore of all natural philosophy is from nature to intelligence; and this, and no other is the true ground and occasion of the instinctive striving to introduce theory into our views of natural phenomena. The highest perfection of natural philosophy would consist in the perfect spiritualization of all the laws of nature into laws of intuition and intellect. The phenomena (the material) must wholly disappear, and the laws alone (the formal) must remain. Thence it comes, that in nature itself the more the principle of law breaks forth, the more does the husk drop off, the phenomena themselves become more spiritual and at length cease altogether in our consciousness. The optical phenomena are but a geometry, the lines of which are drawn by light, and the materiality of this light itself has already become matter of doubt. In the appearances of magnetism all trace of matter is lost, and of the phenomena of gravitation, which not a few among the most illustrious Newtonians have declared no otherwise comprehensible than as an immediate spiritual influence, there remains nothing but its law, the execution of which on a vast scale is the mechanism of the heavenly motions. The theory of natural philosophy would then be completed, when all nature was demonstrated to be identical in essence with that, which in its highest known power exists in man as intelligence and self-consciousness; when the heavens and the earth shall declare not only the power of their maker, but the glory and the presence of their God, even as he appeared to the great prophet during the vision of the mount in the skirts of his divinity.

This may suffice to show, that even natural science, which commences with the material phenomenon as the reality and substance of things existing, does yet by the necessity of theorizing unconsciously, and as it were instinctively, end in nature as an intelligence; and by this tendency the science of nature becomes finally natural philosophy, the one of the two poles of fundamental science.

2. OR THE SUBJECTIVE IS TAKEN AS THE FIRST, AND THE PROBLEM THEN IS, HOW THERE SUPERVENES TO IT A COINCIDENT OBJECTIVE.

In the pursuit of these sciences, our success in each depends on an austere and faithful adherence to its own principles with a careful separation and exclusion of those, which appertain to the opposite science. As the natural philosopher, who directs his views to the objective, avoids above all things the intermixture of the subjective in his knowledge, as for instance, arbitrary suppositions or rather suffictions, occult qualities, spiritual agents, and the substitution of final for efficient causes; so on the other hand, the transcendental or intelligential philosopher is equally anxious to preclude all interpolation of the objective into the subjective principles of his science, as for instance the assumption of impresses or configurations in the brain, correspondent to miniature pictures on the retina painted by rays of light from supposed originals, which are not the immediate and real objects of vision, but deductions from it for the purposes of explanation. This purification of the mind is effected by an absolute and scientific scepticism to which the mind voluntarily determines itself for the specific purpose of future certainty. Descartes who (in his Meditations) himself first, at least of the moderns, gave a beautiful example of this voluntary doubt, this self-determined indetermination, happily expresses its utter difference from the scepticism of vanity or irreligion: 'Nec tamen in eo scepticos imitabar, qui dubitant tantum ut dubitent, et praeter incertitudinem ipsam nihil quaerunt. Nam contra totus in eo eram ut aliquid certi reperirem [Not that I imitated the sceptics, who doubt for the sake of doubting and who seek nothing beyond uncertainty itself. On the contrary, my whole object was to discover certainty]'. Descartes, *de Methodo*. Nor is it less distinct in its motives and final aim, than in its proper objects, which are not as in ordinary scepticism the prejudices of education and circumstance, but those original and innate prejudices which nature herself has planted in all men, and which to all but the philosopher are the first principles of knowledge, and the final test of truth.

Now these essential prejudices are all reducible to the one fundamental presumption, that there exist things without us. As this on the one hand originates, neither in grounds or arguments, and yet on the other hand remains proof against all attempts to remove it by grounds or arguments (*naturam furca expellas tamen usque redibit* [you may drive nature out with a pitchfork, but it will always return]); on the one hand lays claim to immediate certainty as a position at once indemonstrable and irresistible, and yet on the other hand, inasmuch as it refers to something essentially different from ourselves, nay even in opposition to ourselves, leaves it inconceivable how it could possibly become a part of our immediate consciousness; (in other words how that, which *ex hypothesi* [by hypothesis] is and continues to be extrinsic and alien to our being, should become a modification of our being) the philosopher therefore compels himself to treat this faith as nothing more than a prejudice, innate indeed and connatural, but still a prejudice.

The other position, which not only claims but necessitates the admission of its immediate certainty, equally for the scientific reason of the philosopher as for the common sense of mankind at large, namely, I AM, cannot so properly be entitled a prejudice. It is groundless indeed; but then in the very idea it precludes all ground, and separated from the immediate consciousness loses its whole sense and import. It is groundless; but only because it is itself the ground of all other certainty. Now the apparent contradiction, that the former position, namely, the existence of things without us, which from its nature cannot be immediately certain, should be received as blindly and as independently of all grounds as the existence of our own being, the transcendental philosopher can solve only by the supposition, that the former is unconsciously involved in the latter; that it is not only coherent but identical, and one and the same thing with our own immediate self-consciousness. To demonstrate this identity is the office and object of his philosophy.

If it be said, that this is idealism, let it be remembered that it is only so far idealism, as it is at the same time, and on that very account, the truest and most binding realism. For wherein does the realism of mankind properly consist? In the assertion that there exists a something without them, what, or how, or where they know not, which occasions the objects of their perception? Oh no! This is neither connatural or universal. It is what a few have taught and learnt in the schools, and which the many repeat without asking themselves concerning their own meaning. The realism common to all mankind is far elder and lies infinitely deeper than this hypothetical explanation of the origin of our perceptions, an explanation skimmed from the mere surface of mechanical philosophy. It is the table itself, which the man of common sense believes himself to see, not the phantom of a table, from which he may argumentatively deduce the reality of a table, which he does not see. If to destroy the reality of all, that we actually behold, be idealism, what can be more egregiously so, than the system of modern metaphysics, which banishes us to a land of shadows, surrounds us with apparitions, and distinguishes truth from illusion only by the majority of those who dream the same dream? 'I asserted that the world was mad,' exclaimed poor Lee, 'and the world said, that I was mad, and confound them, they outvoted me.'

It is to the true and original realism, that I would direct the attention. This believes and requires neither more nor less, than that the object which it beholds or presents to itself, is the real and very object. In this sense, however much we may strive against it, we are all collectively born idealists, and therefore and only therefore are we at the same time realists. But of this the philosophers of the schools know nothing, or despise the faith as the prejudice of the ignorant vulgar, because they live and move in a crowd of phrases and notions from which human nature has long ago vanished. Oh, ye that reverence yourselves, and walk humbly with the divinity in your own hearts, ye are worthy of a better philosophy! Let the dead bury the dead, but do you preserve your human nature, the depth of which was never yet fathomed by a philosophy made up of notions and mere logical entities.

In the third treatise of my *Logosophia*, announced at the end of this volume, I shall give (*Deo volente* [God willing]) the demonstrations and constructions of the dynamic philosophy scientifically arranged. It is, according to my conviction, no other than the system of Pythagoras and of Plato revived and purified from impure mixtures. 'Doctrina per tot manus tradita tandem in vappam desiit [A doctrine passed down through so many hands has ended up as flat wine]'. The science of arithmetic furnishes instances, that a rule may be useful in practical application, and for the particular purpose may be sufficiently authenticated by the result, before it has itself been fully demonstrated. It is enough, if only it be rendered intelligible. This will, I trust, have been effected in the following theses for those of my readers, who are willing to accompany me through the following chapter, in which the results will be applied to the deduction of the imagination, and with it the principles of production and of genial criticism in the fine arts.

THESIS I

Truth is correlative to being. Knowledge without a correspondent reality is no knowledge; if we know, there must be somewhat known by us. To know is in its very essence a verb active.

THESIS II

All truth is either mediate, that is, derived from some other truth or truths; or immediate and original. The latter is absolute, and its formula A A; the former is of dependent or conditional certainty, and represented in the formula B A. The certainty, which inheres in A, is attributable to B.

Scholium. A chain without a staple, from which all the links derived their stability, or a series without a first, has been not inaptly allegorized, as a string of blind men, each holding the skirt of the man before him, reaching far out of sight, but all moving without the least deviation in one straight line. It would be naturally taken for granted, that there was a guide at the head of the file: what if it were answered, No! Sir, the men are without number, and infinite blindness supplies the place of sight?

Equally inconceivable is a cycle of equal truths without a common and central principle, which prescribes to each its proper sphere in the system of science. That the absurdity does not so immediately strike us, that it does not seem equally unimaginable, is owing to a surreptitious act of the imagination, which, instinctively and without our noticing the same, not only fills out the intervening spaces, and contemplates the cycle (of B C D E F etc.) as a continuous circle (A) giving to all collectively the unity of their common orbit; but likewise supplies by a sort of *subintelligitur* the one central power, which renders the movement harmonious and cyclical.

THESIS III

We are to seek therefore for some absolute truth capable of communicating to other positions a certainty, which it has not itself borrowed; a truth self-grounded, unconditional and known by its own light. In short, we have to find a somewhat which *is*, simply because it *is*. In order to be such, it must be one which is its own predicate, so far at least that all other nominal predicates must be modes and repetitions of itself. Its existence too must be such, as to preclude the possibility of requiring a cause or antecedent without an absurdity.

THESIS IV

That there can be but one such principle, may be proved a priori; for were there two or more, each must refer to some other, by which its equality is affirmed; consequently neither would be self-established, as the hypothesis demands. And a posteriori, it will be proved by the principle itself when it is discovered, as involving universal antecedents in its very conception.

Scholium. If we affirm of a board that it is blue, the predicate (blue) is accidental, and not implied in the subject, board. If we affirm of a circle that it is equiradial, the predicate indeed is implied in the definition of the subject; but the existence of the subject itself is contingent, and supposes both a cause and a percipient. The same reasoning will apply to the indefinite number of supposed indemonstrable truths exempted from the profane approach of philosophic investigation by the amiable Beattie, and other less eloquent and not more profound inaugurators of common sense on the throne of philosophy; a fruitless attempt, were it only that it is the twofold function of philosophy to reconcile reason with common sense, and to elevate common sense into reason.

THESIS V

Such a principle cannot be any thing or object. Each thing is what it is in consequence of some other thing. An infinite, independent *thing*,* is no less a contradiction, than an infinite circle or a sideless triangle. Besides a thing is that, which is capable of being an object of which itself is not the sole percipient. But an object is inconceivable without a subject as its antithesis. *Omne perceptum percipientem supponit* [Everything perceived supposes a perceiver].

But neither can the principle be found in a subject as a subject, contradistinguished from an object: for *unicuique percipienti aliquid objicitur perceptum* [for every perceiver there is an object perceived]. It is to be found therefore neither in object or subject taken separately, and consequently, as no other third is conceivable, it must be found in that which is neither subject nor object exclusively, but which is the identity of both.

THESIS VI

This principle, and so characterized manifests itself in the SUM or I AM; which I shall hereafter indiscriminately express by the words spirit, self, and self-consciousness. In this, and in this alone, object and subject, being and knowing, are identical, each involving and supposing the other. In other words, it is a subject which becomes a subject by the act of constructing itself objectively to itself; but which

* The impossibility of an absolute thing (*substantia unica* [unique substance]) as neither genus, species, nor *individuum* [individual]; as well as its utter unfitness for the fundamental position of a philosophic system will be demonstrated in the critique on Spinozism in the fifth treatise of my Logosophia.

never is an object except for itself, and only so far as by the very same act it becomes a subject. It may be described therefore as a perpetual self-duplication of one and the same power into object and subject, which presuppose each other, and can exist only as antitheses.

Scholium. If a man be asked how he *knows* that he is? he can only answer, *sum quia sum* [I am because I am]. But if (the absoluteness of this certainty having been admitted) he be again asked, how he, the individual person, came to be, then in relation to the ground of his existence, not to the ground of his knowledge of that existence, he might reply, *sum quia deus est* [I am because God is], or still more philosophically, *sum quia in deo sum* [I am, because I am in God].

But if we elevate our conception to the absolute self, the great eternal I AM,° then the principle of being, and of knowledge, of idea, and of reality; the ground of existence, and the ground of the knowledge of existence, are absolutely identical, *Sum quia sum*;* I am, because I affirm myself to be; I affirm myself to be, because I am.

* It is most worthy of notice, that in the first revelation of himself, not confined to individuals; indeed in the very first revelation of his absolute being Jehovah at the same time revealed the fundamental truth of all philosophy, which must either commence with the absolute, or have no fixed commencement; i.e. cease to be philosophy. I cannot but express my regret, that in the equivocal use of the word *that*, for *in that*, or *because*, our admirable version has rendered the passage susceptible of a degraded interpretation in the mind of common readers or hearers, as if it were a mere reproof to an impertinent question. I am what I am, which might be equally affirmed of himself by any existent being.

The Cartesian *Cogito, ergo sum* [I think, therefore I am] is objectionable, because either the *Cogito* [I think] is used *extra gradum* [without reference to degree, absolutely], and then it is involved in the *sum* [I am] and is tautological, or it is taken as a particular mode or dignity, and then it is subordinated to the *sum* [I am] as the species to the genus, or rather as a particular modification to the subject modified; and not preordinated as the arguments seem to require. For *Cogito* [I think] is *sum cogitans* [I am, thinking]. This is clear by the inevidence of the converse. *Cogitat ergo est* [he thinks, therefore he is] is true, because it is a mere application of the logical rule: *Quicquid in genere est, est et in specie* [Whatever is in the genus is also in the species]. *Est* (*cogitans*) *ergo est* [He is (thinking), therefore he is]. It is a cherry tree; therefore it is a tree. But, *est ergo cogitat* [he is, therefore he thinks], is illogical: for *quod est in specie, non* necessario *in genere est* [whatever is in the species is not *necessarily* in the genus]. It may be true. I hold it to be true, that *quicquid vere est, est per veram sui affirmationem* [whatever truly is, is through the true affirmation of itself]; but it is a derivative, not an immediate truth. Here then we have, by anticipation, the distinction between the conditional finite I (which as known in distinct consciousness by occasion of experience is called by Kant's followers the empirical I) and the absolute I AM, and likewise the dependence or rather the inherence of the former in the latter; in whom 'we live, and move, and have our being,' as St Paul divinely asserts, differing widely from the theists of the mechanic school (as Sir I. Newton, Locke, etc.) who must say from *whom* we *had* our being, and with it life and the powers of life.

THESIS VII

If then I know myself only through myself, it is contradictory to require any other predicate of self, but that of self-consciousness. Only in the self-consciousness of a spirit is there the required identity of object and of representation; for herein consists the essence of a spirit, that it is self-representative. If therefore this be the one only immediate truth, in the certainty of which the reality of our collective knowledge is grounded, it must follow that the spirit in all the objects which it views, views only itself. If this could be proved, the immediate reality of all intuitive knowledge would be assured. It has been shown, that a spirit is that, which is its own object, yet not originally an object, but an absolute subject for which all, itself included, may become an object. It must therefore be an act; for every object is, as an object, dead, fixed, incapable in itself of any action, and necessarily finite. Again, the spirit (originally the identity of object and subject) must in some sense dissolve this identity, in order to be conscious of it: *fit alter et idem* [it becomes another and (yet) the same]. But this implies an act, and it follows therefore that intelligence or self-consciousness is impossible, except by and in a will. The self-conscious spirit therefore is a will; and freedom must be assumed as a ground of philosophy, and can never be deduced from it.

THESIS VIII

Whatever in its origin is objective, is likewise as such necessarily finite. Therefore, since the spirit is not originally an object, and as the subject exists in antithesis to an object, the spirit cannot originally be finite. But neither can it be a subject without becoming an object, and as it is originally the identity of both, it can be conceived neither as infinite or finite exclusively, but as the most original union of both. In the existence, in the reconciling, and the recurrence of this contradiction consists the process and mystery of production and life.

THESIS IX

This *principium commune essendi et cognoscendi* [common principle of being and knowing], as subsisting in a will, or primary act of self-duplication, is the mediate or indirect principle of every science; but it is the immediate and direct principle of the ultimate science alone, i.e. of transcendental philosophy alone. For it must be remembered, that all these theses refer solely to one of the two polar sciences, namely, to that which commences with and rigidly confines itself within the subjective, leaving the objective (as far as it is exclusively objective) to natural philosophy, which is its opposite pole. In its very idea therefore as a systematic knowledge of our collective knowing, (*scientia scientiae* [the science of science (i.e. knowledge of knowing)]) it involves the necessity of some one highest principle of knowing, as at once the source and the accompanying form in all particular acts of intellect and perception. This, it has been shown, can be found only in the act and evolution of self-consciousness. We are not investigating an absolute *principium essendi* [principle of being]; for then, I admit, many valid objections might be started against our theory; but an absolute *principium cognoscendi* [principle of knowing]. The result of both the sciences, or their equatorial point, would be the principle of a total and undivided philosophy, as for prudential reasons, I have chosen to anticipate in the scholium to Thesis VI and the note subjoined. In other words, philosophy would pass into religion, and religion become inclusive of philosophy. We begin with the I KNOW MYSELF, in order to end with the absolute I AM. We proceed from the SELF, in order to lose and find all self in GOD.

THESIS X

The transcendental philosopher does not enquire, what ultimate ground of our knowledge there may lie out of our knowing, but what is the last in our knowing itself, beyond which we cannot pass. The principle of our knowing is sought within the sphere of our knowing. It must be something therefore, which can itself be known. It is asserted only, that the act of self-consciousness is for us the source and principle of all our possible knowledge. Whether abstracted from us there exists anything higher and beyond this primary self-knowing, which is for us the form of all our knowing, must be decided by the result.

That the self-consciousness is the fixed point, to which for us all is mortised and annexed, needs no further proof. But that the self-consciousness may be the modification of a higher form of being, perhaps of a higher consciousness, and this again of a yet higher, and so on in an infinite *regressus* [regress]; in short, that self-consciousness may be itself something explicable into something, which must lie beyond the possibility of our knowledge, because the whole synthesis of our intelligence is first formed in and through the self-consciousness, does not at all concern us as transcendental philosophers. For to us the self-consciousness is not a kind of being, but a kind of knowing, and that too the highest and farthest that exists for us. It may however be shown, and has in part already been shown in pages 290–1, that even when the objective is assumed as the first, we yet can never pass beyond the principle of self-consciousness. Should we attempt it, we must be driven back from ground to ground, each of which would cease to be a ground the moment we pressed on it. We must be whirled down the gulf of an infinite series. But this would make our reason baffle the end and purpose of all reason, namely, unity and system. Or we must break off the series arbitrarily, and affirm an absolute something that is in and of itself at once cause and effect (*causa sui* [the cause of itself]) subject and object, or rather the absolute identity of both. But as this is inconceivable, except in a self-consciousness, it follows, that even as natural philosophers we must arrive at the same principle from which as transcendental philosophers we set out; that is, in a self-consciousness in which the *principium essendi* [principle of being] does not stand to the *principium cognoscendi* [principle of knowing] in the relation of cause to effect, but both the one and the other are coinherent and identical. Thus the true system of natural philosophy places the sole reality of things in an absolute, which is at once *causa sui et effectus*, *πατηρ αυτοπατωρ*, *Υιος εαυτου* [self-cause and effect, father of himself, Son of himself]—in the absolute identity of subject and object, which it calls nature, and which in its highest power is nothing else but self-conscious will or intelligence. In this sense the position of Malebranche, that we see all things in God, is a strict philosophical truth; and equally true is the assertion of Hobbes, of Hartley, and of their masters in ancient Greece, that all real knowledge supposes a prior sensation. For sensation itself is but vision nascent, not the cause of intelligence, but intelligence itself revealed as an earlier power in the process of self-construction.

Μάκαρ, ἵλαθί μοι!
Πάτερ, ἵλαθί μοι
Εἰ παρὰ κόσμον,
Εἰ παρὰ μοῖραν
Τῶν σῶν ἔθιγον!

[Be full of goodness unto me, Blessed One, be full of goodness unto me, Father, if beyond what is ordered, beyond what is destined, I touch upon that which is thine.]

Bearing then this in mind, that intelligence is a self-development, not a quality supervening to a substance, we may abstract from all degree, and for the purpose of philosophic construction reduce it to kind, under the idea of an indestructible power with two opposite and counteracting forces, which, by a metaphor borrowed from astronomy, we may call the centrifugal and centripetal forces. The intelligence in the one tends to objectize itself, and in the other to know itself in the object. It will be hereafter my business to construct by a series of intuitions the progressive schemes, that must follow from such a power with such forces, till I arrive at the fulness of the human intelligence. For my present purpose, I assume such a power as my principle, in order to deduce from it a faculty, the generation, agency, and application of which form the contents of the ensuing chapter.

In a preceding page I have justified the use of technical terms in philosophy, whenever they tend to preclude confusion of thought, and when they assist the memory by the exclusive singleness of their meaning more than they may, for a short time, bewilder the attention by their strangeness. I trust, that I have not extended this privilege beyond the grounds on which I have claimed it; namely, the conveniency of the scholastic phrase to distinguish the kind from all degrees, or rather to express the kind with the abstraction of degree, as for instance *multeity* instead of multitude; or secondly, for the sake of correspondence in sound in interdependent or antithetical terms, as subject and object; or lastly, to avoid the wearying recurrence of circumlocutions and definitions. Thus I shall venture to use *potence*, in order to express a specific degree of a power, in imitation of the algebraists. I have even hazarded the new verb *potenziate* with its derivatives in order to express the combination or transfer of powers. It is with new or unusual terms, as with privileges in courts of justice or legislature; there can be no legitimate privilege, where there already exists a positive law adequate to the purpose; and when there is no law in existence, the privilege is to be justified by its accordance with the end, or final cause, of all law. Unusual and new-coined words are doubtless an evil; but vagueness, confusion, and imperfect conveyance of our thoughts, are a far greater. Every system, which is under the necessity of using terms not familiarized by the metaphysics in fashion, will be described as written in an unintelligible style, and the author must expect the charge of having substituted learned jargon for clear conception; while, according to the creed of our modern philosophers, nothing is deemed a clear conception, but what is representable by a distinct image. Thus the conceivable is reduced within the bounds of the picturable. 'Hinc patet, quî fiat ut, *cum irrepraesentabile* et *impossibile* vulgo ejusdem significatûs habeantur, conceptus tam *Continui*, quam *Infiniti*, a plurimis rejiciantur, quippe quorum, *secundum leges cognitionis intuitivae*, repraesentatio est impossibilis. Quanquam autem harum e non paucis scholis explosarum notionum, praesertim prioris, causam hic non gero, maximi tamen momenti erit monuisse: gravissimo illos errore labi, qui tam perversâ argumentandi ratione utuntur. Quicquid enim *repugnat* legibus intellectûs et rationis, utique est impossibile; quod autem, cum rationis purae sit objectum, legibus cognitionis intuitivae tantummodo *non subest*, non item. Nam hinc dissensus inter facultatem *sensitivam* et *intellectualem*, (quarum indolem mox exponam) nihil indigitat, nisi, *quas mens ab intellectu acceptas fert ideas abstractas, illas in concreto exsequi, et in Intuitus commutare saepenumero non posse*. Haec autem reluctantia *subjectiva* mentitur, ut plurimum, repugnantiam aliquam *objectivam*, et incautos facile fallit, limitibus, quibus *mens humana* circumscribitur, pro iis habitis, quibus *ipsa rerum essentia* continetur.*—Kant *de Mundi Sensibilis atque Intelligibilis forma et principiis*, 1770.

* TRANSLATION

'Hence it is clear, from what cause many reject the notion of the continuous and the infinite. They take, namely, the words irrepresentable and impossible in one and the same meaning; and, according to the forms of sensuous evidence, the notion of the continuous and the infinite is doubtless impossible. I am not now pleading the cause of these laws, which not a few schools have thought proper to explode, especially the former (the law of continuity). But it is of the highest importance to admonish the reader, that those, who adopt so perverted a mode of reasoning, are under a grievous error. Whatever opposes the formal principles of the understanding and the reason is confessedly impossible; but not therefore that, which is therefore not amenable to the forms of sensuous evidence, because it is exclusively an object of pure intellect. For this non-coincidence of the sensuous and the intellectual (the nature of which I shall presently lay open) proves nothing more, but that the mind cannot always adequately represent in the concrete, and transform into distinct images, abstract notions derived from the pure intellect. But this contradiction, which is in itself merely subjective (i.e. an incapacity in the nature of man) too often passes for an incongruity or impossibility in the object (i.e. the notions themselves) and seduces the incautious to mistake the limitations of the human faculties for the limits of things, as they really exist.'

I take this occasion to observe, that here and elsewhere Kant uses the terms intuition, and the verb active (*Intueri*, *germanice* Anschauen [to intuit, in German *Anschauen*]) for which we have unfortunately no correspondent word, exclusively for that which can be represented in space and time. He therefore consistently and rightly denies the possibility of intellectual intuitions. But as I see no adequate reason for this exclusive sense of the term, I have reverted to its wider signification authorized by our elder theologians and metaphysicians, according to whom the term comprehends all truths known to us without a medium.

Critics, who are most ready to bring this charge of pedantry and unintelligibility, are the most apt to overlook the important fact, that besides the language of words, there is a language of spirits (*sermo interior* [inner discourse]) and that the former is only the vehicle of the latter. Consequently their assurance, that they do not understand the philosophic writer, instead of proving anything against the philosophy, may furnish an equal and (*caeteris paribus* [other things being equal]) even a stronger presumption against their own philosophic talent.

Great indeed are the obstacles which an English metaphysician has to encounter. Amongst his most respectable and intelligent judges, there will be many who have devoted their attention exclusively to the concerns and interests of human life, and who bring with them to the perusal of a philosophic system an habitual aversion to all speculations, the utility and application of which are not evident and immediate. To these I would in the first instance merely oppose an authority, which they themselves hold venerable, that of Lord Bacon: 'non inutiles scientiae existimandae sunt, quarum in se nullus est usus, si ingenia acuant et ordinent [those sciences ought not to be considered useless that are in themselves without use, if they sharpen and order the wits].'

There are others, whose prejudices are still more formidable, inasmuch as they are grounded in their moral feelings and religious principles, which had been alarmed and shocked by the impious and pernicious tenets defended by Hume, Priestley, and the French fatalists or necessitarians; some of whom had perverted metaphysical reasonings to the denial of the mysteries and indeed of all the peculiar doctrines of Christianity; and others even to the subversion of all distinction between right and wrong. I would request such men to consider what an eminent and successful defender of the Christian faith has observed, that true metaphysics are nothing else but true divinity, and that in fact the writers, who have given them such just offence, were sophists, who had taken advantage of the general neglect into which the science of logic has unhappily fallen, rather than metaphysicians, a name indeed which those writers were the first to explode as unmeaning. Secondly, I would remind them, that as long as there are men in the world to whom the *Γνῶθι σέαυτον* [Know thyself] is an instinct and a command from their own nature, so long will there be metaphysicians and metaphysical speculations; that false metaphysics can be effectually counteracted by true metaphysics alone; and that if the reasoning be clear, solid and pertinent, the truth deduced can never be the less valuable on account of the depth from which it may have been drawn.

A third class profess themselves friendly to metaphysics, and believe that they are themselves metaphysicians. They have no objection to system or terminology, provided it be the method and the nomenclature to which they have been familiarized in the writings of Locke, Hume, Hartley, Condillac, or perhaps Dr Reid, and Professor Stewart. To objections from this cause, it is a sufficient answer, that one main object of my attempt was to demonstrate the vagueness or insufficiency of the terms used in the metaphysical schools of France and Great Britain since the revolution, and that the errors which I propose to attack cannot subsist, except as they are concealed behind the mask of a plausible and indefinite nomenclature.

But the worst and widest impediment still remains. It is the predominance of a popular philosophy, at once the counterfeit and the mortal enemy of all true and manly metaphysical research. It is that corruption, introduced by certain immethodical aphorisming Eclectics, who, dismissing not only all system, but all logical connection, pick and choose whatever is most plausible and showy; who select, whatever words can have some semblance of sense attached to them without the least expenditure of thought, in short whatever may enable men to talk of what they do not understand, with a careful avoidance of everything that might awaken them to a moment's suspicion of their ignorance. This alas! is an irremediable disease, for it brings with it, not so much an indisposition to any particular system, but an utter loss of taste and faculty for all system and for all philosophy. Like echoes that beget each other amongst the mountains, the praise or blame of such men rolls in volleys long after the report from the original blunderbuss. 'Sequacitas est potius et coitio quam consensus: et tamen (quod pessimum est) pusillanimitas ista non sine arrogantiâ et fastidio se offert [It is a following and a going along together rather than a consent; and what is worst of all, this very littleness of spirit comes with a certain air of arrogance and superiority].' *Novum Organum.*

I shall now proceed to the nature and genesis of the imagination; but I must first take leave to notice, that after a more accurate perusal of Mr Wordsworth's remarks on the imagination in his preface to the new edition of his poems, I find that my conclusions are not so consentient with his, as I confess, I had taken for granted. In an article contributed by me to Mr Southey's Omniana, on the soul and its organs of sense, are the following sentences. 'These (the human faculties) I would arrange under the different senses and powers; as the eye, the ear, the touch, etc.; the imitative power, voluntary and automatic; the imagination, or shaping and modifying power; the fancy, or the aggregative and associative power; the understanding, or the regulative, substantiating and realizing power; the speculative reason—*vis theoretica et scientifica* [the capacity for scientific thought], or the power by which we produce, or aim to produce unity, necessity, and universality in all our knowledge by means of principles a priori;* the will, or practical reason; the faculty of choice (*Germanice* [in German], Willkühr) and (distinct both from the moral will and the choice) the *sensation* of volition, which I have found reason to include under the head of single and double touch.' To this, as far as it relates to the subject in question, namely the words 'the aggregative and associative power' Mr Wordsworth's 'only objection is that the definition is too general. To aggregate and to associate, to evoke and combine, belong as well to the imagination as the fancy.' I reply, that if by the power of evoking and combining, Mr W. means the same as, and no more than, I meant by the aggregative and associative, I continue to deny, that it belongs at all to the imagination; and I am disposed to conjecture, that he has mistaken the co-presence of fancy with imagination for the operation of the latter singly. A man may work with two very different tools at the same moment; each has its share in the work, but the work effected by each is distinct and different. But it will probably appear in the next chapter, that deeming it necessary to go back much further than Mr Wordsworth's subject required or permitted, I have attached a meaning to both fancy and imagination, which he had not in view, at least while he was writing that preface. He will judge. Would to heaven, I might meet with many such readers. I will conclude with the words of Bishop Jeremy Taylor: 'he to whom all things are one, who draweth all things to one, and seeth all things in one, may enjoy true peace and rest of spirit.' (J. Taylor's *Via Pacis*.)

* This phrase, *a priori*, is in common most grossly misunderstood, and an absurdity burthened on it, which it does not deserve! By knowledge, *a priori*, we do not mean, that we can know anything previously to experience, which would be a contradiction in terms; but that having once known it by occasion of experience (i.e. something acting upon us from without) we then know, that it must have pre-existed, or the experience itself would have been impossible. By experience only I know, that I have eyes; but then my reason convinces me, that I must have had eyes in order to the experience.

CHAPTER XIII

On the imagination, or esemplastic power.

O Adam! one Almighty is, from whom
All things proceed, and up to him return
If not depraved from good: created all
Such to perfection, one first nature all
Indued with various forms, various degrees
Of substance, and in things that live, of life;
But more refin'd, more spiritous and pure,
As nearer to him plac'd or nearer tending,
Each in their several active spheres assign'd,
Till body up to spirit work, in bounds
Proportion'd to each kind. So from the root
Springs lighter the green stalk: from thence the leaves
More airy: last, the bright consummate flower
Spirits odorous breathes. Flowers and their fruit,
Man's nourishment, by gradual scale sublim'd,
To *vital* spirits aspire: to *animal*:
To *intellectual*!—give both life and sense,
Fancy and understanding: whence the soul
Reason receives. And reason is her *being*,
Discursive or intuitive.

Par. Lost, b. v

Sane si res corporales nil nisi materiale continerent, verissime dicerentur in fluxu consistere neque habere substantiale quicquam, quemadmodum et Platonici olim recte agnovêre.—Hinc igitur, praeter purè mathematica et phantasiae subjecta, collegi quaedam metaphysica solâque mente perceptibilia, esse admittenda: et massae materiali *principium* quoddam superius et, ut sic dicam, *formale* addendum: quandoquidem omnes veritates rerum corporearum ex solis axiomatibus logisticis et geometricis, nempe de magno et parvo, toto et parte, figurâ et situ, colligi non possint; sed alia de causâ et effectu, *actioneque* et *passione*, accedere debeant, quibus ordinis rerum rationes salventur. Id principium rerum, an ἐντελεχείαν an vim appellemus, non refert, modó meminerimus, per solam *Virium* notionem intelligibiliter explicari.

[If indeed corporeal things contained nothing but matter they might truly be said to consist in flux and to have no substance, as the Platonists once rightly recognized.—And so, apart from the purely mathematical and what is subject to the fancy, I have come to the conclusion that certain metaphysical elements perceptible by the mind alone should be admitted, and that some higher and, so to speak, *formal principle* should be added to the material mass, since all the truths about corporeal things cannot be collected from logistic and geometrical axioms alone, i.e. those concerning great and small, whole and part, shape and position, but others must enter into it, i.e. cause and effect, *action* and *passion*, by which the reasons for the order of things are maintained. It does not matter whether we call this principle of things an entelechy or a power so long as we remember that it is intelligibly to be explained only by the idea of *powers*. (*CC*)]

Leibniz: *Op.* T. II. P. II. p. 53.—T. III. p. 321

Σέβομαι Νοερῶν
Κρυφίαν τάξιν
Χωρει ΤΙ ΜΕΣΟΝ
Ου καταχυθέν.

[I venerate the hidden ordering of intellectual things, but there is some medial element that may not be distributed. (*CC*)]

Synesii, *Hymn III.* l. 231.

DESCARTES, speaking as a naturalist, and in imitation of Archimedes, said, give me matter and motion and I will construct you the universe. We must of course understand him to have meant: I will render the construction of the universe intelligible. In the same sense the transcendental philosopher says; grant me a nature having two contrary forces, the one of which tends to expand infinitely, while the other strives to apprehend or find itself in this infinity, and I will cause the world of intelligences with the whole system of their representations to rise up before you. Every other science presupposes intelligence as already existing and complete: the philosopher contemplates it in its growth, and as it were represents its history to the mind from its birth to its maturity.

The venerable Sage of Koenigsberg [Kant] has preceded the march of this master-thought as an effective pioneer in his essay on the introduction of negative quantities into philosophy, published 1763. In this he has shown, that instead of assailing the science of mathematics by metaphysics, as Berkeley did in his Analyst, or of sophisticating it, as Wolff did, by the vain attempt of deducing the first principles of geometry from supposed deeper grounds of ontology, it behoved the metaphysician rather to examine whether the only province of knowledge, which man has succeeded in erecting into a pure science, might not furnish materials or at least hints for establishing and pacifying the unsettled, warring, and embroiled domain of philosophy. An imitation of the mathematical method had indeed been attempted with no better success than attended the essay of David to wear the armour of Saul. Another use however is possible and of far greater promise, namely, the actual application of the positions which had so wonderfully enlarged the discoveries of geometry, *mutatis mutandis* [changing the things that must be changed], to philosophical subjects. Kant having briefly illustrated the utility of such an attempt in the questions of space, motion, and infinitely small quantities, as employed by the mathematician, proceeds to the idea of negative quantities and the transfer of them to metaphysical investigation. Opposites, he well observes, are of two kinds, either logical, i.e. such as are absolutely incompatible; or real without being contradictory. The former he denominates *Nihil negativum irrepraesentabile* [nothing in a negative sense, not representable], the connection of which produces nonsense. A body in motion is something—*Aliquid cogitabile* [something conceivable]; but a body, at one and the same time in motion and not in motion, is nothing, or at most, air articulated into nonsense. But a motory force of a body in one direction, and an equal force of the same body in an opposite direction is not incompatible, and the result, namely rest, is real and representable. For the purposes of mathematical calculus it is indifferent which force we term negative, and which positive, and consequently we appropriate the latter to that, which happens to be the principal object in our thoughts. Thus if a man's capital be ten and his debts eight, the subtraction will be the same, whether we call the capital negative debt, or the debt negative capital. But in as much as the latter stands practically in reference to the former, we of course represent the sum as 10 – 8. It is equally clear that two equal forces acting in opposite directions, both being finite and each distinguished from the other by its direction only, must neutralize or reduce each other to inaction. Now the transcendental philosophy demands; first, that two forces should be conceived which counteract each other by their essential nature; not only not in consequence of the accidental direction of each, but as prior to all direction, nay, as the primary forces from which the conditions of all possible directions are derivative and deducible: secondly, that these forces should be assumed to be both alike infinite, both alike indestructible. The problem will then be to discover the result or product of two such forces, as distinguished from the result of those forces which are finite, and derive their difference solely from the circumstance of their direction. When we have formed a scheme or outline of these two different kinds of force, and of their different results by the process of discursive reasoning, it will then remain for us to elevate the thesis from notional to actual, by contemplating intuitively this one power with its two inherent indestructible yet counteracting forces, and the results or generations to which their interpenetration gives existence, in the living principle and in the process of our own self-consciousness. By what instrument this is possible the solution itself will discover, at the same time that it will reveal, to and for whom it is possible. *Non omnia possumus omnes* [We are not all capable of everything]. There is a philosophic, no less than a poetic genius, which is differenced from the highest perfection of talent, not by degree but by kind.

The counteraction then of the two assumed forces does not depend on their meeting from opposite directions; the power which acts in them is indestructible; it is therefore inexhaustibly re-ebullient; and as something must be the result of these two forces, both alike infinite, and both alike indestructible; and as rest or neutralization cannot be this result; no other conception is possible, but that the product must be a *tertium aliquid* [some third thing], or finite generation. Consequently this conception is necessary. Now this *tertium aliquid* can be no other than an interpenetration of the counteracting powers, partaking of both.

Thus far had the work been transcribed for the press, when I received the following letter from a friend, whose practical judgement I have had ample reason to estimate and revere, and whose taste and sensibility preclude all the excuses which my self-love might possibly have prompted me to set up in plea against the decision of advisers of equal good sense, but with less tact and feeling.

'Dear C.

'You ask my opinion concerning your chapter on the Imagination, both as to the impressions it made on myself, and as to those which I think it will make on the public, i.e. that part of the public, who from the title of the work and from its forming a sort of introduction to a volume of poems, are likely to constitute the great majority of your readers.

'As to myself, and stating in the first place the effect on my understanding, your opinions and method of argument were not only so new to me, but so directly the reverse of all I had ever been accustomed to consider as truth, that even if I had comprehended your premisses sufficiently to have admitted them, and had seen the necessity of your conclusions, I should still have been in that state of mind, which in your note, p. 196, you have so ingeniously evolved, as the antithesis to that in which a man is, when he makes a bull. In your own words, I should have felt as if I had been standing on my head.

'The effect on my feelings, on the other hand, I cannot better represent, than by supposing myself to have known only our light airy modern chapels of ease, and then for the first time to have been placed, and left alone, in one of our largest Gothic cathedrals in a gusty moonlight night of autumn. "Now in glimmer, and now in gloom"; often in palpable darkness not without a chilly sensation of terror; then suddenly emerging into broad yet visionary lights with coloured shadows, of fantastic shapes yet all decked with holy insignia and mystic symbols; and ever and anon coming out full upon pictures and stonework images of great men, with whose names I was familiar, but which looked upon me with countenances and an expression, the most dissimilar to all I had been in the habit of connecting with those names. Those whom I had been taught to venerate as almost superhuman in magnitude of intellect, I found perched in little fretwork niches, as grotesque dwarfs; while the grotesques, in my hitherto belief, stood guarding the high altar with all the characters of apotheosis. In short, what I had supposed substances were thinned away into shadows, while everywhere shadows were deepened into substances:

> If substance may be call'd what shadow seem'd,
> For each seem'd either!
>
> Milton

'Yet after all, I could not but repeat the lines which you had quoted from a MS poem of your own in the Friend, and applied to a work of Mr Wordsworth's though with a few of the words altered:

> . . . An orphic tale indeed,
> A tale *obscure* of high and passionate thoughts
> To *a strange* music chaunted!

'Be assured, however, that I look forward anxiously to your great book on the constructive philosophy, which you have promised and announced: and that I will do my best to understand it. Only I will not promise to descend into the dark cave of Trophonius with you, there to rub my own eyes, in order to make the sparks and figured flashes, which I am required to see.

'So much for myself. But as for the public, I do not hesitate a moment in advising and urging you to withdraw the chapter from the present work, and to reserve it for your announced treatises on the Logos or communicative intellect in Man and Deity. First, because imperfectly as I understand the present chapter, I see clearly that you have done too much, and yet not enough. You have been obliged to omit so many links, from the necessity of compression, that what remains, looks (if I may recur to my former illustration) like the fragments of the winding steps of an old ruined tower. Secondly, a still stronger argument (at least one that I am sure will be more forcible with you) is, that your readers will have both right and reason to complain of you. This chapter, which cannot, when it is printed, amount to so little as an hundred pages, will of necessity greatly increase the expense of the work; and every reader who, like myself, is neither prepared or perhaps calculated for the study of so abstruse a subject so abstrusely treated, will, as I have before hinted, be almost entitled to accuse you of a sort of imposition on him. For who, he might truly observe, could from your title-page, viz. "*My Literary Life and Opinions*," published too as introductory to a volume of miscellaneous poems, have anticipated, or even conjectured, a long treatise on ideal realism, which holds the same relation in abstruseness to Plotinus, as Plotinus does to Plato. It will be well, if already you have not too much of metaphysical disquisition in your work, though as the larger part of the disquisition is historical, it will doubtless be both interesting and instructive to many to whose unprepared minds your speculations on the esemplastic power would be utterly unintelligible. Be assured, if you do publish this chapter in the present work, you will be reminded of Bishop Berkeley's Siris, announced as an Essay on Tar-water, which beginning with Tar ends with the Trinity, the *omne scibile* [everything knowable] forming the interspace. I say in the present work. In that greater work to which you have devoted so many years, and study so intense and various, it will be in its proper place. Your prospectus will have described and announced both its contents and their nature; and if any persons purchase it, who feel no interest in the subjects of which it treats, they will have themselves only to blame.

'I could add to these arguments one derived from pecuniary motives, and particularly from the probable effects on the sale of your present publication; but they would weigh little with you compared with the preceding. Besides, I have long observed, that arguments drawn from your own personal interests more often act on you as narcotics than as stimulants, and that in money concerns you have some small portion of pig nature in your moral idiosyncracy, and like these amiable creatures, must occasionally be pulled backward from the boat in order to make you enter it. All success attend you, for if hard thinking and hard reading are merits, you have deserved it.

Your affectionate, etc.°

In consequence of this very judicious letter, which produced complete conviction on my mind, I shall content myself for the present with stating the main result of the chapter, which I have reserved for that future publication, a detailed prospectus of which the reader will find at the close of the second volume.

The imagination then I consider either as primary, or secondary. The primary imagination I hold to be the living power and prime agent of all human perception, and as a repetition in the finite mind of the eternal act of creation in the infinite I AM. The secondary I consider as an echo of the former, coexisting with the conscious will, yet still as identical with the primary in the kind of its agency, and differing only in degree, and in the mode of its operation. It dissolves, diffuses, dissipates, in order to re-create; or where this process is rendered impossible, yet still at all events it struggles to idealize and to unify. It is essentially vital, even as all objects (*as* objects) are essentially fixed and dead.

Fancy, on the contrary, has no other counters to play with, but fixities and definites. The fancy is indeed no other than a mode of memory emancipated from the order of time and space; and blended with, and modified by that empirical phenomenon of the will, which we express by the word choice. But equally with the ordinary memory it must receive all its materials ready made from the law of association.

Whatever more than this, I shall think it fit to declare concerning the powers and privileges of the imagination in the present work, will be found in the critical essay on the uses of the Supernatural in poetry and the principles that regulate its introduction: which the reader will find prefixed to the poem of The Ancient Mariner.°

CHAPTER XIV

Occasion of the Lyrical Ballads, and the objects originally proposed—Preface to the second edition—The ensuing controversy, its causes and acrimony—Philosophic definitions of a poem and poetry with scholia.

DURING the first year that Mr Wordsworth and I were neighbours, our conversations turned frequently on the two cardinal points of poetry, the power of exciting the sympathy of the reader by a faithful adherence to the truth of nature, and the power of giving the interest of novelty by the modifying colours of imagination. The sudden charm, which accidents of light and shade, which moonlight or sunset diffused over a known and familiar landscape, appeared to represent the practicability of combining both. These are the poetry of nature. The thought suggested itself (to which of us I do not recollect) that a series of poems might be composed of two sorts. In the one, the incidents and agents were to be, in part at least, supernatural; and the excellence aimed at was to consist in the interesting of the affections by the dramatic truth of such emotions, as would naturally accompany such situations, supposing them real. And real in this sense they have been to every human being who, from whatever source of delusion, has at any time believed himself under supernatural agency. For the second class, subjects were to be chosen from ordinary life; the characters and incidents were to be such, as will be found in every village and its vicinity, where there is a meditative and feeling mind to seek after them, or to notice them, when they present themselves.

In this idea originated the plan of the 'Lyrical Ballads'; in which it was agreed, that my endeavours should be directed to persons and characters supernatural, or at least romantic; yet so as to transfer from our inward nature a human interest and a semblance of truth sufficient to procure for these shadows of imagination that willing suspension of disbelief for the moment, which constitutes poetic faith. Mr Wordsworth, on the other hand, was to propose to himself as his object, to give the charm of novelty to things of every day, and to excite a feeling analogous to the supernatural, by awakening the mind's attention from the lethargy of custom, and directing it to the loveliness and the wonders of the world before us; an inexhaustible treasure, but for which in consequence of the film of familiarity and selfish solicitude we have eyes, yet see not, ears that hear not, and hearts that neither feel nor understand.

With this view I wrote the 'Ancient Mariner,' and was preparing among other poems, the 'Dark Ladie,' and the 'Christabel,' in which I should have more nearly realized my ideal, than I had done in my first attempt. But Mr Wordsworth's industry had proved so much more successful, and the number of his poems so much greater, that my compositions, instead of forming a balance, appeared rather an interpolation of heterogeneous matter. Mr Wordsworth added two or three poems written in his own character, in the impassioned, lofty, and sustained diction, which is characteristic of his genius. In this form the 'Lyrical Ballads' were published; and were presented by him, as an experiment, whether subjects, which from their nature rejected the usual ornaments and extra-colloquial style of poems in general, might not be so managed in the language of ordinary life as to produce the pleasurable interest, which it is the peculiar business of poetry to impart. To the second edition he added a preface of considerable length; in which notwithstanding some passages of apparently a contrary import, he was understood to contend for the extension of this style to poetry of all kinds, and to reject as vicious and indefensible all phrases and forms of style that were not included in what he (unfortunately, I think, adopting an equivocal expression) called the language of real life. From this preface, prefixed to poems in which it was impossible to deny the presence of original genius, however mistaken its direction might be deemed, arose the whole long continued controversy. For from the conjunction of perceived power with supposed heresy I explain the inveteracy and in some instances, I grieve to say, the acrimonious passions, with which the controversy has been conducted by the assailants.

Had Mr Wordsworth's poems been the silly, the childish things, which they were for a long time described as being; had they been really distinguished from the compositions of other poets merely by meanness of language and inanity of thought; had they indeed contained nothing more than what is found in the parodies and pretended imitations of them; they must have sunk at once, a dead weight, into the slough of oblivion, and have dragged the preface along with them. But year after year increased the number of Mr Wordsworth's admirers. They were found too not in the lower classes of the reading public, but chiefly among young men of strong sensibility and meditative minds; and their admiration (inflamed perhaps in some degree by opposition) was distinguished by its intensity, I might almost say, by its *religious* fervour. These facts, and the intellectual energy of the author, which was more or less consciously felt, where it was outwardly and even boisterously denied, meeting with sentiments of aversion to his opinions, and of alarm at their consequences, produced an eddy of criticism, which would of itself have borne up the poems by the violence, with which it whirled them round and round. With many parts of this preface in the sense attributed to them and which the words undoubtedly seem to authorize, I never concurred; but on the contrary objected to them as erroneous in principle, and as contradictory (in appearance at least) both to other parts of the same preface, and to the author's own practice in the greater number of the poems themselves. Mr Wordsworth in his recent collection° has, I find, degraded this prefatory disquisition to the end of his second volume, to be read or not at the reader's choice. But he has not, as far as I can discover, announced any change in his poetic creed. All all events, considering it as the source of a controversy, in which I have been honoured more, than I deserve, by the frequent conjunction of my name with his, I think it expedient to declare once for all, in what points I coincide with his opinions, and in what points I altogether differ. But in order to render myself intelligible I must previously, in as few words as possible, explain my ideas, first, of a poem; and secondly, of poetry itself, in kind and in essence.

The office of philosophical disquisition consists in just distinction; while it is the privilege of the philosopher to preserve himself constantly aware, that distinction is not division. In order to obtain adequate notions of any truth, we must intellectually separate its distinguishable parts; and this is the technical process of philosophy. But having so done, we must then restore them in our conceptions to the unity, in which they actually coexist; and this is the result of philosophy. A poem contains the same elements as a prose composition; the difference therefore must consist in a different combination of them, in consequence of a different object proposed. According to the difference of the object will be the difference of the combination. It is possible, that the object may be merely to facilitate the recollection of any given facts or observations by artificial arrangement; and the composition will be a poem, merely because it is distinguished from prose by metre, or by rhyme, or by both conjointly. In this, the lowest sense, a man might attribute the name of a poem to the well-known enumeration of the days in the several months;

> Thirty days hath September,
> April, June, and November, etc.

and others of the same class and purpose. And as a particular pleasure is found in anticipating the recurrence of sounds and quantities, all compositions that have this charm superadded, whatever be their contents, *may* be entitled poems.

So much for the superficial form. A difference of object and contents supplies an additional ground of distinction. The immediate purpose may be the communication of truths; either of truth absolute and demonstrable, as in works of science; or of facts experienced and recorded, as in history. Pleasure, and that of the highest and most permanent kind, may result from the attainment of the end; but it is not itself the immediate end. In other works the communication of pleasure may be the immediate purpose; and though truth either moral or intellectual, ought to be the ultimate end, yet this will distinguish the character of the author, not the class to which the work belongs. Blessed indeed is that state of society, in which the immediate purpose would be baffled by the perversion of the proper ultimate end; in which no charm of diction or imagery could exempt the Bathyllus even of an Anacreon, or the Alexis of Virgil,° from disgust and aversion!

But the communication of pleasure may be the immediate object of a work not metrically composed; and that object may have been in a high degree attained, as in novels and romances. Would then the mere superaddition of metre, with or without rhyme, entitle these to the name of poems? The answer is, that nothing can permanently please, which does not contain in itself the reason why it is so, and not otherwise. If metre be superadded, all other parts must be made consonant with it. They must be such, as to justify the perpetual and distinct attention to each part, which an exact correspondent recurrence of accent and sound are calculated to excite. The final definition then, so deduced, may be thus worded. A poem is that species of composition, which is opposed to works of science, by proposing for its immediate object pleasure, not truth; and from all other species (having this object in common with it) it is discriminated by proposing to itself such delight from the whole, as is compatible with a distinct gratification from each component part.

Controversy is not seldom excited in consequence of the disputants attaching each a different meaning to the same word; and in few instances has this been more striking, than in disputes concerning the present subject. If a man chooses to call every composition a poem, which is rhyme, or measure, or both, I must leave his opinion uncontroverted. The distinction is at least competent to characterize the writer's intention. If it were subjoined, that the whole is likewise entertaining or affecting, as a tale, or as a series of interesting reflections, I of course admit this as another fit ingredient of a poem, and an additional merit. But if the definition sought for be that of a legitimate poem, I answer, it must be one, the parts of which mutually support and explain each other; all in their proportion harmonizing with, and supporting the purpose and known influences of metrical arrangement. The philosophic critics of all ages coincide with the ultimate judgement of all countries, in equally denying the praises of a just poem, on the one hand, to a series of striking lines or distichs, each of which absorbing the whole attention of the reader to itself disjoins it from its context, and makes it a separate whole, instead of an harmonizing part; and on the other hand, to an unsustained composition, from which the reader collects rapidly the general result unattracted by the component parts. The reader should be carried forward, not merely or chiefly by the mechanical impulse of curiosity, or by a restless desire to arrive at the final solution; but by the pleasurable activity of mind excited by the attractions of the journey itself. Like the motion of a serpent, which the Egyptians made the emblem of intellectual power; or like the path of sound through the air; at every step he pauses and half recedes, and from the retrogressive movement collects the force which again carries him onward. 'Precipitandus est *liber* spiritus [the *free* spirit must be hurried onward],' says Petronius Arbiter most happily. The epithet, *liber*, here balances the preceding verb; and it is not easy to conceive more meaning condensed in fewer words.

But if this should be admitted as a satisfactory character of a poem, we have still to seek for a definition of poetry. The writings of Plato, and Bishop Taylor, and the Theoria Sacra of Burnet, furnish undeniable proofs that poetry of the highest kind may exist without metre, and even without the contradistinguishing objects of a poem. The first chapter of Isaiah (indeed a very large proportion of the whole book) is poetry in the most emphatic sense; yet it would be not less irrational than strange to assert, that pleasure, and not truth, was the immediate object of the prophet. In short, whatever specific import we attach to the word, poetry, there will be found involved in it, as a necessary consequence, that a poem of any length neither can be, or ought to be, all poetry. Yet if an harmonious whole is to be produced, the remaining parts must be preserved in keeping with the poetry; and this can be no otherwise effected than by such a studied selection and artificial arrangement, as will partake of one, though not a peculiar, property of poetry. And this again can be no other than the property of exciting a more continuous and equal attention, than the language of prose aims at, whether colloquial or written.

My own conclusions on the nature of poetry, in the strictest use of the word, have been in part anticipated in the preceding disquisition on the fancy and imagination. What is poetry? is so nearly the same question with, what is a poet? that the answer to the one is involved in the solution of the other. For it is a distinction resulting from the poetic genius itself, which sustains and modifies the images, thoughts, and emotions of the poet's own mind. The poet, described in ideal perfection, brings the whole soul of man into activity, with the subordination of its faculties to each other, according to their relative worth and dignity. He diffuses a tone, and spirit of unity, that blends, and (as it were) fuses, each into each, by that synthetic and magical power, to which we have exclusively appropriated the name of imagination. This power, first put in action by the will and understanding, and retained under their irremissive, though gentle and unnoticed, control (*laxis effertur habenis* [it is carried onwards with loose reins]) reveals itself in the balance or reconciliation of opposite or discordant qualities: of sameness, with difference; of the general, with the concrete; the idea, with the image; the individual, with the representative; the sense of novelty and freshness, with old and familiar objects; a more than usual state of emotion, with more than usual order; judgement ever awake and steady self-possession, with enthusiasm and feeling profound or vehement; and while it blends and harmonizes the natural and the artificial, still subordinates art to nature; the manner to the matter; and our admiration of the poet to our sympathy with the poetry. 'Doubtless,' as Sir John Davies observes of the soul (and his words may with slight alteration be applied, and even more appropriately to the poetic imagination)

> Doubtless this could not be, but that she turns
> Bodies to spirit by sublimation strange,
> As fire converts to fire the things it burns,
> As we our food into our nature change.
>
> From their gross matter she abstracts their forms,
> And draws a kind of quintessence from things;
> Which to her proper nature she transforms
> To bear them light, on her celestial wings.

> Thus does she, when from individual states
> She doth abstract the universal kinds;
> Which then re-clothed in divers names and fates
> Steal access through our senses to our minds.

Finally, good sense is the body of poetic genius, fancy its drapery, motion its life, and imagination the soul that is everywhere, and in each; and forms all into one graceful and intelligent whole.

CHAPTER XV

The specific symptoms of poetic power elucidated in a critical analysis of Shakespeare's Venus and Adonis, and Lucrece.

IN the application of these principles to purposes of practical criticism as employed in the appraisal of works more or less imperfect, I have endeavoured to discover what the qualities in a poem are, which may be deemed promises and specific symptoms of poetic power, as distinguished from general talent determined to poetic composition by accidental motives, by an act of the will, rather than by the inspiration of a genial and productive nature. In this investigation, I could not, I thought, do better, than keep before me the earliest work of the greatest genius, that perhaps human nature has yet produced, our *myriad-minded** Shakespeare. I mean the 'Venus and Adonis,' and the 'Lucrece'; works which give at once strong promises of the strength, and yet obvious proofs of the immaturity, of his genius. From these I abstracted the following marks, as characteristics of original poetic genius in general.

1. In the 'Venus and Adonis,' the first and most obvious excellence is the perfect sweetness of the versification; its adaptation to the subject; and the power displayed in varying the march of the words without passing into a loftier and more majestic rhythm, than was demanded by the thoughts, or permitted by the propriety of preserving a sense of melody predominant. The delight in richness

* *Ἄνηρ μυριονοῦς*, a phrase which I have borrowed from a Greek monk, who applies it to a Patriarch of Constantinople. I might have said, that I have *reclaimed*, rather than borrowed it: for it seems to belong to Shakespeare, *de jure singulari, et ex privilegio naturae* [by a law peculiar to himself, and by privilege of nature].

and sweetness of sound, even to a faulty excess, if it be evidently original, and not the result of an easily imitable mechanism, I regard as a highly favourable promise in the compositions of a young man. 'The man that hath not music in his soul' can indeed never be a genuine poet. Imagery (even taken from nature, much more when transplanted from books, as travels, voyages, and works of natural history); affecting incidents; just thoughts; interesting personal or domestic feelings; and with these the art of their combination or intertexture in the form of a poem; may all by incessant effort be acquired as a trade, by a man of talents and much reading, who, as I once before observed, has mistaken an intense desire of poetic reputation for a natural poetic genius; the love of the arbitrary end for a possession of the peculiar means. But the sense of musical delight, with the power of producing it, is a gift of imagination; and this together with the power of reducing multitude into unity of effect, and modifying a series of thoughts by some one predominant thought or feeling, may be cultivated and improved, but can never be learnt. It is in these that 'Poeta nascitur non fit [A poet is born, not made].'

2. A second promise of genius is the choice of subjects very remote from the private interests and circumstances of the writer himself. At least I have found, that where the subject is taken immediately from the author's personal sensations and experiences, the excellence of a particular poem is but an equivocal mark, and often a fallacious pledge, of genuine poetic power. We may perhaps remember the tale of the statuary, who had acquired considerable reputation for the legs of his goddesses, though the rest of the statue accorded but indifferently with ideal beauty; till his wife, elated by her husband's praises, modestly acknowledged, that she herself had been his constant model. In the Venus and Adonis, this proof of poetic power exists even to excess. It is throughout as if a superior spirit more intuitive, more intimately conscious, even than the characters themselves, not only of every outward look and act, but of the flux and reflux of the mind in all its subtlest thoughts and feelings, were placing the whole before our view; himself meanwhile unparticipating in the passions, and actuated only by that pleasurable excitement, which had resulted from the energetic fervour of his own spirit in so vividly exhibiting, what it had so accurately and profoundly contemplated. I think, I should have conjectured from these poems, that even then the great instinct, which impelled the poet to the drama, was secretly working in him, prompting him by a series and never-broken chain of imagery, always vivid and because unbroken, often minute; by the highest effort

of the picturesque in words, of which words are capable, higher perhaps than was ever realized by any other poet, even Dante not excepted; to provide a substitute for that visual language, that constant intervention and running comment by tone, look and gesture, which in his dramatic works he was entitled to expect from the players. His Venus and Adonis seem at once the characters themselves, and the whole representation of those characters by the most consummate actors. You seem to be *told* nothing, but to see and hear everything. Hence it is, that from the perpetual activity of attention required on the part of the reader; from the rapid flow, the quick change, and the playful nature of the thoughts and images; and above all from the alienation, and, if I may hazard such an expression, the utter *aloofness* of the poet's own feelings, from those of which he is at once the painter and the analyst; that though the very subject cannot but detract from the pleasure of a delicate mind, yet never was poem less dangerous on a moral account. Instead of doing as Ariosto, and as, still more offensively, Wieland has done, instead of degrading and deforming passion into appetite, the trials of love into the struggles of concupiscence; Shakespeare has here represented the animal impulse itself, so as to preclude all sympathy with it, by dissipating the reader's notice among the thousand outward images, and now beautiful, now fanciful circumstances, which form its dresses and its scenery; or by diverting our attention from the main subject by those frequent witty or profound reflections, which the poet's ever active mind has deduced from, or connected with, the imagery and the incidents. The reader is forced into too much action to sympathize with the merely passive of our nature. As little can a mind thus roused and awakened be brooded on by mean and indistinct emotion, as the low, lazy mist can creep upon the surface of a lake, while a strong gale is driving it onward in waves and billows.

3. It has been before observed, that images however beautiful, though faithfully copied from nature, and as accurately represented in words, do not of themselves characterize the poet. They become proofs of original genius only as far as they are modified by a predominant passion; or by associated thoughts or images awakened by that passion; or when they have the effect of reducing multitude to unity, or succession to an instant; or lastly, when a human and intellectual life is transferred to them from the poet's own spirit,

> Which shoots its being through earth, sea, and air.

In the two following lines for instance, there is nothing objectionable, nothing which would preclude them from forming, in their proper place, part of a descriptive poem:

> Behold yon row of pines, that shorn and bow'd
> Bend from the sea-blast, seen at twilight eve.

But with the small alteration of rhythm, the same words would be equally in their place in a book of topography, or in a descriptive tour. The same image will rise into a semblance of poetry if thus conveyed:

> Yon row of bleak and visionary pines,
> By twilight-glimpse discerned, mark! how they flee
> From the fierce sea-blast, all their tresses wild
> Streaming before them.

I have given this as an illustration, by no means as an instance, of that particular excellence which I had in view, and in which Shakespeare even in his earliest, as in his latest works, surpasses all other poets. It is by this, that he still gives a dignity and a passion to the objects which he presents. Unaided by any previous excitement, they burst upon us at once in life and in power.

> Full many a glorious morning have I seen
> *Flatter* the mountain tops with sovereign eye.
>
> Shakespeare's Sonnet 33rd

> Not mine own fears, nor the prophetic soul
> Of the wide world dreaming on things to come—
>
> * * * * *
>
> The mortal moon hath her eclipse endur'd,
> And the sad augurs mock their own presage;
> Incertainties now crown themselves assur'd,
> And Peace proclaims olives of endlesss age.
> Now with the drops of this most balmy time
> My Love looks fresh: and Death to me subscribes!
> Since spite of him, I'll live in this poor rhyme,
> While he insults o'er dull and speechless tribes.
> And thou in this shalt find thy monument,
> When tyrant's crests, and tombs of brass are spent.
>
> Sonnet 107

As of higher worth, so doubtless still more characteristic of poetic genius does the imagery become, when it moulds and colours itself to the circumstances, passion, or character, present and foremost in the mind. For unrivalled instances of this excellence, the reader's own memory will refer him to the Lear, Othello, in short to which not of the 'great, ever living, dead man's' dramatic works? *Inopem me copia fecit* [plenty has made me poor]. How true it is to nature, he has himself finely expressed in the instance of love in Sonnet 98.

From you have I been absent in the spring,
When proud pied April drest in all its trim
Hath put a spirit of youth in every thing;
That heavy Saturn laugh'd and leap'd with him.
Yet nor the lays of birds, nor the sweet smell
Of different flowers in odour and in hue,
Could make me any summer's story tell,
Or from their proud lap pluck them, where they grew:
Nor did I wonder at the lilies white,
Nor praise the deep vermillion in the rose;
They were, tho' sweet, but figures of delight,
Drawn after you, you pattern of all those.
Yet seem'd it winter still, and you away,
As with your shadow I with these did play!

Scarcely less sure, or if a less valuable, not less indispensable mark

Γονίμου μέν Ποιητου . . .
. . . οστις ρημα γενναιον λακοι

[of a True Poet, such a one as
utters noble words]

will the imagery supply, when, with more than the power of the painter, the poet gives us the liveliest image of succession with the feeling of simultaneousness!

With this he breaketh from the sweet embrace
Of those fair arms, that held him to her heart,
And homeward through the dark lawns runs apace:
Look how a bright star shooteth from the sky!
So glides he through the night from Venus' eye.

4. The last character I shall mention, which would prove indeed but little, except as taken conjointly with the former; yet without which the former could scarce exist in a high degree, and (even if this were possible) would give promises only of transitory flashes and a meteoric power; is depth, and energy of thought. No man was ever yet a great poet, without being at the same time a profound philosopher.

For poetry is the blossom and the fragrancy of all human knowledge, human thoughts, human passions, emotions, language. In Shakespeare's poems, the creative power, and the intellectual energy wrestle as in a war embrace. Each in its excess of strength seems to threaten the extinction of the other. At length, in the drama they were reconciled, and fought each with its shield before the breast of the other. Or like two rapid streams, that at their first meeting within narrow and rocky banks mutually strive to repel each other, and intermix reluctantly and in tumult; but soon finding a wider channel and more yielding shores blend, and dilate, and flow on in one current and with one voice. The Venus and Adonis did not perhaps allow the display of the deeper passions. But the story of Lucretia seems to favour, and even demand their intensest workings. And yet we find in Shakespeare's management of the tale neither pathos, nor any other dramatic quality. There is the same minute and faithful imagery as in the former poem, in the same vivid colours, inspirited by the same impetuous vigour of thought, and diverging and contracting with the same activity of the assimilative and of the modifying faculties; and with a yet larger display, a yet wider range of knowledge and reflection; and lastly, with the same perfect dominion, often domination, over the whole world of language. What then shall we say? even this; that Shakespeare, no mere child of nature; no automaton of genius; no passive vehicle of inspiration possessed by the spirit, not possessing it; first studied patiently, meditated deeply, understood minutely, till knowledge become habitual and intuitive wedded itself to his habitual feelings, and at length gave birth to that stupendous power, by which he stands alone, with no equal or second in his own class; to that power, which seated him on one of the two glory-smitten summits of the poetic mountain, with Milton as his compeer not rival. While the former darts himself forth, and passes into all the forms of human character and passion, the one Proteus of the fire and the flood; the other attracts all forms and things to himself, into the unity of his own ideal. All things and modes of action shape themselves anew in the being of Milton; while Shakespeare becomes all things, yet for ever remaining himself. O what great men hast thou not produced, England! my country! truly indeed—

> Must *we* be free or die, who speak the tongue,
> Which Shakespeare spake; the faith and morals hold,
> Which Milton held. In every thing we are sprung
> Of earth's first blood, have titles manifold!
>
> Wordsworth

CHAPTER XVI

Striking points of difference between the Poets of the present age and those of the 15th and 16th centuries—Wish expressed for the union of the characteristic merits of both.

CHRISTENDOM, from its first settlement on feudal rights, has been so far one great body, however imperfectly organized, that a similar spirit will be found in each period to have been acting in all its members. The study of Shakespeare's poems (I do not include his dramatic works, eminently as they too deserve that title) led me to a more careful examination of the contemporary poets both in this and in other countries. But my attention was especially fixed on those of Italy, from the birth to the death of Shakespeare; that being the country in which the fine arts had been most sedulously, and hitherto most successfully cultivated. Abstracted from the degrees and peculiarities of individual genius, the properties common to the good writers of each period seem to establish one striking point of difference between the poetry of the fifteenth and sixteenth centuries, and that of the present age. The remark may perhaps be extended to the sister art of painting. At least the latter will serve to illustrate the former. In the present age the poet (I would wish to be understood as speaking generally, and without allusion to individual names) seems to propose to himself as his main object, and as that which is the most characteristic of his art, new and striking images; with incidents that interest the affections or excite the curiosity. Both his characters and his descriptions he renders, as much as possible, specific and individual, even to a degree of portraiture. In his diction and metre, on the other hand, he is comparatively careless. The measure is either constructed on no previous system, and acknowledges no justifying principle but that of the writer's convenience; or else some mechanical movement is adopted, of which one couplet or stanza is so far an adequate specimen, as that the occasional differences appear evidently to arise from accident, or the qualities of the language itself, not from meditation and an intelligent purpose. And the language from Pope's translation of Homer, to Darwin's Temple of Nature, may, notwithstanding some illustrious exceptions, be too faithfully characterized, as claiming to be poetical for no better reason, than that it would be intolerable in conversation or in prose. Though alas! even our prose writings, nay even the style of our more set discourses, strive to be in the fashion, and trick themselves out in the soiled and overworn finery of the meretricious Muse. It is true, that of late a great improvement in this respect is observable in our most popular writers. But it is equally true, that this recurrence to plain sense, and genuine mother English, is far from being general; and that the composition of our novels, magazines, public harangues, etc. is commonly as trivial in thought, and yet enigmatic in expression, as if Echo and Sphinx had laid their heads together to construct it. Nay, even of those who have most rescued themselves from this contagion, I should plead inwardly guilty to the charge of duplicity or cowardice, if I withheld my conviction, that few have guarded the purity of their native tongue with that jealous care, which the sublime Dante in his tract 'De la nobile volgare eloquenza,' declares to be the first duty of a poet. For language is the armoury of the human mind; and at once contains the trophies of its past, and the weapons of its future conquests. 'Animadverte, quam sit ab improprietate verborum pronum hominibus prolabi in errores circa res [Notice, how easily men slip from improper use of words into errors about the things themselves (*CC*)]!' Hobbes: *Exam. et Exmend. hod. Math.*—'Sat vero, in hâc vitae brevitate et naturae obscuritate, rerum est, quibus cognoscendis tempus impendatur, ut confusis et multivocis sermonibus intelligendis illud consumere non opus est. Eheu! quantas strages paravere verba nubila, quae tot dicunt, ut nihil dicunt—nubes potius, e quibus et in rebus politicis et in ecclesiâ turbines et tonitrua erumpunt! Et proinde recte dictum putamus a Platone in Gorgia: *ος αν τα ονοματα ειδει, ισεται και τα πραγματα*: et ab Epicteto, *αρχη παιδευσεως η των ονοματων επισκεψις*: et prudentissime Galenus scribit, *η των ονοματων χρησις παραχθεισα και την των πραγματων επιταραττει γνωσιν*. Egregie vero J. C. Scaliger, in Lib. I. de Plantis: "Est primum, inquit, sapientis officium, bene sentire, ut sibi vivat: proximum, bene loqui, ut patriae vivat" [There are certainly plenty of things in this short life and dark world which are worth time to study, so that we need not spend time in trying to understand confused and ambiguous words. Alas, what great calamities have misty words produced, that say so much that they say nothing—clouds, rather, from which hurricanes burst, both in Church and State. I think that what Plato has said in the Gorgias is indeed true: 'Anyone who knows words will know things too'; and as Epictetus says, 'the study of words is the beginning of education'; and Galen wrote most wisely 'Confusion in our use of words makes confusion in our knowledge of things.' J. C. Scaliger has indeed said excellently, in Book I of his *Plants*: 'A wise man's first duty is to think well so that he can live for himself; the next is to speak well so that he can live for his country' (*CC*)].' Sennertus, *de Puls: Differentiâ*.

Something analogous to the materials and structure of modern poetry I seem to have noticed (but here I beg to be understood as speaking with the utmost diffidence) in our common landscape painters. Their foregrounds and intermediate distances are comparatively unattractive: while the main interest of the landscape is thrown into the background, where mountains and torrents and castles forbid the eye to proceed, and nothing tempts it to trace its way back again. But in the works of the great Italian and Flemish masters, the front and middle objects of the landscape are the most obvious and determinate, the interest gradually dies away in the background, and the charm and peculiar worth of the picture consists, not so much in the specific objects which it conveys to the understanding in a visual language formed by the substitution of figures for words, as in the beauty and harmony of the colours, lines and expression, with which the objects are represented. Hence novelty of subject was rather avoided than sought for. Superior excellence in the manner of treating the same subjects was the trial and test of the artist's merit.

Not otherwise is it with the more polished poets of the 15th and 16th century, especially with those of Italy. The imagery is almost always general: sun, moon, flowers, breezes, murmuring streams, warbling songsters, delicious shades, lovely damsels, cruel as fair, nymphs, naiads, and goddesses, are the materials which are common to all, and which each shaped and arranged according to his judgement or fancy, little solicitous to add or to particularize. If we make an honourable exception in favour of some English poets, the thoughts too are as little novel as the images; and the fable of their narrative poems, for the most part drawn from mythology, or sources of equal notoriety, derive their chief attractions from the manner of treating them; from impassioned flow, or picturesque arrangement. In opposition to the present age, and perhaps in as faulty an extreme, they placed the essence of poetry in the *art*. The excellence, at which they aimed, consisted in the exquisite polish of the diction, combined with perfect simplicity. This their prime object, they attained by the avoidance of every word, which a gentleman would not use in dignified conversation, and of every word and phrase, which none but a learned man would use; by the studied position of words and phrases, so that not only each part should be melodious in itself, but contribute to the harmony of the whole, each note referring and conducing to the melody of all the foregoing and following words of the same period or stanza; and lastly with equal labour, the greater because unbetrayed, by the variation and various harmonies of their metrical movement. Their measures, however, were not indebted for their variety to the introduction of new metres, such as have been attempted of late in the 'Alonzo and Imogen,'° and others borrowed from the German, having in their very mechanism a specific overpowering tune, to which the generous reader humours his voice and emphasis, with more indulgence to the author than attention to the meaning or quantity of the words; but which, to an ear familiar with the *numerous* sounds of the Greek and Roman poets, has an effect not unlike that of galloping over a paved road in a German stage-waggon without springs. On the contrary, our elder bards both of Italy and England produced a far greater, as well as more charming variety by countless modifications, and subtle balances of sound in the common metres of their country. A lasting and enviable reputation awaits that man of genius, who should attempt and realize a union. Who should recall the high finish; the appropriateness; the facility; the delicate proportion; and above all, the perfusive and omnipresent grace; which have preserved, as in a shrine of precious amber, the 'Sparrow' of Catullus, the 'Swallow,' the 'Grasshopper,' and all the other little loves of Anacreon: and which with bright, though diminished glories, revisited the youth and early manhood of Christian Europe, in the vales of Arno,* and the groves of Isis and of

* These thoughts were suggested to me during the perusal of the Madrigals of Giovambatista Strozzi published in Florence (nella Stamperia del Sermartelli) 1st May 1593, by his sons Lorenzo and Filippo Strozzi, with a dedication to their deceased paternal uncle, 'Signor Leone Strozzi, Generale delle battaligie di Santa Chiesa.' As I do not remember to have seen either the poems or their author mentioned in any English work, or have found them in any of the common collections of Italian poetry; and as the little work is of rare occurrence; I will transcribe a few specimens. I have seldom met with compositions that possessed, to my feelings, more of that satisfying *entireness*, that complete adequateness of the manner to the matter which so charms us in Anacreon, joined with the tenderness, and more than the delicacy of Catullus. Trifles as they are, they were probably elaborated with great care; yet in the perusal we refer them to a spontaneous energy rather than to voluntary effort. To a cultivated taste there is a delight in perfection for its own sake, independent of the material in which it is manifested, that none but a cultivated taste can understand or appreciate.

After what I have advanced, it would appear presumption to offer a translation; even if the attempt were not discouraged by the different genius of the English mind and language, which demands a denser body of thought as the condition of a high polish, than the Italian. I cannot but deem it likewise an advantage in the Italian tongue, in

Cam; and who with these should combine the keener interest, deeper pathos, manlier reflection, and the fresher and more various imagery, which give a value and a name that will not pass away to the poets who have done honour to our own times, and to those of our immediate predecessors.

many other respects inferior to our own, that the language of poetry is more distinct from that of prose than with us. From the earlier appearance and established primacy of the Tuscan poets, concurring with the number of independent states, and the diversity of written dialects, the Italians have gained a poetic idiom, as the Greeks before them had obtained from the same causes, with greater and more various discriminations—ex. gr. the ionic for their heroic verses; the attic for their iambic; and the two modes of the doric, the lyric or sacerdotal, and the pastoral, the distinctions of which were doubtless more obvious to the Greeks themselves than they are to us.

I will venture to add one other observation before I proceed to the transcription. I am aware, that the sentiments which I have avowed concerning the points of difference between the poetry of the present age, and that of the period between 1500 and 1650, are the reverse of the opinion commonly entertained. I was conversing on this subject with a friend, when the servant, a worthy and sensible woman, coming in, I placed before her two engravings, the one a pinky-coloured plate of the day, the other a masterly etching by Salvator Rosa, from one of his own pictures. On pressing her to tell us, which she preferred, after a little blushing and flutter of feeling, she replied 'Why, that, Sir! to be sure! (pointing to the ware from the Fleet Street printshops) It's so neat and elegant. T'other is such a scratchy slovenly thing.' An artist, whose writings are scarcely less valuable than his works, and to whose authority more deference will be willingly paid, than I could even wish, should be shown to mine, has told us, and from his own experience too, that good taste must be acquired, and like all other good things, is the result of thought, and the submissive study of the best models. If it be asked, 'But what shall I deem such?' the answer is; presume these to be the best, the reputation of which has been matured into fame by the consent of ages. For wisdom always has a final majority, if not by conviction, yet by acquiescence. In addition to Sir J. Reynolds° I may mention Harris of Salisbury, who in one of his philosophical disquisitions has written on the means of acquiring a just taste with the precision of Aristotle, and the elegance of Quintilian.

(1) MADRIGALE

Gelido suo ruscel chiaro, e tranquillo
M'insegnó Amor, di state a mezzo 'l giorno:
Ardean le selve, ardean le piagge, e i colli.
Ond'io, ch'al piu gran gielo ardo e sfavillo,
Subito corsi; ma si puro adorno
Girsene il vidi, che turbar no 'l volli:
Sol mi specchiava, e 'n dolce ombrosa sponda
Mi stava intento al mormorar dell'onda.

(2) MADRIGALE

Aure dell'angoscioso viver mio
Refrigerio soave,
E dolce si, che piu non mi par grave

Ne 'l arder, ne 'l morir, anz'il desio;
Deh voi 'l ghiaccio, e le nubi, e 'l tempo rio
Discacciatene omai, che l'onda chiara,
E l'ombra non men cara
A scherzare, e cantar per suoi boschetti
E prati Festa ed Allegrezza alletti.

(3) MADRIGALE

Pacifiche, ma spesso in amorosa
Guerra co' fiori, e l'erba
Alla stagione acerba
Verde Insegne del giglio e della rosa
Movete, Aure, pian pian; che tregua o posa,
Se non pace, io ritrove:
E so ben dove — Oh vago, et mansueto
Sguardo, oh labbra d'ambrosia, oh rider lieto!

(4) MADRIGALE

Hor come un Scoglio stassi,
Hor come un Rio se 'n fugge,
Ed hor crud' Orsa rugge,
Hor canta Angelo pio: ma che non fassi?
E che non fammi, O Sassi,
O Rivi, o belve, o Dii, questa mia vaga
Non so, se Ninfa, o Maga,
Non so, se Donna, o Dea,
Non so, se dolce ó rea?

(5) MADRIGALE

Piangendo mi baciaste,
E ridendo il negasté:
In doglia hebbivi pia,
In festa hebbivi ria:
Nacque Gioia di pianti,
Dolor di riso: O amanti
Miseri, habbiate insieme
Ognor Paura e Speme.

(6) MADRIGALE

Bel Fior, tu mi rimembri
La rugiadosa guancia del bel viso;
E si vera l'assembri,
Che 'n te sovente, come in lei m'affiso:
Ed hor dell vago riso,
Hor dell sereno sguardo
Io pur cieco risguardo. Ma qual fugge,
O Rosa, il mattin lieve?
E chi te, come neve,
E 'l mio cor teco, e la mia vita strugge.

[*cont.*]

(7) MADRIGALE

Anna mia, Anna dolce, oh sempre nuovo
E piu chiaro concento,
Quanta dolcezza sento
In sol Anna dicendo? Io mi par pruovo,
Ne quì tra noi ritruovo,
Ne tra cieli armonia,
Che del bel nome suo piu dolce sia:
Altro il Cielo, altro Amore,
Altro non suona l'Eco del mio core.

(8) MADRIGALE

Hor che 'l prato, e la selva si scolora,
Al tuo Sereno ombroso
Muovine, alto Riposo!
Deh ch'io riposi una sol notte, un hora!
Han le fere, e gli augelli, ognun talora
Ha qualche pace; io quando,
Lasso! non vonne errando,
E non piango, e non grido? e qual pur forte?
Ma poiché non sent' egli, odine, Morte!

(9) MADRIGALE

Risi e piansi d'Amor; ne peró mai
Se non in fiamma, ó 'n onda, ó 'n vento scrissi;
Spesso mercè trovai
Crudel; sempre in me morto, in altri vissi!
Hor da' più scuri abyssi al Ciel m'alzai,
Hor ne pur caddi giuso:
Stanco al fin quí son chiuso!

[(1) 'Love showed me his chill stream, clear and tranquil, in summertime at noonday; the woods were burning, the slopes, the hills were burning. So I, who in the coldest frost burn and sparkle, at once hastened to it; but I saw it flowing on so pure and fair that I did not wish to sully it; I only mirrored myself within it, and on its sweet and shady bank I rested, intent upon the murmuring of its wave.' (2) 'Breezes, gentle comfort of my tormented life, and so sweet that no longer does burning or death seem grievous to me, but rather desire alone; pray, drive far away the ice, the clouds, the evil weather, now that the clear wave and the shade, no less dear, entice Festivity and Merriment to sport and sing through their groves and meadows.' (3) 'Oh breezes, peaceful, yet often at amorous war with the flowers and grass, advance softly your green standards of the lily and the rose against the immature season; so that I may find truce or rest, if not peace: and I know well where. Oh charming mild glance, oh ambrosian lips, oh gay laughter!' (4) 'Now she stands fixed like a Rock, now like a River she glides away, now she roars like a savage Bear, now sings like a pitying Angel; but into what does she not transform herself? And into what does she not transform me, Stones or Streams, wild beasts or Gods, this my fair—I know not whether Nymph or Enchantress, whether Lady or Goddess, whether sweet or pitiless?' (5) 'Weeping you kissed me, and laughing you refused; in grief I found you pitiful, in pleasure I found

CHAPTER XVII

Examination of the tenets peculiar to Mr Wordsworth—Rustic life (above all, low and rustic life) especially unfavourable to the formation of a human diction—The best parts of language the product of philosophers, not clowns or shepherds—Poetry essentially ideal and generic—The language of Milton as much the language of real life, yea, incomparably more so than that of the cottager.

As far then as Mr Wordsworth in his preface contended, and most ably contended, for a reformation in our poetic diction, as far as he has evinced the truth of passion, and the dramatic propriety of those figures and metaphors in the original poets, which stripped of their justifying reasons, and converted into mere artifices of connection or ornament, constitute the characteristic falsity in the poetic style of the moderns; and as far as he has, with equal acuteness and clearness, pointed out the process in which this change was effected, and the resemblances between that state into which the reader's mind is thrown by the pleasurable confusion of thought from an unaccustomed train of words and images; and that state which is induced by the natural language of impassioned feeling; he undertook a useful task, and deserves all praise, both for the attempt and for the execution. The provocations to this remonstrance in behalf of truth

you cruel; joy was born of weeping, suffering from laughter. Oh wretched lovers, may you always find together Fear and Hope.' (6) 'Fair flower, you recall to me the dewy cheek of that fair face, and so truly do you resemble it that often I gaze upon you as though upon her: and, blind though I am, contemplate now her charming laugh, now her calm glance. But how lightly, oh Rose, does the morning flee? And who dissolves you like snow, and with you my heart and my very life?' (7) 'My Anna, sweet Anna, oh cadence ever fresh and ever brighter, what sweetness do I feel only in saying Anna? I endeavour indeed, but neither here among us nor in the heavens can I find any harmony which is sweeter than her fair name: Heaven, Love, the Echo of my Heart, plays no other tune.' (8) 'Now that the mead and wood grow dim, beneath your shadowy calm sky move forth, lofty Repose! Ah let me rest one single night, one hour! The wild beasts, the birds, every living thing has sometimes some peace, but I, alas, when do I not wander on, nor weep, nor cry out? and indeed how loudly? But since he does not hear, hear me thou, oh Death.' (9) 'I laughed and wept with Love, yet never did I write except in flame, in water, or in wind; often I found cruel mercy; ever dead to myself, I lived in another; now I rose from the darkest Abyss to Heaven, now I fell down from it again; wearied at last, here have I made my close.'

(translation by Beatrice Corrigan for *CN*)].

and nature were still of perpetual recurrence before and after the publication of this preface. I cannot likewise but add, that the comparison of such poems of merit, as have been given to the public within the last ten or twelve years, with the majority of those produced previously to the appearance of that preface, leave no doubt on my mind, that Mr Wordsworth is fully justified in believing his efforts to have been by no means ineffectual. Not only in the verses of those who have professed their admiration of his genius, but even of those who have distinguished themselves by hostility to his theory, and depreciation of his writings, are the impressions of his principles plainly visible. It is possible, that with these principles others may have been blended, which are not equally evident; and some which are unsteady and subvertible from the narrowness or imperfection of their basis. But it is more than possible, that these errors of defect or exaggeration, by kindling and feeding the controversy, may have conduced not only to the wider propagation of the accompanying truths, but that by their frequent presentation to the mind in an excited state, they may have won for them a more permanent and practical result. A man will borrow a part from his opponent the more easily, if he feels himself justified in continuing to reject a part. While there remain important points in which he can still feel himself in the right, in which he still finds firm footing for continued resistance, he will gradually adopt those opinions, which were the least remote from his own convictions, as not less congruous with his own theory, than with that which he reprobates. In like manner with a kind of instinctive prudence, he will abandon by little and little his weakest posts, till at length he seems to forget that they had ever belonged to him, or affects to consider them at most as accidental and 'petty annexments,' the removal of which leaves the citadel unhurt and unendangered.

My own differences from certain supposed parts of Mr Wordsworth's theory ground themselves on the assumption, that his words had been rightly interpreted, as purporting that the proper diction for poetry in general consists altogether in a language taken, with due exceptions, from the mouths of men in real life, a language which actually constitutes the natural conversation of men under the influence of natural feelings. My objection is, first, that in any sense this rule is applicable only to certain classes of poetry; secondly, that even to these classes it is not applicable, except in such a sense, as hath never by any one (as far as I know or have read) been denied or doubted; and lastly, that as far as, and in that degree in which it is practicable, yet as a rule it is useless, if not injurious, and therefore either need not, or ought not to be practised. The poet informs his reader, that he had generally chosen low and rustic life; but not *as* low and rustic, or in order to repeat that pleasure of doubtful moral effect, which persons of elevated rank and of superior refinement oftentimes derive from a happy imitation of the rude unpolished manners and discourse of their inferiors. For the pleasures so derived may be traced to three exciting causes. The first is the naturalness, in fact, of the things represented. The second is the apparent naturalness of the representation, as raised and qualified by an imperceptible infusion of the author's own knowledge and talent, which infusion does, indeed, constitute it an imitation as distinguished from a mere copy. The third cause may be found in the reader's conscious feeling of his superiority awakened by the contrast presented to him; even as for the same purpose the kings and great barons of yore retained, sometimes actual clowns and fools, but more frequently shrewd and witty fellows in that character. These, however, were not Mr Wordsworth's objects. He chose low and rustic life, 'because in that condition the essential passions of the heart find a better soil, in which they can attain their maturity, are less under restraint, and speak a plainer and more emphatic language; because in that condition of life our elementary feelings coexist in a state of greater simplicity, and consequently may be more accurately contemplated, and more forcibly communicated; because the manners of rural life germinate from those elementary feelings; and from the necessary character of rural occupations are more easily comprehended, and are more durable; and lastly, because in that condition the passions of men are incorporated with the beautiful and permanent forms of nature.'°

Now it is clear to me, that in the most interesting of the poems, in which the author is more or less dramatic, as the 'Brothers,' 'Michael,' 'Ruth,' the 'Mad Mother,' etc. the persons introduced are by no means taken from low or rustic life in the common acceptation of those words; and it is not less clear, that the sentiments and language, as far as they can be conceived to have been really transferred from the minds and conversation of such persons, are attributable to causes and circumstances not necessarily connected with 'their occupations and abode.' The thoughts, feelings, language, and manners of the shepherd-farmers in the vales of Cumberland and Westmoreland, as far as they are actually adopted in those poems, may be accounted for from causes, which will and do produce the same results in every state of life, whether in town or country. As the two principal I rank that independence, which raises a man above servitude, or daily toil for the profit of others, yet not above the necessity of industry and a frugal simplicity of domestic life; and the accompanying unambitious, but solid and religious education, which has rendered few books familiar, but the Bible, and the liturgy or hymnbook. To this latter cause, indeed, which is so far accidental, that it is the blessing of particular countries and a particular age, not the product of particular places or employments, the poet owes the show of probability, that his personages might really feel, think, and talk with any tolerable resemblance to his representation. It is an excellent remark of Dr Henry More's (Enthusiasmus triumphatus, Sec. XXXV) that 'a man of confined education, but of good parts, by constant reading of the bible will naturally form a more winning and commanding rhetoric than those that are learned; the intermixture of tongues and of artificial phrases debasing *their* style.'

It is, moreover, to be considered that to the formation of healthy feelings, and a reflecting mind, negations involve impediments not less formidable, than sophistication and vicious intermixture. I am convinced, that for the human soul to prosper in rustic life, a certain vantage-ground is prerequisite. It is not every man, that is likely to be improved by a country life or by country labours. Education, or original sensibility, or both, must pre-exist, if the changes, forms, and incidents of nature are to prove a sufficient stimulant. And where these are not sufficient, the mind contracts and hardens by want of stimulants; and the man becomes selfish, sensual, gross, and hard-hearted. Let the management of the poor laws in Liverpool, Manchester, or Bristol be compared with the ordinary dispensation of the poor rates in agricultural villages, where the farmers are the overseers and guardians of the poor. If my own experience have not been particularly unfortunate, as well as that of the many respectable country clergymen with whom I have conversed on the subject, the result would engender more than scepticism concerning the desirable influences of low and rustic life in and for itself. Whatever may be concluded on the other side, from the stronger local attachments and enterprising spirit of the Swiss, and other mountaineers, applies to a particular mode of pastoral life, under forms of property, that permit and beget manners truly republican, not to rustic life in general, or to the absence of artificial cultivation. On the contrary the mountaineers, whose manners have been so often eulogized, are in general better educated and greater readers than men of equal rank elsewhere. But where this is not the case, as among the peasantry of North Wales, the ancient mountains, with all their terrors and all their glories, are pictures to the blind, and music to the deaf.

I should not have entered so much into detail upon this passage, but here seems to be the point, to which all the lines of difference converge as to their source and centre. (I mean, as far as, and in whatever respect, my poetic creed does differ from the doctrines promulged in this preface.) I adopt with full faith the principle of Aristotle, that poetry as poetry is essentially ideal,* that it avoids and excludes all accident; that its apparent individualities of rank, character, or occupation must be representative of a class; and that the persons of poetry must be clothed with generic attributes, with the common attributes of the class; not with such as one gifted individual might possibly possess, but such as from his situation it is most probable beforehand, that he *would* possess. If my premisses are right, and my deductions legitimate, it follows that there can be no poetic medium between the swains of Theocritus and those of an imaginary Golden Age.

The characters of the vicar and the shepherd-mariner in the poem of the 'Brothers,' those of the shepherd of Green-head Ghyll in the 'Michael,' have all the verisimilitude and representative quality, that the purposes of poetry can require. They are persons of a known and

* Say not that I am recommending abstractions, for these class characteristics which constitute the instructiveness of a character, are so modified and particularized in each person of the Shakespearian Drama, that life itself does not excite more distinctly that sense of individuality which belongs to real existence. Paradoxical as it may sound, one of the essential properties of geometry is not less essential to dramatic excellence; and Aristotle has accordingly required of the poet an involution of the universal in the individual. The chief differences are, that in geometry it is the universal truth, which is uppermost in the consciousness; in poetry the individual form, in which the truth is clothed. With the ancients, and not less with the elder dramatists of England and France, both comedy and tragedy were considered as kinds of poetry. They neither sought in comedy to make us laugh merely: much less to make us laugh by wry faces, accidents of jargon, slang phrases for the day, or the clothing of commonplace morals in metaphors drawn from the shops or mechanic occupations of their characters. Nor did they condescend in tragedy to wheedle away the applause of the spectators, by representing before them facsimiles of their own mean selves in all their existing meanness, or to work on their sluggish sympathies by a pathos not a whit more respectable than the maudlin tears of drunkenness. Their tragic scenes were meant to affect us indeed; but yet within the bounds of pleasure, and in union with the activity both of our understanding and imagination. They wished to transport the mind to a sense of its possible greatness, and to implant the germs of that greatness, during the temporary oblivion of the worthless 'thing we are,' and of the peculiar state in which each man happens to be, suspending our individual recollections and lulling them to sleep amid the music of nobler thoughts. Friend, Pages 251, 252.

abiding class, and their manners and sentiments the natural product of circumstances common to the class. Take 'Michael' for instance:

> An old man stout of heart, and strong of limb;
> His bodily frame had been from youth to age
> Of an unusual strength: his mind was keen,
> Intense and frugal, apt for all affairs,
> And in his shepherd's calling he was prompt
> And watchful more than ordinary men.
> Hence he had learnt the meaning of all winds,
> Of blasts of every tone, and oftentimes
> When others heeded not, he heard the South
> Make subterraneous music, like the noise
> Of bagpipers on distant highland hills.
> The shepherd, at such warning, of his flock
> Bethought him, and he to himself would say,
> The winds are now devising work for me!
> And truly at all times the storm, that drives
> The traveller to a shelter, summon'd him
> Up to the mountains. He had been alone
> Amid the heart of many thousand mists,
> That came to him and left him on the heights.
> So liv'd he, till his eightieth year was pass'd.
> And grossly that man errs, who should suppose
> That the green valleys, and the streams and rocks,
> Were things indifferent to the shepherd's thoughts.
> Fields, where with cheerful spirits he had breath'd
> The common air; the hills, which he so oft
> Had climb'd with vigorous steps; which had impress'd
> So many incidents upon his mind
> Of hardship, skill or courage, joy or fear;
> Which like a book preserved the memory
> Of the dumb animals, whom he had sav'd,
> Had fed or shelter'd, linking to such acts,
> So grateful in themselves, the certainty
> Of honorable gains; these fields, these hills
> Which were his living being, even more
> Than his own blood—what could they less? had laid
> Strong hold on his affections, were to him
> A pleasurable feeling of blind love,
> The pleasure which there is in life itself.

On the other hand, in the poems which are pitched at a lower note, as the 'Harry Gill,' 'Idiot Boy,' etc. the feelings are those of human nature in general; though the poet has judiciously laid the scene in the country, in order to place himself in the vicinity of interesting images, without the necessity of ascribing a sentimental perception of their beauty to the persons of his drama. In the 'Idiot Boy,' indeed, the mother's character is not so much a real and native product of a 'situation where the essential passions of the heart find a better soil, in which they can attain their maturity and speak a plainer and more emphatic language,' as it is an impersonation of an instinct abandoned by judgement. Hence the two following charges seem to me not wholly groundless: at least, they are the only plausible objections, which I have heard to that fine poem. The one is, that the author has not, in the poem itself, taken sufficient care to preclude from the reader's fancy the disgusting images of ordinary, morbid idiocy, which yet it was by no means his intention to represent. He has even by the 'burr, burr, burr,' uncounteracted by any preceding description of the boy's beauty, assisted in recalling them. The other is, that the idiocy of the boy is so evenly balanced by the folly of the mother, as to present to the general reader rather a laughable burlesque on the blindness of anile dotage, than an analytic display of maternal affection in its ordinary workings.

In the 'Thorn,' the poet himself acknowledges in a note the necessity of an introductory poem, in which he should have portrayed the character of the person from whom the words of the poem are supposed to proceed: a superstitious man moderately imaginative, of slow faculties and deep feelings, 'a captain of a small trading vessel, for example, who being past the middle age of life, had retired upon an annuity, or small independent income, to some village or country town of which he was not a native, or in which he had not been accustomed to live. Such men having nothing to do become credulous and talkative from indolence.' But in a poem, still more in a lyric poem (and the Nurse in Shakespeare's Romeo and Juliet alone prevents me from extending the remark even to dramatic poetry, if indeed the Nurse itself can be deemed altogether a case in point) it is not possible to imitate truly a dull and garrulous discourser, without repeating the effects of dullness and garrulity. However this may be, I dare assert, that the parts (and these form the far larger portion of the whole) which might as well or still better have proceeded from the poet's own imagination, and have been spoken in his own character, are those which have given, and which will continue to give universal delight; and that the passage exclusively appropriate to the supposed narrator, such as the last couplet of the third stanza;* the seven last lines of the

* I've measured it from side to side;
'Tis three feet long, and two feet wide.

tenth;* and the five following stanzas, with the exception of the four admirable lines at the commencement of the fourteenth are felt by many unprejudiced and unsophisticated hearts, as sudden and unpleasant sinkings from the height to which the poet had previously

* Nay, rack your brain—'tis all in vain,
I'll tell you every thing I know;
But to the Thorn, and to the Pond
Which is a little step beyond,
I wish that you would go:
Perhaps, when you are at the place,
You something of her tale may trace.

I'll give you the best help I can:
Before you up the mountain go,
Up to the dreary mountain-top,
I'll tell you all I know.
'Tis now some two-and-twenty years
Since she (her name is Martha Ray)
Gave, with a maiden's true good will,
Her company to Stephen Hill;
And she was blithe and gay,
And she was happy, happy still
Whene'er she thought of Stephen Hill.

And they had fix'd the wedding-day,
The morning that must wed them both;
But Stephen to another maid
Had sworn another oath;
And with this other maid to church
Unthinking Stephen went—
Poor Martha! on that woeful day
A pang of pitiless dismay
Into her soul was sent;
A fire was kindled in her breast,
Which might not burn itself to rest.

They say, full six months after this,
While yet the summer leaves were green,
She to the mountain-top would go,
And there was often seen.
'Tis said, a child was in her womb,
As now to any eye was plain;
She was with child, and she was mad;
Yet often she was sober sad
From her exceeding pain.
Oh me! ten thousand times I'd rather
That he had died, that cruel father!

* * * * *

lifted them, and to which he again re-elevates both himself and his reader.

If then I am compelled to doubt the theory, by which the choice of characters was to be directed, not only a priori, from grounds of reason, but both from the few instances in which the poet himself need be supposed to have been governed by it, and from the comparative inferiority of those instances; still more must I hesitate in my assent to the sentence which immediately follows the former citation; and which I can neither admit as particular fact, or as general rule. 'The language too of these men is adopted (purified indeed from what appear to be its real defects, from all lasting and rational causes of dislike or disgust) because such men hourly communicate with the best objects from which the best part of language is originally derived; and because, from their rank in society, and the sameness and narrow circle of their intercourse, being less under the action of social vanity, they convey their feelings and notions in simple and unelaborated expressions.' To this I reply; that a rustic's language, purified from all provincialism and grossness, and so far reconstructed as to be made consistent with the rules of grammar (which are in essence no other than the laws of universal logic, applied to psychological materials) will not differ from the language of any other man of common sense, however learned or refined he may be, except as far as the notions, which the rustic has to convey, are fewer and more indiscriminate. This will become still clearer, if we add the consideration (equally

Last Christmas when we talked of this,
Old farmer Simpson did maintain,
That in her womb the infant wrought
About its mother's heart, and brought
Her senses back again:
And when at last her time drew near,
Her looks were calm, her senses clear.

No more I know, I wish I did,
And I would tell it all to you;
For what became of this poor child
There's none that ever knew:
And if a child was born or no,
There's no one that could ever tell;
And if 'twas born alive or dead,
There's no one knows, as I have said;
But some remember well,
That Martha Ray about this time
Would up the mountain often climb.

important though less obvious) that the rustic, from the more imperfect development of his faculties, and from the lower state of their cultivation, aims almost solely to convey insulated facts, either those of his scanty experience or his traditional belief; while the educated man chiefly seeks to discover and express those connections of things, or those relative bearings of fact to fact, from which some more or less general law is deducible. For facts are valuable to a wise man, chiefly as they lead to the discovery of the indwelling law, which is the true being of things, and sole solution of their modes of existence, and in the knowledge of which consists our dignity and our power.

As little can I agree with the assertion, that from the objects with which the rustic hourly communicates, the best part of language is formed. For first, if to communicate with an object implies such an acquaintance with it, as renders it capable of being discriminately reflected on; the distinct knowledge of an uneducated rustic would furnish a very scanty vocabulary. The few things, and modes of action, requisite for his bodily conveniences, would alone be individualized; while all the rest of nature would be expressed by a small number of confused, general terms. Secondly, I deny that the words and combinations of words derived from the objects, with which the rustic is familiar, whether with distinct or confused knowledge, can be justly said to form the *best* part of language. It is more than probable, that many classes of the brute creation possess discriminating sounds, by which they can convey to each other notices of such objects as concern their food, shelter, or safety. Yet we hesitate to call the aggregate of such sounds a language, otherwise than metaphorically. The best part of human language, properly so called, is derived from reflection on the acts of the mind itself. It is formed by a voluntary appropriation of fixed symbols to internal acts, to processes and results of imagination, the greater part of which have no place in the consciousness of uneducated man; though in civilized society, by imitation and passive remembrance of what they hear from their religious instructors and other superiors, the most uneducated share in the harvest which they neither sowed or reaped. If the history of the phrases in hourly currency among our peasants were traced, a person not previously aware of the fact would be surprised at finding so large a number, which three or four centuries ago were the exclusive property of the universities and the schools; and at the commencement of the Reformation had been transferred from the school to the pulpit, and thus gradually passed into common life. The extreme difficulty, and often the impossibility, of finding words for the simplest moral and intellectual processes in the languages of uncivilized tribes has proved perhaps the weightiest obstacle to the progress of our most zealous and adroit missionaries. Yet these tribes are surrounded by the same nature, as our peasants are; but in still more impressive forms; and they are, moreover, obliged to particularize many more of them. When therefore Mr Wordsworth adds, 'accordingly such a language' (meaning, as before, the language of rustic life purified from provincialism) 'arising out of repeated experience and regular feelings is a more permanent, and a far more philosophical language, than that which is frequently substituted for it by poets, who think they are conferring honour upon themselves and their art in proportion as they indulge in arbitrary and capricious habits of expression'; it may be answered, that the language, which he has in view, can be attributed to rustics with no greater right, than the style of Hooker or Bacon to Tom Brown or Sir Roger L'Estrange.° Doubtless, if what is peculiar to each were omitted in each, the result must needs be the same. Further, that the poet, who uses an illogical diction, or a style fitted to excite only the low and changeable pleasure of wonder by means of groundless novelty, substitutes a language of folly and vanity, not for that of the rustic, but for that of good sense and natural feeling.

Here let me be permitted to remind the reader, that the positions, which I controvert, are contained in the sentences—'a selection of the real language of men';—'the language of these men (i.e. men in low and rustic life) I propose to myself to imitate, and as far as possible to adopt the very language of men.' 'Between the language of prose and that of metrical composition, there neither is, nor can be any essential difference.' It is against these exclusively, that my opposition is directed.

I object, in the very first instance, to an equivocation in the use of the word 'real.' Every man's language varies, according to the extent of his knowledge, the activity of his faculties, and the depth or quickness of his feelings. Every man's language has, first, its individualities; secondly, the common properties of the class to which he belongs; and thirdly, words and phrases of *universal* use. The language of Hooker, Bacon, Bishop Taylor, and Burke, differ from the common language of the learned class only by the superior number and novelty of the thoughts and relations which they had to convey. The language of Algernon Sidney differs not at all from that, which every well-educated gentleman would wish to write, and (with due allowances for the undeliberateness, and less connected train, of thinking natural and proper to conversation) such as he would wish to talk. Neither one or the other differ half as much from the general language of cultivated society, as the language of Mr Wordsworth's homeliest composition differs from that of a common peasant. For 'real' therefore, we must substitute ordinary, or *lingua communis* [the common tongue]. And this, we have proved, is no more to be found in the phraseology of low and rustic life, than in that of any other class. Omit the peculiarities of each, and the result of course must be common to all. And assuredly the omissions and changes to be made in the language of rustics, before it could be transferred to any species of poem, except the drama or other professed imitation, are at least as numerous and weighty, as would be required in adapting to the same purpose the ordinary language of tradesmen and manufacturers. Not to mention, that the language so highly extolled by Mr Wordsworth varies in every county, nay in every village, according to the accidental character of the clergyman, the existence or non-existence of schools; or even, perhaps, as the exciseman, publican, or barber happen to be, or not to be, zealous politicians, and readers of the weekly newspaper *pro bono publico* [for the public good]. Anterior to cultivation the *lingua communis* of every country, as Dante has well observed, exists everywhere in parts, and nowhere as a whole.

Neither is the case rendered at all more tenable by the addition of the words, 'in a state of excitement.' For the nature of a man's words, when he is strongly affected by joy, grief, or anger, must necessarily depend on the number and quality of the general truths, conceptions and images, and of the words expressing them, with which his mind had been previously stored. For the property of passion is not to create; but to set in increased activity. At least, whatever new connections of thoughts or images, or (which is equally, if not more than equally, the appropriate effect of strong excitement) whatever generalizations of truth or experience, the heat of passion may produce; yet the terms of their conveyance must have pre-existed in his former conversations, and are only collected and crowded together by the unusual stimulation. It is indeed very possible to adopt in a poem the unmeaning repetitions, habitual phrases, and other blank counters, which an unfurnished or confused understanding interposes at short intervals, in order to keep hold of his subject which is still slipping from him, and to give him time for recollection; or in mere aid of vacancy, as in the scanty companies of a country stage the same player pops backwards and forwards, in order to prevent the appearance of empty spaces, in the procession of Macbeth, or Henry

VIIIth. But what assistance to the poet, or ornament to the poem, these can supply, I am at a loss to conjecture. Nothing assuredly can differ either in origin or in mode more widely from the apparent tautologies of intense and turbulent feeling, in which the passion is greater and of longer endurance, than to be exhausted or satisfied by a single representation of the image or incident exciting it. Such repetitions I admit to be a beauty of the highest kind; as illustrated by Mr Wordsworth himself from the song of Deborah. 'At her feet he bowed, he fell, he lay down; at her feet he bowed, he fell; where he bowed, there he fell down dead.'

CHAPTER XVIII

Language of metrical composition, why and wherein essentially different from that of prose—Origin and elements of metre—Its necessary consequences, and the conditions thereby imposed on the metrical writer in the choice of his diction.

I CONCLUDE therefore, that the attempt is impracticable, and that, were it not impracticable, it would still be useless. For the very power of making the selection implies the previous possession of the language selected. Or where can the poet have lived? And by what rules could he direct his choice, which would not have enabled him to select and arrange his words by the light of his own judgement? We do not adopt the language of a class by the mere adoption of such words exclusively, as that class would use, or at least understand; but likewise by following the order, in which the words of such men are wont to succeed each other. Now this order, in the intercourse of uneducated men, is distinguished from the diction of their superiors in knowledge and power, by the greater disjunction and separation in the component parts of that, whatever it be, which they wish to communicate. There is a want of that prospectiveness of mind, that surview, which enables a man to foresee the whole of what he is to convey, appertaining to any one point; and by this means so to subordinate and arrange the different parts according to their relative importance, as to convey it at once, and as an organized whole.

Now I will take the first stanza, on which I have chanced to open, in the Lyrical Ballads. It is one [of] the most simple and the least peculiar in its language.

> In distant countries I have been,
> And yet I have not often seen
> A healthy man, a man full grown,
> Weep in the public road alone.
> But such a one, on English ground,
> And in the broad highway I met;
> Along the broad highway he came,
> His cheeks with tears were wet.
> Sturdy he seem'd, though he was sad,
> And in his arms a lamb he had.

The words here are doubtless such as are current in all ranks of life; and of course not less so, in the hamlet and cottage, than in the shop, manufactory, college, or palace. But is this the order, in which the rustic would have placed the words? I am grievously deceived, if the following less compact mode of commencing the same tale be not a far more faithful copy. 'I have been in a many parts far and near, and I don't know that I ever saw before a man crying by himself in the public road; a grown man I mean, that was neither sick nor hurt,' etc. etc. But when I turn to the following stanza in 'The Thorn':

> At all times of the day and night
> This wretched woman thither goes,
> And she is known to every star
> And every wind that blows:
> And there beside the thorn she sits,
> When the blue day-light's in the skies;
> And when the whirlwind's on the hill,
> Or frosty air is keen and still;
> And to herself she cries,
> Oh misery! Oh misery!
> Oh woe is me! Oh misery!

and compare this with the language of ordinary men; or with that which I can conceive at all likely to proceed, in real life, from such a narrator, as is supposed in the note to the poem; compare it either in the succession of the images or of the sentences, I am reminded of the sublime prayer and hymn of praise, which Milton, in opposition to an established liturgy, presents as a fair specimen of common extemporary devotion, and such as we might expect to hear from every self-inspired minister of a conventicle! And I reflect with delight, how little a mere theory, though of his own workmanship, interferes with the processes of genuine imagination in a man of true poetic genius, who possesses, as Mr Wordsworth, if ever man did, most assuredly does possess,

The vision and the faculty divine.

One point then alone remains, but that the most important; its examination having been, indeed, my chief inducement for the preceding inquisition. 'There neither is or can be any essential difference between the language of prose and metrical composition.' Such is Mr Wordsworth's assertion. Now prose itself, at least, in all argumentative and consecutive works differs, and ought to differ, from the language of conversation; even as reading* ought to differ from talking. Unless therefore the difference denied be that of the mere words, as materials common to all styles of writing, and not of the style itself in the universally admitted sense of the term, it might be naturally presumed that there must exist a still greater between the ordonnance of poetic composition and that of prose, than is expected to distinguish prose from ordinary conversation.

There are not, indeed, examples wanting in the history of literature, of apparent paradoxes that have summoned the public wonder as new and startling truths, but which on examination have shrunk into tame and harmless truisms; as the eyes of a cat, seen in the dark, have been

* It is no less an error in teachers, than a torment to the poor children, to enforce the necessity of reading as they would talk. In order to cure them of *singing* as it is called; that is, of too great a difference. The child is made to repeat the words with his eyes from off the book; and then indeed, his tones resemble talking, as far as his fears, tears and trembling will permit. But as soon as the eye is again directed to the printed page, the spell begins anew; for an instinctive sense tells the child's feelings, that to utter its own momentary thoughts, and to recite the written thoughts of another, as of another, and a far wiser than himself, are two widely different things; and as the two acts are accompanied with widely different feelings, so must they justify different modes of enunciation. Joseph Lancaster, among his other sophistications of the excellent Dr Bell's invaluable system,° cures this fault of *singing*, by hanging fetters and chains on the child, to the music of which, one of his school fellows who walks before, dolefully chants out the child's last speech and confession, birth, parentage, and education. And this soul-benumbing ignominy, this unholy and heart-hardening burlesque on the last fearful infliction of outraged law, in pronouncing the sentence to which the stern and familiarized judge not seldom bursts into tears, has been extolled as a happy and ingenious method of remedying—what? and how?—why, one extreme in order to introduce another, scarce less distant from good sense, and certainly likely to have worse moral effects, by enforcing a semblance of petulant ease and self-sufficiency, in repression, and possible after-perversion of the natural feelings. I have to beg Dr Bell's pardon for his connection of the two names, but he knows that contrast is no less powerful a cause of association than likeness.

mistaken for flames of fire. But Mr Wordsworth is among the last men, to whom a delusion of this kind would be attributed by anyone, who had enjoyed the slightest opportunity of understanding his mind and character. Where an objection has been anticipated by such an author as natural, his answer to it must needs be interpreted in some sense which either is, or has been, or is capable of being controverted. My object then must be to discover some other meaning for the term 'essential difference', in this place, exclusive of the indistinction and community of the words themselves. For whether there ought to exist a class of words in the English, in any degree resembling the poetic dialect of the Greek and Italian, is a question of very subordinate importance. The number of such words would be small indeed, in our language; and even in the Italian and Greek, they consist not so much of different words, as of slight differences in the forms of declining and conjugating the same words; forms, doubtless, which having been, at some period more or less remote, the common grammatic flexions of some tribe or province, had been accidentally appropriated to poetry by the general admiration of certain master intellects, the first established lights of inspiration, to whom that dialect happened to be native.

Essence, in its primary signification, means the principle of individuation, the inmost principle of the possibility of any thing, *as* that particular thing. It is equivalent to the idea of a thing, whenever we use the word *idea*, with philosophic precision. Existence, on the other hand, is distinguished from essence, by the superinduction of reality. Thus we speak of the essence, and essential properties of a circle; but we do not therefore assert, that anything, which really *exists*, is mathematically circular. Thus too, without any tautology we contend for the existence of the Supreme Being; that is, for a reality correspondent to the idea. There is, next, a secondary use of the word *essence*, in which it signifies the point or ground of contradistinction between two modifications of the same substance or subject. Thus we should be allowed to say, that the style of architecture of Westminster Abbey is essentially different from that of Saint Paul's, even though both had been built with blocks cut into the same form, and from the same quarry. Only in this latter sense of the term must it have been denied by Mr Wordsworth (for in this sense alone is it affirmed by the general opinion) that the language of poetry (i.e. the formal construction, or architecture, of the words and phrases) is essentially different from that of prose. Now the burden of the proof lies with the oppugner, not with the supporters of the common belief. Mr

Wordsworth, in consequence, assigns as the proof of his position, 'that not only the language of a large portion of every good poem, even of the most elevated character, must necessarily, except with reference to the metre, in no respect differ from that of good prose; but likewise that some of the most interesting parts of the best poems will be found to be strictly the language of prose, when prose is well written. The truth of this assertion might be demonstrated by innumerable passages from almost all the poetical writings even of Milton himself.' He then quotes Gray's sonnet—

> In vain to me the smiling mornings shine,
> And reddening Phoebus lifts his golden fire;
> The birds in vain their amorous descant join,
> Or cheerful fields resume their green attire;
> These ears alas! for other notes repine;
> *A different object do these eyes require;*
> *My lonely anguish melts no heart but mine,*
> *And in my breast the imperfect joys expire!*
> Yet morning smiles the busy race to cheer,
> And new born pleasure brings to happier men:
> The fields to all their wonted tributes bear,
> To warm their little loves the birds complain.
> *I fruitless mourn to him that cannot hear,*
> *And weep the more because I weep in vain;*

and adds the following remark:—'It will easily be perceived, that the only part of this Sonnet which is of any value, is the lines printed in italics. It is equally obvious, that except in the rhyme, and in the use of the single word "fruitless" for fruitlessly, which is so far a defect, the language of these lines does in no respect differ from that of prose.'

An idealist defending his system by the fact, that when asleep we often believe ourselves awake, was well answered by his plain neighbour, 'Ah, but when awake do we ever believe ourselves asleep?'—Things identical must be convertible. The preceding passage seems to rest on a similar sophism. For the question is not, whether there may not occur in prose an order of words, which would be equally proper in a poem; nor whether there are not beautiful lines and sentences of frequent occurrence in good poems, which would be equally becoming as well as beautiful in good prose; for neither the one or the other has ever been either denied or doubted by anyone. The true question must be, whether there are not modes of expression, a construction, and an order of sentences, which are in their fit and natural place in a serious prose composition, but would be disproportionate and heterogeneous in metrical poetry; and, vice versa, whether in the language of a serious poem there may not be an arrangement both of words and sentences, and a use and selection of (what are called) figures of speech, both as to their kind, their frequency, and their occasions, which on a subject of equal weight would be vicious and alien in correct and manly prose. I contend, that in both cases this unfitness of each for the place of the other frequently will and ought to exist.

And first from the origin of metre. This I would trace to the balance in the mind effected by that spontaneous effort which strives to hold in check the workings of passion. It might be easily explained likewise in what manner this salutary antagonism is assisted by the very state, which it counteracts; and how this balance of antagonists became organized into metre (in the usual acceptation of that term) by a supervening act of the will and judgement, consciously and for the foreseen purpose of pleasure. Assuming these principles, as the data of our argument, we deduce from them two legitimate conditions, which the critic is entitled to expect in every metrical work. First, that as the elements of metre owe their existence to a state of increased excitement, so the metre itself should be accompanied by the natural language of excitement. Secondly, that as these elements are formed into metre artificially, by a voluntary act, with the design and for the purpose of blending delight with emotion, so the traces of present volition should throughout the metrical language be proportionally discernible. Now these two conditions must be reconciled and co-present. There must be not only a partnership, but a union; an interpenetration of passion and of will, of spontaneous impulse and of voluntary purpose. Again, this union can be manifested only in a frequency of forms and figures of speech (originally the offspring of passion, but now the adopted children of power) greater, than would be desired or endured, where the emotion is not voluntarily encouraged, and kept up for the sake of that pleasure, which such emotion so tempered and mastered by the will is found capable of communicating. It not only dictates, but of itself tends to produce, a more frequent employment of picturesque and vivifying language, than would be natural in any other case, in which there did not exist, as there does in the present, a previous and well understood, though tacit, compact between the poet and his reader, that the latter is entitled to expect, and the former bound to supply this species and degree of pleasurable excitement. We may in some measure apply to this union the answer of Polixenes, in the Winter's Tale, to Perdita's neglect of the streaked gilly-flowers, because she had heard it said,

> There is an art which in their piedness shares
> With great creating nature.
> POL: Say there be:
> Yet nature is made better by no mean,
> But nature makes that mean. So ev'n that art,
> Which you say adds to nature, is an art,
> That nature makes! You see, sweet maid, we marry
> *A gentler scyon to the wildest stock*:
> And make conceive a bark of ruder kind
> By bud of nobler race. This is an art,
> Which does mend nature—change it rather; but
> The art itself is nature.

Secondly, I argue from the effects of metre. As far as metre acts in and for itself, it tends to increase the vivacity and susceptibility both of the general feelings and of the attention. This effect it produces by the continued excitement of surprise, and by the quick reciprocations of curiosity still gratified and still re-excited, which are too slight indeed to be at any one moment objects of distinct consciousness, yet become considerable in their aggregate influence. As a medicated atmosphere, or as wine during animated conversation; they act powerfully, though themselves unnoticed. Where, therefore, correspondent food and appropriate matter are not provided for the attention and feelings thus roused, there must needs be a disappointment felt; like that of leaping in the dark from the last step of a staircase, when we had prepared our muscles for a leap of three or four.

The discussion on the powers of metre in the preface is highly ingenious and touches at all points on truth. But I cannot find any statement of its powers considered abstractly and separately. On the contrary Mr Wordsworth seems always to estimate metre by the powers, which it exerts during (and, as I think, in consequence of) its combination with other elements of poetry. Thus the previous difficulty is left unanswered, what the elements are, with which it must be combined in order to produce its own effects to any pleasurable purpose. Double and trisyllable rhymes, indeed, form a lower species of wit, and attended to exclusively for their own sake may become a source of momentary amusement; as in poor Smart's distich to the Welsh Squire who had promised him a hare:

> Tell me thou son of great Cadwallader!
> Hast sent the hare? or hast thou swallow'd her?

But for any poetic purposes, metre resembles (if the aptness of the simile may excuse its meanness) yeast, worthless or disagreeable by itself, but giving vivacity and spirit to the liquor with which it is proportionally combined.

The reference to the 'Children in the Wood' by no means satisfies my judgement.° We all willingly throw ourselves back for a while into the feelings of our childhood. This ballad, therefore, we read under such recollections of our own childish feelings, as would equally endear to us poems, which Mr Wordsworth himself would regard as faulty in the opposite extreme of gaudy and technical ornament. Before the invention of printing, and in a still greater degree, before the introduction of writing, metre, especially alliterative metre, (whether alliterative at the beginning of the words, as in 'Pierce Plouman,' or at the end as in rhymes) possessed an independent value as assisting the recollection, and consequently the preservation, of any series of truths or incidents. But I am not convinced by the collation of facts, that the 'Children in the Wood' owes either its preservation, or its popularity, to its metrical form. Mr Marshall's repository affords a number of tales in prose inferior in pathos and general merit, some of as old a date, and many as widely popular. Tom Hickathrift, Jack the Giant-killer, Goody Two-shoes, and Little Red Riding-hood are formidable rivals. And that they have continued in prose, cannot be fairly explained by the assumption, that the comparative meanness of their thoughts and images precluded even the humblest forms of metre. The scene of Goody Two-shoes in the church is perfectly susceptible of metrical narration; and among the *Θαύματα θαυμαστότατα* [marvels most marvellous] even of the present age, I do not recollect a more astonishing image than that of the 'whole rookery, that flew out of the giant's beard' scared by the tremendous voice, with which this monster answered the challenge of the heroic Tom Hickathrift!

If from these we turn to compositions universally, and independently of all early associations, beloved and admired; would the Maria, the Monk, or the Poor Man's Ass of Sterne,° be read with more delight, or have a better chance of immortality, had they without any change in the diction been composed in rhyme, than in their present state? If I am not grossly mistaken, the general reply would be in the negative. Nay, I will confess, that in Mr Wordsworth's own volumes the Anecdote for Fathers, Simon Lee, Alice Fell, the Beggars, and the Sailor's Mother, notwithstanding the beauties which are to be found in each of them where the poet interposes the music of his own thoughts, would have been more delightful to me in prose, told and managed, as by Mr Wordsworth they would have been, in a moral essay, or pedestrian tour.

Metre in itself is simply a stimulant of the attention, and therefore excites the question: Why is the attention to be thus stimulated? Now the question cannot be answered by the pleasure of the metre itself: for this we have shown to be conditional, and dependent on the appropriateness of the thoughts and expressions, to which the metrical form is superadded. Neither can I conceive any other answer that can be rationally given, short of this: I write in metre, because I am about to use a language different from that of prose. Besides, where the language is not such, how interesting soever the reflections are, that are capable of being drawn by a philosophic mind from the thoughts or incidents of the poem, the metre itself must often become feeble. Take the three last stanzas of the Sailor's Mother, for instance. If I could for a moment abstract from the effect produced on the author's feelings, as a man, by the incident at the time of its real occurrence, I would dare appeal to his own judgement, whether in the metre itself he found a sufficient reason for their being written metrically?

> And thus continuing, she said
> I had a son, who many a day
> Sailed on the seas; but he is dead;
> In Denmark he was cast away:
> And I have travelled far as Hull, to see
> What clothes he might have left, or other property.
>
> The bird and cage, they both were his;
> 'Twas my son's bird; and neat and trim
> He kept it; many voyages
> This singing bird hath gone with him;
> When last he sailed he left the bird behind;
> As it might be, perhaps, from bodings of his mind.
>
> He to a fellow-lodger's care
> Had left it, to be watched and fed,
> Till he came back again; and there
> I found it when my son was dead;
> And now, God help me for my little wit!
> I trail it with me, Sir! he took so much delight in it.

If disproportioning the emphasis we read these stanzas so as to make the rhymes perceptible, even trisyllable rhymes could scarcely produce an equal sense of oddity and strangeness, as we feel here in finding rhymes at all in sentences so exclusively colloquial. I would further ask whether, but for that visionary state, into which the figure of the woman and the susceptibility of his own genius had placed the poet's imagination (a state, which spreads its influence and colouring over all, that coexists with the exciting cause, and in which 'The simplest, and the most familiar things | Gain a strange power of spreading awe around them')* I would ask the poet whether he would not have felt an abrupt downfall in these verses from the preceding stanza?

> The ancient spirit is not dead;
> Old times, thought I, are breathing there!
> Proud was I, that my country bred
> Such strength, a dignity so fair!
> She begged an alms, like one in poor estate;
> I looked at her again, nor did my pride abate.

It must not be omitted, and is besides worthy of notice, that those stanzas furnish the only fair instance that I have been able to discover in all Mr Wordsworth's writings, of an actual adoption, or true imitation, of the real and very language of low and rustic life, freed from provincialisms.

Thirdly, I deduce the position from all the causes elsewhere assigned, which render metre the proper form of poetry, and poetry imperfect and defective without metre. Metre therefore having been connected with poetry most often and by a peculiar fitness, whatever else is combined with metre must, though it be not itself essentially poetic, have nevertheless some property in common with poetry, as an intermedium of affinity, a sort (if I may dare borrow a well-known phrase from technical chemistry) of *mordant*° between it and the superadded metre. Now poetry, Mr Wordsworth truly affirms, does always imply passion; which word must be here understood in its most general sense, as an excited state of the feelings and faculties. And as

* Altered from the description of Night-Mair in the Remorse.

> O Heaven! 'twas frightful! Now run-down and stared at,
> By hideous shapes that cannot be remembered;
> Now seeing nothing and imaging nothing;
> But only being afraid—stifled with fear!
> While every goodly or familiar form
> Had a strange power of spreading terror round me.

NB. Though Shakespeare has for his own all-justifying purposes introduced the night-*mare* with her own foals, yet *mair* means a sister or perhaps a hag.

every passion has its proper pulse, so will it likewise have its characteristic modes of expression. But where there exists that degree of genius and talent which entitles a writer to aim at the honours of a poet, the very act of poetic composition itself is, and is allowed to imply and to produce, an unusual state of excitement, which of course justifies and demands a correspondent difference of language, as truly, though not perhaps in as marked a degree, as the excitement of love, fear, rage, or jealousy. The vividness of the descriptions or declamations in Donne, or Dryden, is as much and as often derived from the force and fervour of the describer, as from the reflections, forms or incidents which constitute their subject and materials. The wheels take fire from the mere rapidity of their motion. To what extent, and under what modifications, this may be admitted to act, I shall attempt to define in an after remark on Mr Wordsworth's reply to this objection, or rather on his objection to this reply, as already anticipated in his preface.

Fourthly, and as intimately connected with this, if not the same argument in a more general form, I adduce the high spiritual instinct of the human being impelling us to seek unity by harmonious adjustment, and thus establishing the principle, that all the parts of an organized whole must be assimilated to the more important and essential parts. This and the preceding arguments may be strengthened by the reflection, that the composition of a poem is among the imitative arts; and that imitation, as opposed to copying, consists either in the interfusion of the same throughout the radically different, or of the different throughout a base radically the same.

Lastly, I appeal to the practice of the best poets, of all countries and in all ages, as authorizing the opinion, (deduced from all the foregoing) that in every import of the word essential, which would not here involve a mere truism, there may be, is, and ought to be, an *essential* difference between the language of prose and of metrical composition.

In Mr Wordsworth's criticism of Gray's Sonnet, the reader's sympathy with his praise or blame of the different parts is taken for granted rather perhaps too easily. He has not, at least, attempted to win or compel it by argumentative analysis. In my conception at least, the lines rejected as of no value do, with the exception of the two first, differ as much and as little from the language of common life, as those which he has printed in italics as possessing genuine excellence. Of the five lines thus honourably distinguished, two of them differ from prose even more widely, than the lines which either precede or follow, in the position of the words.

A different object do these eyes require;
My lonely anguish melts no heart but mine;
And in my breast the imperfect joys expire.

But were it otherwise, what would this prove, but a truth, of which no man ever doubted? *Videlicet* [=viz., 'namely'], that there are sentences, which would be equally in their place both in verse and prose. Assuredly it does not prove the point, which alone requires proof; namely, that there are not passages, which would suit the one, and not suit the other. The first line of this sonnet is distinguished from the ordinary language of men by the epithet to morning. (For we will set aside, at present, the consideration, that the particular word 'smiling' is hackneyed, and (as it involves a sort of personification) not quite congruous with the common and material attribute of *shining*.) And, doubtless, this adjunction of epithets for the purpose of additional description, where no particular attention is demanded for the quality of the thing, would be noticed as giving a poetic cast to a man's conversation. Should the sportsman exclaim, 'Come boys! the rosy morning calls you up,' he will be supposed to have some song in his head. But no one suspects this, when he says, 'A wet morning shall not confine us to our beds.' This then is either a defect in poetry, or it is not. Whoever should decide in the affirmative, I would request him to reperuse any one poem, of any confessedly great poet from Homer to Milton, or from Aeschylus to Shakespeare; and to strike out (in thought I mean) every instance of this kind. If the number of these fancied erasures did not startle him; or if he continued to deem the work improved by their total omission; he must advance reasons of no ordinary strength and evidence, reasons grounded in the essence of human nature. Otherwise I should not hesitate to consider him as a man not so much proof against all authority, as dead to it.

The second line,

And reddening Phoebus lifts his golden fire,

has indeed almost as many faults as words. But then it is a bad line, not because the language is distinct from that of prose; but because it conveys incongruous images, because it confounds the cause and the effect, the real thing with the personified representative of the thing; in short, because it differs from the language of good sense! That the 'Phoebus' is hackneyed, and a schoolboy image, is an accidental fault, dependent on the age in which the author wrote, and not deduced from the nature of the thing. That it is part of an exploded mythology, is an objection more deeply grounded. Yet when the torch of ancient learning was rekindled, so cheering were its beams, that our eldest poets, cut off by Christianity from all accredited machinery, and deprived of all acknowledged guardians and symbols of the great objects of nature, were naturally induced to adopt, as a poetic language, those fabulous personages, those forms of the supernatural* in nature, which had given them such dear delight in the poems of their great masters. Nay, even at this day what scholar of genial taste will not so far sympathize with them, as to read with pleasure in Petrarch, Chaucer, or Spenser, what he would perhaps condemn as puerile in a modern poet?

I remember no poet, whose writings would safelier stand the test of Mr Wordsworth's theory, than Spenser. Yet will Mr Wordsworth say, that the style of the following stanzas is either undistinguished from prose, and the language of ordinary life? Or that it is vicious, and that the stanzas are blots in the Faery Queen?

By this the northern waggoner had set
His sevenfold teme behind the stedfast starre,
That was in ocean waves yet never wet,
But firm is fixt and sendeth light from farre
To all that in the wild deep wandering are.
And chearful chanticleer with his note shrill
Had warned once that Phoebus's fiery carre
In haste was climbing up the easterne hill,
Full envious that night so long his room did fill.

Book I. Can. 2. *St.* 2

At last the golden orientall gate
Of greatest heaven gan to open fayre,
And Phoebus fresh as brydegrome to his mate,
Came dauncing forth, shaking his deawie hayre,
And hurl'd his glist'ring beams through gloomy ayre;
Which when the wakeful elfe perceived, streightway
He started up, and did him selfe prepayre
In sun-bright armes, and battailous array;
For with that pagan proud he combat will that day.

Book I. Can. 5. *St.* 2

* But still more by the mechanical system of philosophy which has needlessly infected our theological opinions, and teaching us to consider the world in its relation to God, as of a building to its mason, leaves the idea of omnipresence a mere abstract notion in the state-room of our reason.

On the contrary to how many passages, both in hymn-books and in blank verse poems, could I (were it not invidious) direct the reader's attention, the style of which is most unpoetic, because, and only because it is the style of prose? He will not suppose me capable of having in my mind such verses, as

> I put my hat upon my head
> And walk'd into the strand;
> And there I met another man,
> Whose hat was in his hand.

To such specimens it would indeed be a fair and full reply, that these lines are not bad, because they are unpoetic; but because they are empty of all sense and feeling; and that it were an idle attempt to prove that an ape is not a Newton, when it is evident that he is not a man. But the sense shall be good and weighty, the language correct and dignified, the subject interesting and treated with feeling; and yet the style shall, notwithstanding all these merits, be justly blameable as prosaic, and solely because the words and the order of the words would find their appropriate place in prose, but are not suitable to metrical composition. The 'Civil Wars' of Daniel is an instructive, and even interesting work; but take the following stanzas (and from the hundred instances which abound I might probably have selected others far more striking)

> And to the end we may with better ease
> Discern the true discourse, vouchsafe to shew
> What were the times foregoing near to these,
> That these we may with better profit know.
> Tell how the world fell into this disease;
> And how so great distemperature did grow;
> So shall we see with what degrees it came;
> How things at full do soon wax out of frame.
>
> Ten kings had from the Norman conqu'ror reign'd
> With intermixt and variable fate,
> When England to her greatest height attain'd
> Of power, dominion, glory, wealth, and state;
> After it had with much ado sustain'd
> The violence of princes with debate
> For titles, and the often mutinies
> Of nobles for their ancient liberties.

For first the Norman, conqu'ring all by might,
By might was forced to keep what he had got;
Mixing our customs and the form of right
With foreign constitutions, he had brought;
Mastering the mighty, humbling the poorer wight,
By all severest means that could be wrought;
And making the succession doubtful rent
His new-got state and left it turbulent.

B. I. St. VII, VIII, and IX

Will it be contended on the one side, that these lines are mean and senseless? Or on the other, that they are not prosaic, and for that reason unpoetic? This poet's well-merited epithet is that of the 'well-languaged Daniel'; but likewise and by the consent of his contemporaries no less than of all succeeding critics, the 'prosaic Daniel.' Yet those, who thus designate this wise and amiable writer from the frequent incorrespondency of his diction to his metre in the majority of his compositions, not only deem them valuable and interesting on other accounts; but willingly admit, that there are to be found throughout his poems, and especially in his *Epistles* and in his *Hymen's Triumph*, many and exquisite specimens of that style which, as the neutral ground, of prose and verse, is common to both. A fine and almost faultless extract, eminent as for other beauties, so for its perfection in this species of diction, may be seen in Lamb's Dramatic Specimens, etc. a work of various interest from the nature of the selections themselves (all from the plays of Shakespeare's contemporaries) and deriving a high additional value from the notes, which are full of just and original criticism, expressed with all the freshness of originality.

Among the possible effects of practical adherence to a theory, that aims to identify the style of prose and verse (if it does not indeed claim for the latter a yet nearer resemblance to the average style of men in the viva voce intercourse of real life) we might anticipate the following as not the least likely to occur. It will happen, as I have indeed before observed, that the metre itself, the sole acknowledged difference, will occasionally become metre to the eye only. The existence of prosaisms, and that they detract from the merit of a poem, must at length be conceded, when a number of successive lines can be rendered, even to the most delicate ear, unrecognizable as verse, or as having even been intended for verse, by simply transcribing them as prose: when if the poem be in blank verse, this can be effected without any alteration, or at most by merely restoring one or two words to their proper places, from which they had been transplanted* for no assignable cause or reason but that of the author's convenience; but if it be in rhyme, by the mere exchange of the final word of each line for some other of the same meaning, equally appropriate, dignified and euphonic.

The answer or objection in the preface to the anticipated remark 'that metre paves the way to other distinctions,' is contained in the following words. 'The distinction of rhyme and metre is voluntary and uniform, and not like that produced by (what is called) poetic diction, arbitrary and subject to infinite caprices, upon which no calculation whatever can be made. In the one case the reader is utterly at the mercy of the poet respecting what imagery or diction he may choose to connect with the passion.' But is this a poet, of whom a poet is speaking? No surely! rather of a fool or madman: or at best of a vain or ignorant fantast! And might not brains so wild and so deficient make just the same havoc with rhymes and metres, as they are supposed to effect with modes and figures of speech? How is the reader at the mercy of such men? If he continue to read their nonsense, is it not his own fault? The ultimate end of criticism is much more to establish the

* As the ingenious gentleman under the influence of the Tragic Muse contrived to dislocate, 'I wish you a good morning, Sir! Thank you, Sir, and I wish you the same,' into two blank verse heroics:—

To you a morning good, good Sir! I wish.
You, Sir! I thank: to you the same wish I.

In those parts of Mr Wordsworth's works which I have thoroughly studied, I find fewer instances in which this would be practicable than I have met in many poems, where an approximation of prose has been sedulously and on system guarded against. Indeed excepting the stanzas already quoted from the *Sailor's Mother*, I can recollect but one instance: viz. a short passage of four or five lines in The Brothers, that model of English pastoral, which I never yet read with unclouded eye.—'James, pointing to its summit, over which they had all purposed to return together, informed them that he would wait for them there. They parted, and his comrades passed that way some two hours after, but they did not find him at the appointed place, a *circumstance of which they took no heed*: but one of them going by chance into the house, which at this time was James's house, learnt *there*, that nobody had seen him all that day.' The only change which has been made is in the position of the little word *there* in two instances, the position in the original being clearly such as is not adopted in ordinary conversation. The other words printed in italics were so marked because, though good and genuine English, they are not the phraseology of common conversation either in the word put in apposition, or in the connection by the genitive pronoun. Men in general would have said, 'but that was a circumstance they paid no attention to, or took no notice of,' and the language is, on the theory of the preface, justified only by the narrator's being the Vicar. Yet if any ear could suspect, that these sentences were ever printed as metre, on those very words alone could the suspicion have been grounded.

principles of writing, than to furnish rules how to pass judgement on what has been written by others; if indeed it were possible that the two could be separated. But if it be asked, by what principles the poet is to regulate his own style, if he do not adhere closely to the sort and order of words which he hears in the market, wake, high-road, or ploughfield? I reply; by principles, the ignorance or neglect of which would convict him of being no poet, but a silly or presumptuous usurper of the name! By the principles of grammar, logic, psychology! In one word by such a knowledge of the facts, material and spiritual, that most appertain to his art, as if it have been governed and applied by good sense, and rendered instinctive by habit, becomes the representative and reward of our past conscious reasonings, insights, and conclusions, and acquires the name of taste. By what rule that does not leave the reader at the poet's mercy, and the poet at his own, is the latter to distinguish between the language suitable to suppressed, and the language, which is characteristic of indulged, anger? Or between that of rage and that of jealousy? Is it obtained by wandering about in search of angry or jealous people in uncultivated society, in order to copy their words? Or not far rather by the power of imagination proceeding upon the *all in each* of human nature? By meditation, rather than by observation? And by the latter in consequence only of the former? As eyes, for which the former has predetermined their field of vision, and to which, as to its organ, it communicates a microscopic power? There is not, I firmly believe, a man now living, who has from his own inward experience a clearer intuition, than Mr Wordsworth himself, that the last mentioned are the true sources of genial discrimination. Through the same process and by the same creative agency will the poet distinguish the degree and kind of the excitement produced by the very act of poetic composition. As intuitively will he know, what differences of style it at once inspires and justifies; what intermixture of conscious volition is natural to that state; and in what instances such figures and colours of speech degenerate into mere creatures of an arbitrary purpose, cold technical artifices of ornament or connection. For even as truth is its own light and evidence, discovering at once itself and falsehood, so is it the prerogative of poetic genius to distinguish by parental instinct its proper offspring from the changelings, which the gnomes of vanity or the fairies of fashion may have laid in its cradle or called by its names. Could a rule be given from without, poetry would cease to be poetry, and sink into a mechanical art. It would be μορφωσις [*morphōsis*, shaping], not ποιησις [*poiēsis*, creating]. The rules of the imagination are themselves the very powers of growth and production. The words, to which they are reducible, present only the outlines and external appearance of the fruit. A deceptive counterfeit of the superficial form and colours may be elaborated; but the marble peach feels cold and heavy, and children only put it to their mouths. We find no difficulty in admitting as excellent, and the legitimate language of poetic fervour self-impassioned, Donne's apostrophe to the Sun in the second stanza of his 'Progress of the Soul.'

> Thee, eye of heaven! this great soul envies not:
> By thy male force is all, we have, begot.
> In the first East thou now beginn'st to shine,
> Suck'st early balm and island spices there;
> And wilt anon in thy loose-rein'd career
> At Tagus, Po, Seine, Thames, and Danow dine,
> And see at night this western world of mine:
> Yet hast thou not more nations seen, than she,
> Who before thee one day began to be,
> And, thy frail light being quenched, shall long, long outlive thee!

Or the next stanza but one:

> Great destiny, the commissary of God,
> That hast marked out a path and period
> For ev'ry thing! Who, where we offspring took,
> Our ways and ends see'st at one instant: thou
> Knot of all causes! Thou, whose changeless brow
> Ne'er smiles or frowns! O vouchsafe thou to look,
> And shew my story in thy eternal book, etc.

As little difficulty do we find in excluding from the honours of unaffected warmth and elevation the madness prepense of pseudo-poesy, or the startling hysteric of weakness over-exerting itself, which bursts on the unprepared reader in sundry odes and apostrophes to abstract terms. Such are the Odes to Jealousy, to Hope, to Oblivion, and the like in Dodsley's collection and the magazines of that day, which seldom fail to remind me of an Oxford copy of verses on the two Suttons, commencing with

> Inoculation, heavenly maid! descend!

It is not to be denied that men of undoubted talents, and even poets of true, though not of first-rate, genius, have from a mistaken theory deluded both themselves and others in the opposite extreme. I once read to a company of sensible and well-educated women the introductory period of Cowley's preface to his 'Pindaric Odes, written in imitation of the style and manner of the odes of Pindar.' 'If (says Cowley) a man should undertake to translate Pindar, word for word, it would be thought that one madman had translated another; as may appear, when he, that understands not the original, reads the verbal traduction of him into Latin prose, than which nothing seems more raving.' I then proceeded with his own free version of the second Olympic composed for the charitable purpose of rationalizing the Theban Eagle.

> Queen of all harmonious things,
> Dancing words and speaking strings,
> What God, what hero, wilt thou sing?
> What happy man to equal glories bring?
> Begin, begin thy noble choice,
> And let the hills around reflect the image of thy voice.
> Pisa does to Jove belong,
> Jove and Pisa claim thy song.
> The fair first-fruits of war, th' Olympic games,
> Alcides offer'd up to Jove;
> Alcides too thy strings may move!
> But oh! what man to join with these can worthy prove?
> Join Theron boldly to their sacred names;
> Theron the next honor claims;
> Theron to no man gives place;
> Is first in Pisa's and in Virtue's race;
> Theron there, and he alone,
> Ev'n his own swift forefathers has outgone.

One of the company exclaimed, with the full assent of the rest, that if the original were madder than this, it must be incurably mad. I then translated the ode from the Greek, and as nearly as possible, word for word; and the impression was, that in the general movement of the periods, in the form of the connections and transitions, and in the sober majesty of lofty sense, it appeared to them to approach more nearly, than any other poetry they had heard, to the style of our Bible in the prophetic books. The first strophe will suffice as a specimen:

> Ye harp-controlling hymns! (or) ye hymns the sovereigns of harps!
> What God? what Hero?
> What Man shall we celebrate?
> Truly Pisa indeed is of Jove,
> But the Olympiad (or the Olympic games) did Hercules establish,

The first-fruits of the spoils of war.
But Theron for the four-horsed car,
That bore victory to him,
It behoves us now to voice aloud:
The Just, the Hospitable,
The Bulwark of Agrigentum,
Of renowned fathers
The Flower, even him
Who preserves his native city erect and safe.

But are such rhetorical caprices condemnable only for their deviation from the language of real life? and are they by no other means to be precluded, but by the rejection of all distinctions between prose and verse, save that of metre? Surely good sense, and a moderate insight into the constitution of the human mind, would be amply sufficient to prove, that such language and such combinations are the native produce neither of the fancy nor of the imagination; that their operation consists in the excitement of surprise by the juxtaposition and apparent reconciliation of widely different or incompatible things. As when, for instance, the hills are made to reflect the image of a voice. Surely, no unusual taste is requisite to see clearly, that this compulsory juxtaposition is not produced by the presentation of impressive or delightful forms to the inward vision, nor by any sympathy with the modifying powers with which the genius of the poet had united and inspirited all the objects of his thought; that it is therefore a species of wit, a pure work of the will, and implies a leisure and self-possession both of thought and of feeling, incompatible with the steady fervour of a mind possessed and filled with the grandeur of its subject. To sum up the whole in one sentence. When a poem, or a part of a poem, shall be adduced, which is evidently vicious in the figures and contexture of its style, yet for the condemnation of which no reason can be assigned, except that it differs from the style in which men actually converse, then, and not till then, can I hold this theory to be either plausible, or practicable, or capable of furnishing either rule, guidance, or precaution, that might not, more easily and more safely, as well as more naturally, have been deduced in the author's own mind from considerations of grammar, logic, and the truth and nature of things, confirmed by the authority of works, whose fame is not of one country, nor of one age.

CHAPTER XIX

Continuation—Concerning the real object which, it is probable, Mr Wordsworth had before him, in his critical preface—Elucidation and application of this.

It might appear from some passages in the former part of Mr Wordsworth's preface, that he meant to confine his theory of style, and the necessity of a close accordance with the actual language of men, to those particular subjects from low and rustic life, which by way of experiment he had purposed to naturalize as a new species in our English poetry. But from the train of argument that follows; from the reference to Milton; and from the spirit of his critique on Gray's sonnet; those sentences appear to have been rather courtesies of modesty, than actual limitations of his system. Yet so groundless does this system appear on a close examination; and so strange and overwhelming* in its consequences, that I cannot, and I do not, believe that the poet did ever himself adopt it in the unqualified sense, in which his expressions have been understood by others, and which indeed according to all the common laws of interpretation they seem to bear. What then did he mean? I apprehend, that in the clear perception, not unaccompanied with disgust or contempt, of the gaudy affectations of a style which passed too current with too many for poetic diction, (though in truth it had as little pretensions to poetry, as to logic or common sense) he narrowed his view for the time; and feeling a justifiable preference for the language of nature, and of good sense, even in its humblest and least ornamented forms, he suffered himself to express, in terms at once too large and too exclusive, his predilection for a style the most remote possible from the false and showy splendour which he wished to explode. It is possible, that this predilection, at first merely comparative, deviated for a time into

* I had in my mind the striking but untranslatable epithet, which the celebrated Mendelssohn applied to the great founder of the Critical Philosophy 'Der *alles-zermalmende* Kant,' i.e. the all-becrushing, or rather the *all-to-nothing-crushing* Kant. In the facility and force of compound epithets, the German from the number of its cases and inflections approaches to the Greek: that language so

Bless'd in the happy marriage of sweet words.

It is in the woeful harshness of its sounds alone that the German need shrink from the comparison.

direct partiality. But the real object, which he had in view, was, I doubt not, a species of excellence which had been long before most happily characterized by the judicious and amiable Garve, whose works are so justly beloved and esteemed by the Germans, in his remarks on Gellert (see Sammlung Einiger Abhandlungen von Christian Garve) from which the following is literally translated. 'The talent, that is required in order to make excellent verses, is perhaps greater than the philosopher is ready to admit, or would find it in his power to acquire: the talent to seek only the apt expression of the thought, and yet to find at the same time with it the rhyme and the metre. Gellert possessed this happy gift, if ever any one of our poets possessed it; and nothing perhaps contributed more to the great and universal impression which his fables made on their first publication, or conduces more to their continued popularity. It was a strange and curious phenomenon, and such as in Germany had been previously unheard of, to read verses in which everything was expressed, just as one would wish to talk, and yet all dignified, attractive, and interesting; and all at the same time perfectly correct as to the measure of the syllables and the rhyme. It is certain, that poetry when it has attained this excellence makes a far greater impression than prose. So much so indeed, that even the gratification which the very rhymes afford, becomes then no longer a contemptible or trifling gratification.'

However novel this phenomenon may have been in Germany at the time of Gellert, it is by no means new, nor yet of recent existence in our language. Spite of the licentiousness with which Spenser occasionally compels the orthography of his words into a subservience to his rhymes, the whole Fairy Queen is an almost continued instance of this beauty. Waller's song 'Go, lovely Rose, etc.' is doubtless familiar to most of my readers; but if I had happened to have had by me the poems of Cotton, more but far less deservedly celebrated as the author of the Virgil Travestied, I should have indulged myself, and I think have gratified many who are not acquainted with his serious works, by selecting some admirable specimens of this style. There are not a few poems in that volume, replete with every excellence of thought, image, and passion, which we expect or desire in the poetry of the milder muse; and yet so worded, that the reader sees no one reason either in the selection or the order of the words, why he might not have said the very same in an appropriate conversation, and cannot conceive how indeed he could have expressed such thoughts otherwise, without loss or injury to his meaning.

But in truth our language is, and from the first dawn of poetry ever has been, particularly rich in compositions distinguished by this excellence. The final *e*, which is now mute, in Chaucer's age was either sounded or dropped indifferently. We ourselves still use either *beloved* or *belov'd* according as the rhyme, or measure, or the purpose of more or less solemnity may require. Let the reader then only adopt the pronunciation of the poet and of the court, at which he lived, both with respect to the final *e* and to the accentuation of the last syllable: I would then venture to ask, what even in the colloquial language of elegant and unaffected women (who are the peculiar mistresses of 'pure English and undefiled,') what could we hear more natural, or seemingly more unstudied, than the following stanzas from Chaucer's Troilus and Creseide.

> And after this forth to the gate he went,
> Ther as Creseide out rode a full gode paas:
> And up and doun there made he many a wente,
> And to himselfe ful oft he said, Alas!
> Fro hennis rode my blisse and my solas:
> As wouldè blisful God now for his joie,
> I might her sene agen come in to Troie!
> And to the yondir hill I gan her guide,
> Alas! and there I toke of her my leave:
> And yond I saw her to her fathir ride;
> For sorrow of which mine hearte shall to-cleve;
> And hithir home I came when it was eve;
> And here I dwel; out-cast from allè joie,
> And shall, til I maie sene her efte in Troie.
> And of himselfe imaginid he ofte
> To ben defaitid, pale and waxen lesse
> Than he was wonte, and that men saidin softe,
> What may it be? who can the sothè guess,
> Why Troilus hath all this heviness?
> And al this n' as but his melancholie,
> That he had of himselfe suche fantasie.
> Another time imaginin he would
> That every wight, that past him by the wey
> Had of him routhe, and that they saien should,
> I am right sorry, Troilus will die!
> And thus he drove a daie yet forth or twey
> As ye have herde: suche life gan he to lede
> As he that stode betwixin hope and drede:
>   For which him likid in his songis shewe
> Th' echeson of his wo as he best might,
> And made a songe of wordis but a fewe,

Somwhat his woefull herté for to light,
And when he was from every mann'is sight
With softé voice he of his lady dere,
That absent was, gan sing as ye may hear:

* * * * *

This song when he thus songin had, full soon
He fell agen into his sighis olde:
And every night, as was his wonte to done,
He stodè the bright moonè to beholde
And all his sorrowe to the moone he tolde,
And said: I wis, when thou art hornid newe,
I shall be glad, if al the world be trewe!

Another exquisite master of this species of style, where the scholar and the poet supplies the material, but the perfect well-bred gentleman the expressions and the arrangement, is George Herbert. As from the nature of the subject, and the too frequent quaintness of the thoughts, his 'Temple; or Sacred Poems and Private Ejaculations' are comparatively but little known, I shall extract two poems. The first is a sonnet, equally admirable for the weight, number, and expression of the thoughts, and for the simple dignity of the language. (Unless indeed a fastidious taste should object to the latter half of the sixth line.) The second is a poem of greater length, which I have chosen not only for the present purpose, but likewise as a striking example and illustration of an assertion hazarded in a former page of these sketches: namely, that the characteristic fault of our elder poets is the reverse of that, which distinguishes too many of our more recent versifiers; the one conveying the most fantastic thoughts in the most correct and natural language; the other in the most fantastic language conveying the most trivial thoughts. The latter is a riddle of words; the former an enigma of thoughts. The one reminds me of an odd passage in Drayton's Ideas:

SONNET IX

As other men, so I myself do muse,
Why in this sort I wrest invention so;
And why these *giddy metaphors* I use,
Leaving the path the greater part do go?
I will resolve you: *I am lunatic!*

The other recalls a still odder passage in the 'Synagogue: or the Shadow of the Temple,' a connected series of poems in imitation of Herbert's 'Temple,' and in some editions annexed to it.

O how my mind
Is gravell'd!
Not a thought,
That I can find,
But's ravell'd
All to nought!
Short ends of threds,
And narrow shreds
Of lists;
Knot's snarled ruffs,
Loose broken tufts
Of twists;
Are my torn meditations ragged cloathing,
Which wound, and woven shape a sute for nothing:
One while I think, and then I am in pain
To think how to unthink that thought again!

Immediately after these burlesque passages I cannot proceed to the extracts promised, without changing the ludicrous tone of feeling by the interposition of the three following stanzas of Herbert's.

VIRTUE

Sweet day so cool, so calm, so bright,
The bridal of the earth and sky:
The dew shall weep thy fall to night,
For thou must dye!

Sweet rose, whose hue angry and brave
Bids the rash gazer wipe his eye:
Thy root is ever in its grave,
And thou must dye!

Sweet spring, full of sweet days and roses,
A nest, where sweets compacted lie:
My musick shews, ye have your closes,
And all must dye!

THE BOSOM SIN

A Sonnet by George Herbert

Lord, with what care hast thou begirt us round!
Parents first season us; then schoolmasters
Deliver us to laws; they send us bound
To rules of reason, holy messengers,

Pulpits and Sundays, sorrow dogging sin,
Afflictions sorted, anguish of all sizes,
Fine nets and stratagems to catch us in,
Bibles laid open, millions of surprizes;
Blessings before hand, ties of gratefulness,
The sound of glory ringing in our ears:
Without, our shame; within our consciences;
Angels and grace, eternal hopes and fears!
Yet all these fences, and their whole array
One cunning bosom-sin blows quite away.

LOVE UNKNOWN

Dear friend, sit down, the tale is long and sad:
And in my faintings, I presume, your love
Will more comply than help. A Lord I had,
And have, of whom some grounds, which may improve,
I hold for two lives, and both lives in me.
To him I brought a dish of fruit one day
And in the middle placed my heart. But he
(I sigh to say)
Lookt on a servant who did know his eye,
Better than you knew me, or (which is one)
Than I myself. The servant instantly,
Quitting the fruit, seiz'd on my heart alone,
And threw it in a font, wherein did fall
A stream of blood, which issued from the side
Of a great rock: I well remember all,
And have good cause: there it was dipt and dy'd,
And washt, and wrung! the very wringing yet
Enforceth tears. *Your heart was foul, I fear.*
Indeed 'tis true. I did and do commit
Many a fault, more than my lease will bear;
Yet still ask'd pardon, and was not deny'd.
But you shall hear. After my heart was well,
And clean and fair, as I one eventide,
(I sigh to tell)
Walkt by myself abroad, I saw a large
And spacious furnace flaming, and thereon
A boiling caldron, round about whose verge
Was in great letters set AFFLICTION.
The greatness shew'd the owner. So I went
To fetch a sacrifice out of my fold,
Thinking with that, which I did thus present,
To warm his love, which, I did fear, grew cold.

But as my heart did tender it, the man
Who was to take it from me, slipt his hand,
And threw my heart into the scalding pan;
My heart that brought it (do you understand?)
The offerer's heart. *Your heart was hard, I fear.*
Indeed 'tis true. I found a callous matter
Began to spread and to expatiate there:
But with a richer drug than scalding water
I bath'd it often, ev'n with holy blood,
Which at a board, while many drank bare wine,
A friend did steal into my cup for good,
Ev'n taken inwardly, and most divine
To supple hardnesses. But at the length
Out of the caldron getting, soon I fled
Unto my house, where to repair the strength
Which I had lost, I hasted to my bed;
But when I thought to sleep out all these faults,
(I sigh to speak)
I found that some had stuff'd the bed with thoughts,
I would say thorns. Dear, could my heart not break,
When with my pleasures even my rest was gone?
Full well I understood who had been there:
For I had given the key to none but one:
It must be he. *Your heart was dull, I fear.*
Indeed a slack and sleepy state of mind
Did oft possess me; so that when I pray'd,
Though my lips went, my heart did stay behind.
But all my scores were by another paid,
Who took my guilt upon him. *Truly, friend;*
For ought I hear, your master shows to you
More favour than you wot of. Mark the end!
The font did only what was old renew:
The caldron suppled what was grown too hard:
The thorns did quicken what was grown too dull:
All did but strive to mend what you had marr'd.
Wherefore be cheer'd, and praise him to the full
Each day, each hour, each moment of the week,
Who fain would have you be new, tender, quick!

CHAPTER XX

The former subject continued—The neutral style, or that common to prose and poetry, exemplified by specimens from Chaucer, Herbert, etc.

I HAVE no fear in declaring my conviction, that the excellence defined and exemplified in the preceding chapter is not the characteristic excellence of Mr Wordsworth's style; because I can add with equal sincerity, that it is precluded by higher powers. The praise of uniform adherence to genuine, logical English is undoubtedly his; nay, laying the main emphasis on the word *uniform* I will dare add that, of all contemporary poets, it is his alone. For in a less absolute sense of the word, I should certainly include Mr Bowles, Lord Byron, and, as to all his later writings, Mr Southey, the exceptions in their works being so few and unimportant. But of the specific excellence described in the quotation from Garve, I appear to find more, and more undoubted specimens in the works of others; for instance, among the minor poems of Mr Thomas Moore, and of our illustrious Laureate. To me it will always remain a singular and noticeable fact; that a theory which would establish this *lingua communis* [common tongue], not only as the best, but as the only commendable style, should have proceeded from a poet, whose diction, next to that of Shakespeare and Milton, appears to me of all others the most individualized and characteristic. And let it be remembered too, that I am now interpreting the controverted passages of Mr W.'s critical preface by the purpose and object, which he may be supposed to have intended, rather than by the sense which the words themselves must convey, if they are taken without this allowance.

A person of any taste, who had but studied three or four of Shakespeare's principal plays, would without the name affixed scarcely fail to recognize as Shakespeare's, a quotation from any other play, though but of a few lines. A similar peculiarity, though in a less degree, attends Mr Wordsworth's style, whenever he speaks in his own person; or whenever, though under a feigned name, it is clear that he himself is still speaking, as in the different dramatis personae of the 'Recluse.' Even in the other poems in which he purposes to be most dramatic, there are few in which it does not occasionally burst forth. The reader might often address the poet in his own words with reference to the persons introduced;

It seems, as I retrace the ballad line by the line
That but half of it is theirs, and the better half is thine.

Who, having been previously acquainted with any considerable portion of Mr Wordsworth's publications, and having studied them with a full feeling of the author's genius, would not at once claim as Wordsworthian the little poem on the rainbow?

The child is father of the man, etc.

Or in the 'Lucy Gray'?

No mate, no comrade Lucy knew;
She dwelt on a wide moor;
The sweetest thing that ever grew
Beside a human door.

Or in the 'Idle Shepherd-boys'?

Along the river's stony marge
The sand-lark chaunts a joyous song;
The thrush is busy in the wood,
And carols loud and strong.
A thousand lambs are on the rock
All newly born! both earth and sky
Keep jubilee, and more than all,
Those boys with their green coronal,
They never hear the cry,
That plaintive cry which up the hill
Comes from the depth of Dungeon Gill.

Need I mention the exquisite description of the sea loch in the 'Blind Highland Boy'? Who but a poet tells a tale in such language to the little ones by the fireside as—

Yet had he many a restless dream
Both when he heard the eagle's scream,
And when he heard the torrents roar,
And heard the water beat the shore
    Near where their cottage stood.

Beside a lake their cottage stood,
Not small like ours a peaceful flood;
But one of mighty size, and strange
That rough or smooth is full of change
    And stirring in its bed.

For to this lake by night and day,
The great sea-water finds its way
Through long, long windings of the hills,
And drinks up all the pretty rills;
And rivers large and strong:

Then hurries back the road it came—
Returns on errand still the same;
This did it when the earth was new;
And this for evermore will do,
As long as earth shall last.

And with the coming of the tide,
Come boats and ships that sweetly ride,
Between the woods and lofty rocks;
And to the shepherd with their flocks
Bring tales of distant lands.

I might quote almost the whole of his 'Ruth,' but take the following stanzas:

But as you have before been told,
This stripling, sportive gay and bold,
And with his dancing crest,
So beautiful, through savage lands
Had roam'd about with vagrant bands
Of Indians in the West.

The wind, the tempest roaring high,
The tumult of a tropic sky,
Might well be dangerous food
For him, a youth to whom was given
So much of earth, so much of heaven,
And such impetuous blood,

Whatever in those climes he found
Irregular in sight or sound,
Did to his mind impart
A kindred impulse; seem'd allied
To his own powers, and justified
The workings of his heart.

Nor less to feed voluptuous thought
The beauteous forms of nature wrought,
Fair trees and lovely flowers;
The breezes their own langour lent,
The stars had feelings, which they sent
Into those magic bowers.

Yet in his worst pursuits, I ween,
That sometimes there did intervene
Pure hopes of high intent:
For passions, link'd to forms so fair
And stately, needs must have their share
Of noble sentiment.

But from Mr Wordsworth's more elevated compositions, which already form three-fourths of his works; and will, I trust, constitute hereafter a still larger proportion;—from these, whether in rhyme or blank verse, it would be difficult and almost superfluous to select instances of a diction peculiarly his own, of a style which cannot be imitated without its being at once recognized, as originating in Mr Wordsworth. It would not be easy to open on any one of his loftier strains, that does not contain examples of this; and more in proportion as the lines are more excellent, and most like the author. For those, who may happen to have been less familiar with his writings, I will give three specimens taken with little choice. The first from the lines on the 'Boy of Winander-Mere,'—who

Blew mimic hootings to the silent owls,
That they might answer him. And they would shout,
Across the watery vale and shout again
With long halloos, and screams, and echoes loud
Redoubled and redoubled, concourse wild
Of mirths and jocund din. And when it chanc'd,
That pauses of deep silence mock'd his skill,
Then sometimes in that silence, while he hung
Listening, a gentle shock of mild surprize
Has carried far into his heart the voice
*Of mountain torrents; or the visible scene**
Would enter unawares into his mind
With all its solemn imagery, its rocks,
Its woods, and that uncertain heaven, received
Into the bosom of the steady lake.

* Mr Wordsworth's having judiciously adopted 'concourse wild' in this passage for 'a wild scene' as it stood in the former edition, encourages me to hazard a remark, which I certainly should not have made in the works of a poet less austerely accurate in the use of words, than he is, to his own great honour. It respects the propriety of the word, 'scene,' even in the sentence in which it is retained. Dryden, and he only in his more careless verses, was the first as far as my researches have discovered, who for the convenience of rhyme used this word in the vague sense, which has been since too current even in our best writers, and which (unfortunately, I think) is given as its first

The second shall be that noble imitation of Drayton* (if it was not rather a coincidence) in the 'Joanna.'

> When I had gazed perhaps two minutes space,
> Joanna, looking in my eyes, beheld
> That ravishment of mine, and laugh'd aloud.
> The rock, like something starting from a sleep,
> Took up the lady's voice, and laugh'd again!
> That ancient woman seated on *Helm-crag*
> Was ready with her cavern! *Hammar-scar*,
> And the tall steep of SILVER-HOW sent forth
> A noise of laughter: southern LOUGHRIGG heard,
> And FAIRFIELD answered with a mountain tone.
> HELVELLYN far into the clear blue sky
> Carried the lady's voice!—old SKIDDAW blew
> His speaking trumpet!—back out of the clouds
> From GLARAMARA southward came the voice:
> And KIRKSTONE tossed it from his misty head!

explanation in Dr Johnson's Dictionary, and therefore would be taken by an incautious reader as its proper sense. In Shakespeare and Milton the word is never used without some clear reference, proper or metaphorical, to the theatre. Thus Milton;

> Cedar and pine, and fir and branching palm
> A Sylvan *scene*; and as the ranks ascend
> Shade above shade, a woody *theatre*
> Of stateliest view.

I object to any extension of its meaning because the word is already more equivocal than might be wished; inasmuch as in the limited use, which I recommend, it may still signify two different things; namely, the scenery, and the characters and actions presented on the stage during the presence of particular scenes. It can therefore be preserved from obscurity only by keeping the original signification full in the mind. Thus Milton again,

> Prepare thou for another scene.

* Which COPLAND scarce had spoke, but quickly every hill
Upon her verge that stands, the neighbouring vallies fill;
HELVILLON from his height, it through the mountains threw.
From whom as soon again, the sound DUNBALRASE drew,
From whose stone-trophied head, it on the WENDROSS went,
Which, tow'rds the sea again, resounded it to DENT.
That BROADWATER, therewith within her banks astound,
In sailing to the sea told it to EGREMOUND,
Whose buildings, walks and streets, with echoes loud and long
Did mightily commend old COPLAND for her song!

Drayton's Polyolbion: *Song XXX*

The third which is in rhyme I take from the 'Song at the feast of Brougham Castle, upon the restoration of Lord Clifford the shepherd to the estates of his ancestors.'

Now another day is come
Fitter hope, and nobler doom:
He hath thrown aside his crook,
And hath buried deep his book;
Armour rusting in the halls
On the blood of Clifford calls:
Quell the Scot, exclaims the lance!
Bear me to the heart of France
Is the longing of the shield—
Tell thy name, thou trembling field!
Field of death, where'er thou be,
Groan thou with our victory!
Happy day, and mighty hour,
When our shepherd, in his power,
Mailed and horsed with lance and sword,
To his ancestors restored,
Like a re-appearing star,
Like a glory from afar,
First shall head the flock of war!
Alas! the fervent harper did not know,
That for a tranquil soul the lay was framed,
Who, long compelled in humble walks to go
Was softened into feeling, soothed, and tamed.
Love had he found in huts where poor men lie:
His daily teachers had been woods and rills,
The silence that is in the starry sky,
The sleep that is among the lonely hills.

The words themselves in the foregoing extracts, are, no doubt, sufficiently common for the greater part. (But in what poem are they not so? if we except a few misadventurous attempts to translate the arts and sciences into verse?) In the 'Excursion' the number of polysyllabic (or what the common people call, *dictionary*) words is more than usually great. And so must it needs be, in proportion to the number and variety of an author's conceptions, and his solicitude to express them with precision. But are those words in those places commonly employed in real life to express the same thought or outward thing? Are they the style used in the ordinary intercourse of spoken words? No! nor are the modes of connections: and still less the breaks and transitions. Would any but a poet—at least could anyone without being conscious that he had expressed himself with noticeable vivacity—have described a bird singing loud by, 'The thrush is *busy* in the wood?' Or have spoken of boys with a string of club-moss round their rusty hats, as the boys 'with their green coronal?' Or have translated a beautiful May-day into 'Both earth and sky keep jubilee?' Or have brought all the different marks and circumstances of a sea-loch before the mind, as the actions of a living and acting power? Or have represented the reflection of the sky in the water, as 'That uncertain heaven received into the bosom of the steady lake?' Even the grammatical construction is not unfrequently peculiar; as 'The wind, the tempest roaring high, the tumult of a tropic sky, might well be *dangerous food to him*, a youth to whom was given, etc.' There is a peculiarity in the frequent use of the ἀσυναρτητὸν [asynartete] (i.e. the omission of the connective particle before the last of several words, or several sentences used grammatically as single words, all being in the same case and governing or governed by the same verb) and not less in the construction of words by apposition (*to him*, *a youth*). In short, were there excluded from Mr Wordsworth's poetic compositions all, that a literal adherence to the theory of his preface *would* exclude, two-thirds at least of the marked beauties of his poetry must be erased. For a far greater number of lines would be sacrificed, than in any other recent poet; because the pleasure received from Wordsworth's poems being less derived either from excitement of curiosity or the rapid flow of narration, the *striking* passages form a larger proportion of their value. I do not adduce it as a fair criterion of comparative excellence, nor do I even think it such; but merely as matter of fact. I affirm, that from no contemporary writer could so many lines be quoted, without reference to the poem in which they are found, for their own independent weight or beauty. From the sphere of my own experience I can bring to my recollection three persons of no everyday powers and acquirements, who had read the poems of others with more and more unallayed pleasure, and had thought more highly of their authors, as poets; who yet have confessed to me, that from no modern work had so many passages started up anew in their minds at different times, and as different occasions had awakened a meditative mood.

CHAPTER XXI

Remarks on the present mode of conducting critical journals.

LONG have I wished to see a fair and philosophical inquisition into the character of Wordsworth, as a poet, on the evidence of his published works; and a positive, not a comparative, appreciation of their characteristic excellencies, deficiencies, and defects. I know no claim, that the mere opinion of any individual can have to weigh down the opinion of the author himself; against the probability of whose parental partiality we ought to set that of his having thought longer and more deeply on the subject. But I should call that investigation fair and philosophical, in which the critic announces and endeavours to establish the principles, which he holds for the foundation of poetry in general, with the specification of these in their application to the different classes of poetry. Having thus prepared his canons of criticism for praise and condemnation, he would proceed to particularize the most striking passages to which he deems them applicable, faithfully noticing the frequent or infrequent recurrence of similar merits or defects, and *as* faithfully distinguishing what is characteristic from what is accidental, or a mere flagging of the wing. Then if his premisses be rational, his deductions legitimate, and his conclusions justly applied, the reader, and possibly the poet himself, may adopt his judgement in the light of judgement and in the independence of free agency. If he has erred, he presents his errors in a definite place and tangible form, and holds the torch and guides the way to their detection.

I most willingly admit and estimate at a high value, the services which the Edinburgh Review,° and others formed afterwards on the same plan, have rendered to society in the diffusion of knowledge. I think the commencement of the Edinburgh Review an important epoch in periodical criticism; and that it has a claim upon the gratitude of the literary republic, and indeed of the reading public at large, for having originated the scheme of reviewing those books only, which are susceptible and deserving of argumentative criticism. Not less meritorious, and far more faithfully and in general far more ably executed, is their plan of supplying the vacant place of the trash or mediocrity, wisely left to sink into oblivion by their own weight, with original essays on the most interesting subjects of the time, religious, or political; in which the titles of the books or pamphlets prefixed furnish only the name and occasion of the disquisition. I do not arraign the keenness, or asperity of its damnatory style, in and for itself, as long as the author is addressed or treated as the mere impersonation of the work then under trial. I have no quarrel with them on this account, as long as no personal allusions are admitted, and no recommitment (for new trial) of juvenile performances, that were published, perhaps forgotten, many years before the commencement of the review: since for the forcing back of such works to public notice no motives are easily assignable, but such as are furnished to the critic by his own personal malignity; or what is still worse, by a habit of malignity in the form of mere wantonness.

> No private grudge they need, no personal spite:
> The *viva sectio* is its own delight!°
> All enmity, all envy, they disclaim,
> Disinterested thieves of our good name:
> Cool, sober murderers of their neighbour's fame!
>
> S.T.C.

Every censure, every sarcasm respecting a publication which the critic, with the criticized work before him, can make good, is the critic's right. The writer is authorized to reply, but not to complain. Neither can anyone prescribe to the critic, how soft or how hard; how friendly, or how bitter, shall be the phrases which he is to select for the expression of each reprehension or ridicule. The critic must know, what effect it is his object to produce; and with a view to this effect must he weigh his words. But as soon as the critic betrays, that he knows more of his author, than the author's publications could have told him; as soon as from this more intimate knowledge, elsewhere obtained, he avails himself of the slightest trait *against* the author; his censure instantly becomes personal injury, his sarcasms personal insults. He ceases to be a critic, and takes on him the most contemptible character to which a rational creature can be degraded, that of a gossip, backbiter, and pasquillant: but with this heavy aggravation, that he steals the unquiet, the deforming passions of the world into the museum; into the very place which, next to the chapel and oratory, should be our sanctuary, and secure place of refuge; offers abominations on the altar of the Muses; and makes its sacred paling the very circle in which he conjures up the lying and profane spirit.

This determination of unlicensed personality, and of permitted and legitimate censure (which I owe in part to the illustrious Lessing,°

himself a model of acute, spirited, sometimes stinging, but always argumentative and honourable, criticism) is beyond controversy the true one: and though I would not myself exercise all the rights of the latter, yet, let but the former be excluded, I submit myself to its exercise in the hands of others, without complaint and without resentment.

Let a communication be formed between any number of learned men in the various branches of science and literature; and whether the president and central committee be in London, or Edinburgh, if only they previously lay aside their individuality, and pledge themselves inwardly, as well as ostensibly, to administer judgement according to a constitution and code of laws; and if by grounding this code on the twofold basis of universal morals and philosophic reason, independent of all foreseen application to particular works and authors, they obtain the right to speak each as the representative of their body corporate; they shall have honour and good wishes from me, and I shall accord to them their fair dignities, though self-assumed, not less cheerfully than if I could enquire concerning them in the herald's office, or turn to them in the book of peerage. However loud may be the outcries for prevented or subverted reputation, however numerous and impatient the complaints of merciless severity and insupportable despotism, I shall neither feel, nor utter aught but to the defence and justification of the critical machine. Should any literary Quixote find himself provoked by its sounds and regular movements, I should admonish him with Sancho Panza, that it is no giant but a windmill; there it stands on its own place, and its own hillock, never goes out of its way to attack anyone, and to none and from none either gives or asks assistance. When the public press has poured in any part of its produce between its millstones, it grinds it off, one man's sack the same as another, and with whatever wind may happen to be then blowing. All the two and thirty winds are alike its friends. Of the whole wide atmosphere it does not desire a single finger-breadth more than what is necessary for its sails to turn round in. But this space must be left free and unimpeded. Gnats, beetles, wasps, butterflies, and the whole tribe of ephemerals and insignificants, may flit in and out and between; may hum, and buzz, and jarr; may shrill their tiny pipes, and wind their puny horns, unchastised and unnoticed. But idlers and bravados of larger size and prouder show must beware, how they place themselves within its sweep. Much less may they presume to lay hands on the sails, the strength of which is neither greater or less than as the wind is, which drives them round. Whomsoever the remorseless arm slings aloft, or whirls along with it in the air, he has himself alone to blame; though when the same arm throws him from it, it will more often double than break the force of his fall.

Putting aside the too manifest and too frequent interference of national party, and even personal predilection or aversion; and reserving for deeper feelings those worse and more criminal intrusions into the sacredness of private life, which not seldom merit legal rather than literary chastisement, the two principal objects and occasions which I find for blame and regret in the conduct of the review in question are: first, its unfaithfulness to its own announced and excellent plan, by subjecting to criticism works neither indecent or immoral, yet of such trifling importance even in point of size and, according to the critic's own verdict, so devoid of all merit, as must excite in the most candid mind the suspicion, either that dislike or vindictive feelings were at work; or that there was a cold prudential predetermination to increase the sale of the review by flattering the malignant passions of human nature. That I may not myself become subject to the charge, which I am bringing against others, by an accusation without proof, I refer to the article on Dr Rennell's sermon in the very first number of the Edinburgh Review as an illustration of my meaning. If in looking through all the succeeding volumes the reader should find this a solitary instance, I must submit to that painful forfeiture of esteem, which awaits a groundless or exaggerated charge.

The second point of objection belongs to this review only in common with all other works of periodical criticism; at least, it applies in common to the general system of all, whatever exception there may be in favour of particular articles. Or if it attaches to the Edinburgh Review, and to its only corrival (the Quarterly) with any peculiar force, this results from the superiority of talent, acquirement, and information which both have so undeniably displayed; and which doubtless deepens the regret though not the blame. I am referring to the substitution of assertion for argument; to the frequency of arbitrary and sometimes petulant verdicts, not seldom unsupported even by a single quotation from the work condemned, which might at least have explained the critic's meaning, if it did not prove the justice of his sentence. Even where this is not the case, the extracts are too often made without reference to any general grounds or rules from which the faultiness or inadmissibility of the qualities attributed may be deduced; and without any attempt to show, that the qualities *are* attributable to the passage extracted. I have met with such extracts from Mr Wordsworth's poems, annexed to such assertions, as led me to imagine, that the reviewer, having written his critique before he had read the work, had then pricked with a pin for passages, wherewith to illustrate the various branches of his preconceived opinions. By what principle of rational choice can we suppose a critic to have been directed (at least in a Christian country, and himself, we hope, a Christian) who gives the following lines, portraying the fervour of solitary devotion excited by the magnificent display of the Almighty's works, as a proof and example of an author's tendency to downright ravings, and absolute unintelligibility?

> O then what soul was his, when on the tops
> Of the high mountains he beheld the sun
> Rise up, and bathe the world in light! He looked—
> Ocean and earth, the solid frame of earth,
> And ocean's liquid mass, beneath him lay
> In gladness and deep joy. The clouds were touch'd,
> And in their silent faces did he read
> Unutterable love! Sound needed none,
> Nor any *voice* of joy: his spirit drank
> The spectacle! sensation, soul, and form,
> All melted into him. They swallowed up
> His animal being: in them did he live,
> And by them did he live: they were his life.
>
> Excursion

Can it be expected, that either the author or his admirers should be induced to pay any serious attention to decisions which prove nothing but the pitiable state of the critic's own taste and sensibility? On opening the review they see a favourite passage, of the force and truth of which they had an intuitive certainty in their own inward experience confirmed, if confirmation it could receive, by the sympathy of their most enlightened friends; some of whom perhaps, even in the world's opinion, hold a higher intellectual rank than the critic himself would presume to claim. And this very passage they find selected, as the characteristic effusion of a mind deserted by reason; as furnishing evidence that the writer was raving, or he could not have thus strung words together without sense or purpose! No diversity of taste seems capable of explaining such a contrast in judgement.

That I had overrated the merit of a passage or poem, that I had erred concerning the degree of its excellence, I might be easily induced to believe or apprehend. But that lines, the sense of which I had analysed and found consonant with all the best convictions of my understanding; and the imagery and diction of which had collected round those convictions my noblest as well as my most delightful feelings; that I should admit such lines to be mere nonsense or lunacy, is too much for the most ingenious arguments to effect. But that such a revolution of taste should be brought about by a few broad assertions, seems little less than impossible. On the contrary, it would require an effort of charity not to dismiss the criticism with the aphorism of the wise man, *in animam malevolam sapientia haud intrare potest* [wisdom cannot enter a malicious soul].

What then if this very critic should have cited a large number of single lines and even of long paragraphs, which he himself acknowledges to possess eminent and original beauty? What if he himself has owned, that beauties as great are scattered in abundance throughout the whole book? And yet, though under this impression, should have commenced his critique in vulgar exultation with a prophecy meant to secure its own fulfilment? With a 'This won't do!' What if after such acknowledgements extorted from his own judgement he should proceed from charge to charge of tameness, and raving; flights and flatness; and at length, consigning the author to the house of incurables, should conclude with a strain of rudest contempt evidently grounded in the distempered state of his own moral associations? Suppose too all this done without a single leading principle established or even announced, and without any one attempt at argumentative deduction, though the poet had presented a more than usual opportunity for it, by having previously made public his own principles of judgement in poetry, and supported them by a connected train of reasoning!

The office and duty of the poet is to select the most dignified as well as

> The happiest, gayest, attitude of things.

The reverse, for in all cases a reverse is possible, is the appropriate business of burlesque and travesty, a predominant taste for which has been always deemed a mark of a low and degraded mind. When I was at Rome, among many other visits to the tomb of Julius II, I went thither once with a Prussian artist, a man of genius and great vivacity of feeling. As we were gazing on Michelangelo's Moses, our conversation turned on the horns and beard of that stupendous statue; of the necessity of each to support the other; of the superhuman effect of the former, and the necessity of the existence of both to give a harmony and integrity both to the image and the feeling excited by it. Conceive them removed, and the statue would become *un*-natural, without being *super*-natural. We called to mind the horns of the rising sun, and I repeated the noble passage from Taylor's Holy Dying. That horns were the emblem of power and sovereignty among the Eastern nations, and are still retained as such in Abyssinia; the Achelous of the ancient Greeks; and the probable ideas and feelings, that originally suggested the mixture of the human and the brute form in the figure, by which they realized the idea of their mysterious Pan, as representing intelligence blended with a darker power, deeper, mightier, and more universal than the conscious intellect of man; than intelligence;—all these thoughts and recollections passed in procession before our minds. My companion, who possessed more than his share of the hatred, which his countrymen bore to the French, had just observed to me, 'a Frenchman, Sir! is the only animal in the human shape, that by no possibility can lift itself up to religion or poetry': when, lo! two French officers of distinction and rank entered the church! 'Mark you,' whispered the Prussian, 'the first thing, which those scoundrels will notice (for they will begin by instantly noticing the statue in parts, without one moment's pause of admiration impressed by the whole) will be the horns and the beard. And the associations, which they will immediately connect with them will be those of a he-goat and a cuckold.' Never did man guess more luckily. Had he inherited a portion of the great legislator's prophetic powers, whose statue we had been contemplating, he could scarcely have uttered words more coincident with the result: for even as he had said, so it came to pass.

In the Excursion the poet has introduced an old man, born in humble but not abject circumstances, who had enjoyed more than usual advantages of education, both from books and from the more awful discipline of nature. This person he represents, as having been driven by the restlessness of fervid feelings, and from a craving intellect, to an itinerant life; and as having in consequence passed the larger portion of his time, from earliest manhood, in villages and hamlets from door to door,

> A vagrant merchant bent beneath his load.

Now whether this be a character appropriate to a lofty didactic poem, is perhaps questionable. It presents a fair subject for controversy; and the question is to be determined by the congruity or incongruity of such a character with what shall be proved to be the essential constituents of poetry. But surely the critic who, passing by all the opportunities which such a mode of life would present to such a man; all the advantages of the liberty of nature, of solitude and of solitary thought; all the varieties of places and seasons, through which his track had lain, with all the varying imagery they bring with them; and lastly, all the observations of men,

> Their manners, their enjoyment and pursuits,
> Their passions and their feelings

which the memory of these yearly journeys must have given and recalled to such a mind—the critic, I say, who from the multitude of possible associations should pass by all these in order to fix his attention exclusively on the pin-papers, and stay-tapes, which might have been among the wares of his pack;° this critic in my opinion cannot be thought to possess a much higher or much healthier state of moral feeling, than the Frenchmen above recorded.

CHAPTER XXII

The characteristic defects of Wordsworth's poetry, with the principles from which the judgement, that they are defects, is deduced—Their proportion to the beauties—For the greatest part characteristic of his theory only.

IF Mr Wordsworth have set forth principles of poetry which his arguments are insufficient to support, let him and those who have adopted his sentiments be set right by the confutation of those arguments, and by the substitution of more philosophical principles. And still let the due credit be given to the portion and importance of the truths, which are blended with his theory: truths, the too exclusive attention to which had occasioned its errors, by tempting him to carry those truths beyond their proper limits. If his mistaken theory have at all influenced his poetic compositions, let the effects be pointed out, and the instances given. But let it likewise be shown, how far the influence has acted; whether diffusively, or only by starts; whether the number and importance of the poems and passages thus infected be great or trifling compared with the sound portion; and lastly, whether they are inwoven into the texture of his works, or are loose and separable. The result of such a trial would evince beyond a doubt, what it is high time to announce decisively and aloud, that the supposed characteristics of Mr Wordsworth's poetry, whether admired or reprobated; whether they are simplicity or simpleness; faithful adherence to essential nature, or wilful selections from human nature of its meanest forms and under the least attractive associations; are as little the real characteristics of his poetry at large, as of his genius and the constitution of his mind.

In a comparatively small number of poems, he chose to try an experiment; and this experiment we will suppose to have failed. Yet even in these poems it is impossible not to perceive, that the natural tendency of the poet's mind is to great objects and elevated conceptions. The poem entitled 'Fidelity' is for the greater part written in language, as unraised and naked as any perhaps in the two volumes. Yet take the following stanza and compare it with the preceding stanzas of the same poem.

> There sometimes does a leaping fish
> Send through the tarn a lonely cheer;
> The crags repeat the Raven's croak
> In symphony austere;
> Thither the rainbow comes—the cloud,
> And mists that spread the flying shroud;
> And sun-beams; and the sounding blast,
> That if it could would hurry past,
> But that enormous barrier binds it fast.

Or compare the four last lines of the concluding stanza with the former half:

> Yet proof was plain that since the day
> On which the traveller thus had died,
> The dog had watch'd about the spot,
> Or by his master's side:
> *How nourish'd there for such long time*
> *He knows who gave that love sublime,*
> *And gave that strength of feeling great*
> *Above all human estimate.*

Can any candid and intelligent mind hesitate in determining, which of these best represents the tendency and native character of the poet's genius? Will he not decide that the one was written because the poet *would* so write, and the other because he could not so entirely repress the force and grandeur of his mind, but that he must in some part or other of every composition write otherwise? In short, that his only disease is the being out of his element; like the swan, that having amused himself, for a while, with crushing the weeds on the river's bank, soon returns to his own majestic movements on its reflecting and sustaining surface. Let it be observed, that I am here supposing the imagined judge, to whom I appeal, to have already decided against the poet's theory, as far as it is different from the principles of the art, generally acknowledged.

I cannot here enter into a detailed examination of Mr Wordsworth's works; but I will attempt to give the main results of my own judgement, after an acquaintance of many years, and repeated perusals. And though, to appreciate the defects of a great mind it is necessary to understand previously its characteristic excellences, yet I have already expressed myself with sufficient fullness, to preclude most of the ill effects that might arise from my pursuing a contrary arrangement. I will therefore commence with what I deem the prominent defects of his poems hitherto published.

The first characteristic, though only occasional defect, which I appear to myself to find in these poems is the inconstancy of the style. Under this name I refer to the sudden and unprepared transitions from lines or sentences of peculiar felicity (at all events striking and original) to a style, not only unimpassioned but undistinguished. He sinks too often and too abruptly to that style, which I should place in the second division of language, dividing it into the three species; first, that which is peculiar to poetry; second, that which is only proper in prose; and third, the neutral or common to both. There have been works, such as Cowley's essay on Cromwell, in which prose and verse are intermixed (not as in the Consolation of Boethius, or the Argenis of Barclay, by the insertion of poems supposed to have been spoken or composed on occasions previously related in prose, but) the poet passing from one to the other as the nature of the thoughts or his own feelings dictated. Yet this mode of composition does not satisfy a cultivated taste. There is something unpleasant in the being thus obliged to alternate states of feeling so dissimilar, and this too in a species of writing, the pleasure from which is in part derived from the preparation and previous expectation of the reader. A portion of that awkwardness is felt which hangs upon the introduction of songs in our modern comic operas; and to prevent which the judicious Metastasio (as to whose exquisite taste there can be no hesitation, whatever doubts may be entertained as to his poetic genius) uniformly placed the aria at the end of the scene, at the same time that he almost always raises and impassions the style of the recitative immediately preceding. Even in real life, the difference is great and evident between words used as the arbitrary marks of thought, our smooth market-coin of intercourse with the image and superscription worn out by currency; and those which convey pictures either borrowed from one outward object to enliven and particularize some other; or used allegorically to body forth the inward state of the person speaking; or such as are at least the exponents of his peculiar turn and unusual extent of faculty. So much so indeed, that in the social circles of private life we often find a striking use of the latter put a stop to the general flow of conversation, and by the excitement arising from concentred attention produce a sort of damp and interruption for some minutes after. But in the perusal of works of literary art, we prepare ourselves for such language; and the business of the writer, like that of a painter whose subject requires unusual splendour and prominence, is so to raise the lower and neutral tints, that what in a different style would be the commanding colours, are here used as the means of that gentle degradation requisite in order to produce the effect of a whole. Where this is not achieved in a poem, the metre merely reminds the reader of his claims in order to disappoint them; and where this defect occurs frequently, his feelings are alternately startled by anticlimax and hyperclimax.

I refer the reader to the exquisite stanzas cited for another purpose from the blind Highland Boy; and then annex as being in my opinion instances of this disharmony in style the two following:

> And one, the rarest, was a shell,
> Which he, poor child, had studied well:
> The shell of a green turtle, thin
> And hollow;—you might sit therein,
> It was so wide, and deep.
>
> Our Highland Boy oft visited
> The house which held this prize, and led
> By choice or chance did thither come
> One day, when no one was at home,
> And found the door unbarred.

Or page 172, vol. I.

> 'Tis gone forgotten, *let me do*
> *My best*. There was a smile or two—
> I can remember them, I see
> The smiles worth all the world to me.

Dear Baby, I must lay thee down:
Thou troublest me with strange alarms!
Smiles hast thou, sweet ones of thine own;
I cannot keep thee in my arms,
For they confound me: *as it is*,
I have forgot those smiles of his!

Or page 269, vol. I.

Thou hast a nest, for thy love and thy rest,
And though little troubled with sloth
Drunken lark! thou would'st be loth
To be such a traveller as I.
Happy, happy liver
With a soul as strong as a mountain river
Pouring out praise to th' Almighty giver!
Joy and jollity be with us both,
Hearing thee or else some other,
As merry a brother
I on earth will go plodding on
By myself chearfully till the day is done.

The incongruity, which I appear to find in this passage, is that of the two noble lines in italics with the preceding and following. So vol. II, page 30.

Close by a pond, upon the further side
He stood alone; a minute's space I guess,
I watch'd him, he continuing motionless;
To the pool's further margin then I drew;
He being all the while before me full in view.

Compare this with the repetition of the same image, in the next stanza but two.

And still as I drew near with gentle pace,
Beside the little pond or moorish flood
Motionless as a cloud the old man stood;
That heareth not the loud winds as they call
And moveth altogether, if it move at all.

Or lastly, the second of the three following stanzas, compared both with the first and the third.

My former thoughts returned, the fear that kills;
And hope that is unwilling to be fed;
Cold, pain, and labour, and all fleshly ills;
And mighty poets in their misery dead.

But now, perplex'd by what the old man had said,
My question eagerly did I renew,
How is it that you live, and what is it you do?

He with a smile did then his tale repeat;
And said, that, gathering leeches far and wide
He travelled; stirring thus about his feet
The waters of the ponds where they abide.
'Once I could meet with them on every side,
But they have dwindled long by slow decay;
Yet still I persevere, and find them where I may.'

While he was talking thus, the lonely place,
The old man's shape, and speech, all troubled me:
In my mind's eye I seemed to see him pace
About the weary moors continually,
Wandering about alone and silently.

Indeed this fine poem is especially characteristic of the author. There is scarce a defect or excellence in his writings of which it would not present a specimen. But it would be unjust not to repeat that this defect is only occasional. From a careful reperusal of the two volumes of poems, I doubt whether the objectionable passages would amount in the whole to one hundred lines; not the eighth part of the number of pages. In the Excursion the feeling of incongruity is seldom excited by the diction of any passage considered in itself, but by the sudden superiority of some other passage forming the context.

The second defect I could generalize with tolerable accuracy, if the reader will pardon an uncouth and new-coined word. There is, I should say, not seldom a *matter-of-factness* in certain poems. This may be divided into, first, a laborious minuteness and fidelity in the representation of objects, and their positions, as they appeared to the poet himself; secondly, the insertion of accidental circumstances, in order to the full explanation of his living characters, their dispositions and actions; which circumstances might be necessary to establish the probability of a statement in real life, where nothing is taken for granted by the hearer, but appear superfluous in poetry, where the reader is willing to believe for his own sake. To this accidentality, I object, as contravening the essence of poetry, which Aristotle pronounces to be σπουδαιότατον καὶ φιλοσοφώτατον γενὸς, the most intense, weighty and philosophical product of human art; adding, as the reason, that it is the most catholic and abstract. The following passage from Davenant's prefatory letter to Hobbes well expresses this truth. 'When I considered the actions which I meant to describe (those inferring the persons) I was again persuaded rather to choose those of a former age, than the present; and in a century so far removed as might preserve me from their improper examinations, who know not the requisites of a poem, nor how much pleasure they lose (and even the pleasures of heroic poesy are not unprofitable) who take away the liberty of a poet, and fetter his feet in the shackles of an historian. For why should a poet doubt in story to mend the intrigues of fortune by more delightful conveyances of probable fictions, because austere historians have entered into bond to truth? An obligation, which were in poets as foolish and unnecessary, as is the bondage of false martyrs, who lie in chains for a mistaken opinion. But by this I would imply, that truth narrative and past is the idol of historians (who worship a dead thing) and truth operative, and by effects continually alive, is the mistress of poets, who hath not her existence in matter, but in reason.'

For this minute accuracy in the painting of local imagery, the lines in the Excursion, p. 96, 97, and 98, may be taken, if not as a striking instance, yet as an illustration of my meaning. It must be some strong motive (as, for instance, that the description was necessary to the intelligibility of the tale) which could induce me to describe in a number of verses what a draughtsman could present to the eye with incomparably greater satisfaction by half a dozen strokes of his pencil, or the painter with as many touches of his brush. Such descriptions too often occasion in the mind of a reader, who is determined to understand his author, a feeling of labour, not very dissimilar to that, with which he would construct a diagram, line by line, for a long geometrical proposition. It seems to be like taking the pieces of a dissected map out of its box. We first look at one part, and then at another, then join and dovetail them; and when the successive acts of attention have been completed, there is a retrogressive effort of mind to behold it as a whole. The poet should paint to the imagination, not to the fancy; and I know no happier case to exemplify the distinction between these two faculties. Masterpieces of the former mode of poetic painting abound in the writings of Milton, ex. gr.

> The fig tree, not that kind for fruit renown'd,
> But such as at this day to Indians known
> In Malabar or Decan, spreads her arms
> Branching so broad and long, that in the ground
> The bended twigs take root, *and daughters grow*

About the mother-tree, a pillar'd shade
High over-arched, and echoing walks between:
There oft the Indian herdsman shunning heat
Shelters in cool, and tends his pasturing herds
At loop holes cut through thickest shade.

Milton, *P. L.* 9, 1100

This is creation rather than painting, or if painting, yet such, and with such co-presence of the whole picture flashed at once upon the eye, as the sun paints in a camera obscura.° But the poet must likewise understand and command what Bacon calls the *vestigia communia* [impressions in common] of the senses, the latency of all in each, and more especially as by a magical *penna duplex* [double pen], the excitement of vision by sound and the exponents of sound. Thus, 'The echoing walks between,' may be almost said to reverse the fable in tradition of the head of Memnon, in the Egyptian statue.° Such may be deservedly entitled the creative words in the world of imagination.

The second division respects an apparent minute adherence to matter-of-fact in character and incidents; a biographical attention to probability, and an anxiety of explanation and retrospect. Under this head I shall deliver, with no feigned diffidence, the results of my best reflection on the great point of controversy between Mr Wordsworth, and his objectors; namely, on the choice of his characters. I have already declared, and I trust justified, my utter dissent from the mode of argument which his critics have hitherto employed. To their question, why did you choose such a character, or a character from such a rank of life? the poet might in my opinion fairly retort: why, with the conception of my character did you make wilful choice of mean or ludicrous associations not furnished by me, but supplied from your own sickly and fastidious feelings? How was it, indeed, probable, that such arguments could have any weight with an author, whose plan, whose guiding principle, and main object it was to attack and subdue that state of association, which leads us to place the chief value on those things on which man differs from man, and to forget or disregard the high dignities, which belong to human nature, the sense and the feeling, which may be, and ought to be, found in all ranks? The feelings with which, as Christians, we contemplate a mixed congregation rising or kneeling before their common maker, Mr Wordsworth would have us entertain at all times as men, and as readers; and by the excitement of this lofty, yet prideless impartiality in poetry, he might hope to have encouraged its continuance in real life. The praise of good men be his! In real life, and, I trust, even in my imagination, I honour a virtuous and wise man, without reference to the presence or absence of artificial advantages. Whether in the person of an armed baron, a laurelled bard, etc. or of an old pedlar, or still older leech-gatherer, the same qualities of head and heart must claim the same reverence. And even in poetry I am not conscious, that I have ever suffered my feelings to be disturbed or offended by any thoughts or images, which the poet himself has not presented.

But yet I object nevertheless, and for the following reasons. First, because the object in view, as an immediate object, belongs to the moral philosopher, and would be pursued, not only more appropriately, but in my opinion with far greater probability of success, in sermons or moral essays, than in an elevated poem. It seems, indeed, to destroy the main fundamental distinction, not only between a poem and prose, but even between philosophy and works of fiction, inasmuch as it proposes truth for its immediate object, instead of pleasure. Now till the blessed time shall come, when truth itself shall be pleasure, and both shall be so united, as to be distinguishable in words only, not in feeling, it will remain the poet's office to proceed upon that state of association, which actually exists as general; instead of attempting first to make it what it ought to be, and then to let the pleasure follow. But here is unfortunately a small hysteron-proteron. For the communication of pleasure is the introductory means by which alone the poet must expect to moralize his readers. Secondly: though I were to admit, for a moment, this argument to be groundless: yet how is the moral effect to be produced, by merely attaching the name of some low profession to powers which are least likely, and to qualities which are assuredly not more likely, to be found in it? The poet, speaking in his own person, may at once delight and improve us by sentiments, which teach us the independence of goodness, of wisdom, and even of genius, on the favours of fortune. And having made a due reverence before the throne of Antonine, he may bow with equal awe before Epictetus among his fellow slaves—

> . . . and rejoice
> In the plain presence of his dignity.

Who is not at once delighted and improved, when the poet Wordsworth himself exclaims,

> O many are the poets that are sown
> By Nature; men endowed with highest gifts,
> The vision and the faculty divine,

Yet wanting the accomplishment of verse,
Nor having e'er, as life advanced, been led
By circumstance to take unto the height
The measure of themselves, these favor'd beings,
All but a scatter'd few, live out their time
Husbanding that which they possess within,
And go to the grave unthought of. Strongest minds
Are often those of whom the noisy world
Hears least.

Excursion, B. I

To use a colloquial phrase, such sentiments, in such language, do one's heart good; though I for my part, have not the fullest faith in the *truth* of the observation. On the contrary I believe the instances to be exceedingly rare; and should feel almost as strong an objection to introduce such a character in a poetic fiction, as a pair of black swans on a lake, in a fancy-landscape. When I think how many, and how much better books, than Homer, or even than Herodotus, Pindar or Aeschylus, could have read, are in the power of almost every man, in a country where almost every man is instructed to read and write; and how restless, how difficultly hidden, the powers of genius are; and yet find even in situations the most favourable, according to Mr Wordsworth, for the formation of a pure and poetic language; in situations which ensure familiarity with the grandest objects of the imagination; but one Burns, among the shepherds of Scotland, and not a single poet of humble life among those of English lakes and mountains; I conclude, that poetic genius is not only a very delicate but a very rare plant.

But be this as it may, the feelings with which,

I think of Chatterton, the marvellous boy,
The sleepless soul, that perish'd in his pride:
Of Burns, that walk'd in glory and in joy
Behind his plough upon the mountain-side—

are widely different from those with which I should read a poem, where the author, having occasion for the character of a poet and a philosopher in the fable of his narration, had chosen to make him a chimney-sweeper; and then, in order to remove all doubts on the subject, had invented an account of his birth, parentage and education, with all the strange and fortunate accidents which had concurred in making him at once poet, philosopher, and sweep! Nothing, but biography, can justify this. If it be admissible even in a novel, it must be one in the manner of Defoe's, that were meant to pass for histories, not in the manner of Fielding's: in the life of Moll Flanders, or Colonel Jack, not in a Tom Jones or even a Joseph Andrews. Much less then can it be legitimately introduced in a poem, the characters of which, amid the strongest individualization, must still remain representative. The precepts of Horace, on this point, are grounded on the nature both of poetry and of the human mind. They are not more peremptory, than wise and prudent. For in the first place a deviation from them perplexes the reader's feelings, and all the circumstances which are feigned in order to make such accidents less improbable, divide and disquiet his faith, rather than aid and support it. Spite of all attempts, the fiction will appear, and unfortunately not as fictitious but as false. The reader not only knows, that the sentiments and language are the poet's own, and his own too in his artificial character, as poet; but by the fruitless endeavours to make him think the contrary, he is not even suffered to forget it. The effect is similar to that produced by an epic poet, when the fable and the characters are derived from Scripture history, as in the *Messiah* of Klopstock, or in Cumberland's *Calvary*: and not merely suggested by it as in the Paradise Lost of Milton. That illusion, contradistinguished from delusion, that negative faith, which simply permits the images presented to work by their own force, without either denial or affirmation of their real existence by the judgement, is rendered impossible by their immediate neighbourhood to words and facts of known and absolute truth. A faith, which transcends even historic belief, must absolutely put out this mere poetic analogon of faith, as the summer sun is said to extinguish our household fires, when it shines full upon them. What would otherwise have been yielded to as pleasing fiction, is repelled as revolting falsehood. The effect produced in this latter case by the solemn belief of the reader, is in a less degree brought about in the instances, to which I have been objecting, by the baffled attempts of the author to make him believe.

Add to all the foregoing the seeming uselessness both of the project and of the anecdotes from which it is to derive support. Is there one word for instance, attributed to the pedlar in the Excursion, characteristic of a pedlar? One sentiment, that might not more plausibly, even without the aid of any previous explanation, have proceeded from any wise and beneficent old man, of a rank or profession in which the language of learning and refinement are natural and to be expected? Need the rank have been at all particularized, where nothing follows which the knowledge of that rank is to explain or illustrate? When on the contrary this information renders the man's language, feelings, sentiments, and information a riddle, which must itself be solved by episodes of anecdote? Finally when this, and this alone, could have induced a genuine poet to inweave in a poem of the loftiest style, and on subjects the loftiest and of most universal interest, such minute matters of fact, (not unlike those furnished for the obituary of a magazine by the friends of some obscure ornament of society lately deceased in some obscure town), as

> Among the hills of Athol he was born.
> There on a small hereditary farm,
> An unproductive slip of rugged ground,
> His Father dwelt; and died in poverty:
> While he, whose lowly fortune I retrace,
> The youngest of three sons, was yet a babe,
> A little one—unconscious of their loss.
> But ere he had outgrown his infant days
> His widowed mother, for a second mate,
> Espoused the teacher of the Village School;
> Who on her offspring zealously bestowed
> Needful instruction.
>
> From his sixth year, the Boy of whom I speak,
> In summer, tended cattle on the hills;
> But through the inclement and the perilous days
> Of long-continuing winter, he repaired
> To his step-father's school.—etc.

For all the admirable passages interposed in this narration, might, with trifling alterations, have been far more appropriately, and with far greater verisimilitude, told of a poet in the character of a poet; and without incurring another defect which I shall now mention, and a sufficient illustration of which will have been here anticipated.

Third; an undue predilection for the dramatic form in certain poems, from which one or other of two evils result. Either the thoughts and diction are different from that of the poet, and then there arises an incongruity of style; or they are the same and indistinguishable, and then it presents a species of ventriloquism, where two are represented as talking, while in truth one man only speaks.

The fourth class of defects is closely connected with the former; but yet are such as arise likewise from an intensity of feeling disproportionate to such knowledge and value of the objects described, as can be fairly anticipated of men in general, even of the most cultivated classes; and with which therefore few only, and those few particularly circumstanced, can be supposed to sympathize: In this class, I comprise occasional prolixity, repetition, and an eddying instead of progression of thought. As instances, see page 27, 28, and 62 of the Poems, Vol. I. and the first eighty lines of the Sixth Book of the Excursion.

Fifth and last; thoughts and images too great for the subject. This is an approximation to what might be called mental bombast, as distinguished from verbal: for, as in the latter there is a disproportion of the expressions to the thoughts so in this there is a disproportion of thought to the circumstance and occasion. This, by the by, is a fault of which none but a man of genius is capable. It is the awkwardness and strength of Hercules with the distaff of Omphale.

It is a well-known fact, that bright colours in motion both make and leave the strongest impressions on the eye. Nothing is more likely too, than that a vivid image or visual spectrum, thus originated, may become the link of association in recalling the feelings and images that had accompanied the original impression. But if we describe this in such lines, as

> They flash upon that inward eye,
> Which is the bliss of solitude!

in what words shall we describe the joy of retrospection, when the images and virtuous actions of a whole well-spent life, pass before that conscience which is indeed the *inward* eye: which is indeed 'the bliss of solitude.' Assuredly we seem to sink most abruptly, not to say burlesquely, and almost as in a medley from this couplet to—

> And then my heart with pleasure fills,
> And dances with the daffodils.
>
> Vol. I. p. 320

The second instance is from Vol. II. page 12, where the poet having gone out for a day's tour of pleasure, meets early in the morning with a knot of gypsies, who had pitched their blanket-tents and straw beds, together with their children and asses, in some field by the roadside. At the close of the day on his return our tourist found them in the same place. 'Twelve hours,' says he,

> Twelve hours, twelve bounteous hours, are gone while I
> Have been a traveller under open sky,
> Much witnessing of change and cheer,
> Yet as I left I find them here!

Whereat the poet, without seeming to reflect that the poor tawny wanderers might probably have been tramping for weeks together through road and lane, over moor and mountain, and consequently must have been right glad to rest themselves, their children and cattle, for one whole day; and overlooking the obvious truth, that such repose might be quite as necessary for them, as a walk of the same continuance was pleasing or healthful for the more fortunate poet; expresses his indignation in a series of lines, the diction and imagery of which would have been rather above, than below the mark, had they been applied to the immense empire of China improgressive for thirty centuries:

> The weary Sun betook himself to rest,
> —Then issued Vesper from the fulgent west,
> Outshining, like a visible God,
> The glorious path in which he trod!
> And now ascending, after one dark hour,
> And one night's diminution of her power,
> Behold the mighty Moon! this way
> She looks, as if at them—but they
> Regard not her:—oh, better wrong and strife,
> Better vain deeds or evil than such life!
> The silent Heavens have goings on:
> The Stars have tasks!—but *these* have none!

The last instance of this defect, (for I know no other than these already cited) is from the Ode, page 351, Vol. II. where, speaking of a child, 'a six years' darling of a pigmy size,' he thus addresses him:

> Thou best philosopher who yet dost keep
> Thy heritage! Thou eye among the blind,
> That, deaf and silent, read'st the eternal deep,
> Haunted for ever by the Eternal Mind—
> Mighty Prophet! Seer blest!
> On whom those truths do rest,
> Which we are toiling all our lives to find!
> Thou, over whom thy immortality
> Broods like the day, a master o'er the slave.
> A presence that is not to be put by!

Now here, not to stop at the daring spirit of metaphor which connects the epithets 'deaf and silent,' with the apostrophized *eye*: or (if we are to refer it to the preceding word, *philosopher*) the faulty and equivocal syntax of the passage; and without examining the propriety of making a 'master *brood* o'er a slave,' or the day brood at all; we will merely ask, what does all this mean? In what sense is a child of that age a philosopher? In what sense does he read 'the eternal deep'? In what sense is he declared to be 'for ever haunted' by the Supreme Being? or so inspired as to deserve the splendid titles of a mighty prophet, a blessed seer? By reflection? by knowledge? by conscious intuition? or by any form or modification of consciousness? These would be tidings indeed; but such as would presuppose an immediate revelation to the inspired communicator, and require miracles to authenticate his inspiration. Children at this age give us no such information of themselves; and at what time were we dipped in the Lethe, which has produced such utter oblivion of a state so godlike? There are many of us that still possess some remembrances, more or less distinct, respecting themselves at six years old; pity that the worthless straws only should float, while treasures, compared with which all the mines of Golconda and Mexico were but straws, should be absorbed by some unknown gulf into some unknown abyss.

But if this be too wild and exorbitant to be suspected as having been the poet's meaning; if these mysterious gifts, faculties, and operations, are not accompanied with consciousness; who else is conscious of them? or how can it be called the child, if it be no part of the child's conscious being? For aught I know, the thinking spirit within me may be substantially one with the principle of life, and of vital operation. For aught I know, it may be employed as a secondary agent in the marvellous organization and organic movements of my body. But, surely, it would be strange language to say, that *I* construct my heart! or that *I* propel the finer influences through my nerves! or that *I* compress my brain, and draw the curtains of sleep round my own eyes! Spinoza and Behmen were on different systems both Pantheists; and among the ancients there were philosophers, teachers of the *EN KAI ΠAN* [ONE AND ALL], who not only taught, that God was All, but that this All constituted God. Yet not even these would confound the part, as a part, with the whole, as the whole. Nay, in no system is the distinction between the individual and God, between the modification, and the one only substance, more sharply drawn, than in that of Spinoza. Jacobi indeed relates of Lessing, that after a conversation with him at the house of the poet, Gleim (the Tyrtaeus and Anacreon of the German Parnassus), in which conversation L. had avowed privately to Jacobi his reluctance to admit any personal existence of the Supreme Being, or the possibility of personality except in a finite intellect, and while they were sitting at table, a shower of rain came on unexpectedly. Gleim expressed his regret at the circumstance, because they had meant to drink their wine in the garden: upon which Lessing in one of his half-earnest, half-joking moods, nodded to Jacobi, and said, 'It is *I*, perhaps, that am doing that,' i.e. raining! and J. answered, 'or perhaps I'; Gleim contented himself with staring at them both, without asking for any explanation.

So with regard to this passage. In what sense can the magnificent attributes, above quoted, be appropriated to a child, which would not make them equally suitable to a bee, or a dog, or a field of corn; or even to a ship, or to the wind and waves that propel it? The omnipresent Spirit works equally in them, as in the child; and the child is equally unconscious of it as they. It cannot surely be, that the four lines, immediately following, are to contain the explanation?

> To whom the grave
> Is but a lonely bed without the sense or sight
> Of day or the warm light,
> A place of thought where we in waiting lie.

Surely, it cannot be that this wonder-rousing apostrophe is but a comment on the little poem of 'We are Seven'? that the whole meaning of the passage is reducible to the assertion, that a child, who by the by at six years old would have been better instructed in most Christian families, has no other notion of death than that of lying in a dark, cold place? And still, I hope, not as in a place of thought! not the frightful notion of lying awake in his grave! The analogy between death and sleep is too simple, too natural, to render so horrid a belief possible for children; even had they not been in the habit, as all Christian children are, of hearing the latter term used to express the former. But if the child's belief be only, that 'he is not dead, but sleepeth': wherein does it differ from that of his father and mother, or any other adult and instructed person? To form an idea of a thing's becoming nothing; or of nothing becoming a thing; is impossible to all finite beings alike, of whatever age, and however educated or uneducated. Thus it is with splendid paradoxes in general. If the words are taken in the common sense, they convey an absurdity; and if, in contempt of dictionaries and custom, they are so interpreted as to avoid the absurdity, the meaning dwindles into some bald truism. Thus you must at once understand the words contrary to their common import, in order to arrive at any sense; and according to their common import, if you are to receive from them any feeling of sublimity or admiration.

Though the instances of this defect in Mr Wordsworth's poems are so few, that for themselves it would have been scarcely just to attract the reader's attention toward them; yet I have dwelt on it, and perhaps the more for this very reason. For being so very few, they cannot sensibly detract from the reputation of an author, who is even characterized by the number of profound truths in his writings, which will stand the severest analysis; and yet few as they are, they are exactly those passages which his blind admirers would be most likely, and best able, to imitate. But Wordsworth, where he is indeed Wordsworth, may be mimicked by copyists, he may be plundered by plagiarists; but he cannot be imitated, except by those who are not born to be imitators. For without his depth of feeling and his imaginative power his sense would want its vital warmth and peculiarity; and without his strong sense, his mysticism would become sickly—mere fog, and dimness!

To these defects which, as appears by the extracts, are only occasional, I may oppose with far less fear of encountering the dissent of any candid and intelligent reader, the following (for the most part correspondent) excellencies. First, an austere purity of language both grammatically and logically; in short a perfect appropriateness of the words to the meaning. Of how high value I deem this, and how particularly estimable I hold the example at the present day, has been already stated: and in part too the reasons on which I ground both the moral and intellectual importance of habituating ourselves to a strict accuracy of expression. It is noticeable, how limited an acquaintance with the masterpieces of art will suffice to form a correct and even a sensitive taste, where none but masterpieces have been seen and admired: while on the other hand, the most correct notions, and the widest acquaintance with the works of excellence of all ages and countries, will not perfectly secure us against the contagious familiarity with the far more numerous offspring of tastelessness or of a perverted taste. If this be the case, as it notoriously is, with the arts of music and painting, much more difficult will it be, to avoid the infection of multiplied and daily examples in the practice of an art, which uses words, and words only, as its instruments. In poetry, in which every line, every phrase, may pass the ordeal of deliberation and deliberate choice, it is possible, and barely possible, to attain that ultimatum which I have ventured to propose as the infallible test of a blameless style; namely, its untranslatableness in words of the same language without injury to the meaning. Be it observed, however, that I include in the meaning of a word not only its correspondent object, but likewise all the associations which it recalls. For language is framed to convey not the object alone, but likewise the character, mood and intentions of the person who is representing it. In poetry it is practicable to preserve the diction uncorrupted by the affectations and misappropriations, which promiscuous authorship, and reading not promiscuous only because it is disproportionally most conversant with the compositions of the day, have rendered general. Yet even to the poet, composing in his own province, it is an arduous work: and as the result and pledge of a watchful good sense, of fine and luminous distinction, and of complete self-possession, may justly claim all the honour which belongs to an attainment equally difficult and valuable, and the more valuable for being rare. It is at all times the proper food of the understanding; but in an age of corrupt eloquence it is both food and antidote.

In prose I doubt whether it be even possible to preserve our style wholly unalloyed by the vicious phraseology which meets us everywhere, from the sermon to the newspaper, from the harangue of the legislator to the speech from the convivial chair, announcing a toast or sentiment. Our chains rattle, even while we are complaining of them. The poems of Boethius rise high in our estimation when we compare them with those of his contemporaries, as Sidonius Apollinaris, etc. They might even be referred to a purer age, but that the prose, in which they are set, as jewels in a crown of lead or iron, betrays the true age of the writer. Much however may be effected by education. I believe not only from grounds of reason, but from having in great measure assured myself of the fact by actual though limited experience, that to a youth led from his first boyhood to investigate the meaning of every word and the reason of its choice and position, logic presents itself as an old acquaintance under new names.

On some future occasion, more especially demanding such disquisition, I shall attempt to prove the close connection between veracity and habits of mental accuracy; the beneficial after-effects of verbal precision in the preclusion of fanaticism, which masters the feelings more especially by indistinct watchwords; and to display the advantages which language alone, at least which language with incomparably greater ease and certainty than any other means, presents to the instructor of impressing modes of intellectual energy so constantly, so imperceptibly, and as it were by such elements and atoms, as to secure in due time the formation of a second nature. When we reflect, that the cultivation of the judgement is a positive command of the moral law, since the reason can give the principle alone, and the conscience bears witness only to the motive, while the application and effects must depend on the judgement: when we consider, that the greater part of our success and comfort in life depends on distinguishing the similar from the same, that which is peculiar in each thing from that which it has in common with others, so as still to select the most probable, instead of the merely possible or positively unfit, we shall learn to value earnestly and with a practical seriousness a mean, already prepared for us by nature and society, of teaching the young mind to think well and wisely by the same unremembered process and with the same never forgotten results, as those by which it is taught to speak and converse. Now how much warmer the interests, how much more genial the feelings of reality and practicability, and thence how much stronger the impulses to imitation are, which a contemporary writer, and especially a contemporary poet, excites in youth and commencing manhood, has been treated of in the earlier pages of these sketches. I have only to add, that all the praise which is due to the exertion of such influence for a purpose so important, joined with that which must be claimed for the infrequency of the same excellence in the same perfection, belongs in full right to Mr Wordsworth. I am far however from denying that we have poets whose general style possesses the same excellence, as Mr Moore, Lord Byron, Mr Bowles, and in all his later and more important works our laurel-honouring Laureate. But there are none, in whose works I do not appear to myself to find more exceptions, than in those of Wordsworth. Quotations or specimens would here be wholly out of place, and must be left for the critic who doubts and would invalidate the justice of this eulogy so applied.

The second characteristic excellence of Mr W.'s works is: a correspondent weight and sanity of the thoughts and sentiments,—won, not from books, but—from the poet's own meditative observation. They are fresh and have the dew upon them. His Muse, at least when in her strength of wing, and when she hovers aloft in her proper element,

> Makes audible a linked lay of truth,
> Of truth profound a sweet continuous lay,
> Not learnt, but native, her own natural notes!
>
> S.T.C.

Even throughout his smaller poems there is scarcely one, which is not rendered valuable by some just and original reflection.

See page 25, vol. 2nd: or the two following passages in one of his humblest compositions.

O Reader! had you in your mind
Such stores as silent thought can bring,
O gentle Reader! you would find
A tale in every thing.

and

I have heard of hearts unkind, kind deeds
With coldness still returning:
Alas! the gratitude of men
Has oftener left me mourning.

or in a still higher strain the six beautiful quatrains, page 134.

Thus fares it still in our decay:
And yet the wiser mind
Mourns less for what age takes away
Than what it leaves behind.

The Blackbird in the summer trees,
The Lark upon the hill,
Let loose their carols when they please,
Are quiet when they will.

With nature never do *they* wage
A foolish strife; they see
A happy youth, and their old age
Is beautiful and free!

But we are pressed by heavy laws;
And often, glad no more,
We wear a face of joy, because
We have been glad of yore.

If there is one, who need bemoan
His kindred laid in earth,
The household hearts that were his own,
It is the man of mirth.

My days, my Friend, are almost gone,
My life has been approved,
And many love me; but by none
Am I enough beloved.

or the sonnet on Bonaparte, page 202, vol. 2; or finally (for a volume would scarce suffice to exhaust the instances,) the last stanza of the poem on the withered celandine, vol. 2, p. 212.

> To be a prodigal's favorite—then, worse truth,
> A miser's pensioner—behold our lot!
> Oh man! that from thy fair and shining youth
> Age might but take the things, youth needed not.

Both in respect of this and of the former excellence, Mr Wordsworth strikingly resembles Samuel Daniel, one of the golden writers of our golden Elizabethan age, now most causelessly neglected: Samuel Daniel, whose diction bears no mark of time, no distinction of age, which has been, and as long as our language shall last will be so far the language of the today and for ever, as that it is more intelligible to us, than the transitory fashions of our own particular age. A similar praise is due to his sentiments. No frequency of perusal can deprive them of their freshness. For though they are brought into the full daylight of every reader's comprehension; yet are they drawn up from depths which few in any age are privileged to visit, into which few in any age have courage or inclination to descend. If Mr Wordsworth is not equally with Daniel alike intelligible to all readers of average understanding in all passages of his works, the comparative difficulty does not arise from the greater impurity of the ore, but from the nature and uses of the metal. A poem is not necessarily obscure, because it does not aim to be popular. It is enough, if a work be perspicuous to those for whom it is written, and,

> Fit audience find, though few.

To the 'Ode on the intimation of immortality from recollections of early childhood' the poet might have prefixed the lines which Dante addresses to one of his own Canzoni—

> Canzon, io credo, che saranno radi
> Che tua regione intendan bene:
> Tanto lor sei faticoso ed alto.

> O lyric song, there will be few, think I,
> Who may thy import understand aright:
> Thou art for *them* so arduous and so high!

But the ode was intended for such readers only as had been accustomed to watch the flux and reflux of their inmost nature, to venture at times into the twilight realms of consciousness, and to feel a deep interest in modes of inmost being, to which they know that the attributes of time and space are inapplicable and alien, but which yet cannot be conveyed, save in symbols of time and space. For such readers the sense is sufficiently plain, and they will be as little disposed to charge Mr Wordsworth with believing the Platonic pre-existence in the ordinary interpretation of the words, as I am to believe, that Plato himself ever meant or taught it.

Πολλα οἱ ὑπ' αγκῶ-
νος ὠκέα βέλη
Ἔνδον εντὶ φαρέτρας
Φωνᾶντα συνετοῖσιν· ες
Δὲ τὸ παν ερμηνέως
Χατίζει. Σοφὸς ὁ πολ-
λα εἴδως φυᾷ·
Μαθόντες δὲ, λάβροι
Παγγλωσσίᾳ, κόρακες ὥς
Ἄκραντα γαρύετον
Διὸς πρὸς ὄρνιχα Θεῖον.

[Full many a swift arrow [has he] beneath [his] arm, within [his] quiver, many an arrow that is vocal to the wise; but for the crowd they need [an interpreter]. The true poet is he who knoweth much by gift of nature, but they that have only learnt the lore of song, and are turbulent and intemperate of tongue, like a pair of crows, chatter in vain against the godlike bird of Zeus. (LCL)]

Third (and wherein he soars far above Daniel) the sinewy strength and originality of single lines and paragraphs: the frequent *curiosa felicitas* [studied felicity] of his diction, of which I need not here give specimens, having anticipated them in a preceding page. This beauty, and as eminently characteristic of Wordsworth's poetry, his rudest assailants have felt themselves compelled to acknowledge and admire.

Fourth; the perfect truth of nature in his images and descriptions as taken immediately from nature, and proving a long and genial intimacy with the very spirit which gives the physiognomic expression to all the works of nature. Like a green field reflected in a calm and perfectly transparent lake, the image is distinguished from the reality only by its greater softness and lustre. Like the moisture or the polish on a pebble, genius neither distorts nor false-colours its objects; but on the contrary brings out many a vein and many a tint, which escape the eye of common observation, thus raising to the rank of gems, what had been often kicked away by the hurrying foot of the traveller on the dusty high road of custom.

Let me refer to the whole description of skating, vol. I, page 42 to 47, especially to the lines

> So through the darkness and the cold we flew,
> And not a voice was idle: with the din
> Meanwhile the precipices rang aloud;
> The leafless trees and every icy crag
> Tinkled like iron; while the distant hills
> Into the tumult sent an alien sound
> Of melancholy, not unnoticed, while the stars
> Eastward were sparkling clear, and in the west
> The orange sky of evening died away.

Or to the poem on the green linnet, vol. I. p. 244. What can be more accurate yet more lovely than the two concluding stanzas?

> Upon yon tuft of hazel trees,
> That twinkle to the gusty breeze,
> Behold him perched in ecstacies,
> Yet seeming still to hover,
> There! where the flutter of his wings
> Upon his back and body flings
> Shadows and sunny glimmerings
> That cover him all over.
>
> While thus before my eyes he gleams,
> A brother of the leaves he seems;
> When in a moment forth he teems
> His little song in gushes:
> As if it pleased him to disdain
> And mock the form when he did feign
> While he was dancing with the train
> Of leaves among the bushes.

Or the description of the blue-cap, and of the noontide silence, p. 284; or the poem to the cuckoo, p. 299; or lastly, though I might multiply the references to ten times the number, to the poem so completely Wordsworth's commencing

> Three years she grew in sun and shower, etc.

Fifth: a meditative pathos, a union of deep and subtle thought with sensibility; a sympathy with man as man; the sympathy indeed of a contemplator, rather than a fellow sufferer or co-mate, (*spectator, haud particeps* [a spectator, not a participant]) but of a contemplator, from whose view no difference of rank conceals the sameness of the nature; no injuries of wind or weather, of toil, or even of ignorance, wholly disguise the human face divine. The superscription and the image of the Creator still remain legible to *him* under the dark lines, with which guilt or calamity had cancelled or cross-barred it. Here the man and the poet lose and find themselves in each other, the one as glorified, the latter as substantiated. In this mild and philosophic pathos, Wordsworth appears to me without a compeer. Such he is; so he writes. See vol. I. page 134 to 136, or that most affecting composition, the 'Affliction of Margaret —— of ——,' page 165 to 168, which no mother, and if I may judge by my own experience, no parent can read without a tear. Or turn to that genuine lyric, in the former edition, entitled, the 'Mad Mother,' page 174 to 178, of which I cannot refrain from quoting two of the stanzas, both of them for their pathos, and the former for the fine transition in the two concluding lines of the stanza, so expressive of that deranged state, in which from the increased sensibility the sufferer's attention is abruptly drawn off by every trifle, and in the same instant plucked back again by the one despotic thought, and bringing home with it, by the blending, fusing power of imagination and passion, the alien object to which it had been so abruptly diverted, no longer an alien but an ally and an inmate.

Suck, little babe, oh suck again!
It cools my blood; it cools my brain:
Thy lips, I feel them, baby! they
Draw from my heart the pain away.
Oh! press me with thy little hand;
It loosens something at my chest;
About that tight and deadly band
I feel thy little fingers prest.
The breeze I see is in the tree!
It comes to cool my babe and me.

Thy father cares not for my breast,
'Tis thine, sweet baby, there to rest,
'Tis all thine own!—and, if its hue,
Be changed, that was so fair to view,
'Tis fair enough for thee, my dove!
My beauty, little child, is flown,
But thou wilt live with me in love,
And what if my poor cheek be brown?
'Tis well for me, thou can'st not see
How pale and wan it else would be.

Last, and pre-eminently I challenge for this poet the gift of imagination in the highest and strictest sense of the word. In the play of fancy, Wordsworth, to my feelings, is not always graceful, and sometimes recondite. The likeness is occasionally too strange, or demands too peculiar a point of view, or is such as appears the creature of predetermined research, rather than spontaneous presentation. Indeed his fancy seldom displays itself, as mere and unmodified fancy. But in imaginative power, he stands nearest of all modern writers to Shakespeare and Milton; and yet in a kind perfectly unborrowed and his own. To employ his own words, which are at once an instance and an illustration, he does indeed to all thoughts and to all objects—

> . . . add the gleam,
> The light that never was on sea or land,
> The consecration, and the poet's dream.

I shall select a few examples as most obviously manifesting this faculty; but if I should ever be fortunate enough to render my analysis of imagination, its origin and characters thoroughly intelligible to the reader, he will scarcely open on a page of this poet's works without recognizing, more or less, the presence and the influences of this faculty.

From the poem on the Yew Trees, vol. I. page 303, 304.

> But worthier still of note
> Are those fraternal four of Borrowdale,
> Joined in one solemn and capacious grove:
> Huge trunks!—and each particular trunk a growth
> Of intertwisted fibres serpentine
> Up-coiling, and inveterately convolved,—
> Not uninformed with phantasy, and looks
> That threaten the prophane;—a pillared shade,
> Upon whose grassless floor of red-brown hue,
> By sheddings from the pinal umbrage tinged
> Perennially—beneath whose sable roof
> Of boughs, as if for festal purpose decked
> With unrejoicing berries, ghostly shapes
> May meet at noontide—FEAR and trembling HOPE,
> SILENCE and FORESIGHT—DEATH, the skeleton,
> And TIME, the shadow—there to celebrate,
> As in a natural temple scattered o'er
> With altars undisturbed of mossy stone,
> United worship; or in mute repose
> To lie, and listen to the mountain flood
> Murmuring from Glaramara's inmost caves.

The effect of the old man's figure in the poem of Resignation and Independence, vol. II. page 33.

> While he was talking thus, the lonely place
> The old man's shape, and speech, all troubled me:
> In my mind's eye I seemed to see him pace
> About the weary moors continually,
> Wandering about alone and silently.

Or the 8th, 9th, 19th, 26th, 31st, and 33d, in the collection of miscellaneous sonnets—the sonnet on the subjugation of Switzerland, page 210, or the last ode from which I especially select the two following stanzas or paragraphs, page 349 to 350.

> Our birth is but a sleep and a forgetting:
> The soul that rises with us, our life's star
> Hath had elsewhere its setting,
>   And cometh from afar.
> Not in entire forgetfulness,
> And not in utter nakedness,
> But trailing clouds of glory do we come
> From God who is our home:
> Heaven lies about us in our infancy!
> Shades of the prison-house begin to close
>   Upon the growing boy;
> But he beholds the light, and whence it flows,
>   He sees it in his joy!
> The youth who daily further from the east
> Must travel, still is nature's priest,
>   And by the vision splendid
>   Is on his way attended;
> At length the man perceives it die away,
> And fade into the light of common day.

And page 352 to 354 of the same ode.

> O joy that in our embers
> Is something that doth live,
> That nature yet remembers
> What was so fugitive!
> The thought of our past years in me doth breed
> Perpetual benediction: not indeed
> For that which is most worthy to be blest
> Delight and liberty the simple creed
> Of childhood, whether busy or at rest,
> With new-fledged hope still fluttering in his breast:—

Not for these I raise
The song of thanks and praise;
But for those obstinate questionings
Of sense and outward things,
Fallings from us, vanishings;
Blank misgivings of a creature
Moving about in worlds not realized,
High instincts, before which our mortal nature
Did tremble like a guilty thing surprised!
But for those first affections,
Those shadowy recollections,
Which, be they what they may,
Are yet the fountain light of all our day,
Are yet a master light of all our seeing;
Uphold us—cherish—and have power to make
Our noisy years seem moments in the being
Of the eternal silence; truths that wake
To perish never:
Which neither listlessness, nor mad endeavour
Nor man nor boy
Nor all that is at enmity with joy
Can utterly abolish or destroy!
Hence, in a season of calm weather,
Though inland far we be,
Our souls have sight of that immortal sea
Which brought us hither,
Can in a moment travel thither—
And see the children sport upon the shore,
And hear the mighty waters rolling evermore.

And since it would be unfair to conclude with an extract, which though highly characteristic must yet from the nature of the thoughts and the subject be interesting, or perhaps intelligible, to but a limited number of readers; I will add from the poet's last published work a passage equally Wordsworthian; of the beauty of which, and of the imaginative power displayed therein, there can be but one opinion, and one feeling. See White Doe, page 5.

Fast the church-yard fills;—anon
Look again and they are gone;
The cluster round the porch, and the folk
Who sate in the shade of the prior's oak!
And scarcely have they disappear'd
Ere the prelusive hymn is heard:—

With one consent the people rejoice,
Filling the church with a lofty voice!
They sing a service which they feel
For 'tis the sun-rise of their zeal
And faith and hope are in their prime
In great Eliza's golden time.
A moment ends the fervent din
And all is hushed without and within;
For though the priest more tranquilly
Recites the holy liturgy,
The only voice which you can hear
Is the river murmuring near.
When soft!—the dusky trees between
And down the path through the open green
Where is no living thing to be seen;
And through yon gateway, where is found,
Beneath the arch with ivy bound,
Free entrance to the church-yard ground;
And right across the verdant sod
Towards the very house of God;
Comes gliding in with lovely gleam,
Comes gliding in serene and slow,
Soft and silent as a dream,
A solitary doe!
White she is as lily of June,
And beauteous as the silver moon
When out of sight the clouds are driven
And she is left alone in heaven!
Or like a ship some gentle day
In sunshine sailing far away—
A glittering ship that hath the plain
Of ocean for her own domain.

* * * * *

What harmonious pensive changes
Wait upon her as she ranges
Round and round this pile of state
Overthrown and desolate!
Now a step or two her way
Is through space of open day,
Where the enamoured sunny light
Brightens her that was so bright:
Now doth a delicate shadow fall,
Falls upon her like a breath
From some lofty arch or Wall,
As she passes underneath.

The following analogy will, I am apprehensive, appear dim and fantastic, but in reading Bartram's Travels I could not help transcribing the following lines as a sort of allegory, or connected simile and metaphor of Wordsworth's intellect and genius.—'The soil is a deep, rich, dark mould, on a deep stratum of tenacious clay; and that on a foundation of rocks, which often break through both strata, lifting their back above the surface. The trees which chiefly grow here are the gigantic, black oak; magnolia magnifloria; fraximus excelsior; platane; and a few stately tulip trees.' What Mr Wordsworth *will* produce, it is not for me to prophesy: but I could pronounce with the liveliest convictions what he is capable of producing. It is the first genuine philosophic poem.

The preceding criticism will not, I am aware, avail to overcome the prejudices of those, who have made it a business to attack and ridicule Mr Wordsworth's compositions.

Truth and prudence might be imaged as concentric circles. The poet may perhaps have passed beyond the latter, but he has confined himself far within the bounds of the former, in designating these critics, as too petulant to be passive to a genuine poet, and too feeble to grapple with him;—'men of palsied imaginations, in whose minds all healthy action is languid;—who, therefore, feel as the many direct them, or with the many are greedy after vicious provocatives.'

Let not Mr Wordsworth be charged with having expressed himself too indignantly, till the wantonness and the systematic and malignant perseverance of the aggressions have been taken into fair consideration. I myself heard the commander-in-chief of this unmanly warfare make a boast of his private admiration of Wordsworth's genius. I have heard him declare, that whoever came into his room would probably find the Lyrical Ballads lying open on his table, and that (speaking exclusively of those written by Mr Wordsworth himself,) he could nearly repeat the whole of them by heart. But a review, in order to be a saleable article, must be personal, sharp, and pointed: and, since then, the poet has made himself, and with himself all who were, or were supposed to be, his friends and admirers, the object of the critic's revenge —how? by having spoken of a work so conducted in the terms which it deserved! I once heard a clergyman in boots and buckskin avow, that he would cheat his own father in a horse. A moral system of a similar nature seems to have been adopted by too many anonymous critics. As we used to say at school, in reviewing they *make* [play at] being rogues: and he, who complains, is to be laughed at for his ignorance of the game. With the pen out of their hand they are honourable men. They exert indeed power (which is to that of the injured party who should attempt to expose their glaring perversions and misstatements, as twenty to one) to write down, and (where the author's circumstances permit) to impoverish the man, whose learning and genius they themselves in private have repeatedly admitted. They knowingly strive to make it impossible for the man even to publish* any future work without exposing himself to all the wretchedness of debt and embarrassment. But this is all in their vocation: and bating what they do in their vocation, 'who can say that black is the white of their eye?'°

So much for the detractors from Wordsworth's merits. On the other hand, much as I might wish for their fuller sympathy, I dare not flatter myself, that the freedom with which I have declared my opinions concerning both his theory and his defects, most of which are more or less connected with his theory either as cause or effect, will be satisfactory or pleasing to *all* the poet's admirers and advocates. More indiscriminate than mine their admiration may be: deeper and more sincere it cannot be. But I have advanced no opinion either for praise or censure, other than as texts introductory to the reasons which compel me to form it. Above all, I was fully convinced that such a criticism was not only wanted; but that, if executed with adequate ability, it must conduce in no mean degree to Mr Wordsworth's reputation. His fame belongs to another age, and can neither be accelerated or retarded. How small the proportion of the defects are to the beauties, I have repeatedly declared; and that no one of them originates in deficiency of poetic genius. Had they been more and greater, I should still, as a friend to his literary character in the present age, consider an analytic display of them as pure gain; if only it removed, as surely to all reflecting minds even the foregoing analysis must have removed, the strange mistake so slightly grounded, yet so widely and industriously propagated, of Mr Wordsworth's turn for simplicity! I am not half as much irritated by hearing his enemies abuse him for vulgarity of style, subject, and conception; as I am disgusted with the gilded side of the same meaning, as displayed by some affected admirers with whom he is, forsooth, a sweet, simple poet! and so natural, that little master Charles, and his younger sister,

* Not many months ago an eminent bookseller was asked what he thought of ——? The answer was: 'I have heard his powers very highly spoken of by some of our first-rate men; but I would not have a work of his if anyone would give it me: for he is spoken but slightly of, or not at all in the Quarterly Review: and the Edinburgh, you know, is decided, to cut him up!'—

are so charmed with them, that they play at 'Goody Blake,' or at 'Johnny and Betty Foy!'

Were the collection of poems published with these biographical sketches, important enough (which I am not vain enough to believe) to deserve such a distinction, even as I have done, so would I be done unto.

For more than eighteen months have the volume of poems, entitled Sibylline Leaves, and the present volumes up to this page been printed, and ready for publication. But ere I speak of myself in the tones, which are alone natural to me under the circumstances of late years, I would fain present myself to the reader as I was in the first dawn of my literary life:

> When Hope grew round me, like the climbing vine,
> And fruits and foliage not my own seem'd mine!

For this purpose I have selected from the letters which I wrote home from Germany, those which appeared likely to be most interesting, and at the same time most pertinent to the title of this work.

SATYRANE'S LETTERS.

LETTER I

On Sunday morning, September 16, 1798, the Hamburg packet set sail from Yarmouth: and I, for the first time in my life, beheld my native land retiring from me. At the moment of its disappearance—in all the kirks, churches, chapels, and meeting-houses, in which the greater number, I hope, of my countrymen were at that time assembled, I will dare question whether there was one more ardent prayer offered up to heaven, than that which I then preferred for my country. Now then (said I to a gentleman who was standing near me) we are out of our country. Not yet, not yet! he replied, and pointed to the sea; 'This, too, is a Briton's country.' This *bon mot* gave a fillip to my spirits, I rose and looked round on my fellow passengers, who were all on the deck. We were eighteen in number, *videlicet* [= viz., 'namely'], five Englishmen, an English lady, a French gentleman and his servant, an Hanoverian and his servant, a Prussian, a Swede, two Danes, and a Mulatto boy, a German tailor and his wife (the smallest couple I ever beheld) and a Jew. We were all on the deck; but in a short time I observed marks of dismay. The lady retired to the cabin in some confusion, and many of the faces round me assumed a very doleful and frog-coloured appearance; and within an hour the number of those on deck was lessened by one half. I was giddy, but not sick, and the giddiness soon went away, but left a feverishness and want of appetite, which I attributed, in great measure, to the *saeva Mephitis* [dreadful exhalation] of the bilge-water; and it was certainly not decreased by the exportations from the cabin. However, I was well enough to join the able-bodied passengers, one of whom observed not inaptly, that Momus might have discovered an easier way to see a man's inside, than by placing a window in his breast. He needed only have taken a salt-water trip in a packet-boat.

I am inclined to believe, that a packet is far superior to a stage-coach, as a means of making men open out to each other. In the latter the uniformity of posture disposes to dozing, and the definiteness of the period at which the company will separate, makes each individual think more of those, to whom he is going, than of those with whom he is going. But at sea, more curiosity is excited, if only on this account, that the pleasant or unpleasant qualities of your companions are of greater importance to you, from the uncertainty how long you may be obliged to house with them. Besides, if you are countrymen, that now begins to form a distinction and a bond of brotherhood; and if of different countries, there are new incitements of conversation, more to ask and more to communicate. I found that I had interested the Danes in no common degree. I had crept into the boat on the deck and fallen asleep; but was awaked by one of them about three o'clock in the afternoon, who told me that they had been seeking me in every hole and corner, and insisted that I should join their party and drink with them. He talked English with such fluency, as left me wholly unable to account for the singular and even ludicrous incorrectness with which he spoke it. I went, and found some excellent wines and a dessert of grapes with a pineapple. The Danes had christened me Doctor Teology, and dressed as I was all in black, with large shoes and black worsted stockings, I might certainly have passed very well for a Methodist missionary. However I disclaimed my title. What then may you be? A man of fortune? No!—A merchant? No! A merchant's traveller? No!—A clerk? No! *un philosophe*,° perhaps? It was at that time in my life, in which of all possible names and characters I had the greatest disgust to that of 'un philosophe.' But I was weary of being questioned, and rather than be nothing, or at best only the abstract idea of a man, I submitted by a bow, even to the aspersion implied in the word 'un philosophe.'—The Dane then informed me, that all in the present party were philosophers likewise. Certes we were not of the Stoic school. For we drank and talked and sung, till we talked and sung all together; and then we rose and danced on the deck a set of dances, which in one sense of the word at least, were very intelligibly and appropriately intitled reels. The passengers who lay in the cabin below in all the agonies of sea-sickness, must have found our bacchanalian merriment

> . . . a tune
> Harsh and of dissonant mood for their complaint.

I thought so at the time; and (by way, I suppose, of supporting my newly assumed philosophical character) I thought too, how closely the greater number of our virtues are connected with the fear of death, and how little sympathy we bestow on pain, where there is no danger.

The two Danes were brothers. The one was a man with a clear white complexion, white hair, and white eyebrows, looked silly, and nothing that he uttered gave the lie to his looks. The other, whom, by way of eminence I have called The Dane, had likewise white hair, but was much shorter than his brother, with slender limbs, and a very thin face slightly pock-fretten. This man convinced me of the justice of an old remark, that many a faithful portrait in our novels and farces has been rashly censured for an outrageous caricature, or perhaps nonentity. I had retired to my station in the boat—he came and seated himself by my side, and appeared not a little tipsy. He commenced the conversation in the most magnific style, and as a sort of pioneering to his own vanity, he flattered me with *such* grossness! The parasites of the old comedy were modest in the comparison. His language and accentuation were so exceedingly singular, that I determined for once in my life to take notes of a conversation. Here it follows, somewhat abridged indeed, but in all other respects as accurately as my memory permitted.

THE DANE. Vat imagination! vat language! vat vast science! and vat eyes! vat a milk-vite forehead!—O my heafen! vy, you're a Got!

Answer. You do me too much honour, Sir.

THE DANE. O me! if you should dink I is flattering you!—No, no, no! I haf ten tousand a year—yes, ten tousand a year—yes, ten tousand pound a year! Vell—and vat is dhat? a mere trifle! I 'ouldn't gif my sincere heart for ten times dhe money.—Yes, you're a Got! I a mere man! But, my dear friend! dhink of me, as a man! Is, is—I mean to ask you now, my dear friend—is I not very eloquent? Is I not speak English very fine?

Answ. Most admirably! Believe me, Sir! I have seldom heard even a native talk so fluently.

THE DANE (*squeezing my hand with great vehemence*). My *dear* friend! vat an affection and fidelity we have for each odher! But tell me, do tell me,—Is I not, now and den, speak some fault? Is I not in some wrong?

Answ. Why, Sir! perhaps it might be observed by nice critics in the English language, that you occasionally use the word 'is' instead of 'am.' In our best companies we generally say I am, and not I is or Ise. Excuse me, Sir! it is a mere trifle.

THE DANE. O!—is, is, am, am, am. Yes, yes—I know, I know.

Answ. I am, thou art, he is, we are, ye are, they are.

THE DANE. Yes, yes—I know, I know—Am, am, am, is dhe presens, and Is is dhe perfectum [perfect tense]—yes, yes—and are is dhe plusquam perfectum [pluperfect].

Answ. And 'art,' Sir! is . . .?

THE DANE. My dear friend! it is dhe plusquam perfectum, no, no—dhat is a great lie. 'Are' is the plusquam perfectum—and 'art' is dhe plusquam plueperfectum—(*then swinging my hand to and fro, and cocking his little bright hazel eyes at me, that danced with vanity and wine*) You see, my dear friend! that I too have some lehrning.

Answ. Learning, Sir? Who dares suspect it? Who can listen to you for a minute, who can even look at you, without perceiving the extent of it?

THE DANE. My *dear* friend!—(*then with a would-be humble look, and in a tone of voice as if he was reasoning*) I could not talk so of presens and imperfectum, and futurum and plusquamplue perfectum, and all dhat, my dear friend! without *some* lehrning?

Answ. Sir! a man like you cannot talk on any subject without discovering the depth of his information.

THE DANE. Dhe grammatic Greek, my friend! ha! ha! ha! (*laughing, and swinging my hand to and fro—then with a sudden transition to great solemnity*) Now I will tell you, my dear friend! Dhere did happen about me vat de whole historia of Denmark record no instance about nobody else. Dhe bishop did ask me all dhe questions about all dhe religion in dhe Latin grammar.

Answ. The grammar, Sir? The language, I presume—

THE DANE (*a little offended*). Grammar is language, and language is grammar—

Answ. Ten thousand pardons!

THE DANE. Vell, and I was only fourteen years—

Answ. Only fourteen years old?

THE DANE. No more. I was fourteen years old—and he asked me all questions, religion and philosophy, and all in dhe Latin language—and I answered him all every one, my dear friend! all in dhe Latin language.

Answ. A prodigy! an absolute prodigy!

THE DANE. No, no, no! he was a bishop, a great superintendant.

Answ. Yes! a bishop.

THE DANE. A bishop—not a mere predicant, not a prediger [preacher]—

Answ. My dear Sir! we have misunderstood each other. I said that your answering in Latin at so early an age was a prodigy, that is, a thing that is wonderful, that does not often happen.

THE DANE. Often! Dhere is not von instance recorded in dhe whole historia of Denmark.

Answ. And since then Sir—?

THE DANE. I was sent ofer to dhe Vest Indies—to our Island, and dhere I had no more to do vid books. No! no! I put my genius another way—and I haf made ten tousand pound a year. Is not dhat *ghenius*, my dear friend!—But vat is money! I dhink the poorest man alive my equal. Yes, my dear friend! my little fortune is pleasant to my generous heart, because I can do good—no man with so little a fortune ever did so much generosity—no person, no man person, no woman person ever denies it. But we are all Got's children.

Here the Hanoverian interrupted him, and the other Dane, the Swede, and the Prussian, joined us, together with a young Englishman who spoke the German fluently, and interpreted to me many of the Prussian's jokes. The Prussian was a travelling merchant, turned of three score, a hale man, tall, strong, and stout, full of stories, gesticulations, and buffoonery with the soul as well as the look of a mountebank, who, while he is making you laugh, picks your pocket. Amid all his droll looks and droll gestures, there remained one look untouched by laughter; and that one look was the true face, the others were but its mask. The Hanoverian was a pale, fat, bloated young man, whose father had made a large fortune in London, as an army contractor. He seemed to emulate the manners of young Englishmen of fortune. He was a good-natured fellow, not without information or literature; but a most egregious coxcomb. He had been in the habit of attending the House of Commons, and had once spoken, as he informed me, with great applause in a debating society. For this he appeared to have qualified himself with laudable industry: for he was perfect in Walker's Pronouncing Dictionary, and with an accent, which forcibly reminded me of the Scotchman in Roderick Random, who professed to teach the English pronunciation, he was constantly deferring to my superior judgement, whether or no I had pronounced this or that word with propriety, or 'the true delicacy.' When he spoke, though it were only half a dozen sentences, he always rose; for which I could detect no other motive, than his partiality to that elegant phrase so liberally introduced in the orations of our British legislators, 'While I am on my legs.' The Swede, whom for reasons that will soon appear, I shall distinguish by the name of 'Nobility,' was a strong-featured, scurvy-faced man, his complexion resembling, in colour, a red-hot poker beginning to cool. He appeared miserably dependent on the Dane; but was however incomparably the best-informed and most rational of the party. Indeed his manners and conversation discovered him to be both a man of the world and a gentleman. The Jew was in the hold: the French gentleman was lying on the deck so ill, that I could observe nothing concerning him, except the affectionate attentions of his servant to him. The poor fellow was very sick himself, and every now and then ran to the side of the vessel, still keeping his eye on his master, but returned in a moment and seated himself again by him, now supporting his head, now wiping his forehead and talking to him all the while in the most soothing tones. There had been a matrimonial squabble of a very ludicrous kind in the cabin, between the little German tailor and his little wife. He had secured two beds, one for himself, and one for her. This had struck the little woman as a very cruel action; she insisted upon their having but one, and assured the mate in the most piteous tones, that she was his lawful wife. The mate and the cabin-boy decided in her favour, abused the little man for his want of tenderness with much humour, and hoisted him into the same compartment with his sea-sick wife. This quarrel was interesting to me, as it procured me a bed, which I otherwise should not have had.

In the evening, at 7 o'clock, the sea rolled higher, and the Dane, by means of the greater agitation, eliminated enough of what he had been swallowing to make room for a great deal more. His favourite potation was sugar and brandy, i.e. a very little warm water with a large quantity of brandy, sugar, and nutmeg. His servant boy, a black-eyed Mulatto, had a good-natured round face, exactly the colour of the skin of the walnut kernel. The Dane and I were again seated, tête-à-tête, in the ship's boat. The conversation, which was now indeed rather an oration than a dialogue, became extravagant beyond all that I ever heard. He told me that he had made a large fortune in the island of Santa Cruz, and was now returning to Denmark to enjoy it. He expatiated on the style in which he meant to live, and the great undertakings which he proposed to himself to commence, till the brandy aiding his vanity, and his vanity and garrulity aiding the brandy, he talked like a madman—entreated me to accompany him to Denmark—there I should see his influence with the Government, and he would introduce me to the king, etc. etc. Thus he went on dreaming aloud, and then passing with a very lyrical transition to the subject of general politics, he declaimed, like a member of the Corresponding Society,° *about* (not concerning) the Rights of Man, and assured me that notwithstanding his fortune, he thought the poorest man alive his equal. 'All are equal, my dear friend! all are equal! Ve are all Got's children. The poorest man haf the same rights with me. Jack! Jack! some more sugar and brandy. Dhere is dhat fellow now! He is a Mulatto—but he is my equal.—That's right, Jack! (*taking the sugar and brandy*) Here you Sir! shake hands with dhis gentleman! Shake hands with me, you dog! Dhere, dhere!—We are all equal my dear friend!—Do I not speak like Socrates, and Plato, and Cato—they were all philosophers, my dear *philosophe*! all very great men!—and so was Homer and Virgil—but they were poets, yes, yes! I know all about it!—But what can anybody say more than this? we are all equal, all Got's children. I haf ten thousand a year, but I am no more than the meanest man alive. I haf no pride; and yet, my dear friend! I can say, do! and it is done. Ha! ha! ha! my dear friend! Now dhere is dhat gentleman (*pointing to* '*Nobility*') he is a Swedish baron—you shall see. Ho! (*calling to the Swede*) get me, will you, a bottle of wine from the cabin.'

SWEDE.—Here, Jack! go and get your master a bottle of wine from the cabin.

DANE. No, no, no! do *you* go now—you go yourself—*you* go now!

SWEDE. Pah!—

DANE. Now go! Go, I pray you.

And the Swede went!

After this the Dane commenced an harangue on religion, and mistaking me for 'un philosophe' in the continental sense of the word, he talked of Deity in a declamatory style, very much resembling the devotional rants of that rude blunderer, Mr Thomas Paine, in his Age of Reason, and whispered in my ear, what damned *hypocrism* all Jesus

Christ's business was. I dare aver, that few men have less reason to charge themselves with indulging in persiflage than myself. I should hate it if it were only that it is a Frenchman's vice, and feel a pride in avoiding it because our own language is too honest to have a word to express it by. But in this instance the temptation had been too powerful, and I have placed it on the list of my offences. Pericles answered one of his dearest friends who had solicited him on a case of life and death, to take an equivocal oath for his preservation: *Debeo amicis opitulari, sed usque ad Deos.** Friendship herself must place her last and boldest step on this side the altar. What Pericles would not do to save a friend's life, you may be assured I would not hazard merely to mill the chocolate-pot of a drunken fool's vanity till it frothed over. Assuming a serious look, I professed myself a believer, and sunk at once an hundred fathoms in his good graces. He retired to his cabin, and I wrapped myself up in my greatcoat, and looked at the water. A beautiful white cloud of foam at momently intervals coursed by the side of the vessel with a roar, and little stars of flame danced and sparkled and went out in it: and every now and then light detachments of this white cloud-like foam darted off from the vessel's side, each with its own small constellation, over the sea, and scoured out of sight like a Tartar troop over a wilderness.

It was cold, the cabin was at open war with my olfactories, and I found reason to rejoice in my greatcoat, a weighty high-caped, respectable rug, the collar of which turned over, and played the part of a night-cap very passably. In looking up at two or three bright stars, which oscillated with the motion of the sails, I fell asleep, but was awakened at one o'clock, Monday morning, by a shower of rain. I found myself compelled to go down into the cabin, where I slept very soundly, and awoke with a very good appetite at breakfast time, my nostrils, the most placable of all the senses, reconciled to or indeed insensible of the mephitis.

Monday, September 17th, I had a long conversation with the Swede, who spoke with the most poignant contempt of the Dane, whom he described as a fool, purse-mad; but he confirmed the boasts of the Dane respecting the largeness of his fortune, which he had acquired in the first instance as an advocate, and afterwards as a planter. From the Dane and from himself I collected that he was indeed a Swedish nobleman, who had squandered a fortune, that was never very large, and had made over his property to the Dane, on whom he was now

* *Translation.* It behoves me to side with my friends but only as far as the gods.

utterly dependent. He seemed to suffer very little pain from the Dane's insolence. He was in high degree humane and attentive to the English lady, who suffered most fearfully, and for whom he performed many little offices with a tenderness and delicacy which seemed to prove real goodness of heart. Indeed, his general manners and conversation were not only pleasing, but even interesting; and I struggled to believe his insensibility respecting the Dane philosophical fortitude. For though the Dane was now quite sober, his character oozed out of him at every pore. And after dinner, when he was again flushed with wine, every quarter of an hour or perhaps oftener he would shout out to the Swede, 'Ho! Nobility, go—do such a thing! Mr Nobility!—tell the gentlemen such a story,' and so forth, with an insolence which must have excited disgust and detestation, if his vulgar rants on the sacred rights of equality, joined to his wild havoc of general grammar no less than of the English language, had not rendered it so irresistibly laughable.

At four o'clock I observed a wild duck swimming on the waves, a single solitary wild duck. It is not easy to conceive, how interesting a thing it looked in that round objectless desert of waters. I had associated such a feeling of immensity with the ocean, that I felt exceedingly disappointed, when I was out of sight of all land, at the narrowness and nearness, as it were, of the circle of the horizon. So little are images capable of satisfying the obscure feelings connected with words. In the evening the sails were lowered, lest we should run foul of the land, which can be seen only at a small distance. And at four o'clock, on Tuesday morning, I was awakened by the cry of land! land! It was an ugly island rock at a distance on our left, called Heiligeland, well known to many passengers from Yarmouth to Hamburg, who have been obliged by stormy weather to pass weeks and weeks in weary captivity on it, stripped of all their money by the exorbitant demands of the wretches who inhabit it. So at least the sailors informed me.—About nine o'clock we saw the mainland, which seemed scarcely able to hold its head above water, low, flat, and dreary, with lighthouses and landmarks which seemed to give a character and language to the dreariness. We entered the mouth of the Elbe, passing Neuwerk; though as yet the right bank only of the river was visible to us. On this I saw a church, and thanked God for my safe voyage, not without affectionate thoughts of those I had left in England. At eleven o'clock on the same morning we arrived in Cuxhaven, the ship dropped anchor, and the boat was hoisted out, to carry the Hanoverian and a few others on shore. The captain agreed to take us, who remained, to Hamburg for ten guineas, to which the Dane contributed so largely, that the other passengers paid but half a guinea each. Accordingly we hauled anchor, and passed gently up the river. At Cuxhaven both sides of the river may be seen in clear weather; we could now see the right bank only. We passed a multitude of English traders that had been waiting many weeks for a wind. In a short time both banks became visible, both flat and evidencing the labour of human hands by their extreme neatness. On the left bank I saw a church or two in the distance; on the right bank we passed by steeple and windmill and cottage, and windmill and single house, windmill and windmill, and neat single house, and steeple. These were the objects and in the succession. The shores were very green and planted with trees not inelegantly. Thirty-five miles from Cuxhaven, the night came on us, and as the navigation of the Elbe is perilous, we dropped anchor.

Over what place, thought I, does the moon hang to *your* eye, my dearest friend? To me it hung over the left bank of the Elbe. Close above the moon was a huge volume of deep black cloud, while a very thin fillet crossed the middle of the orb, as narrow and thin and black as a ribbon of crape. The long trembling road of moonlight, which lay on the water and reached to the stern of our vessel, glimmered dimly and obscurely. We saw two or three lights from the right bank, probably from bedrooms. I felt the striking contrast between the silence of this majestic stream, whose banks are populous with men and women and children, and flocks and herds—between the silence by night of this peopled river, and the ceaseless noise, and uproar, and loud agitations of the desolate solitude of the ocean. The passengers below had all retired to their beds; and I felt the interest of this quiet scene the more deeply from the circumstance of having just quitted them. For the Prussian had during the whole of the evening displayed all his talents to captivate the Dane, who had admitted him into the train of his dependants. The young Englishman continued to interpret the Prussian's jokes to me. They were all without exception profane and abominable, but some sufficiently witty, and a few incidents, which he related in his own person, were valuable as illustrating the manners of the countries in which they had taken place.

Five o'clock on Wednesday morning we hauled the anchor, but were soon obliged to drop it again in consequence of a thick fog, which our captain feared would continue the whole day; but about nine it cleared off, and we sailed slowly along, close by the shore of a very beautiful island, forty miles from Cuxhaven, the wind continuing slack. This holme or island is about a mile and a half in length, wedge-shaped, well wooded, with glades of the liveliest green, and rendered more interesting by the remarkably neat farmhouse on it. It seemed made for retirement without solitude—a place that would allure one's friends while it precluded the impertinent calls of mere visitors. The shores of the Elbe now became more beautiful, with rich meadows and trees running like a low wall along the river's edge; and peering over them, neat houses and (especially on the right bank) a profusion of steeple-spires, white, black, or red. An instinctive taste teaches men to build their churches in flat countries with spire-steeples, which as they cannot be referred to any other object, point as with silent finger to the sky and stars, and sometimes when they reflect the brazen light of a rich though rainy sunset, appear like a pyramid of flame burning heavenward. I remember once, and once only, to have seen a spire in a narrow valley of a mountainous country. The effect was not only mean but ludicrous, and reminded me against my will of an extinguisher;° the close neighbourhood of the high mountain, at the foot of which it stood, had so completely dwarfed it, and deprived it of all connection with the sky or clouds. Forty-six English miles from Cuxhaven, and sixteen from Hamburg, the Danish village Veder ornaments the left bank with its black steeple, and close by it the wild and pastoral hamlet of Schulau. Hitherto both the right and left bank, green to the very brink, and level with the river, resembled the shores of a park canal. The trees and houses were alike low, sometimes the low trees overtopping the yet lower houses, sometimes the low houses rising above the yet lower trees. But at Schulau the left bank rises at once forty or fifty feet, and stares on the river with its perpendicular fassade [façade (German)] of sand, thinly patched with tufts of green. The Elbe continued to present a more and more lively spectacle from the multitude of fishing boats and the flocks of sea-gulls wheeling round them, the clamorous rivals and companions of the fishermen; till we came to Blankaness, a most interesting village scattered amid scattered trees, over three hills in three divisions. Each of the three hills stares upon the river, with faces of bare sand, with which the boats with their bare poles, standing in files along the banks, made a sort of fantastic harmony. Between each fassade lies a green and woody dell, each deeper than the other. In short it is a large village made up of individual cottages, each cottage in the centre of its own little wood or orchard, and each with its own separate path: a village with a labyrinth of paths, or rather a neighbourhood of houses! It is inhabited by fishermen and boat-makers, the Blankanese boats being in great request through the whole navigation of the Elbe. Here first we saw the spires of Hamburg, and from hence as far as Altona the left bank of the Elbe is uncommonly pleasing, considered as the vicinity of an industrious and republican city—in that style of beauty, or rather prettiness, that might tempt the citizen into the country, and yet gratify the taste which he had acquired in the town. Summerhouses and Chinese show-work are everywhere scattered along the high and green banks; the boards of the farmhouses left unplastered and gaily painted with green and yellow; and scarcely a tree not cut into shapes and made to remind the human being of his own power and intelligence instead of the wisdom of nature. Still, however, these are links of connection between town and country, and far better than the affectation of tastes and enjoyments for which men's habits have disqualified them. Pass them by on Saturdays and Sundays with the burgers of Hamburg smoking their pipes, the women and children feasting in the alcoves of box and yew, and it becomes a nature of its own. On Wednesday, four o'clock, we left the vessel, and passing with trouble through the huge masses of shipping that seemed to choke the wide Elbe from Altona upward, we were at length landed at the Boom House, Hamburg.

LETTER II

(To a lady)

Ratzeburg

Meine liebe Freundin [My dear (lady-)friend],

See how natural the German comes from me, though I have not yet been six weeks in the country!—almost as fluently as English from my neighbour the Amptschreiber (or public secretary) who as often as we meet, though it should be half a dozen times in the same day, never fails to greet me with—'**ddam your ploot unt eyes, my dearest Englander! vhee goes it!'—which is certainly a proof of great generosity on his part, these words being his whole stock of English. I had, however, a better reason than the desire of displaying my proficiency: for I wished to put you in good humour with a language, from the acquirement of which I have promised myself much edification and the means too of communicating a new pleasure to you and your sister, during our winter readings. And how can I do this better than by pointing out its gallant attention to the ladies? Our English affix, *ess*, is, I believe, confined either to words derived from the Latin, as *actress*, *directress*, etc. or from the French, as *mistress*, *duchess*, and the like. But the German, *in*, enables us to designate the sex in every possible relation of life. Thus the Amptman's lady is the Frau Amptman*in* [Mrs Wife-of-the-District-Administrator]—the secretary's wife (by the by the handsomest woman I have yet seen in Germany) is Die allerliebste Frau Amptschreiber*in* [the very charming Mrs Wife-of-the-Public-Secretary]—the colonel's lady, Die Frau Obrist*in* or colonel*lin*—and even the parson's wife, die frau pastor*in*. But I am especially pleased with their *freundin*, which, unlike the *amica* [friend (female)] of the Romans, is seldom used but in its best and purest sense. Now, I know, it will be said, that a friend is already something more than a friend, when a man feels an anxiety to express to himself that this friend is a female; but this I deny—in that sense at least in which the objection will be made. I would hazard the impeachment of heresy, rather than abandon my belief that there is a sex in our souls as well as in their perishable garments; and he who does not feel it, never truly loved a sister—nay, is not capable even of loving a wife as she deserves to be loved, if she indeed be worthy of that holy name.

Now I know, my gentle friend, what you are murmuring to yourself—'This is so like him! running away after the first bubble, that chance has blown off from the surface of his fancy; when one is anxious to learn where he is and what he has seen.' Well then! that I am settled at Ratzeburg, with my motives and the particulars of my journey hither, —— will inform you. My first letter to him, with which doubtless he has edified your whole fireside, left me safely landed at Hamburg on the Elbe Stairs, at the Boom House. While standing on the stairs, I was amused by the contents of the passage-boat which crosses the river once or twice a day from Hamburg to Haarburg. It was stowed close with all people of all nations, in all sorts of dresses; the men all with pipes in their mouths, and these pipes of all shapes and fancies—straight and wreathed, simple and complex, long and short, cane, clay, porcelain, wood, tin, silver, and ivory; most of them with silver chains and silver bowl-covers. Pipes and boots are the first universal characteristic of the male Hamburgers that would strike the eye of a raw traveller. But I forget my promise of journalizing as much as possible.—Therefore, *Septr.* 19*th afternoon*. My companion who, you recollect, speaks the French language with unusual propriety, had formed a kind of confidential acquaintance with the emigrant, who appeared to be a man of sense, and whose manners were those of a perfect gentleman. He seemed about fifty or rather more. Whatever is unpleasant in French manners from excess in the degree, had been softened down by age or affliction; and all that is delightful in the kind, alacrity and delicacy in little attentions, etc. remained, and without bustle, gesticulation, or disproportionate eagerness. His demeanour exhibited the minute philanthropy of a polished Frenchman, tempered by the sobriety of the English character disunited from its reserve. There is something strangely attractive in the character of a gentleman when you apply the word emphatically, and yet in that sense of the term which it is more easy to feel than to define. It neither includes the possession of high moral excellence, nor of necessity even the ornamental graces of manner. I have now in my mind's eye a parson whose life would scarcely stand scrutiny even in the court of honour, much less in that of conscience; and his manners, if nicely observed, would of the two excite an idea of awkwardness rather than of elegance: and yet everyone who conversed with him felt and acknowledged *the gentleman*. The secret of the matter, I believe to be this—we feel the gentlemanly character present to us, whenever under all the circumstances of social intercourse, the trivial not less than the important, through the whole detail of his manners and deportment, and with the ease of a habit, a person shows respect to others in such a way, as at the same time implies in his own feelings an habitual and assured anticipation of reciprocal respect from them to himself. In short, the gentlemanly character arises out of the feeling of equality acting, as a habit, yet flexible to the varieties of rank, and modified without being disturbed or superseded by them. This description will perhaps explain to you the ground of one of your own remarks, as I was englishing to you the interesting dialogue concerning the causes of the corruption of eloquence. 'What perfect gentlemen these old Romans must have been! I was impressed, I remember, with the same feeling at the time I was reading a translation of Cicero's philosophical dialogues and of his epistolary correspondence: while in Pliny's Letters I seemed to have a different feeling—he gave me the notion of a very *fine* gentleman.'—You uttered the words as if you had felt that the adjunct had injured the substance and the increased degree altered the kind. Pliny was the courtier of an absolute monarch—Cicero an aristocratic republican. For this reason the character of gentleman, in the sense to which I have confined it, is frequent in England, rare in France, and found, where it is found, in age or the latest period of manhood; while in Germany the character is almost unknown. But the proper antipode of a gentleman is to be sought for among the Anglo-American democrats.

I owe this digression, as an act of justice, to this amiable Frenchman, and of humiliation for myself. For in a little controversy between us on the subject of French poetry, he made me feel my own ill behaviour by the silent reproof of contrast, and when I afterwards apologized to him for the warmth of my language, he answered me with a cheerful expression of surprise, and an immediate compliment, which a gentleman might both make with dignity and receive with pleasure. I was pleased, therefore, to find it agreed on, that we should, if possible, take up our quarters in the same house. My friend went with him in search of an hotel, and I to deliver my letters of recommendation.

I walked onward at a brisk pace, enlivened not so much by anything I actually saw, as by the confused sense that I was for the first time in my life on the *continent* of our planet. I seemed to myself like a liberated bird that had been hatched in an aviary, who now after his first soar of freedom poises himself in the upper air. Very naturally I began to wonder at all things, some for being so like and some for being so unlike the things in England—Dutch women with large umbrella hats shooting out half a yard before them, with a prodigal plumpness of petticoat behind—the women of Hamburg with caps plaited on the caul with silver or gold, or both, bordered round with stiffened lace, which stood out before their eyes, but not lower, so that the eyes sparkled through it—the Hanoverian women with the forepart of the head bare, then a stiff lace standing up like a wall perpendicular on the cap, and the cap behind tailed with an enormous quantity of ribbon which lies or tosses on the back:

> Their visnomies seem'd like a goodly banner
> Spread in defiance of all enemies.
>
> Spenser

—The ladies all in English dresses, all rouged, and all with bad teeth: which you notice instantly from their contrast to the almost animal, too glossy mother-of-pearl whiteness and the regularity of the teeth of the laughing, loud-talking country-women and servant-girls, who with their clean white stockings and with slippers without heel-quarters tripped along the dirty streets, as if they were secured by a charm from the dirt: with a lightness too, which surprised me, who had always considered it as one of the annoyances of sleeping in an inn, that I had to clatter upstairs in a pair of them. The streets narrow; to my English nose sufficiently offensive, and explaining at first sight the universal use of boots; without any appropriate path for the foot-passengers; the gable ends of the houses all towards the street, some in the ordinary triangular form and *entire* as the botanists say, but the greater number notched and scolloped with more than Chinese grotesqueness. Above all, I was struck with the profusion of windows, so large and so many, that the houses look all glass. Mr Pitt's window tax, with its pretty little *additionals* sprouting out from it like young toadlets on the back of a Surinam toad,° would certainly improve the appearance of the Hamburg houses, which have a slight summer look, not in keeping with their size, incongruous with the climate, and precluding that feeling of retirement and self-content, which one wishes to associate with a house in a noisy city. But a conflagration would, I fear, be the previous requisite to the production of any architectural beauty in Hamburg: for verily it is a filthy town. I moved on and crossed a multitude of ugly bridges, with huge black deformities of water-wheels close by them. The water intersects the city everywhere, and would have furnished to the genius of Italy the capabilities of all that is most beautiful and magnificent in architecture. It might have been the rival of Venice, and it is huddle and ugliness, stench and stagnation. The Jungfer Stieg (i.e. Young Ladies' Walk) to which my letters directed me, made an exception. It is a walk or promenade planted with treble rows of elm-trees, which being yearly pruned and cropped remain slim and dwarf-like. This walk occupies one side of a square piece of water, and with many swans on it perfectly tame, and moving among the swans showy pleasure-boats with ladies in them, rowed by their husbands or lovers. . . .

(*Some paragraphs have been here omitted.*)

. . . thus embarrassed by sad and solemn politeness still more than by broken English, it sounded like the voice of an old friend when I heard the emigrant's servant enquiring after me. He had come for the purpose of guiding me to our hotel. Through streets and streets I pressed on as happy as a child, and, I doubt not, with a childish expression of wonderment in my busy eyes, amused by the wicker waggons with movable benches across them, one behind the other, (these were the hackney coaches); amused by the signboards of the shops, on which all the articles sold within are painted, and that too very exactly, though in a grotesque confusion (a useful substitute for language in this great mart of nations) amused with the incessant tinkling of the shop and house doorbells, the bell hanging over each door and struck with a small iron rod at every entrance and exit;—and finally, amused by looking in at the windows, as I passed along; the ladies and gentlemen drinking coffee or playing cards, and the gentlemen all smoking. I wished myself a painter, that I might have sent you a sketch of one of the card parties. The long pipe of one gentleman rested on the table, its bowl half a yard from his mouth fuming like a censer by the fish-pool—the other gentleman, who was dealing the cards, and of course had both hands employed, held his pipe in his teeth, which hanging down between his knees, smoked beside his ankles. Hogarth himself never drew a more ludicrous distortion both of attitude and physiognomy, than this effort occasioned: nor was there wanting beside it one of those beautiful female faces which the same Hogarth, in whom the satirist never extinguished that love of beauty which belonged to him as a poet, so often and so gladly introduces as the central figure in a crowd of humorous deformities, which figure (such is the power of true genius!) neither acts, nor is meant to act as a contrast; but diffuses through all, and over each of the group, a spirit of reconciliation and human kindness; and even when the attention is no longer consciously directed to the cause of this feeling, still blends its tenderness with our laughter: and thus prevents the instructive merriment at the whims of nature or the foibles or humours of our fellow men from degenerating into the heart-poison of contempt or hatred.

Our hotel *Die wilde Mann* [The Wild Man], (the sign of which was no bad likeness of the landlord, who had engrafted on a very grim face a restless grin, that was at every man's service, and which indeed, like an actor rehearsing to himself, he kept playing in expectation of an occasion for it)—neither our hotel, I say, nor its landlord were of the genteelest class. But it has one great advantage for a stranger, by being in the market-place, and the next neighbour of the huge church of St Nicholas: a church with shops and houses built up against it, out of which wens and warts its high massy steeple rises, necklaced near the top with a round of large gilt balls. A better pole-star could scarcely be desired. Long shall I retain the impression made on my mind by the awful echo, so loud and long and tremulous, of the deep-toned clock within this church, which awoke me at two in the morning from a distressful dream, occasioned, I believe, by the feather bed, which is used here instead of bed-clothes. I will rather carry my blanket about with me like a wild Indian, than submit to this abominable custom. Our emigrant acquaintance was, we found, an intimate friend of the celebrated Abbé de Lisle: and from the large fortune which he possessed under the monarchy, had rescued sufficient not only for independence, but for respectability. He had offended some of his fellow emigrants in London, whom he had obliged with considerable sums, by a refusal to make further advances, and in consequence of their intrigues had received an order to quit the kingdom. I thought it one proof of his innocence, that he attached no blame either to the Alien Act,° or to the Minister who had exerted it against him; and a still greater, that he spoke of London with rapture, and of his favourite niece, who had married and settled in England, with all the fervour and all the pride of a fond parent. A man sent by force out of a country, obliged to sell out of the stocks at a great loss, and exiled from those pleasures and that style of society which habit had rendered essential to his happiness, whose predominant feelings were yet all of a private nature, resentment for friendship outraged, and anguish for domestic affections interrupted—such a man, I think, I could dare warrant guiltless of espionage in any service, most of all in that of the present French Directory. He spoke with extasy of Paris under the monarchy: and yet the particular facts, which made up his description, left as deep a conviction on my mind, of French worthlessness, as his own tale had done of emigrant ingratitude. Since my arrival in Germany, I have not met a single person, even among those who abhor the Revolution, that spoke with favour, or even charity, of the French emigrants. Though the belief of their influence in the origination of this disastrous war, (from the horrors of which, North Germany deems itself only reprieved, not secured) may have some share in the general aversion with which they are regarded; yet I am deeply persuaded that the far greater part is owing to their own profligacy, to their treachery and hard-heartedness to each other, and the domestic misery or corrupt principles which so many of them have carried into the families of their protectors. My heart dilated with honest pride, as I recalled to mind the stern yet amiable characters of the English patriots, who sought refuge on the Continent at the Restoration! O let not our civil war under the first Charles be paralleled with the French Revolution! In the former, the chalice overflowed from excess of principle; in the latter, from the fermentation of the dregs! The former, was a civil war between the virtues and virtuous prejudices of the two parties; the latter, between the vices. The Venetian glass of the French monarchy shivered and flew asunder with the working of a double poison.°

Sept. 20th. I was introduced to Mr Klopstock, the brother of the poet, who again introduced me to Professor Ebeling, an intelligent and lively man, though deaf: so deaf, indeed, that it was a painful effort to talk with him, as we were obliged to drop all our pearls into a huge ear-trumpet. From this courteous and kind-hearted man of letters, (I hope, the German literati in general may resemble this first specimen) I heard a tolerable Italian pun, and an interesting anecdote. When Buonaparte was in Italy, having been irritated by some instance of perfidy, he said in a loud and vehement tone, in a public company—"'tis a true proverb, *gli Italiani tutti ladroni* (i.e. the Italians all plunderers). A lady had the courage to reply 'Non tutti; ma buona parte,' (not all, but a good part, or Buonaparte). This, I confess, sounded to my ears, as one of the many good things that *might have been* said. The anecdote is more valuable; for it instances the ways and means of French insinuation. Hoche had received much information concerning the face of the country from a map of unusual fullness and accuracy, the maker of which, he heard, resided at Düsseldorf. At the storming of Düsseldorf by the French army, Hoche previously ordered, that the house and property of this man should be preserved, and entrusted the performance of the order to an officer on whose troop he could rely. Finding afterwards that the man had escaped before the storming commenced, Hoche exclaimed, 'He had no reason to flee! it is *for* such men, not *against* them, that the French nation makes war, and consents to shed the blood of its children.' You remember Milton's sonnet—

> The great Emathian conqueror bid spare
> The house of Pindarus when temple and tower
> Went to the ground . . .

Now though the Düsseldorf map-maker may stand in the same relation to the Theban bard, as the snail that marks its path by lines of film on the wall it creeps over, to the eagle that soars sunward and beats the tempest with its wings; it does not therefore follow, that the Jacobin of France may not be as valiant a general and as good a politician, as the madman of Macedon.

From Professor Ebeling's, Mr Klopstock accompanied my friend and me to his own house, where I saw a fine bust of his brother. There was a solemn and heavy greatness in his countenance which corresponded to my preconceptions of his style and genius.—I saw there, likewise, a very fine portrait of Lessing, whose works are at present the chief object of my admiration. His eyes were uncommonly like mine, if anything, rather larger and more prominent. But the lower part of his face and his nose—O what an exquisite expression of elegance and sensibility!—There appeared no depth, weight, or comprehensiveness, in the forehead.—The whole face seemed to say, that Lessing was a man of quick and voluptuous feelings; of an active but light fancy; acute; yet acute not in the observation of actual life, but in the arrangements and management of the ideal world, i.e. in taste, and in metaphysics. I assure you, that I wrote these very words in my memorandum book with the portrait before my eyes, and when I knew nothing of Lessing but his name, and that he was a German writer of eminence.

We consumed two hours and more over a bad dinner, at the table d'hôte. 'Patience at a German ordinary, smiling at time.' The Germans are the worst cooks in Europe. There is placed for every two persons a bottle of common wine—Rhenish and Claret alternately; but in the houses of the opulent during the many and long intervals of the dinner, the servants hand round glasses of richer wines. At the Lord of Culpin's they came in this order. Burgundy—Madeira—Port—Frontiniac—Pacchiaretti—Old Hock—Mountain—Champagne—Hock again—Bishop, and lastly, Punch. A tolerable quantum [amount], methinks! The last dish at the ordinary, viz. slices of roast pork (for all the larger dishes are brought in, cut up, and first handed round and then set on the table) with stewed prunes and other sweet fruits, and this followed by cheese and butter, with plates of apples, reminded me of Shakespeare* and Shakespeare put it in my head to go to the French comedy.

Bless me! why it is worse than our modern English plays! The first act informed me, that a court martial is to be held on a Count Vatron, who had drawn his sword on the Colonel, his brother-in-law. The officers plead in his behalf—in vain! His wife, the Colonel's sister, pleads with most tempestuous agonies—in vain! She falls into hysterics and faints away, to the dropping of the inner curtain! In the second act sentence of death is passed on the Count—his wife, as frantic and hysterical as before: more so (good industrious creature!) she could not be. The third and last act, the wife still frantic, very frantic indeed! the soldiers just about to fire, the handkerchief actually dropped, when reprieve! reprieve! is heard from behind the scenes: and in comes Prince somebody, pardons the Count, and the wife is still frantic, only with joy; that was all!

O dear lady! this is one of the cases, in which laughter is followed by melancholy: for such is the kind of drama, which is now substituted

* *Slender*. 'I bruised my shin with playing with sword and dagger for a dish of stewed prunes, and by my troth I cannot abide the smell of hot meat since.' So again, *Evans*. 'I will make an end of my dinner: there's pippins and cheese yet to come.'

everywhere for Shakespeare and Racine. You well know, that I offer violence to my own feelings in joining these names. But however meanly I may think of the French serious drama, even in its most perfect specimens; and with whatever right I may complain of its perpetual falsification of the language, and of the connections and transitions of thought, which nature has appropriated to states of passion; still, however, the French tragedies are consistent works of art, and the offspring of great intellectual power. Preserving a fitness in the parts, and a harmony in the whole, they form a nature of their own, though a false nature. Still they excite the minds of the spectators to active thought, to a striving after ideal excellence. The soul is not stupefied into mere sensations, by a worthless sympathy with our own ordinary sufferings, or an empty curiosity for the surprising, undignified by the language or the situations which awe and delight the imagination. What (I would ask of the crowd, that press forward to the pantomimic tragedies and weeping comedies of Kotzebue and his imitators) what are you seeking? Is it comedy? But in the comedy of Shakespeare and Molière the more accurate my knowledge, and the more profoundly I think, the greater is the satisfaction that mingles with my laughter. For though the qualities which these writers portray are ludicrous indeed, either from the kind or the excess, and exquisitely ludicrous, yet are they the natural growth of the human mind and such as, with more or less change in the drapery, I can apply to my own heart, or at least to whole classes of my fellow creatures. How often are not the moralist and the metaphysician obliged for the happiest illustrations of general truths and the subordinate laws of human thought and action to quotations not only from the tragic characters but equally from the Jaques, Falstaff, and even from the fools and clowns of Shakespeare, or from the Miser, Hypochondriast, and Hypocrite, of Molière! Say not, that I am recommending abstractions: for these class characteristics, which constitute the instructiveness of a character, are so modified and particularized in each person of the Shakespearian drama, that life itself does not excite more distinctly that sense of individuality which belongs to real existence. Paradoxical as it may sound, one of the essential properties of geometry is not less essential to dramatic excellence, and (if I may mention his name without pedantry to a lady) Aristotle has accordingly required of the poet an involution of the universal in the individual. The chief differences are, that in geometry it is the universal truth itself, which is uppermost in the consciousness, in poetry the individual form in which the truth is clothed. With the ancients, and not less with the elder dramatists of England and France, both comedy and tragedy were considered as kinds of poetry. They neither sought in comedy to make us laugh merely, much less to make us laugh by wry faces, accidents of jargon, slang phrases for the day, or the clothing of commonplace morals in metaphors drawn from the shops or mechanic occupations of their characters; nor did they condescend in tragedy to wheedle away the applause of the spectators, by representing before them facsimiles of their own mean selves in all their existing meanness, or to work on their sluggish sympathies by a pathos not a whit more respectable than the maudlin tears of drunkenness. Their tragic scenes were meant to affect us indeed, but within the bounds of pleasure, and in union with the activity both of our understanding and imagination. They wished to transport the mind to a sense of its possible greatness, and to implant the germs of that greatness during the temporary oblivion of the worthless 'thing, we are' and of the peculiar state, in which each man happens to be; suspending our individual recollections and lulling them to sleep amid the music of nobler thoughts.

Hold! (methinks I hear the spokesman of the crowd reply, and we will listen to him. I am the plaintiff, and be he the defendant.)

DEFENDANT. Hold! are not our modern sentimental plays filled with the best Christian morality?

PLAINTIFF. Yes! just as much of it, and just that part of it which you can exercise without a single Christian virtue—without a single sacrifice that is really painful to you!—just as much as flatters you, sends you away pleased with your own hearts, and quite reconciled to your vices, which can never be thought very ill of, when they keep such good company, and walk hand in hand with so much compassion and generosity; adulation so loathsome, that you would spit in the man's face who dared offer it to you in a private company, unless you interpreted it as insulting irony, you appropriate with infinite satisfaction, when you share the garbage with the whole sty, and gobble it out of a common trough. No Caesar must pace your boards—no Antony, no royal Dane, no Orestes, no Andromache!—

D. No: or as few of them as possible. What has a plain citizen of London, or Hamburg, to do with your kings and queens, and your old schoolboy pagan heroes? Besides, everybody knows the stories: and what curiosity can we feel——

P. What, Sir, not for the manner? not for the delightful language of the poet? not for the situations, the action and reaction of the passions?

D. You are hasty, Sir! the only curiosity, we feel, is in the story: and how can we be anxious concerning the end of a play, or be surprised by it, when we know how it will turn out?

P. Your pardon, for having interrupted you! we now understand each other. You seek then, in a tragedy, which wise men of old held for the highest effort of human genius, the same gratification, as that you receive from a new novel, the last German romance, and other dainties of the day, which can be enjoyed but once. If you carry these feelings to the sister art of painting, Michelangelo's Sistine Chapel, and the Scripture Gallery of Raphael, can expect no favour from you. You know all about them beforehand; and are, doubtless, more familiar with the subjects of those paintings, than with the tragic tales of the historic or heroic ages. There is a consistency, therefore, in your preference of contemporary writers: for the great men of former times, those at least who were deemed great by our ancestors, sought so little to gratify this kind of curiosity, that they seemed to have regarded the story in a not much higher light, than the painter regards his canvas: as that on, not by, which they were to display their appropriate excellence. No work, resembling a tale or romance, can well show less variety of invention in the incidents, or less anxiety in weaving them together, than the Don Quixote of Cervantes. Its admirers feel the disposition to go back and reperuse some preceding chapter, at least ten times for once that they find any eagerness to hurry forwards: or open the book on those parts which they best recollect, even as we visit those friends oftenest whom we love most, and with whose characters and actions we are the most intimately acquainted. In the divine Ariosto, (as his countrymen call this, their darling poet) I question whether there be a single tale of his own invention, or the elements of which, were not familiar to the readers of 'old romance.' I will pass by the ancient Greeks, who thought it even necessary to the fable of a tragedy, that its substance should be previously known. That there had been at least fifty tragedies with the same title, would be one of the motives which determined Sophocles and Euripides, in the choice of Electra, as a subject. But Milton—

D. Aye Milton, indeed! but do not Dr Johnson, and other great men tell us, that nobody now reads Milton but as a task?

P. So much the worse for them, of whom this can be truly said! But why then do you pretend to admire Shakespeare? The greater part, if not all, of his dramas were, as far as the names and the main incidents are concerned, already stock plays. All the stories, at least, on which they are built, pre-existed in the chronicles, ballads, or translations of contemporary or preceding English writers. Why, I repeat, do you pretend to admire Shakespeare? Is it, perhaps, that you only pretend to admire him? However, as once for all, you have dismissed the well-known events and personages of history, or the epic Muse, what have you taken in their stead? Whom has your tragic Muse armed with her bowl and dagger? the sentimental Muse I should have said, whom you have seated in the throne of tragedy? What heroes has she reared on her buskins?

D. O! our good friends and next-door neighbours—honest tradesmen, valiant tars, high spirited half-pay officers, philanthropic Jews, virtuous courtesans, tender-hearted braziers, and sentimental rat-catchers! (a little bluff or so, but all our very generous, tender-hearted characters *are* a little rude or misanthropic, and all our misanthropes very tender-hearted.)

P. But I pray you, friend, in what actions great or interesting, can such men be engaged?

D. They give away a great deal of money: find rich dowries for young men and maidens who have all other good qualities; they browbeat lords, baronets, and Justices of the Peace, (for they are as bold as Hector!)—they rescue stage-coaches at the instant they are falling down precipices; carry away infants in the sight of opposing armies; and some of our performers act a muscular able-bodied man to such perfection, that our dramatic poets, who always have the actors in their eye, seldom fail to make their favourite male character as strong as Samson. And then they take such prodigious leaps! And what is *done* on the stage is more striking even than what is acted. I once remember such a deafening explosion, that I could not hear a word of the play for half an act after it: and a little real gunpowder being set fire to at the same time, and smelt by all the spectators, the naturalness of the scene was quite astonishing!

P. But how can you connect with such men and such actions that dependence of thousands on the fate of one, which gives so lofty an interest to the personages of Shakespeare, and the Greek tragedians? How can you connect with them that sublimest of all feelings, the power of destiny and the controlling might of heaven, which seems to elevate the characters which sink beneath its irresistible blow?

D. O mere fancies! We seek and find on the present stage our own wants and passions, our own vexations, losses, and embarrassments.

P. It is your own poor pettifogging nature then, which you desire to have represented before you? not human nature in its height and vigour? But surely you might find the former with all its joys and sorrows, more conveniently in your own houses and parishes.

D. True! but here comes a difference. Fortune is blind, but the poet has his eyes open, and is besides as complaisant as fortune is capricious. He makes every thing turn out exactly as we would wish it. He gratifies us by representing those as hateful or contemptible whom we hate and wish to despise.

P. (*aside*) That is, he gratifies your envy by libelling your superiors.

D. He makes all those precise moralists, who affect to be better than their neighbours, turn out at last abject hypocrites, traitors, and hard-hearted villains; and your men of spirit, who take their girl and their glass with equal freedom, prove the true men of honour, and (that no part of the audience may remain unsatisfied) reform in the last scene, and leave no doubt on the minds of the ladies, that they will make most faithful and excellent husbands: though it does seem a pity, that they should be obliged to get rid of qualities which had made them so interesting! Besides, the poor become rich all at once; and in the final matrimonial choice the opulent and high-born themselves are made to confess, that virtue is the only true nobility, and that a lovely woman is a dowry of herself!

P. Excellent! But you have forgotten those brilliant flashes of loyalty, those patriotic praises of the king and old England, which, especially if conveyed in a metaphor from the ship or the shop, so often solicit and so unfailingly receive the public plaudit! I give your prudence credit for the omission. For the whole system of your drama is a moral and intellectual Jacobinism of the most dangerous kind, and those commonplace rants of loyalty are no better than hypocrisy in your playwrights, and your own sympathy with them a gross self-delusion. For the whole secret of dramatic popularity consists with you, in the confusion and subversion of the natural order of things, their causes and their effects; in the excitement of surprise, by representing the qualities of liberality, refined feeling, and a nice sense of honour (those things rather, which pass among you for such) in persons and in classes of life where experience teaches us least to expect them; and in rewarding with all the sympathies that are the dues of virtue, those criminals whom law, reason, and religion, have excommunicated from our esteem!

And now good night! Truly! I might have written this last sheet without having gone to Germany, but I fancied myself talking to you by your own fireside, and can you think it a small pleasure to me to forget now and then, that I am not there. Besides, you and my other good friends have made up your minds to me as I am, and from whatever place I write you will expect that part of my 'Travels' will consist of the excursions in my own mind.

LETTER III

Ratzeburg

No little fish thrown back again into the water, no fly unimprisoned from a child's hand, could more buoyantly enjoy its element, than I this clean and peaceful house, with this lovely view of the town, groves, and lake of Ratzeburg, from the window at which I am writing. My spirits certainly, and my health I fancied, were beginning to sink under the noise, dirt, and unwholesome air of our Hamburg hotel. I left it on Sunday, Sept. 23d. with a letter of introduction from the poet Klopstock, to the Amptman [District Administrator] of Ratzeburg. The Amptman received me with kindness, and introduced me to the worthy pastor, who agreed to board and lodge me for any length of time not less than a month. The vehicle, in which I took my place, was considerably larger than an English stage-coach, to which it bore much the same proportion and rude resemblance, that an elephant's ear does to the human. Its top was composed of naked boards of different colours, and seeming to have been parts of different wainscots. Instead of windows there were leathern curtains with a little eye of glass in each: they perfectly answered the purpose of keeping out the prospect and letting in the cold. I could observe little, therefore, but the inns and farmhouses at which we stopped. They were all alike, except in size: one great room, like a barn, with a hayloft over it, the straw and hay dangling in tufts through the boards which formed the ceiling of the room, and the floor of the loft. From this room, which is paved like a street, sometimes one, sometimes two smaller ones, are enclosed at one end. These are commonly floored. In the large room the cattle, pigs, poultry, men, women, and children, live in amicable community: yet there was an appearance of cleanliness and rustic comfort. One of these houses I measured. It was an hundred feet in length. The apartments were taken off from one corner. Between these and the stalls there was a small interspace, and here the breadth was forty-eight feet, but thirty-two where the stalls were; of course, the stalls were on each side eight feet in depth. The faces of the cows, etc. were turned towards the room; indeed they were in it, so that they had at least the comfort of seeing each other's faces. Stall-feeding is universal in this part of Germany, a practice concerning which the agriculturalist and the poet are likely to entertain opposite opinions—or at least, to have very different feelings. The woodwork of these buildings on the outside is left unplastered, as in old houses among us, and being painted red and green, it cuts and tesselates the buildings very gaily. From within three miles of Hamburg almost to Molln, which is thirty miles from it, the country as far as I could see it, was a dead flat, only varied by woods. At Molln it became more beautiful. I observed a small lake nearly surrounded with groves, and a palace in view belonging to the King of Great Britain, and inhabited by the Inspector of the Forests. We were nearly the same time in travelling the thirty-five miles from Hamburg to Ratzeburg, as we had been in going from London to Yarmouth, one hundred and twenty-six miles.

The lake of Ratzeburg runs from south to north, about nine miles in length, and varying in breadth from three miles to half a mile. About a mile from the southernmost point it is divided into two, of course very unequal, parts by an island, which being connected by a bridge and a narrow slip of land with the one shore, and by another bridge of immense length with the other shore, forms a complete isthmus. On this island the town of Ratzeburg is built. The pastor's house or vicarage, together with the Amptman's [District Administrator's], Amptschreiber's [Public Secretary's], and the church, stands near the summit of a hill, which slopes down to the slip of land and the little bridge, from which, through a superb military gate, you step into the island town of Ratzeburg. This again is itself a little hill, by ascending and descending which, you arrive at the long bridge, and so to the other shore. The water to the south of the town is called the Little Lake, which however, almost engrosses the beauties of the whole: the shores being just often enough green and bare to give the proper effect to the magnificent groves which occupy the greater part of their circumference. From the turnings, windings, and indentations of the shore, the views vary almost every ten steps, and the whole has a sort of majestic beauty, a feminine grandeur. At the north of the Great Lake, and peeping over it, I see the seven church towers of Lubec, at the distance of twelve or thirteen miles, yet as distinctly as if they were not three. The only defect in the view is, that Ratzeburg is built entirely of red bricks, and all the houses roofed with red tiles. To the eye, therefore, it presents a clump of brick-dust red. Yet this evening, Oct. 10th. twenty minutes past five, I saw the town perfectly beautiful, and the whole softened down into *complete keeping*, if I may borrow a term from the painters. The sky over Ratzeburg and all the east, was a pure evening blue, while over the west it was covered with light sandy clouds. Hence a deep red light spread over the whole prospect, in undisturbed harmony with the red town, the brown-red woods, and the yellow-red reeds on the skirts of the lake. Two or three boats, with single persons paddling them, floated up and down in the rich light, which not only was itself in harmony with all, but brought all into harmony.

I should have told you that I went back to Hamburg on Thursday (Sept. 27th.) to take leave of my friend,° who travels southward, and returned hither on the Monday following. From Empfelde, a village half-way from Ratzeburg, I walked to Hamburg through deep sandy roads and a dreary flat: the soil everywhere white, hungry, and excessively pulverized; but the approach to the city is pleasing. Light cool country houses, which you can look through and see the gardens behind them, with arbours and trellis-work, and thick vegetable walls, and trees in cloisters and piazzas, each house with neat rails before it, and green seats within the rails. Every object, whether the growth of nature or the work of man, was neat and artificial. It pleased me far better, than if the houses and gardens, and pleasure fields, had been in a nobler taste: for this nobler taste would have been mere apery. The busy, anxious, money-loving merchant of Hamburg could only have adopted, he could not have enjoyed the simplicity of nature. The mind begins to love nature by imitating human conveniences in nature; but this is a step in intellect, though a low one—and were it not so, yet all around me spoke of innocent enjoyment and sensitive comforts, and I entered with unscrupulous sympathy into the enjoyments and comforts even of the busy, anxious, money-loving merchants of Hamburg. In this charitable and catholic mood I reached the vast ramparts of the city. These are huge green cushions, one rising above the other, with trees growing in the interspaces, pledges and symbols of a long peace. Of my return I have nothing worth communicating, except that I took extra post, which answers to posting in England. These North German post-chaises are uncovered wicker carts. An English dust-cart is a piece of finery, a chef-d'oeuvre of mechanism, compared with them: and the horses! a savage might use their ribs instead of his fingers for a numeration table. Wherever we stopped, the postilion fed his cattle with the brown rye bread of which he ate himself, all breakfasting together, only the horses had no gin to their water, and the postilion no water to his gin. Now and henceforward for subjects of more interest to you, and to the objects in search of which I left you: namely, the literati and literature of Germany.

Believe me, I walked with an impression of awe on my spirits, as W—— and myself accompanied Mr Klopstock to the house of his brother, the poet, which stands about a quarter of a mile from the city gate. It is one of a row of little commonplace summerhouses, (for so they looked) with four or five rows of young meagre elm-trees before the windows, beyond which is a green, and then a dead flat intersected with several roads. Whatever beauty (thought I) may be before the poet's eyes at present, it must certainly be purely of his own creation. We waited a few minutes in a neat little parlour, ornamented with the figures of two of the Muses and with prints, the subjects of which were from Klopstock's odes. The poet entered. I was much disappointed in his countenance, and recognized in it no likeness to the bust. There was no comprehension in the forehead, no weight over the eyebrows, no expression of peculiarity, moral or intellectual on the eyes, no massiveness in the general countenance. He is if anything rather below the middle size. He wore very large half-boots which his legs filled, so fearfully were they swollen. However, though neither W—— nor myself could discover any indications of sublimity or enthusiasm in his physiognomy, we were both equally impressed with his liveliness, and his kind and ready courtesy. He talked in French with my friend, and with difficulty spoke a few sentences to me in English. His enunciation was not in the least affected by the entire want of his upper teeth. The conversation began on his part by the expression of his rapture at the surrender of the detachment of French troops under General Humbert. Their proceedings in Ireland with regard to the committee which they had appointed, with the rest of their organizing system, seemed to have given the poet great entertainment. He then declared his sanguine belief in Nelson's victory, and anticipated its confirmation with a keen and triumphant pleasure. His words, tones, looks, implied the most vehement anti-Gallicanism. The subject changed to literature, and I enquired in Latin concerning the history of German poetry and the elder German poets. To my great astonishment he confessed, that he knew very little on the subject. He had indeed occasionally read one or two of their elder writers, but not so as to enable him to speak of their merits. Professor Ebeling, he said, would probably give me every information of this kind: the subject had not particularly excited his curiosity. He then talked of Milton and Glover, and thought Glover's blank verse superior to Milton's. W—— and myself expressed our surprise: and my friend gave his definition and notion of harmonious verse, that it consisted (the English iambic blank verse above all) in the apt arrangement of pauses and cadences, and the sweep of whole paragraphs,

> . . . with many a winding bout
> Of linked sweetness long drawn out,

and not in the even flow, much less in the prominence or antithetic vigour, of single lines, which were indeed injurious to the total effect, except where they were introduced for some specific purpose. Klopstock assented, and said that he meant to confine Glover's superiority to single lines. He told us that he had read Milton, in a prose translation, when he was fourteen.* I understood him thus myself, and W—— interpreted Klopstock's French as I had already construed it. He appeared to know very little of Milton—or indeed of our poets in general. He spoke with great indignation of the English prose translation of his Messiah. All the translations had been bad, very bad—but the English was *no* translation—there were pages on pages not in the original:—and half the original was not to be found in the translation. W—— told him that I intended to translate a few of his odes as specimens of German lyrics—he then said to me in English, 'I wish you would render into English some select passages of the Messiah, and *revenge* me of your countryman!' It was the liveliest thing which he produced in the whole conversation. He told us, that his first ode was fifty years older than his last. I looked at him with much emotion—I considered him as the venerable father of German poetry; as a good man; as a Christian; seventy-four years old; with legs enormously swollen; yet active, lively, cheerful, and kind, and communicative. My eyes felt as if a tear were swelling into them. In the portrait of Lessing there was a toupee periwig, which enormously injured the effect of his physiognomy—Klopstock wore the same, powdered and frizzled. By the by, old men ought never to wear powder—the contrast between a large snow-white wig and the colour of an old man's skin is disgusting, and wrinkles in such a neighbourhood appear only channels for dirt. It is an honour to poets and great men, that you think of them as parts of nature; and anything of trick and fashion wounds you in them as much as when you see venerable

* This was accidentally confirmed to me by an old German gentleman at Helmstadt, who had been Klopstock's school- and bedfellow. Among other boyish anecdotes, he related that the young poet set a particular value on a translation of the Paradise Lost, and always slept with it under his pillow.

yews clipped into miserable peacocks.—The author of the Messiah should have worn his own grey hair.—His powder and periwig were to the eye what Mr Virgil would be to the ear.

Klopstock dwelt much on the superior power which the German language possessed of concentrating meaning. He said, he had often translated parts of Homer and Virgil, line by line, and a German line proved always sufficient for a Greek or Latin one. In English you cannot do this. I answered, that in English we could commonly render one Greek heroic line in a line and a half of our common heroic metre, and I conjectured that this line and a half would be found to contain no more syllables than one German or Greek hexameter. He did not understand me:* and I who wished to hear his opinions, not to correct them, was glad that he did not.

* Klopstock's observation was partly true and partly erroneous. In the literal sense of his words, and if we confine the comparison to the average of space required for the expression of the same thought in the two languages, it is erroneous. I have translated some German hexameters into English hexameters, and find, that on the average three lines English will express four lines German. The reason is evident: our language abounds in monosyllables and disyllables. The German, not less than the Greek, is a polysyllable language. But in another point of view the remark was not without foundation. For the German possessing the same unlimited privilege of forming compounds, both with prepositions and with epithets as the Greek, it can express the richest single Greek word in a single German one, and is thus freed from the necessity of weak or ungraceful paraphrases. I will content myself with one example at present, viz. the use of the prefixed particles *ver*, *zer*, *ent*, and *weg*: thus, *reissen* to rend, *verreissen* to rend away, *zerreissen* to rend to pieces, *entreissen* to rend off or out of a thing, in the active sense: or *schmelzen* to melt—*ver*, *zer*, *ent*, *schmelzen*—and in like manner through all the verbs neuter and active. If you consider only how much we should feel the loss of the prefix *be*, as in *bedropt*, *besprinkle*, *besot*, especially in our poetical language, and then think that this same mode of composition is carried through all their simple and compound prepositions, and many of their adverbs; and that with most of these the Germans have the same privilege as we have of dividing them from the verb and placing them at the end of the sentence; you will have no difficulty in comprehending the reality and the cause of this superior power in the German of condensing meaning, in which its great poet exulted. It is impossible to read half a dozen pages of Wieland without perceiving that in this respect the German has no rival but the Greek. And yet I seem to feel, that concentration or condensation is not the happiest mode of expressing this excellence, which seems to consist not so much in the less time required for conveying an impression, as in the unity and simultaneousness with which the impression is conveyed. It tends to make their language more picturesque: it *depictures* images better. We have obtained this power in part by our compound verbs derived from the Latin: and the sense of its great effect no doubt induced our Milton both to the use and the abuse of Latin derivatives. But still these prefixed particles, conveying no separate or separable meaning to the mere English reader, cannot possibly act on the mind with the force or liveliness of an original and homogeneous language such as the German is, and besides are confined to certain words.

We now took our leave. At the beginning of the French Revolution Klopstock wrote odes of congratulation. He received some honorary presents from the French Republic (a golden crown I believe) and, like our Priestley, was invited to a seat in the legislature, which he declined. But when French liberty metamorphosed herself into a fury, he sent back these presents with a *palinodia* [recantation], declaring his abhorrence of their proceedings: and since then he has been perhaps more than enough an anti-Gallican. I mean, that in his just contempt and detestation of the crimes and follies of the Revolutionists, he suffers himself to forget that the Revolution itself is a process of the divine providence; and that as the folly of men is the wisdom of God, so are their iniquities instruments of his goodness. From Klopstock's house we walked to the ramparts, discoursing together on the poet and his conversation, till our attention was diverted to the beauty and singularity of the sunset and its effects on the objects round us. There were woods in the distance. A rich sandy light (nay, of a much deeper colour than sandy) lay over these woods that blackened in the blaze. Over that part of the woods which lay immediately under the intenser light, a brassy mist floated. The trees on the ramparts, and the people moving to and fro between them, were cut or divided into equal segments of deep shade and brassy light. Had the trees, and the bodies of the men and women, been divided into equal segments by a rule or pair of compasses, the portions could not have been more regular. All else was obscure. It was a fairy scene! and to increase its romantic character among the moving objects thus divided into alternate shade and brightness, was a beautiful child, dressed with the elegant simplicity of an English child, riding on a stately goat, the saddle, bridle, and other accoutrements of which were in a high degree costly and splendid. Before I quit the subject of Hamburg, let me say, that I remained a day or two longer than I otherwise should have done, in order to be present at the feast of St Michael, the patron saint of Hamburg, expecting to see the civic pomp of this commercial republic. I was however disappointed. There were no processions, two or three sermons were preached to two or three old women in two or three churches, and St Michael and his patronage wished elsewhere by the higher classes, all places of entertainment, theatre, etc. being shut up on this day. In Hamburg, there seems to be no religion at all: in Lubec it is confined to the women. The men seem determined to be divorced from their wives in the other world, if they cannot in this. You will not easily conceive a more singular sight, than is presented by the vast aisle of the principal church at Lubec seen from the organ-loft: for being filled with female servants and persons in the same class of life, and all their caps having gold and silver cauls, it appears like a rich pavement of gold and silver.

I will conclude this letter with the mere transcription of notes, which my friend W—— made of his conversations with Klopstock, during the interviews that took place after my departure. On these I shall make but one remark at present, and that will appear a presumptuous one, namely, that Klopstock's remarks on the venerable sage of Koenigsberg are to my own knowledge injurious and mistaken; and so far is it from being true, that his system is now given up, that throughout the universities of Germany there is not a single professor who is not, either a Kantean; or a disciple of Fichte, whose system is built on the Kantean, and presupposes its truth; or lastly who, though an antagonist of Kant as to his theoretical work, has not embraced wholly or in part his moral system, and adopted part of his nomenclature.

'Klopstock having wished to see the Calvary of Cumberland, and asked what was thought of it in England, I went to Remnant's (the English bookseller) where I procured the Analytical Review, in which is contained the review of Cumberland's Calvary. I remembered to have read there some specimens of a blank verse translation of the Messiah. I had mentioned this to Klopstock, and he had a great desire to see them. I walked over to his house and put the book into his hands. On adverting to his own poem, he told me he began the Messiah when he was seventeen: he devoted three entire years to the plan without composing a single line. He was greatly at a loss in what manner to execute his work. There were no successful specimens of versification in the German language before this time. The first three cantos he wrote in a species of measured or numerous prose. This, though done with much labour and some success, was far from satisfying him. He had composed hexameters both Latin and Greek as a school exercise, and there had been also in the German language attempts in that style of versification. These were only of very moderate merit.—One day he was struck with the idea of what could be done in this way—he kept his room a whole day, even went without his dinner, and found that in the evening he had written twenty-three hexameters, versifying a part of what he had before written in prose. From that time, pleased with his efforts, he composed no more in prose. Today he informed me that he had finished his plan before he read Milton. He was enchanted to see an author who before him had trod the same path. This is a contradiction of what he said before. He did not wish to speak of his poem to anyone till it was finished: but some of his friends who had seen what he had finished, tormented him till he had consented to publish a few books in a journal. He was then I believe very young, about twenty-five. The rest was printed at different periods, four books at a time. The reception given to the first specimens was highly flattering. He was nearly thirty years in finishing the whole poem, but of these thirty years not more than two were employed in the composition. He only composed in favourable moments; besides he had other occupations. He values himself upon the plan of his odes, and accuses the modern lyrical writers of gross deficiency in this respect. I laid the same accusation against Horace: he would not hear of it—but waived the discussion. He called Rousseau's Ode to Fortune a moral dissertation in stanzas. I spoke of Dryden's St Cecilia; but he did not seem familiar with our writers. He wished to know the distinctions between our dramatic and epic blank verse. He recommended me to read his Herman before I read either the Messiah or the odes. He flattered himself that some time or other his dramatic poems would be known in England. He had not heard of Cowper. He thought that Voss in his translation of the Iliad had done violence to the idiom of the Germans, and had sacrificed it to the Greek, not remembering sufficiently that each language has its particular spirit and genius. He said Lessing was the first of their dramatic writers. I complained of Nathan as tedious. He said there was not enough of action in it; but that Lessing was the most chaste of their writers. He spoke favourably of Goethe; but said that his "Sorrows of Werter" was his best work, better than any of his dramas: he preferred the first written to the rest of Goethe's dramas, Schiller's "Robbers" he found so extravagant, that he could not read it. I spoke of the scene of the setting sun. He did not know it. He said Schiller could not live. He thought Don Carlos the best of his dramas; but said that the plot was inextricable.—It was evident, he knew little of Schiller's works: indeed he said, he could not read them. Bürger he said was a true poet, and would live; that Schiller, on the contrary, must soon be forgotten; that he gave himself up to the imitation of Shakespeare, who often was extravagant, but that Schiller was ten thousand times more so. He spoke very slightingly of Kotzebue, as an immoral author in the first place, and next, as deficient in power. At Vienna, said he, they are transported with him; but we do not reckon the people of Vienna either the wisest or the wittiest people of Germany. He said Wieland was a charming author, and a sovereign master of his own language:

that in this respect Goethe could not be compared to him, or indeed could anybody else. He said that his fault was to be fertile to exuberance. I told him the Oberon had just been translated into English. He asked me, if I was not delighted with the poem. I answered, that I thought the story began to flag about the seventh or eighth book; and observed that it was unworthy of a man of genius to make the interest of a long poem turn entirely upon animal gratification. He seemed at first disposed to excuse this by saying, that there are different subjects for poetry, and that poets are not willing to be restricted in their choice. I answered, that I thought the *passion* of love as well suited to the purposes of poetry as any other passion; but that it was a cheap way of pleasing to fix the attention of the reader through a long poem on the mere *appetite*. Well! but, said he, you see, that such poems please everybody. I answered, that it was the province of a great poet to raise people up to his own level, not to descend to theirs. He agreed, and confessed, that on no account whatsoever would he have written a work like the Oberon. He spoke in raptures of Wieland's style, and pointed out the passage where Retzia is delivered of her child, as exquisitely beautiful. I said that I did not perceive any very striking passages; but that I made allowance for the imperfections of a translation. Of the thefts of Wieland, he said, they were so exquisitely managed, that the greatest writers might be proud to steal as he did. He considered the books and fables of old romance writers in the light of the ancient mythology, as a sort of common property, from which a man was free to take whatever he could make a good use of. An Englishman had presented him with the odes of Collins, which he had read with pleasure. He knew little or nothing of Gray, except his Essay in the churchyard. He complained of the fool in Lear. I observed, that he seemed to give a terrible wildness to the distress; but still he complained. He asked whether it was not allowed, that Pope had written rhyme poetry with more skill than any of our writers—I said, I preferred Dryden, because his couplets had greater variety in their movement. He thought my reason a good one; but asked whether the rhyme of Pope were not more exact. This question I understood as applying to the final terminations, and observed to him that I believed it was the case; but that I thought it was easy to excuse some inaccuracy in the final sounds, if the general sweep of the verse was superior. I told him that we were not so exact with regard to the final endings of lines as the French. He did not seem to know that we made no distinction between masculine and feminine (i.e. single or double,) rhymes: at least he put inquiries to me on this subject. He seemed to think, that no language could ever be so far formed as that it might not be enriched by idioms borrowed from another tongue. I said this was a very dangerous practice; and added that I thought Milton had often injured both his prose and verse by taking this liberty too frequently. I recommended to him the prose works of Dryden as models of pure and native English. I was treading upon tender ground, as I have reason to suppose that he has himself liberally indulged in the practice.

'The same day I dined at Mr Klopstock's, where I had the pleasure of a third interview with the poet. We talked principally about indifferent things. I asked him what he thought of Kant. He said that his reputation was much on the decline in Germany. That for his own part he was not surprised to find it so, as the works of Kant were to him utterly incomprehensible—that he had often been pestered by the Kanteans; but was rarely in the practice of arguing with them. His custom was to produce the book, open it and point to a passage, and beg they would explain it. This they ordinarily attempted to do by substituting their own ideas. "I do not want, I say, an explanation of your own ideas, but of the passage which is before us. In this way I generally bring the dispute to an immediate conclusion." He spoke of Wolff as the first metaphysician they had in Germany. Wolff had followers; but they could hardly be called a sect, and luckily till the appearance of Kant, about fifteen years ago, Germany had not been pestered by any sect of philosophers whatsoever; but that each man had separately pursued his enquiries uncontrolled by the dogmas of a master. Kant had appeared ambitious to be the founder of a sect, that he had succeeded: but that the Germans were now coming to their senses again. That Nicolai and Engel had in different ways contributed to disenchant the nation; but above all the incomprehensibility of the philosopher and his philosophy. He seemed pleased to hear, that as yet Kant's doctrines had not met with many admirers in England—did not doubt but that we had too much wisdom to be duped by a writer who set at defiance the common sense and common understandings of men. We talked of tragedy. He seemed to rate highly the power of exciting tears—I said that nothing was more easy than to deluge an audience, that it was done every day by the meanest writers.'

I must remind you, my friend, first, that these notes, etc. are not intended as specimens of Klopstock's intellectual power, or even 'colloquial prowess,' to judge of which by an accidental conversation, and this with strangers, and those too foreigners, would be not only unreasonable, but calumnious. Secondly, I attribute little other interest to the remarks than what is derived from the celebrity of the person who made them. Lastly, if you ask me, whether I have read the Messiah, and what I think of it? I answer—as yet the first four books only: and as to my opinion (the reasons of which hereafter) you may guess it from what I could not help muttering to myself, when the good pastor this morning told me, that Klopstock was the German Milton—'a very *German* Milton indeed!'—Heaven preserve you, and

S. T. Coleridge

CHAPTER XXIII

Quid quod praefatione praemunierim libellum, quâ conor omnem offendiculi ansam praecidere? Neque quicquam addubito, quin ea candidis omnibus faciat satis. Quid autem facias istis, qui vel ob ingenii pertinaciam sibi satisfieri nolint, vel stupidiores sint quam ut satisfactionem intelligant? Nam quem ad modum Simonides dixit, Thessalos hebetiores esse quam ut possint a se decipi, ita quosdam videas stupidiores quam ut placari queant. Adhaec, non mirum est, invenire quod calumnietur qui nihil aliud quaerit nisi quod calumnietur.

[What has been the use of forearming my little book with a preface in which I try to cut off any handle for the least offence? I have no doubt that it will satisfy everyone who is sincere. What, however, are you to do about those who either refuse to be satisfied out of natural stubbornness or who are too stupid to appreciate their own satisfaction? For even as Simonides said, 'The Thessalians are too dim-witted to be able to be deceived by me,' so you may see that some people are too stupid to be capable of being placated. Besides, it is not surprising that a man who searches only for something to abuse finds nothing else. (*CN*)]

Erasmus, *ad Dorpium*, *Theologum*

IN the rifacciamento [remodelling] of The Friend, I have inserted extracts from the Conciones ad Populum, printed, though scarcely published, in the year 1795, in the very heat and height of my anti-Ministerial enthusiasm: these in proof that my principles of politics have sustained no change.—In the present chapter, I have annexed to my letters from Germany, with particular reference to that, which contains a disquisition on the modern drama, a critique on the tragedy of Bertram, written within the last twelve months:° in proof, that I have been as falsely charged with any fickleness in my principles of taste.—The letter was written to a friend: and the apparent abruptness with which it begins, is owing to the omission of the introductory sentences.

You remember, my dear Sir, that Mr Whitbread, shortly before his death, proposed to the assembled subscribers of Drury Lane Theatre, that the concern should be farmed to some responsible individual under certain conditions and limitations: and that his proposal was rejected, not without indignation, as subversive of the main object, for the attainment of which the enlightened and patriotic assemblage of philodramatists had been induced to risk their subscriptions. Now this object was avowed to be no less than the redemption of the British stage not only from horses, dogs, elephants, and the like zoological rarities, but also from the more pernicious barbarisms and Kotzebuisms in morals and taste.° Drury Lane was to be restored to its former classical renown; Shakespeare, Jonson, and Otway, with the expurgated Muses of Vanburgh, Congreve, and Wycherley, were to be reinaugurated in their rightful dominion over British audiences; and the herculean process was to commence, by exterminating the speaking monsters imported from the banks of the Danube, compared with which their mute relations, the emigrants from Exeter Change, and Polito (late Pidcock's) show-carts, were tame and inoffensive. Could an heroic project, at once so refined and so arduous, be consistently entrusted to, could its success be rationally expected from, a mercenary manager, at whose critical quarantine the *lucri bonus odor* [sweet smell of money] would conciliate a bill of health to the plague in person? No! As the work proposed, such must be the work-masters. Rank, fortune, liberal education, and (their natural accompaniments, or consequences) critical discernment, delicate tact, disinterestedness, unsuspected morals, notorious patriotism, and tried Maecenasship, these were the recommendations that influenced the votes of the proprietary subscribers of Drury Lane Theatre, these the motives that occasioned the election of its Supreme Committee of Management. This circumstance alone would have excited a strong interest in the public mind, respecting the first production of the tragic Muse which had been announced under such auspices, and had passed the ordeal of such judgements: and the tragedy, on which you have requested my judgement, was the work on which the great expectations, justified by so many causes, were doomed at length to settle.

But before I enter on the examination of *Bertram, or the Castle of St Aldobrand*, I shall interpose a few words, on the phrase *German drama*, which I hold to be altogether a misnomer. At the time of Lessing, the German stage, such as it was, appears to have been a flat and servile copy of the French. It was Lessing who first introduced the name and the works of Shakespeare to the admiration of the Germans; and I should not perhaps go too far, if I add, that it was Lessing who first proved to all thinking men, even to Shakespeare's own countrymen, the true nature of his apparent irregularities. These, he demonstrated, were deviations only from the *accidents* of the Greek tragedy; and from such accidents as hung a heavy weight on the wings of the Greek poets, and narrowed their flight within the limits of what we may call the *heroic opera*. He proved, that in all the essentials of art, no less than in the truth of nature, the plays of Shakespeare were incomparably more coincident with the principles of Aristotle, than the productions of Corneille and Racine, notwithstanding the boasted regularity of the latter. Under these convictions, were Lessing's own dramatic works composed. Their deficiency is in depth and in imagination: their excellence is in the construction of the plot; the good sense of the sentiments; the sobriety of the morals; and the high polish of the diction and dialogue. In short, his dramas are the very antipodes of all those which it has been the fashion of late years at once to abuse and to enjoy, under the name of the German drama. Of this latter, Schiller's *Robbers* was the earliest specimen; the first-fruits of his youth (I had almost said of his boyhood) and as such, the pledge, and promise of no ordinary genius. Only as such, did the maturer judgement of the author tolerate the play. During his whole life he expressed himself concerning this production with more than needful asperity, as a monster not less offensive to good taste, than to sound morals; and in his latter years his indignation at the unwonted popularity of the *Robbers* seduced him into the contrary extremes, viz. a studied feebleness of interest (as far as the interest was to be derived from incidents and the excitement of curiosity); a diction elaborately metrical; the affectation of rhymes; and the pedantry of the chorus.

But to understand the true character of the *Robbers*, and of the countless imitations which were its spawn, I must inform you, or at least call to your recollection, that about that time, and for some years before it, three of the most popular books in the German language were, the translations of Young's *Night Thoughts*, Harvey's *Meditations*, and Richardson's *Clarissa Harlowe*. Now we have only to combine the bloated style and peculiar rhythm of Harvey, which is poetic only on account of its utter unfitness for prose, and might as appropriately be called prosaic, from its utter unfitness for poetry; we have only, I repeat, to combine these Harveyisms with the strained thoughts, the figurative metaphysics and solemn epigrams of Young on the one hand; and with the loaded sensibility, the minute detail, the morbid consciousness of every thought and feeling in the whole flux and reflux of the mind, in short the self-involution and dreamlike continuity of Richardson on the other hand; and then to add the horrific incidents, and mysterious villains, (geniuses of supernatural intellect, if you will take the authors' words for it, but on a level with the meanest ruffians of the condemned cells, if we are to judge by their actions and contrivances)—to add the ruined castles, the dungeons, the trapdoors, the skeletons, the flesh-and-blood ghosts, and the perpetual moonshine of a modern author, (themselves the literary brood of the *Castle of Otranto*, the translations of which, with the imitations and improvements aforesaid, were about that time beginning to make as much noise in Germany as their originals were making in England),—and as the compound of these ingredients duly mixed, you will recognize the so called German drama. The *olla podrida* [stew, hotchpotch] thus cooked up, was denounced, by the best critics in Germany, as the mere cramps of weakness, and orgasms of a sickly imagination on the part of the author, and the lowest provocation of torpid feeling on that of the readers. The old blunder however, concerning the irregularity and wildness of Shakespeare, in which the German did but echo the French, who again were but the echoes of our own critics, was still in vogue, and Shakespeare was quoted as authority for the most anti-Shakespearean drama. We have indeed two poets who wrote as one, near the age of Shakespeare, to whom (as the worst characteristic of their writings), the Coryphaeus [leader] of the present drama may challenge the honour of being a poor relation, or impoverished descendant. For if we would charitably consent to forget the comic humour, the wit, the felicities of style, in other words, all the poetry, and nine-tenths of all the genius of Beaumont and Fletcher, that which would remain becomes a Kotzebue.

The so-called German drama, therefore, is English in its origin, English in its materials, and English by readoption; and till we can prove that Kotzebue, or any of the whole breed of Kotzebues, whether dramatists or romantic writers, or writers of romantic dramas, were ever admitted to any other shelf in the libraries of well-educated Germans than were occupied by their originals, and apes' apes in their mother country, we should submit to carry our own brat on our own shoulders; or rather consider it as a lack-grace returned from transportation with such improvements only in growth and manners as young transported convicts usually come home with.

I know nothing that contributes more to a clear insight into the true nature of any literary phenomenon, than the comparison of it with some elder production, the likeness of which is striking, yet only apparent: while the difference is real. In the present case this opportunity is furnished us, by the old Spanish play, entitled *Atheista Fulminato*, formerly, and perhaps still, acted in the churches and monasteries of Spain, and which, under various names (*Don Juan*, *the Libertine*, etc.) has had its day of favour in every country throughout Europe. A popularity so extensive, and of a work so grotesque and extravagant, claims and merits philosophical attention and investigation. The first point to be noticed is, that the play is throughout imaginative. Nothing of it belongs to the real world, but the names of the places and persons. The comic parts, equally with the tragic; the living, equally with the defunct characters, are creatures of the brain; as little amenable to the rules of ordinary probability, as the Satan of *Paradise Lost*, or the Caliban of the *Tempest*, and therefore to be understood and judged of as impersonated abstractions. Rank, fortune, wit, talent, acquired knowledge, and liberal accomplishments, with beauty of person, vigorous health, and constitutional hardihood,—all these advantages, elevated by the habits and sympathies of noble birth and national character, are supposed to have combined in Don Juan, so as to give him the means of carrying into all its practical consequences the doctrine of a godless nature, as the sole ground and efficient cause not only of all things, events, and appearances, but likewise of all our thoughts, sensations, impulses, and actions. Obedience to nature is the only virtue: the gratification of the passions and appetites her only dictate: each individual's self-will the sole organ through which nature utters her commands, and

> Self-contradiction is the only wrong!
> For by the laws of spirit, in the right
> Is every individual character
> That acts in strict consistence with itself.

That speculative opinions, however impious and daring they may be, are not always followed by correspondent conduct, is most true, as well as that they can scarcely in any instance be systematically realized, on account of their unsuitableness to human nature and to the institutions of society. It can be hell, only where it is all hell: and a separate world of devils is necessary for the existence of any one complete devil. But on the other hand it is no less clear, nor, with the biography of Carrier° and his fellow atheists before us, can it be denied without wilful blindness, that the (so-called) system of nature, (i.e. materialism, with the utter rejection of moral responsibility, of a present providence, and of both present and future retribution) may influence the characters and actions of individuals, and even of communities, to a degree that almost does away the distinction between men and devils, and will make the page of the future historian resemble the narration of a madman's dreams. It is not the wickedness of Don Juan, therefore, which constitutes the character an abstraction, and removes it from the rules of probability; but the rapid succession of the correspondent acts and incidents, his intellectual superiority, and the splendid accumulation of his gifts and desirable qualities, as coexistent with entire wickedness in one and the same person. But this likewise is the very circumstance which gives to this strange play its charm and universal interest. Don Juan is, from beginning to end, an intelligible character: as much so as the Satan of Milton. The poet asks only of the reader, what as a poet he is privileged to ask: viz. that sort of negative faith in the existence of such a being, which we willingly give to productions professedly ideal, and a disposition to the same state of feeling, as that with which we contemplate the idealized figures of the Apollo Belvedere, and the Farnese Hercules. What the Hercules is to the eye in corporeal strength, Don Juan is to the mind in strength of character. The ideal consists in the happy balance of the generic with the individual. The former makes the character representative and symbolical, therefore instructive; because, *mutatis mutandis* [changing the things that must be changed], it is applicable to whole classes of men. The latter gives its living interest; for nothing lives or is real, but as definite and individual. To understand this completely, the reader need only recollect the specific state of his feelings, when in looking at a picture of the historic (more properly of the poetic or heroic) class, he objects to a particular figure as being too much of a portrait; and this interruption of his complacency he feels without the least reference to, or the least acquaintance with, any person in real life whom he might recognize in this figure. It is enough that such a figure is not ideal: and therefore not ideal, because one of the two factors or elements of the ideal is in excess. A similar and more powerful objection he would feel towards a set of figures which were mere abstractions, like those of Cipriani, and what have been called Greek forms and faces, i.e. outlines drawn according to a recipe.

These again are not ideal; because in these the other element is in excess. *Forma formans per formam formatam translucens* [the forming form shining through the formed form], is the definition and perfection of ideal art.

This excellence is so happily achieved in the *Don Juan*, that it is capable of interesting without poetry, nay, even without words, as in our pantomime of that name. We see clearly how the character is formed; and the very extravagance of the incidents, and the superhuman entireness of Don Juan's agency, prevents the wickedness from shocking our minds to any painful degree. (We do not believe it enough for this effect; no, not even with that kind of temporary and negative belief or acquiescence which I have described above.) Meantime the qualities of his character are too desirable, too flattering to our pride and our wishes, not to make up on this side as much additional faith as was lost on the other. There is no danger (thinks the spectator or reader) of my becoming such a monster of iniquity as Don Juan! *I* never shall be an atheist! *I* shall never disallow all distinction between right and wrong! *I* have not the least inclination to be so outrageous a drawcansir in my love affairs! But to possess such a power of captivating and enchanting the affections of the other sex! to be capable of inspiring in a charming and even a virtuous woman, a love so deep, and so entirely personal to me! that even my worst vices, (if I *were* vicious) even my cruelty and perfidy, (if I *were* cruel and perfidious) could not eradicate the passion! To be so loved for my own self, that even with a distinct knowledge of my character, she yet died to save me! this, sir, takes hold of two sides of our nature, the better and the worse. For the heroic disinterestedness, to which love can transport a woman, cannot be contemplated without an honourable emotion of reverence towards womanhood: and on the other hand, it is among the mysteries, and abides in the dark groundwork of our nature, to crave an outward confirmation of that something within us, which is our very self, that something, not made up of our qualities and relations, but itself the supporter and substantial basis of all these. Love me, and not my qualities, may be a vicious and an insane wish, but it is not a wish wholly without a meaning.

Without power, virtue would be insufficient and incapable of revealing its being. It would resemble the magic transformation of Tasso's heroine into a tree, in which she could only groan and bleed. (Hence power is necessarily an object of our desire and of our admiration.) But of all power, that of the mind is, on every account, the grand desideratum of human ambition. We shall be as gods in knowledge, was and must have been the first temptation: and the coexistence of great intellectual lordship with guilt has never been adequately represented without exciting the strongest interest, and for this reason, that in this bad and heterogeneous co-ordination we can contemplate the intellect of man more exclusively as a separate self-subsistence, than in its proper state of subordination to his own conscience, or to the will of an infinitely superior being.

This is the sacred charm of Shakespeare's male characters in general. They are all cast in the mould of Shakespeare's own gigantic intellect; and this is the open attraction of his Richard, Iago, Edmund, etc. in particular. But again; of all intellectual power, that of superiority to the fear of the invisible world is the most dazzling. Its influence is abundantly proved by the one circumstance, that it can bribe us into a voluntary submission of our better knowledge, into suspension of all our judgement derived from constant experience, and enable us to peruse with the liveliest interest the wildest tales of ghosts, wizards, genii, and secret talismans. On this propensity, so deeply rooted in our nature, a specific dramatic probability may be raised by a true poet, if the whole of his work be in harmony: a dramatic probability, sufficient for dramatic pleasure, even when the component characters and incidents border on impossibility. The poet does not require us to be awake and believe; he solicits us only to yield ourselves to a dream; and this too with our eyes open, and with our judgement *perdue* [hidden away] behind the curtain, ready to awaken us at the first motion of our will: and meantime, only, not to *dis*believe. And in such a state of mind, who but must be impressed with the cool intrepidity of Don John on the appearance of his father's ghost:°

GHOST. Monster! behold these wounds!
D. JOHN. I do! They were well meant and well performed, I see.
GHOST. Repent, repent of all thy villanies.
My clamorous blood to heaven for vengeance cries,
Heaven will pour out his judgments on you all.
Hell gapes for you, for you each fiend doth call,
And hourly waits your unrepenting fall.
You with eternal horrors they'll torment,
Except of all your crimes you suddenly repent. (*Ghost sinks*)
D. JOHN. Farewell, thou art a foolish ghost. Repent, quoth he! what could this mean? our senses are all in a mist sure.
D. ANTONIO (*one of D. Juan's reprobate companions*). They are not! 'Twas a ghost.

D. LOPEZ (*another reprobate*). I ne'er believed those foolish tales before.

D. JOHN. Come! 'Tis no matter. Let it be what it will, it must be natural.

D. ANT. And nature is unalterable in us too.

D. JOHN. 'Tis true! The nature of a ghost can not change ours.

Who also can deny a portion of sublimity to the tremendous consistency with which he stands out the fearful trial, like a second Prometheus?

Chorus of Devils

STATUE-GHOST. Will you not relent and feel remorse?

D. JOHN. Could'st thou bestow another heart on me I might. But with this heart I have, I can not.

D. LOPEZ. These things are prodigious.

D. ANTON. I have a sort of grudging to relent, but something holds me back.

D. LOP. If we could, 'tis now too late. I will not.

D. ANT. We defy thee!

GHOST. Perish ye impious wretches, go and find the punishments laid up in store for you!

(*Thunder and lightning. D. Lop. and D. Ant. are swallowed up*)

GHOST (*to D. John*). Behold their dreadful fates, and know that thy last moment's come!

D. JOHN. Think not to fright me, foolish ghost; I'll break your marble body in pieces and pull down your horse.

(*Thunder and lightning—chorus of devils, etc.*)

D. JOHN. These things I see with wonder, but no fear.
Were all the elements to be confounded,
And shuffled all into their former chaos;
Were seas of sulphur flaming round about me,
And all mankind roaring within those fires,
I could not fear, or feel the least remorse.
To the last instant I would dare thy power.
Here I stand firm, and all thy threats condemn.
Thy murderer (*to the ghost of one whom he had murdered*) stands here! Now do thy worst!

(*He is swallowed up in a cloud of fire*)

In fine the character of Don John consists in the union of everything desirable to human nature, as means, and which therefore by the well-known law of association become at length desirable on their own account. On their own account, and in their own dignity they are here displayed, as being employed to ends so unhuman, that in the effect, they appear almost as means without an end. The ingredients too are mixed in the happiest proportion, so as to uphold and relieve each other—more especially in that constant interpoise of wit, gaiety, and social generosity, which prevents the criminal, even in his most atrocious moments, from sinking into the mere ruffian, as far at least, as our imagination sits in judgement. Above all, the fine suffusion through the whole, with the characteristic manners and feelings, of a highly bred gentleman gives life to the drama. Thus having invited the statue-ghost of the governor whom he had murdered, to supper, which invitation the marble ghost accepted by a nod of the head, Don John has prepared a banquet.

D. JOHN. Some wine, sirrah! Here's to Don Pedro's ghost—he should have been welcome.

D. LOP. The rascal is afraid of you after death.

(*One knocks hard at the door*)

D. JOHN (*to the servant*). Rise and do your duty.

SERV. Oh the devil, the devil! (*Marble ghost enters*)

D. JOHN. Ha! 'tis the ghost! Let's rise and receive him! Come Governor you are welcome, sit there; if we had thought you would have come, we would have staid for you.

* * * * *

Here Governor, your health! Friends put it about! Here's excellent meat, taste of this ragout. Come, I'll help you, come eat and let old quarrels be forgotten. (*The ghost threatens him with vengeance*)

D. JOHN. We are too much confirmed—curse on this dry discourse. Come here's to your mistress, you had one when you were living: not forgetting your sweet sister. (*Devils enter*)

D. JOHN. Are these some of your retinue? Devils say you? I'm sorry I have no burnt brandy to treat 'em with, that's drink fit for devils. etc.

Nor is the scene from which we quote interesting, in dramatic probability alone; it is susceptible likewise of a sound moral; of a moral that has more than common claims on the notice of a too numerous class, who are ready to receive the qualities of gentlemanly courage, and scrupulous honour (in all the recognized laws of honour,) as the substitutes of virtue, instead of its ornaments. This, indeed, is the moral value of the play at large, and that which places it at a world's distance from the spirit of modern Jacobinism. The latter introduces to us clumsy copies of these showy instrumental qualities, in order to reconcile us to vice and want of principle; while the *Atheista Fulminato* presents an exquisite portraiture of the same qualities, in all their gloss and glow, but presents them for the sole purpose of displaying their hollowness, and in order to put us on our guard by demonstrating their utter indifference to vice and virtue, whenever these, and the like accomplishments are contemplated for themselves alone.

Eighteen years ago I observed,° that the whole secret of the modern Jacobinical drama, (which, and not the German, is its appropriate designation,) and of all its popularity, consists in the confusion and subversion of the natural order of things in their causes and effects: namely, in the excitement of surprise by representing the qualities of liberality, refined feeling, and a nice sense of honour (those things rather which pass amongst us for such) in persons and in classes where experience teaches us least to expect them; and by rewarding with all the sympathies which are the due of virtue, those criminals whom law, reason, and religion have excommunicated from our esteem.

This of itself would lead me back to *Bertram, or the Castle of St Aldobrand*; but, in my own mind, this tragedy was brought into connection with the *Libertine*, (Shadwell's adaptation of the *Atheista Fulminato* to the English stage in the reign of Charles II,) by the fact, that our modern drama is taken, in the substance of it, from the first scene of the third act of the *Libertine*. But with what palpable superiority of judgement in the original! Earth and hell, men and spirits, are up in arms against Don John: the two former acts of the play have not only prepared us for the supernatural, but accustomed us to the prodigious. It is, therefore, neither more nor less than we anticipate when the Captain exclaims: 'In all the dangers I have been, such horrors I never knew. I am quite unmanned'; and when the Hermit says, 'that he had beheld the ocean in wildest rage, yet ne'er before saw a storm so dreadful, such horrid flashes of lightning, and such claps of thunder, were never in my remembrance.' And Don John's burst of startling impiety is equally intelligible in its motive, as dramatic in its effect.

But what is there to account for the prodigy of the tempest at Bertram's shipwreck? It is a mere supernatural effect without even a hint of any supernatural agency; a prodigy without any circumstance mentioned that is prodigious; and a miracle introduced without a ground, and ending without a result. Every event and every scene of the play might have taken place as well if Bertram and his vessel had been driven in by a common hard gale, or from want of provisions. The first act would have indeed lost its greatest and most sonorous picture; a scene for the sake of a scene, without a word spoken; as such, therefore, (a rarity without a precedent) we must take it, and be thankful! In the opinion of not a few, it was, in every sense of the word, the best scene in the play. I am quite certain it was the most innocent; and the steady, quiet uprightness of the flame of the wax candles which the monks held over the roaring billows amid the storm of wind and rain, was really miraculous.

The Sicilian sea-coast: a convent of monks: night: a most portentous, unearthly storm: a vessel is wrecked: contrary to all human expectation, one man saves himself by his prodigious powers as a swimmer, aided by the peculiarity of his destination—

PRIOR. All, all did perish—
1ST MONK. Change, change those drenched weeds—
PRIOR. I wist not of them—every soul did perish—
Enter 3d Monk hastily
3D MONK. No, there was one did battle with the storm
With careless desperate force; full many times
His life was won and lost, as tho' he recked not—
No hand did aid him, and he aided none—
Alone he breasted the broad wave, alone
That man was saved.

Well! This man is led in by the monks, supposed dripping wet, and to very natural enquiries he either remains silent, or gives most brief and surly answers, and after three or four of these half-line courtesies, 'dashing off the monks' who had saved him, he exclaims in the true sublimity of our modern misanthropic heroism—

Off! ye are men—there's poison in your touch.
But I must yield, for this (*What?*) hath left me strengthless.

So end the three first scenes. In the next (the Castle of St Aldobrand), we find the servants there equally frightened with this unearthly storm, though wherein it differed from other violent storms we are not told, except that Hugo informs us, page 9—

PIET. Hugo, well met. Does e'en thy age bear
Memory of so terrible a storm?
HUGO. They have been frequent lately.
PIET. They are ever so in Sicily.
HUGO. So it is said. But storms when I was young
Would still pass o'er like Nature's fitful fevers,
And rendered all more wholesome. Now their rage
Sent thus unseasonable and profitless
Speaks like the threats of heaven.

A most perplexing theory of Sicilian storms is this of old Hugo! and what is very remarkable, not apparently founded on any great familiarity of his own with this troublesome article. For when Pietro asserts the 'ever more frequency' of tempests in Sicily, the old man professes to know nothing more of the fact, but by hearsay. 'So it is said.'—But why he assumed this storm to be unseasonable, and on what he grounded his prophecy (for the storm is still in full fury) that it would be profitless, and without the physical powers common to all other violent seawinds in purifying the atmosphere, we are left in the dark; as well concerning the particular points in which he knew it (during its continuance) to differ from those that he had been acquainted with in his youth. We are at length introduced to the Lady Imogine, who, we learn, had not rested 'through' the night, not on account of the tempest, for

> Long ere the storm arose, her restless gestures
> Forbade all hope to see her blest with sleep.

Sitting at a table, and looking at a portrait, she informs us—First, that portrait-painters may make a portrait from memory—

> The limner's art may trace the absent feature.

For surely these words could never mean, that a painter may have a person sit to him who afterwards may leave the room or perhaps the country? Second, that a portrait-painter can enable a mourning lady to possess a good likeness of her absent lover, but that the portrait-painter cannot, and who shall—

> Restore the *scenes* in which they met and parted?

The natural answer would have been—Why the scene-painter to be sure! But this unreasonable lady requires in addition sundry things to be painted that have neither lines nor colours—

> The thoughts, the recollections sweet and bitter,
> Or the Elysian dreams of lovers when they loved.

Which last sentence must be supposed to mean; when they were present, and making love to each other.—Then, if this portrait could speak, it would 'acquit the faith of womankind.' How? Had she remained constant? No, she has been married to another man, whose wife she now is. How then? Why, that, in spite of her marriage vow, she had continued to yearn and crave for her former lover—

> This has her body, that her mind:
> Which has the better bargain?

The lover, however, was not contented with this precious

arrangement, as we shall soon find. The lady proceeds to inform us, that during the many years of their separation, there have happened in the different parts of the world, a number of 'such things'; even such, as in a course of years always have, and till the Millennium, doubtless always will happen somewhere or other. Yet this passage, both in language and in metre, is perhaps among the best parts of the play. The lady's loved companion and most esteemed attendant, Clotilda, now enters and explains this love and esteem by proving herself a most passive and dispassionate listener, as well as a brief and lucky querist, who asks by chance, questions that we should have thought made for the very sake of the answers. In short, she very much reminds us of those puppet heroines, for whom the showman contrives to dialogue without any skill in ventriloquism. This, notwithstanding, is the best scene in the play, and though crowded with solecisms, corrupt diction, and offences against metre, would possess merits sufficient to outweigh them, if we could suspend the moral sense during the perusal. It tells well and passionately the preliminary circumstances, and thus overcomes the main difficulty of most first acts, viz. that of retrospective narration. It tells us of her having been honourably addressed by a noble youth, of rank and fortune vastly superior to her own: of their mutual love, heightened on her part by gratitude; of his loss of his sovereign's favour: his disgrace; attainder; and flight; that he (thus degraded) sank into a vile ruffian, the chieftain of a murderous banditti [company of bandits]; and that from the habitual indulgence of the most reprobate habits and ferocious passions, he had become so changed, even in his appearance and features,

> That she who bore him had recoiled from him,
> Nor known the alien visage of her child,
> Yet still *she* (Imogine) lov'd him.

She is compelled by the silent entreaties of a father, perishing with 'bitter shameful want on the cold earth,' to give her hand, with a heart thus irrecoverably pre-engaged, to Lord Aldobrand, the enemy of her lover, even to the very man who had baffled his ambitious schemes, and was, at the present time, entrusted with the execution of the sentence of death which had been passed on Bertram. Now, the proof of 'woman's love,' so industriously held forth for the sympathy, if not the esteem of the audience, consists in this, that though Bertram had become a robber and a murderer by trade, a ruffian in manners, yea, with form and features at which his own mother could not but 'recoil,' yet she (Lady Imogine) 'the wife of a most noble, honoured

Lord,' estimable as a man, exemplary and affectionate as a husband, and the fond father of her only child—that she, notwithstanding all this, striking her heart, dares to say to it—

> But thou art Bertram's still, and Bertram's ever.

A monk now enters, and entreats in his prior's name for the wonted hospitality, and 'free noble usage' of the Castle of St Aldobrand for some wretched shipwrecked souls, and from this we learn, for the first time, to our infinite surprise, that notwithstanding the supernaturalness of the storm aforesaid, not only Bertram, but the whole of his gang, had been saved, by what means we are left to conjecture, and can only conclude that they had all the same desperate swimming powers, and the same saving destiny as the hero, Bertram himself. So ends the first act, and with it the tale of the events, both those with which the tragedy begins, and those which had occurred previous to the date of its commencement. The second displays Bertram in disturbed sleep, which the Prior who hangs over him prefers calling a 'starting trance,' and with a strained voice, that would have awakened one of the seven sleepers, observes to the audience—

> How the lip works! How the bare teeth *do* grind!
> And beaded drops course* down his writhen brow!

The dramatic effect of which passage we not only concede to the admirers of this tragedy, but acknowledge the further advantage of preparing the audience for the most surprising series of wry faces, proflated mouths, and lunatic gestures that were ever 'launched' on an audience to 'sear the sense.'†

* . . . The big round tears
Coursed one another down his innocent nose
In piteous chase,

says Shakespeare of a wounded stag hanging its head over a stream: naturally, from the position of the head, and most beautifully, from the association of the preceding image, of the chase, in which 'the poor sequester'd stag from the hunter's aim had ta'en a hurt.' In the supposed position of Bertram, the metaphor, if not false, loses all the propriety of the original.

† Among a number of other instances of words chosen without reason, Imogine in the first act declares, that thunderstorms were not able to intercept her prayers for 'the desperate man, in desperate *ways* who *dealt*'—

> Yea, when the launched bolt did sear her sense,
> Her soul's deep orisons were breathed for him;

i.e. when a red-hot bolt launched at her from a thunder-cloud had cauterized her sense, in plain English, burnt her eyes out of her head, she kept still praying on.

> Was not *this* love? Yes, thus doth woman love!

PRIOR. I will awake him from this *horrid trance*,
This is no natural sleep! Ho, *wake thee*, stranger!

This is rather a whimsical application of the verb reflex we must confess, though we remember a similar transfer of the agent to the patient in a manuscript tragedy, in which the Bertram of the piece, prostrating a man with a single blow of his fist exclaims—'Knock me thee down, then ask thee if thou liv'st.'—Well; the stranger obeys, and whatever his sleep might have been, his waking was perfectly natural, for lethargy itself could not withstand the scolding stentorship of Mr Holland, the Prior. We next learn from the best authority, his own confession, that the misanthropic hero, whose destiny was incompatible with drowning, is Count Bertram, who not only reveals his past fortunes, but avows with open atrocity, his satanic hatred of Imogine's lord, and his frantic thirst of revenge; and so the raving character raves, and the scolding character scolds—and what else? Does not the Prior act? Does he not send for a posse of constables or thief-takers to handcuff the villain, and take him either to Bedlam or Newgate? Nothing of the kind; the author preserves the unity of character, and the scolding Prior from first to last does nothing but scold, with the exception indeed of the last scene of the last act, in which with a most surprising revolution he whines, weeps and kneels to the condemned blaspheming assassin out of pure affection to the high-hearted man, the sublimity of whose angel-sin rivals the star-bright apostate, (i.e. who was as proud as Lucifer, and as wicked as the Devil) and, 'had thrilled him,' (Prior Holland aforesaid) with wild admiration.

Accordingly in the very next scene, we have this tragic Macheath,° with his whole gang, in the castle of St Aldobrand, without any attempt on the Prior's part either to prevent him, or to put the mistress and servants of the castle on their guard against their new inmates, though he (the Prior) knew, and confesses that he knew that Bertram's 'fearful mates' were assassins so habituated and naturalized to guilt, that—

> When their drenched hold forsook both gold and gear,
> They gripped their daggers with a murderer's instinct;

and though he also knew, that Bertram was the leader of a band whose trade was blood. To the castle however he goes, thus with the holy Prior's consent, if not with his assistance; and thither let us follow him.

No sooner is our hero safely housed in the castle of St Aldobrand, than he attracts the notice of the lady and her confidante, by his 'wild and terrible dark eyes,' 'muffled form,' 'fearful form,'* 'darkly wild,' 'proudly stern,' and the like commonplace indefinites, seasoned by merely verbal antitheses, and at best, copied with very slight change, from the Conrade of Southey's Joan of Arc. The lady Imogine, who has been (as is the case, she tells us, with all soft and solemn spirits,) worshipping the moon on a terrace or rampart within view of the castle, insists on having an interview with our hero, and this too tête-à-tête. Would the reader learn why and wherefore the confidante is excluded, who very properly remonstrates against such 'conference, alone, at night, with one who bears such fearful form,' the reason follows—'why, *therefore* send him!' I say, follows, because the next line, 'all things of fear have lost their power over me,' is separated from the former by a break or pause, and besides that it is a very poor answer to the danger, is no answer at all to the gross indelicacy of this wilful exposure. We must therefore regard it as a mere afterthought, that a little softens the rudeness, but adds nothing to the weight of that exquisite woman's reason aforesaid. And so exit Clotilda and enter Bertram, who 'stands without looking at her,' that is, with his lower limbs forked, his arms akimbo, his side to the lady's front, the whole figure resembling an inverted Y. He is soon however roused from the state surly to the state frantic, and then follow raving, yelling, cursing, she fainting, he relenting, in runs Imogine's child, squeaks 'mother!' He snatches it up, and with a 'God bless thee, child! Bertram has kissed thy child,'—the curtain drops. The third act is short, and short be our account of it. It introduces Lord St Aldobrand on his road homeward, and next Imogine in the convent, confessing the foulness of her heart to the Prior, who first indulges his old humour with a fit of senseless scolding, then leaves her alone with her ruffian paramour, with whom she makes at once an infamous appointment, and the curtain drops, that it may be carried into act and consummation.

* This sort of repetition is one of this writer's peculiarities, and there is scarce a page which does not furnish one or more instances—Ex. gr. in the first page or two. Act I, line 7th, 'and *deemed* that I might sleep.'—Line 10, 'Did rock and *quiver* in the bickering *glare*.'—Lines 14, 15, 16, 'But by the momently *gleams* of sheeted blue, Did the pale marbles *glare* so *sternly* on me, I almost *deemed* they lived.'—Line 37, 'The *glare* of Hell.—Line 35, 'O holy Prior, this is no *earthly storm*.'—Line 38, 'This is no *earthly storm*.'—Line 42, '*Dealing* with us.'—Line 43, '*Deal* thus sternly.'—Line 44, 'Speak! thou hast *something seen!*'—'A *fearful sight!*'—Line 45, 'What hast thou *seen?* A piteous, *fearful sight*.'—Line 48, '*quivering gleams*.'—Line 50, 'In the hollow *pauses of the storm*.'—Line 61, 'The *pauses of the storm*, etc.'

I want words to describe the mingled horror and disgust, with which I witnessed the opening of the fourth act, considering it as a melancholy proof of the depravation of the public mind. The shocking spirit of Jacobinism seemed no longer confined to politics. The familiarity with atrocious events and characters appeared to have poisoned the taste, even where it had not directly disorganized the moral principles, and left the feelings callous to all the mild appeals, and craving alone for the grossest and most outrageous stimulants. The very fact then present to our senses, that a British audience could remain passive under such an insult to common decency, nay, receive with a thunder of applause, a human being supposed to have come reeking from the consummation of this complex foulness and baseness, these and the like reflections so pressed as with the weight of lead upon my heart, that actor, author, and tragedy would have been forgotten, had it not been for a plain elderly man sitting beside me, who with a very serious face, that at once expressed surprise and aversion, touched my elbow, and pointing to the actor, said to me in a half-whisper—'Do you see that little fellow there? he has just been committing adultery!' Somewhat relieved by the laugh which this droll address occasioned, I forced back my attention to the stage sufficiently to learn, that Bertram is recovered from a transient fit of remorse, by the information that St Aldobrand was commissioned (to do, what every honest man must have done without commission, if he did his duty) to seize him and deliver him to the just vengeance of the law; an information which (as he had long known himself to be an attainted traitor and proclaimed outlaw, and not only a trader in blood himself, but notoriously the captain of a gang of thieves, pirates and assassins) assuredly could not have been new to him. It is this, however, which alone and instantly restores him to his accustomed state of raving, blasphemy, and nonsense. Next follows Imogine's constrained interview with her injured husband, and his sudden departure again, all in love and kindness, in order to attend the feast of St Anselm at the convent. This was, it must be owned, a very strange engagement for so tender a husband to make within a few minutes after so long an absence. But first his lady has told him that she has 'a vow on her,' and wishes 'that black perdition may gulf her perjured soul,'—(Note: she is lying at the very time)—if she ascends his bed, till her penance is accomplished. How, therefore, is the poor husband to amuse himself in this interval of her penance? But do not be distressed, reader, on account of the St Aldobrand's absence! As the author has contrived to send him out of the house, when a husband would be in his, and the lover's way, so he will doubtless not be at a loss to bring him back again as soon as he is wanted. Well! the husband gone in on the one side, out pops the lover from the other, and for the fiendish purpose of harrowing up the soul of his wretched accomplice in guilt, by announcing to her with most brutal and blasphemous execrations his fixed and deliberate resolve to assassinate her husband; all this too is for no discoverable purpose on the part of the author, but that of introducing a series of super-tragic starts, pauses, screams, struggling, dagger-throwing, falling on the ground, starting up again wildly, swearing, outcries for help, falling again on the ground, rising again, faintly tottering towards the door, and, to end the scene, a most convenient fainting fit of our lady's, just in time to give Bertram an opportunity of seeking the object of his hatred, before she alarms the house, which indeed she has had full time to have done before, but that the author rather chose she should amuse herself and the audience by the above-described ravings and startings. She recovers slowly, and to her enter Clotilda, the confidante and mother confessor; then commences, what in theatrical language is called the madness, but which the author more accurately entitles, delirium, it appearing indeed a sort of intermittent fever with fits of light-headedness off and on, whenever occasion and stage effect happen to call for it. A convenient return of the storm (we told the reader beforehand how it would be) had changed—

> The rivulet, that bathed the Convent walls,
> Into a foaming flood: upon its brink
> The Lord and his small train *do* stand appalled.
> With torch and bell from their high battlements
> The monks *do* summon to the pass in vain;
> He must return to-night.—

Talk of the devil, and his horns appear, says the proverb: and sure enough, within ten lines of the exit of the messenger, sent to stop him, the arrival of Lord St Aldobrand is announced. Bertram's ruffian band now enter, and range themselves across the stage, giving fresh cause for Imogine's screams and madness. St Aldobrand having received his mortal wound behind the scenes, totters in to welter in his blood, and to die at the feet of this double-damned adultress.

Of her, as far as she is concerned in this 4th act, we have two additional points to notice: first, the low cunning and Jesuitical trick with which she deludes her husband into words of forgiveness, which he himself does not understand; and secondly, that everywhere she is made the object of interest and sympathy, and it is not the author's fault, if at any moment she excites feelings less gentle, than those we are accustomed to associate with the self-accusations of a sincere, religious penitent. And did a British audience endure all this?—They received it with plaudits, which, but for the rivalry of the carts and hackney coaches, might have disturbed the evening prayers of the scanty week-day congregation at St Paul's cathedral.

Tempora mutantur, nos et mutamur in illis.

[Times change, and we change with them.]

Of the fifth act, the only thing noticeable (for rant and nonsense, though abundant as ever, have long before the last act become things of course), is the profane representation of the high altar in a chapel, with all the vessels and other preparations for the Holy Sacrament. A hymn is actually sung on the stage by the chorister boys! For the rest, Imogine, who now and then talks deliriously, but who is always light-headed as far as her gown and hair can make her so, wanders about in dark woods with cavern rocks and precipices in the back-scene; and a number of mute dramatis personae move in and out continually, for whose presence, there is always at least this reason, that they afford something to be *seen*, by that very large part of a Drury Lane audience who have small chance of *hearing* a word. She had, it appears, taken her child with her, but what becomes of the child, whether she murdered it or not, nobody can tell, nobody can learn; it was a riddle at the representation, and after a most attentive perusal of the play, a riddle it remains.

No more I know, I wish I did,
And I would tell it all to you;
For what became of this poor child
There's none that ever knew.

Wordsworth's Thorn

Our whole information* is derived from the following words—

PRIOR. Where is thy child?
CLOTIL. (*pointing to the cavern into which she has looked*). Oh he lies cold within his cavern-tomb!
Why dost thou urge her with the horrid theme?

* The child is an important personage, for I see not by what possible means the author could have ended the second and third acts but for its timely appearance. How ungrateful then not further to notice its fate?

PRIOR (who will not, the reader may observe, be disappointed of his dose of scolding).

It was to make (*quere* wake)° one living cord o' th' heart,
And I will try, tho' my own breaks at it.
Where is thy child?

IMOG. (*with a frantic laugh*). The forest-fiend hath snatched him—
He (who? the fiend or the child?) rides the night-mare thro' the wizzard woods.

Now these two lines consist in a senseless plagiarism from the counterfeited madness of Edgar in Lear, who, in imitation of the gipsy incantations, puns on the old word Mair, a hag; and the no less senseless adoption of Dryden's forest-fiend, and the wizzard-stream by which Milton, in his Lycidas, so finely characterizes the spreading Deva [Dee], *fabulosus Amnis* [the fabled Torrent]. Observe too these images stand unique in the speeches of Imogine, without the slightest resemblance to anything she says before or after. But we are weary. The characters in this act frisk about, here, there, and everywhere, as teasingly as the Jack o' Lanthorn-lights which mischievous boys, from across a narrow street, throw with a looking-glass on the faces of their opposite neighbours. Bertram disarmed, out-heroding Charles de Moor in the Robbers, befaces the collected knights of St Anselm (all in complete armour,) and so, by pure dint of black looks, he outdares them into passive poltroons. The sudden revolution in the Prior's manners we have before noticed, and it is indeed so *outré*, that a number of the audience imagined a great secret was to come out, viz.: that the Prior was one of the many instances of a youthful sinner metamorphosed into an old scold, and that this Bertram would appear at last to be his son. Imogine reappears at the convent, and dies of her own accord. Bertram stabs himself, and dies by her side, and that the play may conclude as it began, viz. in a superfetation of blasphemy upon nonsense, because he had snatched a sword from a despicable coward, who retreats in terror when it is pointed towards him in sport; this *felo de se* [self-murderer], and thief-captain, this loathsome and leprous confluence of robbery, adultery, murder, and cowardly assassination, this monster whose best deed is, the having saved his betters from the degradation of hanging him, by turning jack ketch [executioner] to himself, first recommends the charitable monks and holy Prior to pray for his soul, and then has the folly and impudence to exclaim—

I died no felon's death,
A warrior's weapon freed a warrior's soul!—

CHAPTER XXIV

CONCLUSION

It sometimes happens that we are punished for our faults by incidents, in the causation of which these faults had no share: and this I have always felt the severest punishment. The wound indeed is of the same dimensions; but the edges are jagged, and there is a dull underpain that survives the smart which it had aggravated. For there is always a consolatory feeling that accompanies the sense of a proportion between antecedents and consequents. The sense of before and after becomes both intelligible and intellectual when, and only when, we contemplate the succession in the relations of cause and effect, which like the two poles of the magnet manifest the being and unity of the one power by relative opposites, and give, as it were, a substratum of permanence, of identity, and therefore of reality, to the shadowy flux of time. It is eternity revealing itself in the phenomena of time: and the perception and acknowledgement of the proportionality and appropriateness of the present to the past, prove to the afflicted soul, that it has not yet been deprived of the sight of God, that it can still recognize the effective presence of a Father, though through a darkened glass and a turbid atmosphere, though of a Father that is chastising it. And for this cause, doubtless, are we so framed in mind, and even so organized in brain and nerve, that all confusion is painful.—It is within the experience of many medical practitioners, that a patient, with strange and unusual symptoms of disease, has been more distressed in mind, more wretched, from the fact of being unintelligible to himself and others, than from the pain or danger of the disease: nay, that the patient has received the most solid comfort, and resumed a genial and enduring cheerfulness, from some new symptom or product, that had at once determined the name and nature of his complaint, and rendered it an intelligible effect of an intelligible cause: even though the discovery did at the same moment preclude all hope of restoration. Hence the mystic theologians, whose delusions we may more confidently hope to separate from their actual intuitions, when we condescend to read their works without the presumption that whatever our fancy (always the ape, and too often the adulterator and counterfeit of our memory) has not made or cannot make a picture of, must be nonsense,—hence, I say, the mystics have joined in representing the state of the reprobate spirits as a dreadful dream in which there is no sense of reality, not even of the pangs they are enduring—an eternity without time, and as it were below it—God present without manifestation of his presence. But these are depths, which we dare not linger over. Let us turn to an instance more on a level with the ordinary sympathies of mankind. Here then, and in this same healing influence of light and distinct beholding, we may detect the final cause of that instinct which in the great majority of instances leads and almost compels the afflicted to communicate their sorrows. Hence too flows the alleviation that results from 'opening out our griefs': which are thus presented in distinguishable forms instead of the mist, through which whatever is shapeless becomes magnified and (literally) enormous. Casimir, in the fifth Ode of his third Book, has happily* expressed this thought.

Me longus silendi
Edit amor; facilesque Luctus

Hausit medullas. Fugerit ocius,
Simul negantem visere jusseris
Aures amicorum, et loquacem
Questibus evacuâris iram.

Olim querendo desinimus queri,
Ipsoque fletu lacryma perditur,
Nec fortis aequè, si per omnes
Cura volet residetque ramos.

Vires amicis perdit in auribus
Minorque semper dividitur dolor
Per multa permissus vagari
Pectora.—

* Classically too, as far as consists with the allegorizing fancy of the modern, that still striving to project the inward, contradistinguishes itself from the seeming ease with which the poetry of the ancients reflects the world without. Casimir affords, perhaps, the most striking instance of this characteristic difference.—For his style and diction are really classical: while Cowley, who resembles Casimir in many respects, completely barbarizes his Latinity, and even his metre, by the heterogeneous nature of his thoughts. That Dr Johnson should have passed a contrary judgement, and have even preferred Cowley's Latin poems to Milton's, is a caprice that has, if I mistake not, excited the surprise of all scholars. I was much amused last summer with the laughable affright, with which an Italian poet perused a page of Cowley's Davideis, contrasted with the enthusiasm with which he first ran through, and then read aloud, Milton's Mansus and Ad Patrem.

[A long-lasting passion for silence has devoured [you] and grief has consumed [your] soft marrows. It will swiftly flee as soon as you command it, refuse though it may, to visit the ears of your friends and empty out your anger in talk and lamentations. Often by complaining we cease to complain, the tear dries in the very act of weeping, nor is care so strong if it takes wing and settles on all the branches. Pain loses strength in loving ears and ever grows less by division, allowed to wander through many breasts. (*CC*)]

Id. Lib. iii. Od. 5

I shall not make this an excuse, however, for troubling my readers with any complaints or explanations, with which, as readers, they have little or no concern. It may suffice (for the present at least) to declare that the causes that have delayed the publication of these volumes for so long a period after they had been printed off, were not connected with any neglect of my own; and that they would form an instructive comment on the chapter [ch. XI] concerning authorship as a trade, addressed to young men of genius in the first volume of this work. I remember the ludicrous effect which the first sentence of an autobiography, which happily for the writer was as meagre in incidents as it is well possible for the life of an individual to be—'The *eventful* life which I am about to record, from the hour in which I rose into existence on this planet, etc.' Yet when, notwithstanding this warning example of self-importance before me, I review my own life, I cannot refrain from applying the same epithet to it, and with more than ordinary emphasis—and no private feeling, that affected myself only, should prevent me from publishing the same, (for write it I assuredly shall, should life and leisure be granted me) if continued reflection should strengthen my present belief, that my history would add its contingent to the enforcement of one important truth, viz. that we must not only love our neighbours as ourselves, but ourselves likewise as our neighbours; and that we can do neither unless we love God above both.

Who lives, that's not
Depraved or depraves? Who dies, that bears
Not one spurn to the grave—of their friends' gift?

Strange as the delusion may appear, yet it is most true that three years ago I did not know or believe that I had an enemy in the world: and now even my strongest sensations of gratitude are mingled with fear, and I reproach myself for being too often disposed to ask,—Have I one friend?—During the many years which intervened between the composition and the publication of the Christabel, it became almost as well known among literary men as if it had been on common sale, the same references were made to it, and the same liberties taken with it, even to the very names of the imaginary persons in the poem. From almost all of our most celebrated poets, and from some with whom I had no personal acquaintance, I either received or heard of expressions of admiration that (I can truly say) appeared to myself utterly disproportionate to a work, that pretended to be nothing more than a common fairy-tale. Many, who had allowed no merit to my other poems, whether printed or manuscript, and who have frankly told me as much, uniformly made an exception in favour of the Christabel and the poem, entitled Love. Year after year, and in societies of the most different kinds, I had been entreated to recite it: and the result was still the same in all, and altogether different in this respect from the effect produced by the occasional recitation of any other poems I had composed.—This before the publication. And since then, with very few exceptions, I have heard nothing but abuse, and this too in a spirit of bitterness at least as disproportionate to the pretensions of the poem, had it been the most pitiably below mediocrity, as the previous eulogies, and far more inexplicable. In the Edinburgh Review it was assailed with a malignity and a spirit of personal hatred that ought to have injured only the work in which such a tirade was suffered to appear: and this review was generally attributed (whether rightly or no I know not) to a man, who both in my presence and in my absence, has repeatedly pronounced it the finest poem of its kind in the language.—This may serve as a warning to authors, that in their calculations on the probable reception of a poem, they must subtract to a large amount from the panegyric, which may have encouraged them to publish it, however unsuspicious and however various the sources of this panegyric may have been. And, first, allowances must be made for private enmity, of the very existence of which they had perhaps entertained no suspicion—for personal enmity behind the mask of anonymous criticism: secondly, for the necessity of a certain proportion of abuse and ridicule in a review, in order to make it saleable, in consequence of which, if they have no friends behind the scenes, the chance must needs be against them; but lastly and chiefly, for the excitement and temporary sympathy of feeling, which the recitation of the poem by an admirer, especially if he be at once a warm admirer and a man of acknowledged celebrity, calls forth in the audience. For this is really a species of animal magnetism,° in which the enkindling reciter, by perpetual comment of looks and tones, lends his own will and apprehensive faculty to his auditors. They live for the time within the dilated sphere of his intellectual being. It is equally possible, though not equally common, that a reader left to himself should sink below the poem, as that the poem left to itself should flag beneath the feelings of the reader.—But in my own instance, I had the additional misfortune of having been gossiped about, as devoted to metaphysics, and worse than all to a system incomparably nearer to the visionary flights of Plato, and even to the jargon of the mystics, than to the established tenets of Locke. Whatever therefore appeared with my name was condemned beforehand, as predestined metaphysics. In a dramatic poem, which had been submitted by me to a gentleman of great influence in the theatrical world, occurred the following passage.—

> O we are querulous creatures! Little less
> Then all things can suffice to make us happy:
> And little more than nothing is enough
> To make us wretched.

Aye, here now! (exclaimed the critic) here come Coleridge's metaphysics! And the very same motive (that is, not that the lines were unfit for the present state of our immense theatres; but that they were metaphysics*) was assigned elsewhere for the rejection of the two following passages. The first is spoken in answer to a usurper, who had rested his plea on the circumstance, that he had been chosen by the acclamations of the people.—

> What people? How conven'd? Or if conven'd,
> Must not that magic power that charms together
> Millions of men in council, needs have power
> To win or wield them? Rather, O far rather,
> Shout forth thy titles to yon circling mountains,
> And with a thousandfold reverberation
> Make the rocks flatter thee, and the volleying air,
> Unbribed, shout back to thee, King Emerich!
> By wholesome laws to embank the Sovereign Power;
> To deepen by restraint; and by prevention
> Of lawlesss will to amass and guide the flood
> In its majestic channel, is man's task

* Poor unlucky metaphysics! and what are they? A single sentence expresses the object and thereby the contents of this science. *Γνῶθι σέαυτον*: *et Deum quantum licet et in Deo omnia scibis*. Know thyself: and so shalt thou know God, as far as is permitted to a creature, and in God all things.—Surely, there is a strange—nay, rather a too natural—aversion in many to know themselves.

And the true patriot's glory! In all else
Men safelier trust to heaven, than to themselves
When least themselves: even in those whirling crowds
Where folly is contagious, and too oft
Even wise men leave their better sense at home
To chide and wonder at them, when return'd.

The second passage is in the mouth of an old and experienced courtier, betrayed by the man in whom he had most trusted.

And yet Sarolta, simple, inexperienced,
Could see him as he was and oft has warn'd me.
Whence learnt she this? O she was innocent.
And to be innocent is Nature's wisdom.
The fledge dove knows the prowlers of the air
Fear'd soon as seen, and flutters back to shelter!
And the young steed recoils upon his haunches,
The never-yet-seen adder's hiss first heard!
Ah! surer than suspicion's hundred eyes
Is that fine sense, which to the pure in heart
By mere oppugnancy of their own goodness
Reveals the approach of evil!

As therefore my character as a writer could not easily be more injured by an overt act than it was already in consequence of the report, I published a work, a large portion of which was professedly metaphysical. A long delay occurred between its first annunciation and its appearance; it was reviewed therefore by anticipation with a malignity, so avowedly and exclusively personal, as is, I believe, unprecedented even in the present contempt of all common humanity that disgraces and endangers the liberty of the press. After its appearance, the author of this lampoon was chosen to review it in the Edinburgh Review: and under the single condition, that he should have written what he himself really thought, and have criticized the work as he would have done had its author been indifferent to him, I should have chosen that man myself both from the vigour and the originality of his mind, and from his particular acuteness in speculative reasoning, before all others.—I remembered Catullus's lines,

Desine de quoquam quicquam bene velle mereri,
  Aut aliquem fieri posse putare pium.
Omnia sunt ingrata: nihil fecisse benigne est:
  Imo, etiam taedet, taedet obestque magis.

Ut mihi, quem nemo gravius nec acerbius urget
Quam modo qui me unum atque unicum amicum habuit.

[Leave off wishing to deserve any thanks from anyone, or thinking that anyone can ever become grateful. All this wins no thanks; to have acted kindly is nothing, rather it is wearisome, wearisome and harmful; so is it now with me, who am vexed and troubled by no one so bitterly as by him who but now held me for his one and only friend. (LCL)]

But I can truly say, that the grief with which I read this rhapsody of predetermined insult, had the rhapsodist himself for its whole and sole object: and that the indignant contempt which it excited in me, was as exclusively confined to his employer and suborner. I refer to this review at present, in consequence of information having been given me, that the innuendo of my 'potential infidelity,' grounded on one passage of my first Lay Sermon, has been received and propagated with a degree of credence, of which I can safely acquit the originator of the calumny. I give the sentences as they stand in the sermon, premising only that I was speaking exclusively of miracles worked for the outward senses of men. 'It was only to overthrow the usurpation exercised in and through the senses, that the senses were miraculously appealed to. Reason and religion are their own evidence. The natural sun is in this respect a symbol of the spiritual. Ere he is fully arisen, and while his glories are still under veil, he calls up the breeze to chase away the usurping vapours of the night-season, and thus converts the air itself into the minister of its own purification: not surely in proof or elucidation of the light from heaven, but to prevent its interception.

'Wherever, therefore, similar circumstances coexist with the same moral causes, the principles revealed, and the examples recorded, in the inspired writings render miracles superfluous: and if we neglect to apply truths in expectation of wonders, or under pretext of the cessation of the latter, we tempt God and merit the same reply which our Lord gave to the Pharisees on a like occasion.'

In the sermon and the notes both the historical truth and the necessity of the miracles are strongly and frequently asserted. 'The testimony of books of history (i.e. relatively to the signs and wonders, with which Christ came) is one of the strong and stately pillars of the Church; but it is not the foundation!' Instead, therefore, of defending myself, which I could easily effect by a series of passages, expressing the same opinion, from the Fathers and the most eminent Protestant divines, from the Reformation to the Revolution, I shall merely state what my belief is, concerning the true evidences of Christianity. 1. Its consistency with right reason, I consider as the outer court of the temple—the common area, within which it stands. 2. The miracles, with and through which the religion was first revealed and attested, I regard as the steps, the vestibule, and the portal of the temple. 3. The sense, the inward feeling, in the soul of each believer of its exceeding desirableness—the experience, that he needs something, joined with the strong foretokening, that the redemption and the graces propounded to us in Christ are what he needs—this I hold to be the true foundation of the spiritual edifice. With the strong a priori probability that flows in from 1 and 3 on the correspondent historical evidence of 2, no man can refuse or neglect to make the experiment without guilt. But, 4, it is the experience derived from a practical conformity to the conditions of the Gospel—it is the opening eye; the dawning light; the terrors and the promises of spiritual growth; the blessedness of loving God as God, the nascent sense of sin hated as sin, and of the incapability of attaining to either without Christ; it is the sorrow that still rises up from beneath and the consolation that meets it from above; the bosom treacheries of the principal in the warfare and the exceeding faithfulness and long suffering of the uninterested ally;—in a word, it is the actual trial of the faith in Christ, with its accompaniments and results, that must form the arched roof, and the faith itself is the completing keystone. In order to an efficient belief in Christianity, a man must have been a Christian, and this is the seeming *argumentum in circulo* [circular argument], incident to all spiritual truths, to every subject not presentable under the forms of time and space, as long as we attempt to master by the reflex acts of the understanding what we can only know by the act of becoming. 'Do the will of my father, and ye shall know whether I am of God.' These four evidences I believe to have been and still to be, for the world, for the whole Church, all necessary, all equally necessary; but that at present, and for the majority of Christians born in Christian countries, I believe the third and the fourth evidences to be the most operative, not as superseding but as involving a glad undoubting faith in the two former. *Credidi, ideóque intellexi* [I believed, and therefore I understood], appears to me the dictate equally of philosophy and religion, even as I believe redemption to be the antecedent of sanctification, and not its consequent. All spiritual predicates may be construed indifferently as modes of action or as states of being. Thus holiness and blessedness are the same idea, now seen in relation to act and now to existence. The ready belief which has been yielded to the slander of my 'potential infidelity,' I attribute in part to the openness with which

I have avowed my doubts, whether the heavy interdict, under which the name of Benedict Spinoza lies, is merited on the whole or to the whole extent. Be this as it may, I wish, however, that I could find in the books of philosophy, theoretical or moral, which are alone recommended to the present students of theology in our established schools, a few passages as thoroughly Pauline, as completely accordant with the doctrines of the established Church, as the following sentences in the concluding page of Spinoza's Ethics. 'Deinde quó mens amore divino seu beatitudine magis gaudet, eó plus *intelligit*, eó majorem in affectus habet potentiam, et eó minus ab affectibus, qui mali sunt, patitur: atque adeò ex eo, quód mens hoc amore divino seu beatitudine gaudet, potestatem habet libidines coercendi, nemo beatitudine gaudet quia affectus coercuit; sed contra potestas libidines coercendi ex ipsâ beatitudine oritur [Again, in proportion as the mind rejoices more in this divine love or blessedness, so does it the more *understand*, so much the more power has it over the emotions, and so much the less is it subject to those emotions which are evil; and so much is it true that from the fact that the mind rejoices in this divine love or blessedness, it therefore has the power of controlling lusts, that no one rejoices in the blessedness because it has controlled the emotions, but, on the contrary, the power of controlling lusts arises from the blessedness itself]'.

With regard to the Unitarians, it has been shamelessly asserted, that I have denied them to be Christians. God forbid! For how should I know, what the piety of the heart may be, or what quantum [amount] of error in the understanding may consist with a saving faith in the intentions and actual dispositions of the whole moral being in any one individual? Never will God reject a soul that sincerely loves him: be his speculative opinions what they may: and whether in any given instance certain opinions, be they unbelief, or misbelief, are compatible with a sincere love of God, God only can know.—But this I have said, and shall continue to say: that if the doctrines, the sum of which I believe to constitute the truth in Christ, be Christianity, then Unitarian*ism* is not, and vice versa: and that in speaking theologically and impersonally, i.e. of psilanthropism and theanthropism as schemes of belief, without reference to individuals who profess either the one or the other, it will be absurd to use a different language as long as it is the dictate of common sense, that two opposites cannot properly be called by the same name. I should feel no offence if a Unitarian applied the same to me, any more than if he were to say, that 2 and 2 being 4, 4 and 4 must be 8.

Αλλα βροτων
Τον μεν κενοφρονες αυχαι
Εξ αγαθων εβαλον.
Τον δ'αυ καταμεμφθεντ' αγαν
Ισχυν οικειων κατεσφαλεν καλων,
Χειρος ελκων οπισσω, Θυμος ατολμος.

[But, among mortals, *one* is cast down from his blessings by empty-headed conceit, whereas *another*, underrating his strength too far, hath been thwarted from winning the honours within his reach, by an uncourageous spirit that draggeth him back by the hand. (LCL)]

Pindar, Nem. Ode xi

This has been my object, and this alone can be my defence—and O! that with this my personal as well as my literary life might conclude! the unquenched desire I mean, not without the consciousness of having earnestly endeavoured to kindle young minds, and to guard them against the temptations of scorners, by showing that the scheme of Christianity, as taught in the liturgy and homilies of our Church, though not discoverable by human reason, is yet in accordance with it; that link follows link by necessary consequence; that religion passes out of the ken of reason only where the eye of reason has reached its own horizon; and that faith is then but its continuation: even as the day softens away into the sweet twilight, and twilight, hushed and breathless, steals into the darkness. It is night, sacred night! the upraised eye views only the starry heaven which manifests itself alone: and the outward beholding is fixed on the sparks twinkling in the aweful depth, though suns of other worlds, only to preserve the soul steady and collected in its pure act of inward adoration to the great I AM, and to the filial Word° that reaffirmeth it from eternity to eternity, whose choral echo is the universe.

ΘΕΩ ΜΟΝΩ ΔΟΞΑ [GLORY TO GOD ALONE]

LETTERS°

[*To Robert Southey*, 18 September 1794]

Sept—18th— 10 o clock Thursday Morning

Well, my dear Southey! I am at last arrived at Jesus [College]. My God! how tumultuous are the movements of my Heart—Since I quitted this room what and how important Events have been evolved! America! Southey! Miss Fricker!—Yes—Southey—you are right—Even Love is the creature of strong Motive—I certainly love her. I think of her incessantly & with unspeakable tenderness—with that inward melting away of Soul that symptomatizes it.

Pantisocracy°—O I shall have such a scheme of it! My head, my heart are all alive—I have drawn up my arguments in battle array—they shall have the *Tactician* Excellence of the Mathematician with the Enthusiasm of the Poet—The Head shall be the Mass—the Heart the fiery Spirit, that fills, informs, and agitates the whole—Harwood!—Pish! I say nothing of him——

SHAD GOES WITH US. HE IS MY BROTHER!
I am longing to be with you—Make Edith my Sister—Surely, Southey! we shall be frendotatoi meta frendous. Most friendly where all are friends. She must therefore be more emphatically my Sister.

Brookes & Berdmore, as I suspected, have spread my Opinions in mangled forms at Cambridge—Caldwell the most excellent, the most pantisocratic of Aristocrats, has been laughing at me—Up I arose terrible in Reasoning—he fled from me—because 'he could not answer for his own Sanity sitting so near a madman of Genius!' He told me, that the Strength of my Imagination had intoxicated my Reason—and that the acuteness of my Reason had given a directing Influence to my Imagination.—Four months ago the Remark would not have been more elegant than Just—. Now it is Nothing.—

I like your Sonnets exceedingly—the best of any I have yet seen.—tho' to the eye Fair is the extended Vale—should be To the Eye Tho' fair the extended Vale—I by no means disapprove of Discord introduced to produce *effect*—nor is my Ear so fastidious as to be angry with it where it could not have been avoided without weakening the Sense—But Discord for Discord's sake is rather too licentious.—

'Wild wind' has no other but alliterative beauty—it applies to a storm, not to the Autumnal Breeze that makes the trees rustle mournfully—Alter it to

That rustle to the sad wind moaning by.

''Twas a long way & tedious'—& the three last lines are marked Beauties—unlaboured Strains poured soothingly along from the feeling Simplicity of Heart.—The next Sonnet is altogether exquisite—the circumstance common yet new to Poetry—the moral accurate & full of Soul. '*I never saw*['] &c is most exquisite——I am almost ashamed to write the following—it is so inferior—Ashamed! No—Southey—God knows my heart—I am *delighted* to feel you superior to me in Genius as in Virtue.

No more my Visionary Soul shall dwell
On Joys, that were! No more endure to weigh
The Shame and Anguish of the evil Day,
Wisely forgetful! O'er the Ocean swell
Sublime of Hope I seek the cottag'd Dell,
Where Virtue calm with careless step may stray,
And dancing to the moonlight Roundelay
The Wizard Passions weave an holy Spell.
Eyes that have ach'd with Sorrow! ye shall weep
Tears of doubt-mingled Joy, like theirs who start
From Precipices of distemper'd Sleep,
On which the fierce-eyed Fiends their Revels k[eep,]
And see the rising Sun, & feel it dart
New Rays of Pleasance trembling to the Heart.

I have heard from Allen—and write the *third* Letter to him. Yours is the *second*.—Perhaps you would like two Sonnets I have written to my Sally.——

When I have received an answer from Allen, I will tell you the contents of his first Letter.—

My Comp— to Heath——

I will write you a huge big Letter next week—at present I have to transact the Tragedy Business, to wait on the Master, to write to Mrs Southey, Lovell, &c &c—

God love you—&

S. T. Coleridge

[*To 'Citizen' John Thelwall,*° 19 November 1796]

My dear Thelwall

Ah me! literary *Adventure* is but bread and cheese *by chance!* I keenly sympathize with you—sympathy, the only poor consolation I can offer you. Can no plan be suggested? I mention one not as myself approving it; but because it was mentioned to me—Briefly thus—If the Lovers of Freedom in the principal towns would join together by eights or tens, to send for what *books* they want directly to you, & if you could place yourself in such a line that you might have books from the different Publishers at Booksellers' price.—Suppose now, that 12 or 14 people should agree together that a little order book should be kept in the *Shop* of one of them—& when any one of these wanted a book, to write it down. And as soon as enough were ordered, to make it worth the carriage, to write up to you for them?——I repeat, that I mention the plan merely because it was mentioned to me. Shame fall on the friends of Freedom if they will do nothing better! If they will do nothing better, they will not do even this!—And the plan would disgust the Country Booksellers, who ought not to be alienated.

Have you any connection with the Corresponding Society Magazine—I have not seen it yet—Robert Southey is one of its benefactors—Of course, you have read the Joan of Arc. Homer is the Poet for the Warrior—Milton for the Religionist—Tasso for Women—Robert Southey for the Patriot. The first & fourth books of the Joan of Arc are to me more interesting than the same number of Lines in any poem whatsoever.—But you, & I, my dear Thelwall! hold different *creeds* in poetry as well as religion. N'importe [it doesn't matter].—By the bye, of your works I have now all, except your essay on animal vitality which I never had, & your *poems* which I bought on their first publication, & lost them. From those poems I should have supposed our poetical *tastes* more nearly alike, than I find, they are.—The poem on the Sols flashes Genius thro' Strophe 1. Antistrophe I. & Epode I.—the rest I do not perhaps understand——only I *love* these two lines—

> Yet sure the Verse that shews the friendly mind
> To Friendship's ear not harshly flows.—

Your larger *Narrative* affected me greatly. It is admirably written—& displays strong Sense animated by Feeling, & illumined by

Imagination—& neither in the thoughts or rhythm does it encroach on poetry.——

There have been two poems of mine in the New Monthly magazine—with my name—indeed, I make it a scruple of conscience never to publish any thing, however trifling, without it. Did you like them? The first was written at the desire of a beautiful little Aristocrat—Consider it therefore, as a Lady's Poem. Bowles (the bard of my idolatry) has written a poem lately without plan or meaning—but the component parts are divine. It is entitled—Hope, an allegorical Sketch. I will copy two of the Stanzas, which must be peculiarly interesting to you, virtuous High-Treasonist, & your friends, the other Acquitted Felons!—

> But see as one awaked from deadly Trance
> With hollow and dim eyes and stony stare
> CAPTIVITY with faltering step advance!
> Dripping & knotted was her coal-black Hair,
> For she had long been hid, as in the Grave:
> No sounds the silence of her prison broke,
> Nor one Companion had she in her Cave
> Save TERROR's dismal Shape, that no word broke,*
> But to a stony Coffin on the Floor
> With lean and hideous finger pointed evermore.
>
> The lark's shrill song, the early Village Chime,
> The upland echo of the winding Horn,
> The far-heard Clock that spoke the passing time,
> Had never pierc'd her Solitude forlorn:
> At length releas'd from the deep Dungeon's gloom
> She feels the fragrance of the vernal Gale,
> She sees more sweet the living Landscape bloom
> And whilst she listens to Hope's tender tale,
> She thinks, her long-lost Friends shall bless her sight,
> And almost faints with Joy amidst the broad Day-light!

The last line is indeed exquisite.—

Your portrait of yourself interested me—As to me, my face, unless when animated by immediate eloquence, expresses great Sloth, & great, indeed almost ideotic, good nature. 'Tis a mere carcase of a face: fat, flabby, & expressive chiefly of inexpression.—Yet, I am told, that my eyes, eyebrows, & forehead are physiognomically good—; but of

* for broke read spoke, as in Bowles.

this the Deponent knoweth not. As to my shape, 'tis a good shape enough, if measured—but my gait is awkward, & the walk, & the *Whole man* indicates *indolence capable of energies.*—I am, & ever have been, a great reader—& have read almost every thing—a library-cormorant—I am *deep* in all out of the way books, whether of the monkish times, or of the puritanical aera—I have read & digested most of the Historical Writers—; but I do not *like* History. Metaphysics, & Poetry, & 'Facts of mind'—(i.e. Accounts of all the strange phantasms that ever possessed your philosophy-dreamers from Tauth [Thoth], the Egyptian to Taylor, the English Pagan,) are my darling Studies.—In short, I seldom read except to amuse myself—& I am almost always reading.——Of useful knowledge, I am a so-so chemist, & I love chemistry——all else is *blank*,—but I *will* be (please God) an Horticulturist & a Farmer. I compose very little—& I absolutely hate composition. Such is my dislike, that even a sense of Duty is sometimes too weak to overpower it.

I cannot breathe thro' my nose—so my mouth, with sensual thick lips, is almost always open. In conversation I am impassioned, and oppose what I deem [error] with an eagerness, which is often mistaken for personal asperity——but I am ever so swallowed up in the *thing*, that I perfectly forget my *opponent*. Such am I. I am just about to read Dupuis' 12 octavos, which I have got from London. I shall read only one Octavo a week—for I cannot *speak* French at all, & I read it slowly.—— . . .

[*To Benjamin Flower*,° 11 December 1796]

My much esteemed Friend

I truly sympathize with you in your severe Loss, and pray to God that he may give you a sanctified use of your Affliction. The death of a young person of high hopes and opening faculties impresses me less gloomily, than the Departure of the Old. To my mere natural Reason, the former *appears* like a *transition*; there seems an *incompleteness* in the life of such a person, contrary to the general order of nature; and it makes the heart say, 'this is not all.' But when an old man sinks into the grave, we have seen the bud, the blossom, and the fruit; and the unassisted mind droops in melancholy, as if *the Whole* had come and gone.—But God hath been merciful to us, and strengthened our eyes thro' faith, and Hope may cast her anchor in a certain bottom, and the young and old may rejoice before God and the Lamb, weeping as tho' they wept not, and crying in the Spirit of faith, Art thou not from everlasting, O Lord God my Holy One? We shall not die!——I have known affliction, yea, my friend! I have been myself sorely afflicted, and have rolled my dreary eye from earth to Heaven, and found no comfort, till it pleased the Unimaginable High & Lofty One to make my Heart more tender in regard of religious feelings. My philosophical refinements, & metaphysical Theories lay by me in the hour of anguish, as toys by the bedside of a Child deadly-sick. May God continue his visitations to my soul, bowing it down, till the pride & Laodicean self-confidence of human Reason be utterly done away; and I cry with deeper & yet deeper feelings, O my Soul! thou art wretched, and miserable, & poor, and blind, and naked!——The young Lady, who in a fit of frenzy killed her own mother, was the Sister of my dearest Friend, and herself dear to me as an only Sister. She is recovered, and is acquainted with what she has done, and is very calm. She was a truly pious young woman; and her Brother, whose soul is almost wrapped up in her, hath had his heart purified by this horror of desolation, and prostrates his Spirit at the throne of God in believing Silence. The Terrors of the Almighty are the whirlwind, the earthquake, and the Fire that precede the still small voice of his Love. The pestilence of our lusts must be scattered, the strong-layed Foundations of our Pride blown up, & the stubble & chaff of our Vanities burnt, ere we can give ear to the inspeaking Voice of Mercy, 'Why *will* ye die?'——

My answer to Godwin° will be a six shilling Octavo; and is designed to shew not only the absurdities and wickedness of *his* System, but to detect what appear to me the defects of all the systems of morality before & since Christ, & to show that wherein they have been right, they have exactly coincided with the Gospel, and that each has erred exactly where & in proportion as, he has deviated from that perfect canon. My last Chapter will attack the credulity, superstition, calumnies, and hypocrisy of the present race of Infidels. Many things have fallen out to retard the work; but I hope, that it will appear shortly after Christmas, at the farthest. I have endeavoured to make it a cheap book; and it will contain as much matter as is usually sold for eight shillings. . . .

[*To John Thelwall*, 17 December 1796]

My dear Thelwall

I should have written you long ere this, had not the settlement of my affairs previous to my leaving Bristol, and the organization of my *new plan* occupied me with bulky anxieties that almost excluded every thing but Self from my thoughts. And, besides, my Health has been very bad, and remains so—A nervous Affection from my right temple to the extremity of my right shoulder almost distracted me, & made the frequent use of Laudanum absolutely necessary. And since I have subdued this, a Rheumatic Complaint in the back of my Head & Shoulders, accompanied with sore throat, and depression of the animal Spirits, has convinced me that a man may change bad Lodgers without bettering himself. I write these things not so much to apologize for my silence, or for the pleasure of complaining, as that you may know the reason why I have not given you 'a strict account' how I have disposed of your Books. This I will shortly do, with all the Veracity, which, that solemn Incantation '*upon your honor*' must necessarily have conjur'd up.—

Your second & third part promise great things—I have counted the subjects, and by a nice calculation find that eighteen Scotch Doctors would write fifty four Quarto Volumes; each chusing his Thesis out of your Syllabus. May you do good by them; and moreover enable yourself to do more good—I *should* say—to continue to do good. *My farm* will be a garden of one acre & an half; in which I mean to raise vegetables & corn enough for myself & Wife, and feed a couple of snouted & grunting Cousins from the refuse. My evenings I shall devote to Literature; and by Reviews, the Magazine, and other shilling-scavenger Employments shall probably gain 40£ a year—which Economy & Self-Denial, Goldbeaters, shall hammer till it cover my annual Expences. Now in favor of this scheme I shall say nothing: for the more vehement my ratiocinations were previous to the experiment, the more ridiculous my failure would appear; and if the Scheme deserve the said ratiocination, I shall *live down* all your objections. I doubt not, that the time will come when all our Utilities will be directed in one simple path. That Time however is not come; and imperious circumstances point out to each one his particular Road. Much good may be done in all. I am not *fit* for *public* Life; yet the Light shall stream to a far distance from the taper in my cottage window. Meantime, do *you* uplift the *torch* dreadlessly, and shew to mankind the face of that Idol, which they have worshipped in Darkness!

And now, my dear fellow! for a little sparring about Poetry. My first *Sonnet is obscure*; but you ought to distinguish between obscurity residing in the uncommonness of the thought, and that which proceeds from thoughts unconnected & language not adapted to the expression of them. When you *do* find out the meaning of my poetry, can you (in general, I mean) alter the language so as to make it more perspicuous—the thought remaining the same?—By 'dreamy semblance' I *did* mean semblance of some unknown Past, like to a dream—and not 'a semblance *presented* in a dream.'—I meant to express, that oftimes, for a second or two, it flashed upon my mind, that the then company, conversation, & every thing, had occurred before, with all the precise circumstances; so as to make Reality appear a Semblance, and the Present like a dream in Sleep. Now this thought is obscure; because few people have experienced the same feeling. Yet several have—& they were proportionably delighted with the lines as expressing some strange sensations, which they themselves had never ventured to communicate, much less had ever seen developed in poetry. (—The lines I have since altered to—

> Oft o'er my Brain does that strange Rapture roll
> Which makes the Present (while its brief fits last)
> Seem a mere Semblance of some Unknown Past,
> Mix'd with such feelings, as distress the Soul
> When dreaming that she dreams.)

Next as to 'mystical'—Now that the thinking part of Man, i.e. the Soul, existed previously to its appearance in its present body, may be very wild philosophy; but it is very intelligible poetry, inasmuch as Soul is an orthodox word in all our poets; they meaning by 'Soul' a being inhabiting our body, & playing upon it, like a Musician inclosed in an Organ whose keys were placed inwards.—Now this opinion I do not hold—not that I am a Materialist; but because I am a Berkleian——Yet as you who are not a Christian wished you were, that we might meet in Heaven, so I, who do not believe in this descending, & incarcerated Soul, yet said, if my Baby had died before I had seen him, I should have *struggled* to believe it.——Bless me! a commentary of 35 lines in defence of a Sonnet!—And I do not like the Sonnet much myself——. In some (indeed in many of my poems,) there is a garishness & swell of diction, which I hope, that my poems in future, if I write any, will be clear of—; but seldom, I think, any *conceits*.—In the second Edition now printing I have swept the book with the expurgation Besom to a fine tune—having omitted nearly one third.—As to Bowles, I affirm, that the manner of his accentuation in the words 'brōad dāy-līght['] (thrēē lōng S̄yllables) is a beauty, as it admirably expresses the Captive's *dwelling* on the sight of Noon—with rapture & a kind of Wonder.

> The common Sun, the Air, the Skies,
> To Him are opening Paradise.
>
> Gray.

But supposing my defence not tenable, yet how a blunder in metre stamps a man, Italian or Della Cruscan,° I cannot perceive.——As to my own poetry I do confess that it frequently both in thought & language deviates from 'nature & simplicity.' But that Bowles, the most tender, and, with the exception of Burns, the only *always-natural* poet in our Language, that *he* should not escape the charge of Della Cruscanism / this cuts the skin & *surface* of my Heart.——'Poetry to have its highest relish must be *impassioned*!' true! but first, Poetry ought not always to have its *highest* relish, & secondly, judging of the cause from its effect, Poetry, though treating on lofty & abstract truths, ought to be deemed *impassioned* by him, who reads it with impassioned feelings.—Now Collins' Ode on the poetical character—that part of it, I should say, beginning with—'The Band (as faery Legends say) Was wove on that creating Day,' has inspired & whirled *me* along with greater agitations of enthusiasm than any the most *impassioned* Scene in Schiller or Shakespeare——using 'impasssioned' in its confined sense for writings in which the human passions of Pity, Fear, Anger, Revenge, Jealousy, or Love are brought into view with their workings.——Yet I consider the latter poetry as more valuable, because it gives *more general* pleasure—& I judge of all things by their Utility.——I feel strongly, and I think strongly; but I seldom feel without thinking, or think without feeling. Hence tho' my poetry has in general a *hue* of tenderness, or Passion over it, yet it seldom exhibits unmixed & simple tenderness or Passion. My philosophical opinions are blended with, or deduced from, my feelings: & this, I think, peculiarizes my style of Writing. And like every thing else, it is sometimes a beauty, and sometimes a fault. But do not let us introduce an act of Uniformity° against Poets—I have room enough in *my* brain to admire, aye & almost equally, the *head* and fancy of Akenside, and the *heart* and fancy of Bowles, the solemn Lordliness of Milton, & the divine Chit chat of Cowper: and whatever a man's excellence is, that will be likewise his fault. . . .

[*To John Thelwall*, 31 December 1796]

Enough, my dear Thelwall, of Theology. In my book on Godwin I compare the two Systems—his & Jesus's—& that book I am sure you will read with attention.—I entirely accord with your opinion of Southey's Joan—the 9th book is execrable—and the poem tho' it frequently reach the *sentimental*, does not display, the *poetical*, *Sublime*. In language at once natural, perspicuous, & dignified, in manly pathos, in soothing & sonnet-like description, and above all, in character, & *dramatic* dialogue, Southey is unrivalled; but as certainly he does not possess opulence of Imagination, lofty-paced Harmony, or that *toil* of thinking, which is necessary in order to plan *a Whole*. Dismissing mock humility, & hanging your mind as a looking-glass over my Idea-pot, so as to image on the said mind all the bubbles that boil in the said Idea-pot, (there's a damn'd long-winded Metaphor for you) I think, that an admirable Poet might be made by *amalgamating him & me*. I *think* too much for a *Poet*; he too little for a *great* Poet. But he abjures *thinking*—& lays the whole stress of excellence—on *feeling*.—Now (as you say) they must go together.—Between ourselves, the *Enthusiasm* of Friendship is not with S. & me. We quarreled—& the quarrel lasted for a twelvemonth—We are now reconciled; but the cause of the Difference was solemn—& 'the blasted oak puts not forth its buds anew'—we are *acquaintances*—& feel *kindliness* towards each other; but I do not *esteem*, or LOVE Southey, as I must esteem & love the man whom I dared call by the holy name of FRIEND!—and vice versâ Southey of me—I say no more—it is a painful subject—& do you say nothing—I mention this, for obvious reasons—but let it go no farther.——It is a painful subject. Southey's direction at present is—R. Southey, No. 8, Westgate Buildings, Bath, but he leaves Bath for London in the course of a week.

You imagine that I know Bowles personally—I never *saw* him but once; & when I was a boy, & in Salisbury *market-place*.

The passage in your letter respecting your Mother affected me greatly.—Well, true or false, Heaven is a less gloomy idea than Annihilation!—Dr Beddoes, & Dr Darwin think that *Life* is utterly inexplicable, writing as Materialists—You, I understand, have adopted the idea that it is the result of organized matter acted on by external Stimuli.—As likely as any other system; but you *assume* the thing to be proved—the '*capability* of being stimulated into sensation' *as* a *property* of organized matter—now 'the Capab.' &c is *my* definition of *animal Life*——Monro believes in a plastic immaterial Nature—all-pervading—

> And what if all of animated Nature
> Be but organic harps diversely fram'd
> That tremble into *thought* as o'er them sweeps
> Plastic & vast &c—

(by the bye—that is my favorite of *my* poems—do *you* like it?) Hunter that the *Blood* is the Life—which is saying nothing at all—for if the blood were *Life*, it could never be otherwise than Life—and to say, it is *alive*, is saying nothing—& Ferriar believes in a *Soul*, like an orthodox Churchman—So much for Physicians & Surgeons—Now as to the Metaphysicians, Plato says, it is *Harmony*—he might as well have said, a fiddle stick's end—but I love Plato—his dear *gorgeous* Nonsense! And *I*, *tho' last not least*, *I* do not know what to think about it—on the whole, I have rather made up my mind that I am a mere *apparition*—a naked Spirit!—And that Life is I myself I! which is a mighty clear account of it. Now I have written all this not to expose my ignorance (that is an accidental effect, not the final cause) but to shew you, that I want to see your Essay on Animal Vitality—of which Bowles, the Surgeon, spoke in high Terms—Yet *he* believes in a *body* & a *soul*. Any book may be left at Robinson's for *me*, 'to be put into the next parcel sent to Joseph Cottle, Bookseller, Bristol.'——Have you received an Ode of mine from Parsons's? In your next letter tell me what you think of the *scattered* poems, I sent you—send me any poems, and I will be minute in Criticism——for, O Thelwall! even a long-winded Abuse is more consolatory to an *Author's* feelings than a short-breathed, asthma-lunged Panegyric.——Joking apart, I would to God we could sit by a fireside & joke vivâ voce, face to face—Stella & Sara, Jack Thelwall, & I!——As I once wrote to my dear *friend*, T. Poole, 'repeating'

> Such Verse as Bowles, heart-honour'd Poet, sang,
> That wakes the Tear yet steals away the Pang,
> Then or with Berkeley or with Hobbes romance it
> Dissecting Truth with metaphysic lancet.
> Or drawn from up those dark unfathom'd Wells
> In wiser folly clink the Cap & Bells.
> How many tales we told! What jokes we made!
> Conundrum, Crambo, Rebus, or Charade;

Ænigmas, that had driven the Theban* mad,
And Puns then best when exquisitely bad;
And I, if aught of archer vein I hit,
With my own Laughter stifled my own Wit.

[*To Thomas Poole*, 6 February 1797]

My dear Poole

I could inform the dullest author how he might write an interesting book—let him relate the events of his own Life with honesty, not disguising the feelings that accompanied them.—I never yet read even a Methodist's 'Experience' in the Gospel Magazine without receiving instruction & amusement: & I should almost despair of that Man, who could peruse the Life of John Woolman without an amelioration of Heart.—As to my Life, it has all the charms of variety: high Life, & low Life, Vices & Virtues, great Folly & some Wisdom. However what I am depends on what I have been; and you MY BEST FRIEND! have a right to the narration.—To me the task will be a useful one; it will renew and deepen *my* reflections on the past; and it will perhaps make you behold with no unforgiving or impatient eye those weaknesses and defects in my character, which so many untoward circumstances have concurred to plant there.——

My family on my Mother's side can be traced up, I know not, how far—The Bowdens inherited a house-stye & a pig-stye in the Exmore Country, in the reign of Elizabeth, as I have been told—& to my own knowledge, they have inherited nothing better since that time.—On my father's side I can rise no higher than my Grandfather, who was dropped, when a child, in the Hundred of Coleridge in the County of Devon; christened, educated, & apprenticed by the parish.—He afterwards became a respectable Woolen-draper in the town of South Molton. / I have mentioned these particulars, as the time may come in which it will be useful to be able to prove myself a genuine Sans culotte, my veins uncontaminated with one drop of Gentility. My father received a better education than the others of his Family in consequence of his own exertions, not of his superior advantages. When he was not quite 16 years old, my Grandfather became bankrupt; and by a series of misfortunes was reduced to extreme poverty. My father received the half of his last crown & his blessing;

* Œdipus.

and walked off to seek his fortune. After he had proceeded a few miles, he sate him down on the side of the road, so overwhelmed with painful thoughts that he wept audibly. A Gentleman passed by, who knew him: & enquiring into his distresses took my father with him, & settled him in a neighb'ring town as a schoolmaster. His school increased; and he got money & knowledge: for he commenced a severe & ardent student. Here too he married his first wife, by whom he had three daughters; all now alive. While his first wife lived, having scraped up money enough, at the age of 20 he walked to Cambridge, entered at Sidney College, distinguished himself for Hebrew & Mathematics, & might have had a fellowship: if he had not been married.—He returned—his wife died—Judge Buller's Father gave him the living of Ottery St Mary, & put the present Judge to school with him—he married my Mother, by whom he had ten children of whom I am the youngest, born October 20th [21], 1772.

These sketches I received from my mother & Aunt; but I am utterly unable to fill them up by any particularity of times, or places, or names. Here I shall conclude my first Letter, because I cannot pledge myself for the accuracy of the accounts, & I will not therefore mingle them with those, for the accuracy of which in the minutest parts I shall hold myself amenable to the Tribunal of Truth.—You must regard this Letter, as the first chapter of an history; which is devoted to dim traditions of times too remote to be pierced by the eye of investigation.——

Yours affectionately
S. T. Coleridge

[*To Thomas Poole*, March 1797]

My dear Poole

My Father, (Vicar of, and Schoolmaster at, Ottery St. Mary, Devon) was a profound Mathematician, and well-versed in the Latin, Greek, & Oriental Languages. He published, or rather attempted to publish, several works: 1st, Miscellaneous Dissertations arising from the 17th and 18th Chapters of the Book of Judges; II. Sententiae excerptae [selected sayings], for the use of his own School; 3rd (& his best work) a Critical Latin Grammar; in the preface to which he proposes a bold Innovation in the names of the Cases. My father's new nomenclature was not likely to become popular, altho' it must be allowed to be both sonorous and expressive—exempli gratiâ [for example]—he calls the ablative the Quippe-quare-quale-quia-quidditive Case!—My Father made the world his confidant with respect to his Learning & ingenuity: & the world seems to have kept the secret very faithfully.—His various works, uncut, unthumbed, have been preserved free from all pollution, except that of his Family's Tails.—This piece of good-luck promises to be hereditary: for all *my* compositions have the same amiable *homestaying* propensity.—The truth is, My Father was not a first-rate Genius—he was however a first-rate Christian. I need not detain you with his Character—in learning, good-heartedness, absentness of mind, & excessive ignorance of the world, he was a perfect *Parson Adams.*°—My Mother was an admirable Economist, and managed exclusively.—My eldest Brother's name was John: he went over to the East Indies in the Company's Service; he was a successful Officer, & a brave one, I have heard: he died of a consumption there about 8 years ago. My second Brother was called William—he went to Pembroke College, Oxford; and afterwards was assistant to Mr Newcome's School, at Hackney. He died of a putrid fever the year before my Father's death, & just as he was on the eve of marriage with Miss Jane Hart, the eldest Daughter of a very wealthy Druggist in Exeter.—My third Brother, James, has been in the army since the age of sixteen—has married a woman of fortune—and now lives at Ottery St Mary, a respectable Man. My Brother Edward, the wit of the Family, went to Pembroke College; & afterwards, to Salisbury, as assistant to Dr Skinner: he married a woman 20 years older than his Mother. She is dead: & he now lives at Ottery St Mary, an idle Parson. My fifth Brother, George, was educated at Pembroke College, Oxford; and from thence went to Mr Newcome's, Hackney, on the death of William. He stayed there fourteen years: when the living of Ottery St Mary was given him—there he now has a fine school, and has lately married Miss Jane Hart; who with beauty, & wealth, had remained a faithful Widow to the memory of William for 16 years.—My Brother George is a man of reflective mind & elegant Genius. He possesses Learning in a greater degree than any of the Family, excepting myself. His manners are grave, & hued over with a tender sadness. In his moral character he approaches every way nearer to Perfection than any man I ever yet knew—indeed, he is worth the whole family in a Lump. My sixth Brother, Luke (indeed the seventh, for one Brother, the second, died in his Infancy, & I had forgot to mention him) was bred as a medical Man—he married Miss Sara Hart: and died at the age of 22, leaving one child, a lovely Boy, still alive. My Brother Luke was a man of uncommon Genius,—a severe student, & a good man.——The 8th Child was a Sister, Anne—she died a little after my Brother Luke—aged 21.

> Rest, gentle Shade! & wait thy Maker's will;
> Then rise *unchang'd*, and be an Angel still!

The 9th Child was called Francis: he went out as a Midshipman, under Admiral Graves—his Ship lay on the Bengal Coast—& he accidentally met his Brother John—who took him to Land, & procured him a Commission in the Army.—He shot himself (having been left carelessly by his attendant) in a delirious fever brought on by his excessive exertions at the siege of Seringapatam: at which his conduct had been so gallant, that Lord Cornwallis payed him a high compliment in the presence of the army, & presented him with a valuable gold Watch, which my Mother now has.—All my Brothers are remarkably handsome; but they were as inferior to Francis as I am to them. He went by the name of 'the handsome Coleridge.' The tenth & last Child was S. T. Coleridge, the subject of these Epistles: born (as I told you in my last) October 20th, 1772.

From October 20th, 1772 to October 20th, 1773.——Christened Samuel Taylor Coleridge—my Godfather's name being Samuel Taylor Esq. I had another Godfather, his name was Evans: & two Godmothers; both called 'Monday' [Mundy].—

From October 20th, 1778 to October 20th 1774.——In this year I was carelessly left by my Nurse—ran to the Fire, and pulled out a live coal—burnt myself dreadfully—while my hand was being Drest by a Mr Young, I spoke for the first time (so my Mother informs me) & said—'Nasty Doctor Young'!—The snatching at fire, & the circumstance of my first words expressing hatred to professional men, are they at all *ominous*? This Year, I went to School—My Schoolmistress, the very image of Shenstone's,° was named, Old Dame Key—she was nearly related to Sir Joshua Reynolds.—

From October 20th 1774 to October 1775. I was inoculated; which I mention, because I distinctly remember it: & that my eyes were bound—at which I manifested so much obstinate indignation, that at last they removed the bandage—and unaffrighted I looked at the lancet & suffered the scratch.—At the close of this Year I could read a Chapter in the Bible.

Here I shall end; because the remaining years of my life *all* assisted to form *my particular mind*—the three first years had nothing in them that seems to relate to it.

[*To Joseph Cottle*,° April 1797]

My dearest Cottle

I love & respect you, as a Brother. And my memory deceives me woefully, if I have not evidenced by the animated turn of my conversation, when we have been tête à tête, how much your company interested me.—But when last in Bristol the day I meant to have devoted to you, was such a day of sadness, that I could *do nothing*—On the Saturday, the Sunday, & the ten days after my arrival at Stowey I felt a depression too dreadful to be described—

> So much I felt my genial spirits droop!
> My Hopes all flat, Nature within me seem'd
> In all her functions weary of herself.

Wordsworth's conversation &c rous'd me somewhat; but even now I am not the man, I have been—& I think, never shall.—A sort of calm hopelessness diffuses itself over my heart.—Indeed every mode of life, which has promised me bread & cheese, has been, one after another, torn away from me—but God remains! I have no immediate pecuniary distress, having received ten pound from Lloyd's Father at Birmingham.—I employ myself now on a book of morals in answer to Godwin, & on my Tragedy . . . .

David Hartley is well & grows—Sara is well, and desires a Sister's Love to you—

Tom Poole desires to be kindly remembered to you—I see they have reviewed Southey's Poems, & my Ode in the Monthly Review—Notwithstanding the Reviews, I, who in the sincerity of my heart am *jealous for* Robert Southey's fame, regret the publication of that Volume. Wordsworth complains with justice, that Southey writes *too much at his ease*—that he too seldom 'feels his burthen'd breast

> Heaving beneath th' incumbent Deity.'

He certainly will make literature more *profitable to him* from the fluency with which he writes, & the facility, with which he pleases himself. But I fear, that to Posterity his Wreath will look unseemly—here an ever living Amaranth, & close by its side some Weed of an hour, sere, yellow, & shapeless—his exquisite Beauties will lose half their effect from the bad company, they keep.—Besides, I am fearful that he will begin to rely too much on *story* & *event* in his poems to the neglect of those *lofty imaginings*, that are peculiar to, & definitive of, *the* Poet. The *story* of Milton might be told in *two pages*—it is this [whic]h distinguishes *an* Epic *Poem* from a *Romance in metre*. Observe the march of Milton—his severe application, his laborious polish, his deep metaphysical researches, his *prayers to God* before he began his great poem—all, that could lift & swell his intellect, became his daily food.—I should not think of devoting less than 20 years to an Epic Poem. Ten to collect materials, & warm my mind with universal Science—I would be a tolerable mathematician, I would thoroughly know mechanics, hydrostatics, optics, & Astronomy—Botany, Metallurgy, fossillism, chemistry, geology, Anatomy, Medicine—then *the mind of man*—then the *minds of men*—in *all* Travels, Voyages, & Histories. So I would spend ten years—the next five in the composition of the poem—& the five last in the correction of it—So I would write, haply not unhearing of that divine and nightly-whispering Voice, which speaks to mighty Minds of predestinated Garlands starry & unwithering!—

God love you & S. T. Coleridge

[*To Thomas Poole*, 9 October 1797]

My dearest Poole

From March to October—a long silence! but [as] it is possible, that I may have been preparing materials for future letters, the time cannot be considered as altogether subtracted from you.

From October 1775 to October 1778.

These three years I continued at the reading-school—because I was too little to be trusted among my Father's School-boys—. After breakfast I had a halfpenny given me, with which I bought three cakes at the Baker's close by the school of my old mistress—& these were my dinner on every day except Saturday & Sunday—when I used to dine at home, and wallowed in a beef & pudding dinner.—I am remarkably fond of Beans & Bacon—and this fondness I attribute to my father's having given me a penny for having eat a large quantity of beans, one Saturday—for the other boys did not like them, and as it was an economic food, my father thought, that my attachment & penchant for it ought to be encouraged.——My Father was very fond of me, and I was my mother's darling—in consequence, I was very miserable. For Molly, who had nursed my Brother Francis, and was immoderately fond of him, hated me because my mother took more notice of me than of Frank—and Frank hated me, because my mother gave me now & then a bit of cake, when he had none—quite forgetting that for one bit of cake which I had & he had not, he had twenty sops in the pan & pieces of bread & butter with sugar on them from Molly, from whom I received only thumps & ill names.—So I became fretful, & timorous, & a tell-tale—& the School-boys drove me from play, & were always tormenting me—& hence I took no pleasure in boyish sports—but read incessantly. My Father's Sister kept an *every-thing* Shop at Crediton—and there I read thro' all the gilt-cover little books that could be had at that time, & likewise all the uncovered tales of Tom Hickathrift, Jack the Giant-killer, &c &c &c &c &c—/—and I used to lie by the wall, and *mope*—and my spirits used to come upon me suddenly, & in a flood—& then I was accustomed to run up and down the church-yard, and act over all I had been reading on the docks, the nettles, and the rank-grass.—At six years old I remember to have read Belisarius, Robinson Crusoe, & Philip Quarll—and then I found the Arabian Nights' entertainments—one tale of which (the tale of a man who was compelled to seek for a pure virgin) made so deep an impression on me (I had read it in the evening while my mother was mending stockings) that I was haunted by spectres, whenever I was in the dark—and I distinctly remember the anxious & fearful eagerness, with which I used to watch the window, in which the books lay—& whenever the Sun lay upon them, I would seize it, carry it by the wall, & bask, & read—. My Father found out the effect, which these books had produced—and burnt them.—So I became a *dreamer*—and acquired an indisposition to all bodily activity—and I was fretful, and inordinately passionate, and as I could not play at any thing, and was slothful, I was despised & hated by the boys; and because I could read & spell, & had, I may truly say, a memory & understanding forced into almost an unnatural ripeness, I was flattered & wondered at by all the old women—& so I became very vain, and despised most of the boys, that were at all near my own age—and before I was eight years old, I was a *character*—sensibility, imagination, vanity, sloth, & feelings of deep & bitter contempt for almost all who traversed the orbit of my understanding, were even then prominent & manifest.

From October 1778 to 1779.—That which I began to be from 3 to 6, I continued from 6 to 9.—In this year I was admitted into the grammer school, and soon outstripped all of my age.— I had a dangerous putrid fever this year—My Brother George lay ill of the same fever in the next room.——My poor Brother Francis, I remember, stole up in spite of orders to the contrary, & sate by my bedside, & read Pope's Homer to me—Frank had a violent love of beating me—but whenever that was superseded by any humour or circumstance, he was always very fond of me—& used to regard me with a strange mixture of admiration & contempt—strange it was not—: for he hated books, and loved climbing, fighting, playing, & robbing orchards, to distraction.—

My mother relates a story of me, which I repeat here—because it must be regarded as my first piece of wit.—During my fever I asked why Lady Northcote (our neighbour) did not come & see me.—My mother said, She was afraid of catching the fever—I was piqued & answered—Ah—Mamma! the four Angels round my bed an't afraid of catching it.—I suppose, you know the old prayer—

Matthew! Mark! Luke! & John!
God bless the bed which I lie on.
Four Angels round me spread,
Two at my foot & two at my bed [head]—

This prayer I said nightly—& most firmly believed the truth of it.—Frequently have I, half-awake & half-asleep, my body diseased & fevered by my imagination, seen armies of ugly Things bursting in upon me, & these four angels keeping them off.—In my next I shall carry on my life to my Father's Death.—

God bless you, my dear Poole! | & your affectionate
S. T. Coleridge.

[*To Thomas Poole*, 16 October 1797]

Dear Poole

From October 1779 to Oct. 1781.——I had asked my mother one evening to cut my cheese *entire*, so that I might toast it: this was no easy matter, it being a *crumbly* cheese—My mother however did it—/ I went into the garden for some thing or other, and in the mean time my Brother Frank *minced* my cheese, 'to disappoint the favorite'. I returned, saw the exploit, and in an agony of passion flew at Frank—he pretended to have been seriously hurt by my blow, flung himself on the ground, and there lay with outstretched limbs——I hung over him moaning & in a great fright—he leaped up, & with a horse-laugh gave me a severe blow in the face—I seized a knife, and was running at him, when my Mother came in & took me by the arm— / I expected a flogging—& struggling from her I ran away, to a hill at the bottom of which the Otter flows—about one mile from Ottery.—There I stayed; my rage died away; but my obstinacy vanquished my fears—& taking out a little shilling book which had, at the end, morning & evening prayers, I very devoutly repeated them—thinking *at the same time* with inward & gloomy satisfaction, how miserable my Mother must be!—I distinctly remember my feelings when I saw a Mr Vaughan pass over the Bridge, at about a furlong's distance—and how I watched the Calves in the fields beyond the river. It grew dark—& I fell asleep—it was towards the latter end of October—& it proved a dreadful stormy night— / I felt the cold in my sleep, and dreamt that I was pulling the blanket over me, & actually pulled over me a dry thorn bush, which lay on the hill—in my sleep I had rolled from the top of the hill to within three yards of the River, which flowed by the unfenced edge of the bottom.—I awoke several times, and finding myself wet & stiff, and cold, closed my eyes again that I might forget it.——In the mean time my Mother waited about half an hour, expecting my return, when the *Sulks* had evaporated—I not returning, she sent into the Church-yard, & round the town—not found!—Several men & all the boys were sent to ramble about & seek me—in vain! My Mother was almost distracted—and at ten o'clock at night I was *cry'd* by the crier in Ottery, and in two villages near it—with a reward offered for me.—No one went to bed—indeed, I believe, half the town were up all one night! To return to myself—About five in the morning or a little after, I was broad awake; and attempted to get up & walk—but I could not move—I saw the Shepherds & Workmen at a distance—& cryed but so faintly, that it was impossible to hear me 80 yards off——and there I might have lain & died—for I was now almost given over, the ponds & even the river near which I was lying, having been dragged.—But by good luck Sir Stafford Northcote, who had been out all night, resolved to make one other trial, and came so near that he heard my crying—He carried me in his arms, for near a quarter of a mile; when we met my father & Sir Stafford's Servants.—I remember, & never shall forget, my father's face as he looked upon me while I lay in the servant's arms—so calm, and the tears stealing down his face: for I was the child of his old age.——My Mother, as you may suppose, was outrageous with joy—in rushed a *young Lady*, crying out—'I hope, you'll whip him, Mrs Coleridge!'—This woman still lives at Ottery—& neither Philosophy or Religion have been able to conquer the antipathy which I *feel* towards her, whenever I see her.—I was put to bed—& recovered in a day or so—but I was certainly injured—For I was weakly, & subject to the ague for many years after—.—

My Father (who had so little of parental ambition in him, that he had destined his children to be Blacksmiths &c, & had accomplished his intention but for my Mother's pride & spirit of aggrandizing her family) my father had however resolved, that I should be a Parson. I read every book that came in my way without distinction—and my father was fond of me, & used to take me on his knee, and hold long conversations with me. I remember, that at eight years old I walked with him one winter evening from a farmer's house, a mile from Ottery——& he told me the names of the stars—and how Jupiter was a thousand times larger than our world—and that the other twinkling stars were Suns that had worlds rolling round them—& when I came home, he shewed me how they rolled round—/. I heard him with a profound delight & admiration; but without the least mixture of wonder or incredulity. For from my early reading of Faery Tales, & Genii &c &c—my mind had been habituated *to the Vast*——& I never regarded my senses in any way as the criteria of my belief. I regulated all my creeds by my conceptions not by my *sight*—even at that age. Should children be permitted to read Romances, & Relations of Giants & Magicians, & Genii?——I know all that has been said against it; but I have formed my faith in the affirmative.—I know no other way of giving the mind a love of 'the Great', & 'the Whole'.—Those who have been led to the same truths step by step thro' the constant testimony of their senses, seem to me to want a sense which I possess—They contemplate nothing but *parts*—and all *parts* are necessarily little—and the Universe to them is but a mass of *little things.*—It is true, that the mind *may* become credulous & prone to superstition by the former method—but are not the Experimentalists credulous even to madness in believing any absurdity, rather than believe the grandest truths, if they have not the testimony of their own senses in their favor?—I have known some who have been *rationally* educated, as it is styled. They were marked by a microscopic acuteness; but when they looked at great things, all became a blank & they saw nothing—and denied (very illogically) that any thing could be seen; and uniformly put the negation of a power for the possession of a power—& called the want of imagination Judgment, & the never being moved to Rapture Philosophy!——

Towards the latter end of September 1781 my Father went to Plymouth with my Brother Francis, who was to go as Midshipman under Admiral Graves; the Admiral was a friend of my Father's.—My Father settled my Brother; & returned Oct. 4th, 1781—. He arrived at Exeter about six o'clock—& was pressed to take a bed there by the Harts—but he refused—and to avoid their intreaties he told them—that he had never been superstitious—but that the night before he had had a dream which had made a deep impression. He dreamt that Death had appeared to him, as he is commonly painted, & touched him with his Dart. Well he returned home—& all his family, I excepted, were up. He told my mother his dream—; but he was in high health & good spirits—& there was a bowl of Punch made—& my Father gave a long & particular account of his Travel, and that he had placed Frank under a religious Captain &c—/ At length, he went to bed, very well, & in high Spirits.—A short time after he had lain down he complained of a pain in his bowells, which he was subject to, from the wind—my mother got him some peppermint water—and after a pause, he said—'I am much better now, my dear!'—and lay down again. In a minute my mother heard a noise in his throat—and spoke to him—but he did not answer—and she spoke repeatedly in vain. Her *shriek* awaked me—& I said, 'Papa is dead.'—I did not know my Father's return, but I knew that he was expected. How I came to think of his Death, I cannot tell; but so it was.—Dead he was—some said it was the Gout in the Heart—probably, it was a fit of Apoplexy / —He was an Israelite without guile; simple, generous, and, taking some scripture texts in their literal sense; he was conscientiously indifferent to the good & the evil of this world.—

God love you & S. T. Coleridge

[*To Thomas Poole*, 19 February 1798]

From October 1781 to October 1782.

After the death of my father we, of course, changed houses, & I remained with my mother till the spring of 1782, and was a day-scholar to Parson Warren, my Father's successor— / He was a booby, I believe; and I used to delight my poor mother by relating little instances of his deficiency in grammar knowlege — every detraction from his merits seemed an oblation to the memory of my Father, especially as Parson Warren did certainly *pulpitize* much better.—

Somewhere, I think, about April 1792, [1782] Judge Buller, who had been educated by my Father, sent for me, having procured a Christ's Hospital Presentation.—I accordingly went to London, and was received by my mother's Brother, Mr Bowden, a Tobacconist & (at the same [time]) clerk to an Underwriter. My Uncle lived at the corner of the Stock exchange, & carried on his shop by means of a confidential Servant, who, I suppose, fleeced him most unmercifully.—He was a widower, & had one daughter who lived with a Miss Cabriere, an old Maid of great sensibilities & a taste for literature——Betsy Bowden had obtained an unlimited influence over her mind, which she still retains—Mrs Holt (for this is her name now) was, when I knew her, an ugly & an artful woman & not the kindest of Daughters—but indeed, my poor Uncle would have wearied the patience & affection of an Euphrasia.—He was generous as the air & a man of very considerable talents—but he was a Sot.—He received me with great affection, and I stayed ten weeks at his house, during which time I went occasionally to Judge Buller's. My Uncle was very proud of me, & used to carry me from Coffee-house to Coffee-house, and Tavern to Tavern, where I drank, & talked & disputed, as if I had been a man— /. Nothing was more common than for a large party to exclaim in my hearing, that I *was a prodigy*, &c &c &c—so that, while I remained at my Uncle's, I was most completely spoilt & pampered, both mind & body. At length the time came, & I donned the *Blue* coat & yellow stockings, & was sent down to Hertford, a town 20 miles from London, where there are about 300 of the younger Blue coat boys—At Hertford I was very happy, on the whole; for I had plenty to eat & drink, & pudding & vegetables almost every day. I stayed there six weeks; and then was drafted up to the great school at London, where I arrived in September, 1792 [1782]—and was placed in the second ward, then called Jefferies's ward; & in the under Grammar School. There are twelve Wards, or dormitories, of unequal sizes, beside the Sick Ward, in the great School—& they contained, all together, 700 boys; of whom I think nearly one third were the Sons of Clergymen. There are 5 Schools, a Mathematical, a Grammar, a drawing, a reading, & a writing School—all very large Buildings.—When a boy is admitted, if he read very badly, he is either sent to Hertford or to the Reading-School—(N.B. Boys are admissible from 7 to 12 years old)—If he learn to read tolerably well before 9, he is drafted into the lower Grammar-school—if not, into the writing-school, as having given proof of unfitness for classical attainment.—If before he is eleven he climbs up to the first form of the lower

Grammar-school, he is drafted into the head Grammar School—if not, at 11 years old he is sent into the writing School, where he continues till 14 or 15—and is then either apprenticed, & articled as clerk, or whatever else his turn of mind, or of fortune shall have provided for him. Two or three times a year the Mathematical Master beats up for recruits for the King's boys, as they are called—and all, who like the navy, are drafted into the Mathematical & Drawing Schools—where they continue till 16 or 17, & go out as Midshipmen & Schoolmasters in the Navy.—The Boys, who are drafted into the head Grammar School, remain there till 13—& then if not chosen for the university, go into the writing school. Each dormitory has a Nurse, or Matron—& there is a head Matron to superintend all these Nurses.—The boys were, when I was admitted, under excessive subordination to each other, according to rank in School—& every ward was governed by four Monitors, (appointed by the *Steward*, who was the supreme Governor out of School—our Temporal Lord) and by four *Markers*, who wore silver medals, & were appointed by the head Grammar Master, who was our supreme Spiritual Lord. The same boys were commonly both Monitors & Markers—We read in classes on Sundays to our *Markers*, & were catechized by them, & under their sole authority during prayers, &c—all other authority was in the monitors; but, as I said, the same boys were ordinarily both the one & the other.—Our diet was very scanty—Every morning a bit of dry bread & some bad small beer—every evening a larger piece of bread, & cheese or butter, whichever we liked—For dinner—on Sunday, boiled beef & broth—Monday, Bread & butter, & milk & water—on Tuesday, roast mutton, Wednesday, bread & butter & rice milk, Thursday, boiled beef & broth—Friday, boiled mutton & broth—Saturday, bread & butter, & pease porritch—Our food was portioned—& excepting on Wednesdays I never had a belly full. Our appetites were *damped* never satisfied—and we had no vegetables.—

S. T. Coleridge

[*To Thomas Poole*, after receiving news of the death of Berkeley Coleridge, 6 April 1799]

My dearest Poole

Your two letters, dated, Jan. 24th and March 15th, followed close on each other. I was still enjoying 'the livelier impulse and the dance of thought' which the first had given me, when I received the second.—At the time, in which I read Sara's lively account of the miseries which herself and the infant had undergone, all was over & well—there was nothing to *think* of—only a mass of Pain was brought suddenly and closely within the sphere of my perception, and I was made to suffer it over again. For this bodily frame is an imitative Thing, and touched by the imagination gives the hour that is past, as faithfully as a repeating watch.—But death—the death of an Infant—of one's own Infant!—I read your letter in calmness, and walked out into the open fields, oppressed, not by my feelings, but by the riddles, which the Thought so easily proposes, and solves—never! A Parent—in the strict and exclusive sense a *Parent*—! to me it is a *fable* wholly without meaning except in the *moral* which it suggests—a fable, of which the Moral is God. Be it so—my dear dear Friend! O let it be so! La nature (says Pascal) 'La Nature confond les Pyrrhoniens, et la raison confond les Dogmatistes. Nous avons une impuissance à prouver, invincible à tout le Dogmatisme: nous avons une idée de la vérité, invincible à tout le Pyrrhonisme [Nature confounds the Pyrrhonists, and reason the Dogmatists. Our inability to prove things cannot be conquered by Dogmatism; our idea of truth cannot be conquered by Pyrrhonism].' I find it wise and human to believe, even on slight evidence, opinions, the contrary of which cannot be proved, & which promote our happiness without hampering our Intellect.—My Baby has not lived in vain—this life has been to him what it is to all of us, education & developement! Fling yourself forward into your immortality only a few thousand years, & how small will not the difference between one year old & sixty years appear!—Consciousness—! it is not otherwise necessary to our conceptions of future Continuance than as connecting the *present link* of our Being with the one *immediately* preceding it; & *that* degree of Consciousness, *that* small portion of *memory*, it would not only be arrogant, but in the highest degree absurd, to deny even to a much younger Infant.—'Tis a strange assertion, that the Essence of Identity lies in *recollective* Consciousness—'twere scarcely less ridiculous to affirm, that the 8 miles from Stowey to Bridgewater consist in the 8 mile stones. Death in a doting old age falls upon my feelings ever as a more hopeless Phaenomenon than Death in Infancy / ; but *nothing* is hopeless.—What if the vital force which I sent from my arm into the stone, as I flung it in the air & skimm'd it upon the water—what if even that did not perish!—It was *life*—! it was a particle of *Being*—! it was *Power*!—& *how could* it perish—? *Life*, *Power*, *Being*!—organization may [be] &

probably *is*, their *effect*; their *cause* it *cannot* be!—I have indulged very curious fancies concerning that force, that *swarm* of motive Powers which I sent out of my body into that Stone; & which, one by one, left the untractable or already possessed Mass, and—but the German Ocean lies between us.—It is all too far to send you such fancies as these!——'Grief' indeed,

> Doth love to dally with fantastic thoughts,
> And smiling, like a sickly Moralist,
> Finds some resemblance to her own Concerns
> In the Straws of Chance, & Things Inanimate!

But I cannot truly say that I grieve—I am perplexed—I am sad—and a little thing, a very trifle would make me weep; but for the death of the Baby I have *not* wept!—Oh! this strange, strange, strange Scene-shifter, Death! that giddies one with insecurity, & so unsubstantiates the living Things that one has grasped and handled!—/ Some months ago Wordsworth transmitted to me a most sublime Epitaph / whether it had any reality, I cannot say.—Most probably, in some gloomier moment he had fancied the moment in which his Sister might die.

Epitaph

> A Slumber did my spirit seal,
> I had no human fears:
> She seem'd a Thing, that could not feel
> The touch of earthly years.
>
> No motion has she now, no force;
> She neither hears nor sees,
> Mov'd round in Earth's diurnal course
> With rocks, & stones, and trees! . . .

[*To William Godwin*, 25 March 1801]

Dear Godwin

I fear, your Tragedy will find me in a very unfit state of mind to sit in Judgement on it. I have been, during the last 3 months, undergoing a process of intellectual *exsiccation*. In my long Illness I had compelled into hours of Delight many a sleepless, painful hour of Darkness by chasing down metaphysical Game—and since then I have continued the Hunt, till I found myself unaware at the Root of Pure Mathematics—and up that tall smooth Tree, whose few poor Branches are all at its very summit, am I climbing by pure adhesive strength of arms and thighs—still slipping down, still renewing my ascent.—You would not know me—! all sounds of similitude keep at such a distance from each other in my mind, that I have *forgotten* how to make a rhyme—I look at the Mountains (that visible God Almighty that looks in at my windows) I look at the Mountains only for the Curves of their outlines; the Stars, as I behold them, form themselves into Triangles—and my hands are scarred with scratches from a Cat, whose back I was rubbing in the Dark in order to see whether the sparks from it were refrangible by a Prism. The Poet is dead in me—my imagination (or rather the Somewhat that had been imaginative) lies, like a Cold Snuff on the circular Rim of a Brass Candle-stick, without even a stink of Tallow to remind you that it was once cloathed & mitred with Flame. That is past by!—I was once a Volume of Gold Leaf, rising & riding on every breath of Fancy—but I have beaten myself back into weight & density, & now I sink in quicksilver, yea, remain squat and square on the earth amid the hurricane, that makes Oaks and Straws join in one Dance, fifty yards high in the Element.

However, I will do what I can—Taste & Feeling have I none, but what I have, give I unto thee.—But I repeat, that I am unfit to decide on any but works of severe Logic.

I write now to beg, that, if you have not sent your Tragedy, you may remember to send Antonio with it, which I have not yet seen—& likewise my Campbell's Pleasures of Hope, which Wordsworth wishes to see.

Have you seen the second Volume of the Lyrical Ballads, & the Preface prefixed to the First?—I should judge of a man's Heart, and Intellect precisely according to the degree & intensity of the admiration, with which he read those poems—Perhaps, instead of Heart I should have said Taste, but when I think of The Brothers, of Ruth, and of Michael, I recur to the expression, & am enforced to say *Heart*. If I die, and the Booksellers will give you any thing for my Life, be sure to say—'Wordsworth descended on him, like the *Γνῶθι σεαυτόν* [Know thyself] from Heaven; by shewing to him what true Poetry was, he made him know, that he himself was no Poet.'

In your next Letter you will perhaps give me some hints respecting your prose Plans.—.

God bless you
& S. T. Coleridge

[*To William Sotheby*,° 13 July 1802]

My dear Sir

... I will acknowledge to you, that your very, very kind Letter was not only a Pleasure to me, but a Relief to my mind / for after I had left you on the Road between Ambleside & Grasmere, I was dejected by the apprehension, that I had been unpardonably loquacious, and had oppressed you, & still more Mrs Sotheby, with my many words so impetuously uttered. But in simple truth you were yourselves in part the innocent causes of it / for the meeting with you; the manner of the meeting; your kind attentions to me; the deep & healthful delight, which every impressive & beautiful object seemed to pour out upon you; kindred opinions, kindred pursuits, kindred feelings, in persons whose Habits & as it were *Walk* of Life, have been so different from my own—; these, and more than these which I would but cannot say, all flowed in upon me with unusually strong Impulses of Pleasure / and Pleasure, in a body & soul such as I happen to possess, 'intoxicates more than strong Wine.'—However, *I promise to be a much more subdued creature—when you next meet me* / for I had but just recovered from a state of extreme dejection brought on in part by Ill-health, partly by other circumstances / and Solitude and solitary Musings do of themselves impregnate our Thoughts perhaps with more Life & Sensation, than will leave the Balance quite even.—But you, my dear Sir! looked [at a] Brother Poet with a Brother's Eyes—O that you were now in my study, & saw what is now before the window, at which I am writing, that rich mulberry-purple which a floating Cloud has thrown on the Lake—& that quiet Boat making its way thro' it to the Shore!—We have had little else but Rain & squally weather since you left us, till within the last three Days—but showery weather is no evil to us—& even that most oppressive of all weathers, hot small *Drizzle*, exhibits the Mountains the best of any. It produced such new combinations of Ridges in the Lodore & Borrodale Mountains, on Saturday morning, that, I declare, had I been blindfolded & so brought to the Prospect, I should scarcely have known them again. It was a Dream, such as Lovers have—a wild & transfiguring, yet enchantingly lovely, Dream of an Object lying by the side of the Sleeper. Wordsworth, who has walked thro' Switzerland, declared that he never saw any thing superior—perhaps nothing equal—in the Alps.— . . . On my return to Keswick I reperused the erste Schiffer° with great attention; & the result was an increasing Disinclination to the business of translating it / . . . It is easy to cloathe Imaginary Beings with our own Thoughts & Feelings; but to send ourselves out of ourselves, to *think* ourselves in to the Thoughts and Feelings of Beings in circumstances wholly & strangely different from our own / hoc labor, hoc opus [this is the work, this is the task] / and who has atchieved it? Perhaps only Shakespeare. Metaphisics is a word, that you, my dear Sir! are no great Friend to / but yet you will agree, that a great Poet must be, implicitè [implicitly] if not explicitè [explicitly], a profound Metaphysician. He may not have it in logical coherence, in his Brain & Tongue; but he must have it by *Tact* / for all sounds, & forms of human nature he must have the *ear* of a wild Arab listening in the silent Desart, the eye of a North American Indian tracing the footsteps of an Enemy upon the Leaves that strew the Forest—; the *Touch* of a Blind Man feeling the face of a darling Child— / and do not think me a Bigot, if I say, that I have read no French or German Writer, who appears to me to have had a *heart* sufficiently pure & simple to be capable of this or any thing like it. . . .

[*To Sara Hutchinson*, 6 August 1802]

There is one sort of Gambling, to which I am much addicted; and that not of the least criminal kind for a man who has children & a Concern.—It is this. When I find it convenient to descend from a mountain, I am too confident & too indolent to look round about & wind about 'till I find a track or other symptom of safety; but I wander on, & where it is first *possible* to descend, there I go—relying upon fortune for how far down this possibility will continue. So it was yesterday afternoon, I passed down from Broadcrag, skirted the Precipices, and found myself cut off from a most sublime Crag-summit, that seemed to rival Sca' Fell Man in height, & to outdo it in fierceness. A Ridge of Hill lay low down, & divided this Crag (called Doe-crag) & Broad-crag—even as the Hyphen divides the words broad & crag. I determined to go thither; the first place I came to, that was not direct Rock, I slipped down, & went on for a while with tolerable ease—but now I came (it was midway down) to a smooth perpendicular Rock about 7 feet high—this was nothing—I put my hands on the Ledge, & dropped down / in a few yards came just such another / I *dropped* that too / and yet another, seemed not higher—I would not stand for a trifle / so I dropped that too / but the stretching of the muscle[s] of my hands & arms, & the jolt of the Fall on my Feet, put my whole Limbs in a *Tremble*, and I paused, & looking down, saw that I had little else to encounter but a succession of these little Precipices—it was in truth a Path that in a very hard Rain is, no doubt, the channel of a most splendid Waterfall.—So I began to suspect that I ought not to go on / but then unfortunately tho' I could with ease drop down a smooth Rock 7 feet high, I could not *climb* it / so go on I must / and on I went / the next 3 drops were not half a Foot, at least not a foot more than my own height / but every Drop increased the Palsy of my Limbs—I shook all over, Heaven knows without the least influence of Fear / and now I had only two more to drop down / to return was impossible—but of these two the first was tremendous / it was twice my own height, & the Ledge at the bottom was [so] exceedingly narrow, that if I dropt down upon it I must of necessity have fallen backwards & of course killed myself. My Limbs were all in a tremble—I lay upon my Back to rest myself, & was beginning according to my Custom to laugh at myself for a Madman, when the sight of the Crags above me on each side, & the impetuous Clouds just over them, posting so luridly & so rapidly northward, overawed me / I lay in a state of almost prophetic Trance & Delight—& blessed God aloud, for the powers of Reason & the Will, which remaining no Danger can overpower us! O God, I exclaimed aloud—how calm, how blessed am I now / I know not how to proceed, how to return / but I am calm & fearless & confident / if this Reality were a Dream, if I were asleep, what agonies had I suffered! what screams!—When the Reason & the Will are away, what remain to us but Darkness & Dimness & a bewildering Shame, and Pain that is utterly Lord over us, or fantastic Pleasure, that draws the Soul along swimming through the air in many shapes, even as a Flight of Starlings in a Wind.—I arose, & looking down saw at the bottom a heap of Stones—which had fallen abroad—and rendered the narrow Ledge on which they had been piled, doubly dangerous / at the bottom of the third Rock that I dropt from, I met a dead Sheep quite rotten—This heap of Stones, I guessed, & have since found that I guessed aright, had been piled up by the Shepherd to enable him to climb up & free the poor creature whom he had observed to be crag-fast—but seeing nothing but rock over rock, he had desisted & gone for help—& in the mean time the poor creature had fallen down & killed itself.—As I was looking at these I glanced my eye to my left, & observed that the Rock was rent from top to bottom—I measured the breadth of the Rent, and found that there was no danger of my being *wedged* in / so I put my Knap-sack round to my side, & slipped down as between two walls, without any danger or difficulty——the next Drop brought me down on the Ridge called the How / I hunted out my Besom Stick, which I had flung before me when I first came to the Rocks—and wisely gave over all thoughts of ascending Doe-Crag—for now the Clouds were again coming in most tumultuously—so I began to descend / when I felt an odd sensation across my whole Breast—not pain nor itching—& putting my hand on it I found it all bumpy—and on looking saw the whole of my Breast from my Neck to my Navel—& exactly all that my Kamell-hair Breast-shield covers, filled with great red heat-bumps, so thick that no hair could lie between them. They still remain / but are evidently less—& I have no doubt will wholly disappear in a few Days. . . .

[*To William Sotheby*, 10 September 1802]

. . . Bowles's Stanzas on Navigation are among the best in that second Volume / but the whole volume is woefully inferior to its Predecessor. There reigns thro' all the blank verse poems such a perpetual trick of *moralizing* every thing—which is very well, occasionally—but never to see or describe any interesting appearance in nature, without connecting it by dim analogies with the moral world, proves faintness of Impression. Nature has her proper interest; & he will know what it is, who believes & feels, that every Thing has a Life of its own, & that we are all *one Life*. A Poet's *Heart* & *Intellect* should be *combined*, *intimately* combined & *unified*, with the great appearances in Nature—& not merely held in solution & loose mixture with them, in the shape of formal Similies. I do not mean to *exclude* these formal Similies—there are moods of mind, in which they are natural—pleasing moods of mind, & such as a Poet will often have, & sometimes express; but they are not his highest, & most appropriate moods. They are 'Sermoni propiora [nearer to discourse]' which I once translated—'*Properer for a Sermon*.' The truth is—Bowles has indeed the *sensibility* of a poet; but he has not the *Passion* of a great Poet. His latter Writings all want *native* Passion—Milton here & there supplies him with an appearance of it—but he has no native Passion, because he is not a Thinker—& has probably weakened his Intellect by the haunting Fear of becoming extravagant / Young somewhere in one of his prose works remarks that there is as profound a Logic in the most daring & dithyrambic parts of Pindar, as in the Ὄργανον [*Organon*] of Aristotle—the remark is a valuable one /

> Poetic Feelings, like the flexuous Boughs
> Of mighty Oaks, yield homage to the Gale,
> Toss in the strong winds, drive before the Gust,
> Themselves one giddy storm of fluttering Leaves;
> Yet all the while, self-limited, remain
> Equally near the fix'd and parent Trunk
> Of Truth & Nature, in the howling Blast
> As in the Calm that stills the Aspen Grove.—

That this is deep in our Nature, I felt when I was on Sca' fell—. I involuntarily poured forth a Hymn in the manner of the *Psalms*, tho' afterwards I thought the Ideas &c disproportionate to our humble mountains—& accidentally lighting on a short Note in some swiss Poems, concerning the Vale of Chamouny, & its Mountain, I transferred myself thither, in the Spirit, & adapted my former feelings to these grander external objects. You will soon see it in the Morning Post—& I should be glad to know whether & how far it pleased you.—It has struck [me] with great force lately, that the Psalms afford a most compleat answer to those, who state the Jehovah of the Jews, as a personal & national God—& the Jews, as differing from the Greeks, only in calling the minor Gods, Cherubim & Seraphim—& confining the word God to their Jupiter. It must occur to every Reader that the Greeks in their religious poems address always the Numina Loci [Spirits of the Place], the Genii, the Dryads, the Naiads, &c &c—All natural Objects were *dead*—mere hollow Statues—but there was a Godkin or Goddessling *included* in each—In the Hebrew Poetry you find nothing of this poor Stuff—as poor in genuine Imagination, as it is mean in Intellect— / At best, it is but Fancy, or the aggregating Faculty of the mind—not *Imagination*, or the *modifying*, and *co-adunating* Faculty. This the Hebrew Poets appear to me to have possessed beyond all others—& next to them the English. In the Hebrew Poets each Thing has a Life of its own, & yet they are all one Life. In God they move & live, & *have* their Being—not *had*, as the cold System of Newtonian Theology represents / but *have*. Great pleasure indeed, my dear Sir! did I receive from the latter part of your Letter. If there be any two subjects which have in the very depth of my Nature interested me, it has been the Hebrew & Christian Theology, & the Theology of Plato. Last winter I read the Parmenides & the Timaeus with great care—and O! that you were here, even in this howling Rain-Storm that dashes itself against my windows, on the other side of my blazing Fire, in that great Arm Chair there—I guess, we should encroach on the morning before we parted. How little the Commentators of Milton have availed themselves of the writings of Plato / Milton's Darling! But alas! commentators only hunt out verbal Parallelisms—*numen abest* [the spirit is absent]. I was much impressed with this in all the many Notes on that beautiful Passage in Comus from l. 629 to 641—all the puzzle is to find out what Plant Haemony is—which they discover to be the English Spleenwort—& decked out, as a mere play & licence of poetic Fancy, with all the strange properties suited to the purpose of the Drama—They thought little of Milton's platonizing Spirit—who wrote nothing without an interior meaning. 'Where more is meant, than meets the ear' is true of himself beyond all writers. He was so great a Man, that he seems to have considered Fiction as profane, unless where it is consecrated by being emblematic of some Truth / What an unthinking & ignorant man we must have supposed Milton to be, if without any hidden meaning, he had described [it] as growing in such abundance that the dull Swain treads on it daily—& yet as never *flowering*—Such blunders Milton, of all others, was least likely to commit—Do look at the passage—apply it as an Allegory of Christianity, or to speak more precisely of the Redemption by the Cross—every syllable is full of Light!— . . .

[*To Robert Southey*, 7 August 1803]

My dear Southey

The last 3 days I have been fighting against a restless wish to write to you. I am afraid, lest I should infect you with my fears rather than furnish you with any new arguments—give you impulses rather than motives—and prick you with *spurs*, that had been dipt in the vaccine matter of my own cowardliness—. While I wrote that last sentence, I had a vivid recollection—indeed an ocular Spectrum—of our room in College Street—/ a curious instance of association / you remember how incessantly in that room I used to be compounding these half-verbal, half-visual metaphors. It argues, I am persuaded, a particular state of general feeling—& I hold, that association depends in a much greater degree on the recurrence of resembling states of Feeling, than on Trains of Idea / that the recollection of early childhood in latest old age depends on, & is explicable by this—& if this be true, Hartley's System totters.—If I were asked, how it is that very old People remember *visually* only the events of early childhood—& remember the intervening Spaces either not at all, or only verbally—I should think it a perfectly philosophical answer / that old age remembers childhood by becoming 'a second childhood.' This explanation will derive some additional value if you would look into Hartley's solution of the phaenomena / how flat, how wretched!—Believe me, Southey! a metaphysical Solution, that does not instantly *tell* for something in the Heart, is grievously to be suspected as apocry[p]hal. I almost think, that Ideas *never* recall Ideas, as far as they are Ideas—any more than Leaves in a forest create each other's motion—The Breeze it is that runs thro' them / it is the Soul, the state of Feeling—. If I had said, no *one* Idea ever recalls another, I am confident that I could support the assertion.——And this is a Digression.—My dear Southey, again & again I say, that whatever your Plan be, I will continue to work for you with equal zeal if not with equal pleasure / —But the arguments against your plan weigh upon me the more heavily, the more I reflect—& it could not be otherwise than that I should feel a confirmation of them from Wordsworth's compleat coincidence—I having requested his deliberate opinion without having communicated an Iota of my own.—You seem to me, dear friend! to hold the dearness of a scarce work for a proof, that the work would have a general Sale—if not scarce.—Nothing can be more fallacious than this. Burton's anatomy used to sell for a guinea to two guineas—it was republished / has it payed the expence of reprinting? Scarcely.—Literary History informs us, that most of those great continental Bibliographies &c were published by the munificence of princes, or nobles, or great monasteries.—A Book from having had little or no sale, except among great Libraries, may become so scarce, that the number of competitors for it, tho' few, may be proportionally very great. I have observed, that great works are now a days bought, not for curiosity, or the amor proprius [self-love]—but under the notion that they contain all the *knowledge*, a man may ever want/ and that if he has it on his *Shelf*, why there it is, as snug as if it were in his *Brain*. . . .

[*To George Coleridge*, 2 April 1806]

My dear Brother

The omniscience of the supreme Being has always appeared to me among the most tremendous thoughts, of which an imperfect rational Being is capable; and to the very best of men one of the most awful attributes of God is, the Searcher of Hearts. As he knows us, we are not capable of knowing ourselves—it is not impossible, that this perfect (as far as in a creature can be) Self-knowlege may be among the spiritual punishments of the abandoned, as among the joys of the redeemed Spirits. Yet there are occasions, when it would be both a comfort and advantage to us, if with regard to a particular conduct & the feelings & impulses connected with it, we could make known to another and with the same degree of vividness the state of our own Hearts, even as it exists in our own consciousness. Sure am I at least, that I should rejoice if without the pain & struggles of communication (pain referent not to any delicacy or self-reproach of my own) there could be conveyed to you a fair Abstract of all that has passed within me, concerning yourself and Ottery, and the place of my future residence, & the nature of my future employments (all more or less connected with you)—but after I have been with you awhile, in proportion as I gain your confidence & confident esteem, so I shall be able to pour my whole Heart into you—I leave this place (a seat of Sir G. Beaumont's) on Saturday, March [April] the 4th—& proceed to Bristol—where I am to meet Mrs Coleridge, & the two children (for Hartley is with me) and immediately proceed to Ottery.—If you find reason to believe, that I should be an assistance or a comfort to you by settling there in any connection with you, I am prepared to strike root in my native place; and if you knew the depth of the friendship, I have now for ten years (without the least fluctuation amid the tenderest and yet always respectful Intimacy) felt toward, and enjoyed from, Mr W. Wordsworth, as well as the mutual Love between me and his immediate House-hold, you would not think the less of my affection and sense of duty towards you, my paternal Brother, when I confess that the resolution to settle myself at so great a distance from him has occasioned one among the two or three *very severe* struggles of my life. Previously however to my meeting you, and at the time of thus communicating to you my resolve, provided it should be satisfactory to you— it is absolutely necessary that I should put you in possession of the true state of my domestic Affairs—the agony, which I feel on the very thought of the subject and the very attempt to write concerning it, has been a principal cause not only of the infrequency & omission of my correspondence with you, but of the distraction of all settled pursuits hitherto—

In short, with many excellent qualities, of strict modesty, attention to her children, and economy, Mrs Coleridge has a temper & general tone of feeling, which after a long—& for six years at least—a patient Trial I have found wholly incompatible with even an endurable Life, & such as to preclude all chance of my ever developing the talents, which my Maker has entrusted to me—or of applying the acquirements, which I have been making one after the other, because I could not be doing nothing, & was too sick at heart to exert myself in drawing from the sources of my own mind to any perseverance in any regular plan. The few friends, who have been Witnesses of my domestic Life, have long advised separation, as the necessary condition of every thing desirable for me—nor does Mrs Coleridge herself state or pretend to any objection on the score of attachment to me;—that it will not look *respectable* for her, is the sum into which all her objections resolve themselves.—At length however, it is settled (indeed, the state of my Health joined with that of my circumstances, and the duty of providing what I can, for my three Children, would of themselves dictate the measure, tho' we were only indifferent to each other) but Mrs Coleridge wishes—& very naturally—to accompany me into Devonshire, that our separation may appear free from all shadow of suspicion of any other cause than that of unfitness & unconquerable difference of Temper. O that those, who have been Witnesses of the Truth, could but add for me that commentary on my last Words, which my very respect for Mrs Coleridge's many estimable qualities would make it little less than torture to me to attempt.—However, we part as Friends—the boys of course will be with me. What more need be said, I shall have an opportunity of saying when we are together.—If you wish to write to me, before my arrival, my address will be—Mr Wade's, Aggs' Printing-office, St Augustin's Back, Bristol.

Make my apologies to my dear Nephews; and assure them, that it will be a great Joy to me to endeavour to compensate for my epistolary neglect by my conversation with them—and that any valuable Knowledge, which it should be in my power to communicate to them, will on their account become more valuable to me.—My Love & my Duty to all, who have to claim it from me. I am, my dear Brother, with grateful & affectionate esteem

your friend & brother,
S. T. Coleridge

[*To John Morgan*, 14 May 1814]

My dear Morgan

If it could be said with as little *appearance* of profaneness, as there is feeling or intention in my mind, I might affirm; that I had been crucified, dead, and buried, descended into *Hell*, and am now, I humbly trust, rising again, tho' slowly and gradually. I thank you from my heart for your far too kind Letter to Mr Hood—so much of it is true that such as you described I always wished to be. I know, it will be vain to attempt to persuade Mrs Morgan or Charlotte, that a man, whose moral feelings, reason, understanding, and senses are perfectly sane and vigorous, may yet have been *mad*—And yet nothing is more true. By the long long Habit of the accursed Poison my Volition (by which I mean the faculty *instrumental* to the Will, and by which alone the Will can realize itself—its Hands, Legs, & Feet, as it were) was compleatly deranged, at times frenzied, dissevered itself from the Will, & became an independent faculty: so that I was perpetually in the state, in which you may have seen paralytic Persons, who attempting to push a step forward in one direction are violently forced round to the opposite. I was sure that no ease, much less pleasure, would ensue: nay, was certain of an accumulation of pain. But tho' there was no prospect, no gleam of Light before, an indefinite indescribable Terror as with a scourge of ever restless, ever coiling and uncoiling Serpents, drove me on from behind.—The worst was, that in *exact proportion* to the *importance* and *urgency* of any Duty was it, as of a fatal necessity, sure to be neglected: because it added to the Terror above described. In exact proportion, as I *loved* any person or persons more than others, & would have sacrificed my Life for them, were *they* sure to be the most barbarously mistreated by silence, absence, or breach of promise.—I used to think St James's Text, 'He who offendeth in one point of the Law, offendeth in all', very harsh; but my own sad experience has taught me its aweful, dreadful Truth.—What crime is there scarcely which has not been included in or followed from the one guilt of taking opium? Not to speak of ingratitude to my maker for the wasted Talents; of ingratitude to so many friends who have loved me I know not why; of barbarous neglect of my family; excess of cruelty to Mary & Charlotte, when at Box, and both ill—(a vision of Hell to me when I think of it!) I have in this one dirty business of Laudanum an hundred times deceived, tricked, nay, actually & consciously LIED.—And yet *all* these vices are so opposite to my nature, that but for this *free-agency-annihilating* Poison, I verily believe that I should have suffered myself to have been cut to pieces rather than have committed any one of them.

At length, it became too bad. I used to take 4 to 5 ounces a day of Laudanum, once [*manuscript torn here*] . . . [ou]nces, i.e. near a Pint—besides great quantities [of liquo]r. From the Sole of my foot to the Crown of [my h]eart there was not an Inch in which I was not [contin]ually in torture: for more than a fortnight no [sleep] ever visited my Eye lids—but the agonies of [remor]se were far worse than all!—Letters past between Cottle, Hood, & myself—& our kind Friend, Hood, sent Mr Daniel to me. At his second Call I told him plainly (for I had sculked out the night before & got Laudanum) that while I was in my own power, all would be in vain—I should inevitably cheat & trick *him*, just as I had done Dr Tuthill—that I must either be removed to a place of confinement, or at all events have a Keeper.—Daniel saw the truth of my observations, & my most faithful excellent friend, Wade, procured a strong-bodied, but decent, meek, elderly man, to superintend me, under the name of my Valet—All in the House were forbidden to fetch any thing but by the Doctor's order.—Daniel generally spends two or three hours a day with me—and already from 4 & 5 ounces has brought me down to four tea-spoonfuls in the 24 Hours—The terror & the indefinite craving are gone—& he expects to drop it altogether by the middle of next week—Till a day or two after that I would rather not see you.

[*To John Morgan*, 15 May 1814]

My dear Morgan

To continue from my last—Such was the direful state of my mind, that (I tell it you with horror) the razors, penknife, & every possible instrument of Suicide it was found necessary to remove from my room! My faithful, my *inexhaustibly patient* Friend, WADE, has caused a person to sleep by my bed side, on a bed on the floor: so that I might never be altogether alone—O Good God! why do such good men love me! At times, it would be more delightful to me to lie in the Kennel, & (as Southey said) 'unfit to be pulled out by any honest man except with a pair of Tongs.'—What *he* then said (perhaps) rather unkindly of me, was prophetically true! Often have I wished to have been thus trodden & spit upon, if by any means it might be an atonement for the direful guilt, that (like all others) first *smiled* on me, like Innocence! then crept closer, & yet closer, till it had thrown its serpent folds round & round me, and I was no longer in my own power!—*Something* even the most wretched of Beings (*human* Beings at least) owes to himself—& this I *will* say & *dare* with truth say—that never was I led to this wicked direful practice of taking Opium or Laudanum by any desire or expectation of exciting *pleasurable* sensations; but purely by *terror*, by cowardice of pain, first of mental pain, & afterwards as my System became weakened, even of bodily Pain.

My Prayers have been fervent, in agony of Spirit, and for hours together, incessant! still ending, O! only for the merits, for the agonies, for the cross of my blessed Redeemer! For I am nothing, but evil—I can do nothing, but evil! Help, Help!—I believe! help thou my unbelief!— . . .

[*To William Wordsworth*, 30 May 1815]

My honored Friend

On my return from Devizes, whither I had gone to procure some Vaccine matter (the Small Pox having appeared in Calne, and Mrs M's Sister believing herself never to have had it) I found your Letter: and I will answer it immediately, tho' to answer it as I could wish to do would require more recollection and arrangement of Thought than is always to be commanded on the Instant. But I dare not trust my own habit of procrastination—and do what I would, it would be impossible in a single Letter to give more than *general* convictions. But even after a tenth or twentieth Letter I should still be disquieted as knowing how poor a substitute must Letters be for a vivâ voce examination of a Work with its Author, Line by Line. It is most uncomfortable from many, many Causes, to express anything but sympathy, and gratulation to an absent friend, to whom for the more substantial Third of a Life we have been habituated to look up: especially, where our Love, tho' increased by many and different influences, yet begun and throve and knit its Joints in the perception of his Superiority. It is not in *written words*, but by the hundred modifications that Looks make, and Tone, and denial of the FULL sense of the very words used, that one can reconcile the struggle between sincerity and diffidence, between the Persuasion, that I am in the Right, and that as deep tho' not so vivid conviction, that it may be the positiveness of Ignorance rather than the Certainty of Insight. Then come the Human Frailties—the dread of giving pain, or exciting suspicions of alteration and Dyspathy—in short, the almost inevitable Insincerities between imperfect Beings, however sincerely attached to each other. It is hard (and I am *Protestant* enough to doubt whether it is right) to confess the whole Truth even of one's Self—Human Nature scarce endures it even *to* one's Self!—but to me it is still harder to do this of and to a revered Friend.—

But to your Letter. First, I had never determined to print the Lines addressed to you—I lent them to L. Beaumont on her promise that they should be copied and returned—& not knowing of any copy in my own possession I sent for them, because I was making a *Mss* Collection of *all* my poems, publishable or unpublishable—& still more perhaps, for the Handwriting of the only perfect Copy, that entrusted to her Ladyship.—Most assuredly, I never once thought of printing them without having consulted you—and since I lit on the first rude draught, and corrected it as well as I could, I wanted no additional reason for its not being published in my Life Time, than its *personality* respecting myself—After the opinions, I had given publicly, for the preference of the Lycidas (moral no less than poetical) to Cowley's Monody, I could not have printed it consistently—. It is for the Biographer, not the Poet, to give the *accidents* of *individual* Life. Whatever is not representative, generic, may be indeed most poetically exprest, but is not Poetry. Otherwise, I confess, your prudential Reasons would not have weighed with me, except as far as my name might haply injure your reputation—for there is nothing in the Lines as far [as] your Powers are concerned, which I have not as fully expressed elsewhere—and I hold it a miserable cowardice to withhold a deliberate opinion only because the Man is alive.

2ndly. for the EXCURSION. I feared that had I been silent concerning the Excursion, Lady B. would have drawn some strange inference—& yet I had scarcely sent off the Letter before I repented that I had not run that risk rather than have approach to Dispraise communicated to you by a third person—. But what did my criticism amount to, reduced to its full and naked Sense?—This: that *comparatively* with the *former* Poem [the unpublished *Prelude*] the excursion, as far as it was new to me, had disappointed my expectations—that the Excellences were so many and of so high a class, that it was impossible to attribute the inferiority, if any such really existed, to any flagging of the Writer's own genius—and that I conjectured that it might have been occasioned by the influence of self-established Convictions having given to certain Thoughts and Expressions a depth & force which they had not for readers in general.—In order therefore to explain the *disappointment* I must recall to your mind what my *expectations* were: and as these again were founded on the supposition, that (in whatever order it might be published) the Poem on the growth of your own mind was as the ground-plat and the Roots, out of which the Recluse was to have sprung up as the Tree—as far as the same Sap in both, I expected them doubtless to have formed one compleat Whole, but in matter, form, and product to be different, each not only a distinct but a different Work.—In the first I had found 'themes by thee first sung aright'—

Of Smiles spontaneous and mysterious fears
(The first-born they of Reason and Twin-birth),
Of Tides obedient to external force,
And Currents self-determin'd, as might seem,
Or by some central Breath; of moments aweful,
Now in thy inner Life, and now abroad,
When Power stream'd from thee, and thy Soul received
The Light reflected as a Light bestow'd!—
Of Fancies fair, and milder Hours of Youth,
Hyblaean murmurs of poetic Thought
Industrious in its Joy, in vales and Glens
Native or outland, Lakes and famous Hills!
Or on the lonely High-road, when the stars
Were rising; or by secret mountain streams,
The Guides and the Companions of thy Way.

Of more than *Fancy*—of the *social sense*
Distending wide and man beloved as man:
Where France in all her towns lay vibrating
Ev'n as a Bark becalm'd beneath the Burst
Of Heaven's immediate Thunder, when no Cloud
Is visible, or Shadow on the Main!—
For Thou wert there, thy own Brows garlanded,
Amid the tremor of a realm aglow,
Amid a mighty Nation jubilant,
When from the general Heart of Human Kind
HOPE sprang forth, like a full-born Deity!
Of that dear Hope afflicted, and amaz'd,
So homeward summon'd! thenceforth calm & sure
From the dread Watch tower of man's absolute Self
With Light unwaning on her Eyes, to look

> Far on!—herself a Glory to behold,
> The Angel of the Vision!—Then (last strain!)
> Of Duty! Chosen Laws controlling choice!
> Action and Joy!—AN ORPHIC SONG INDEED,
> A SONG DIVINE OF HIGH AND PASSIONATE TRUTHS
> TO THEIR OWN MUSIC CHAUNTED!

Indeed thro' the whole of that Poem 'με Αὔρα τις εἰσέπνευσε μυστικωτάτη [a most mystical Breeze blew upon me].' *This* I considered as 'the EXCURSION'; and the second as 'THE RECLUSE' I had (from what I had at different times gathered from your conversation on the Plan) anticipated as commencing with you set down and settled in an abiding Home, and that with the Description of that Home you were to begin a *Philosophical Poem*, the result and fruits of a Spirit so fram'd & so disciplin'd, as had been told in the former. Whatever in Lucretius is Poetry is not philosophical, whatever is philosophical is not Poetry: and in the very Pride of confident Hope I looked forward to the Recluse, as the *first* and *only* true Phil. Poem in existence. Of course, I expected the Colors, Music, imaginative Life, and Passion of *Poetry*; but the matter and arrangement of *Philosophy*—not doubting from the advantages of the Subject that the Totality of a System was not only capable of being harmonized with, but even calculated to aid, the unity (Beginning, Middle, and End) of a *Poem*. Thus, whatever the Length of the Work might be, still it was a *determinate* Length: of the subjects announced each would have its own appointed place, and excluding repetitions each would relieve & rise in interest above the other—. I supposed you first to have meditated the faculties of Man in the abstract, in their correspondence with his Sphere of action, and first, in the Feeling, Touch, and Taste, then in the Eye, & last in the Ear, to have laid a solid and immoveable foundation for the Edifice by removing the sandy Sophisms of Locke, and the Mechanic Dogmatists, and demonstrating that the Senses were living growths and developements of the Mind & Spirit in a much juster as well as higher sense, than the mind can be said to be formed by the Senses—. Next, I understood that you would take the Human Race in the concrete, have exploded the absurd notion of Pope's Essay on Man, Darwin, and all the countless Believers—even (strange to say) among Xtians of Man's having progressed from an Ouran Outang state—so contrary to all History, to all Religion, nay, to all Possibility—to have affirmed a Fall in some sense, as a fact, the possibility of which cannot be understood from the nature of the Will, but the reality of which is attested by Experience & Conscience—Fallen men contemplated in the different ages of the World, and in the different states—Savage—Barbarous—Civilized—the lonely Cot, or Borderer's Wigwam—the Village—the Manufacturing Town—Sea-port—City—Universities—and not disguising the sore evils, under which the whole Creation groans, to point out however a manifest Scheme of Redemption from this Slavery, of Reconciliation from this Enmity with Nature—what are the Obstacles, the *Antichrist* that must be & already is—and to conclude by a grand didactic swell on the necessary identity of a true Philosophy with true Religion, agreeing in the results and differing only as the analytic and synthetic process, as discursive from intuitive, the former chiefly useful as perfecting the latter—in short, the necessity of a general revolution in the modes of developing & disciplining the human mind by the substitution of Life, and Intelligence (considered in its different powers from the Plant up to that state in which the difference of Degree becomes a new kind (man, self-consciousness) but yet not by essential opposition) for the philosophy of mechanism which in every thing that is most worthy of the human Intellect strikes *Death*, and cheats itself by mistaking clear Images for distinct conceptions, and which idly demands Conceptions where Intuitions alone are possible or adequate to the majesty of the Truth.—In short, Facts elevated into Theory—Theory into Laws—& Laws into living & intelligent Powers—true Idealism necessarily perfecting itself in Realism, & Realism refining itself into Idealism.—

Such or something like this was the Plan, I had supposed that you were engaged on—. Your own words will therefore explain my feelings—viz—that your object 'was not to convey recondite or refined truths but to place commonplace Truths in an interesting point of View.' Now this I supposed to have been in your two Volumes of Poems, as far as was desirable, or p[ossible,] without an insight into the whole Truth—. How can common [trut]hs be made permanently interesting but by being *bottomed* in our common nature—it is only by the profoundest Insight into Numbers and Quantity that a sublimity & even religious Wonder become attached to the simplest operations of Arithmetic, the most evident properties of the Circle or Triangle—.

I have only to finish a Preface which I shall have done in two or at farthest three days—and I will then, dismissing all comparison either with the Poem on the Growth of your own Support [*sic*], or with the imagined Plan of the Recluse, state fairly my main Objections to the Excursion as it is—But it would have been alike unjust both to you and to myself, if I had led you to suppose that any disappointment, I may have felt, arose wholly or chiefly from the Passages, I do not like—or from the Poem considered irrelatively.—

Allston lives at 8, Buckingham Place, Fitzroy Square—He has lost his wife—& been most unkindly treated—& most unfortunate—I hope, you will call on him.—Good God! to think of such a Grub as *Dawe* with more than he can do—and such a Genius as Allston, without a Single Patron!

God bless you!—I am & never have been other than your | most affectionate

S. T. Coleridge.—

[*To Thomas Allsop*,° 30 March 1820]

My dear young Friend

The only impression left by you on my mind, of which I am aware, is an increased desire to see you again and at shorter intervals. Were you my son by nature, I could not hold you dearer or more earnestly desire to retain you the adopted of whatever within me will remain when the dross and alloy of infirmity shall have been purged away. I feel the most entire confidence, that no prosperous change of my outward circumstances would add to *your* faith in the sincerity of this assurance: still, however, the average of men being what it is, and it being neither possible nor desirable to be fully conscious in our understanding of the habits of thinking and judging in the world around us, and yet to be wholly impassive and unaffected by them in our feelings, it would endear, and give a new value to, an honorable competence, that I should be able to evince the true nature and degree of my esteem and attachment, beyond the suspicion even of the sordid, and separate from all that is accidental and adventitious. But yet the gratitude, I feel to you, is so genial a Warmth, and blends so undistinguishably with my affections, is so perfectly one of the family in the Household of Love, that I would not be otherwise than obliged and indebted to you: and God is my witness, that my wish for an easier and less embarrassed lot is *chiefly* (I think, I might have said, *exclusively*) grounded in the deep conviction, that exposed to a less bleak aspect I should bring forth flowers and fruits, both more abundant and more worthy of the unexampled kindness of your *faith* in me.—Interpreting the '*wine*' and the 'ivy garland' as figures of poetry signifying competence & the removal of the petty needs of the body that plug up the pipes of the playing Fountain (and such, it is too well known, was the intent and meaning of the hardly used Poet)—and O! how often, when my heart has begun to swell from the genial warmth of thought as our northern Lakes from the (so called) bottom-winds when all above and around is Stillness and Sunshine—how often have I repeated in my own name the sweet Stanza of Edmund Spenser—

> Thou kenst not, Percie, how the rhyme should rage
> O! if my temples were bedew'd with wine,
> And girt in garlands of wild ivy twine—
> How I could rear the Muse on stately stage
> And teach her tread aloft in buskin fine
> With queint Bellona in her equipage!

Read this as you would a note at the bottom of a page.

> But ah! Mecaenas is ywrapt in clay
> And great Augustus long ago is dead—

this is a natural sigh, & natural too is the reflection that follows—

> And if that any buddes of Poesy
> Yet of the old stock gin to shoot again,
> 'Tis or *self*-lost the worldling's meed to gain,
> And with the rest to breathe its ribauldry:
> Or as it sprung, it wither must again—
> Tom Piper makes them better melody.

but tho' natural, the complaint is not equally philosophical, were it only on this account, that I know of no age in which the same has not been advanced, & with the same grounds. Nay, I retract. There never was a time, in which the *complaint* would be so little wise, tho' perhaps none in which the *fact* is more *prominent*. Neither Philosophy or Poetry ever did, nor as long as they are terms of comparative excellence & contradistinction, ever can be *popular*, nor honored with the praise and favor of Contemporaries. But on the other hand, there never was a time, in which either books, that were *held* for excellent as poetic or philosophic, had so extensive and rapid a sale, or men reputed Poets and Philosophers of a high rank were so much *looked up to* in Society or so munificently, almost profusely, rewarded.—Walter Scott's Poems & Novels (except only the two wretched Abortions, Ivanhoe & the Bride of Ravensmuir or whatever its name be) supply both instance & solution of the *present* conditions & components of popularity—viz—to amuse without requiring any effort of thought, & without exciting any deep emotion. The age seems *sore* from excess of stimulation, just as a day or two after a thorough Debauch & long sustained Drinking-match a man feels all over like a Bruise. Even to *admire* otherwise than *on the whole* and where 'I admire' is but a synonyme for 'I remember, I *liked* it very much *when I was reading it*', is too much an effort, would be too disquieting an emotion! Compare Waverley, Guy Mannering, &c with works that had an *immediate run* in the last generation—Tristram Shandy, Roderick Random, Sir Ch. Grandison, Clarissa Harlow, & Tom Jones (all which became popular as soon as published & therefore instances fairly in point) and you will be convinced, that the difference of Taste is real & not any fancy or croaking of my own.

But enough of these Generals. It was my purpose to open myself out to you in detail.—My health, I have reason to believe, is so intimately connected with the state of my Spirits, and these again so dependent on my thoughts, prospective and retrospective, that I should not doubt the being favored with a sufficiency for my noblest undertakings, had I the ease of heart requisite for the necessary abstraction of the Thoughts, and such a reprieve from the goading of the immediate exigencies as might make tranquillity possible. But alas! I know by experience (and this knowledge is not the less, because the regret is not unmixed with self-blame and the consciousness of want of exertion and fortitude) that my health will continue to decline, as long as the pain from reviewing the barrenness of the Past is great in an inverse proportion to any rational anticipations of the Future. As I now am, however, from 5 to 6 hours devoted to actual writing and composition in the day is the utmost, that my strength, not to speak of my nervous system, will permit: and the invasions on this portion of my time from applications, often of the most senseless kind, are such and so many, as to be almost as ludicrous even to myself as they are vexatious. In less than a week I have not seldom received half a dozen packets or parcels, of works printed or manuscript, urgently requesting my *candid judgement*, or my correcting hand—add to these Letters from Lords & Ladies urging me to write reviews & puffs of heaven-born Geniuses, whose whole merit consists in their being Ploughmen or Shoemakers—Ditto from Actors—Ditto, Intreaties—for money or recommendations to Publishers from Ushers out of place, &c &c—and to *me*, who have neither influence, interest, or money—and what is still more apropos, can neither bring myself to tell smooth falshoods or harsh truths, and in the struggle too often do both in the anxiety to do neither—. I have already the *written* materials and contents, requiring only to be put together, from the loose papers and numerous Common-place or Memorandum Books, & needing no other change whether of omission, addition, or correction, than the mere act of arranging & the opportunity of seeing the whole collectively, bring with them of course—the following Works. I. Characteristics of Shakespeare's Dramatic Works, with a critical Review of each Play—together with a relative and comparative Critique on the kind and degree of the merits & demerits of the Dramatic Works of Ben Jonson, Beaumont & Fletcher, and Massinger. The history of the English Drama, the accidental advantages, it afforded to Shakespeare, without in the least detracting from the perfect originality, or proper creation of the Shakespearian Drama; the contra-distinction of the Latter from the Greek Drama, and its still remaining *Uniqueness*, with the causes of this from the combined influences of Shakespeare himself, as Man, Poet, Philosopher, and finally, by conjunction of all these, *Dramatic Poet*; and of the age, events, manners and state of the English Language. This work, with every art of compression, amounts to three Volumes Oct. of about 500 pages each.—II. Philosophical Analysis of the Genius and Works of Dante, Spenser, Milton, Cervantes, and Calderon—with similar but more compressed Criticisms of Chaucer, Ariosto, Donne, Rabelais, and others, during the predominance of the romantic Poesy.—In one large Volume.—These two works will, I flatter myself, form a complete Code of the Principles of Judgement & Feeling applied to Works of Taste—and not of *Poetry* only, but of *Poesy* in all its forms, Painting, Statuary, Music &c.—

III. The History of Philosophy, considered as a tendency of the Human Mind to exhibit the powers of the Human Reason—to discover by its own strength the origin & laws of Man and the world, from Pythagoras to Locke & Condillac—2 Volumes.

IV. Letters on the Old and New Testament, and on the doctrines and principles held in common by the Fathers and Founders of the Reformation, addressed to a Candidate for Holy Orders—including advice on the plan and subjects of Preaching, proper to a Minister of the Established Church.

To the completion of these four Works I have literally nothing more to do, than *to transcribe*; but, as I before hinted, from so many scraps & *sibylline* leaves, including Margins of Books & blank Pages, that unfortunately I must be my own Scribe—& not done by myself, they will be all but lost—or perhaps (as has been too often the case already) furnish feathers for the Caps of others—some for this purpose, and some to plume the arrows of detraction to be let fly against the luckless Bird, from whom they had been plucked or moulted!

In addition to these,—of my GREAT WORK, to the preparation of which more than twenty years of my life have been devoted, and on which my hopes of extensive and permanent Utility, of Fame in the noblest* sense of the word, mainly rest—that, by which I might

> As now by thee, by all the Good be known,
> When this weak frame lies moulder'd in its grave,
> Which self-surviving I might call my own,
> Which Folly can not mar, nor Hate deprave—
> The Incense of those Powers which risen in flame
> Might make me dear to Him from whom they came!

of this work, to which all my other writings (unless I except my poems, and these I can exclude in part only) are introductory and preparative; and the result of which (if the premises be, as I with the most tranquil assurance am convinced, they are—insubvertible, the deductions legitimate, and the conclusions commensurate & only commensurate with both) must finally be a revolution of all that has been called *Philosophy* or Metaphysics in England and France since the aera of the commencing predominance of the mechanical system at the Restoration of our second Charles, and with this the present fashionable Views not only of Religion, Morals and Politics but even of the modern Physics and Physiology—You will not blame the earnestness of my expressions or the high importance which I attach to this work: for how with less noble objects & less faith in their attainment could I stand acquitted of folly and abuse of Time, Talent, and Learning in a Labor of 3 fourths of my *intellectual* Life?—of this work something more than a Volume has been dictated by me, so as to exist fit for the Press, to my friend and enlightened Pupil, Mr Green—and more than as much again would have been evolved & delivered to paper, but that for the last six or 8 months I have been compelled to break off our weekly Meetings from the necessity of writing (alas! alas! of *attempting* to write) for purposes and on the subjects of the passing Day.—Of my poetic works I would fain finish

* Turn to Milton's Lycidas, vith Stanza—'Alas! what boots it with incessant care' to the end of that paragraph. The sweetest music does not fall sweeter on my ear, than this Stanza on both mind & ear as often as I repeat it aloud.

the Christabel—Alas! for the proud times when I planned, when I had present to my mind the materials as well as the Scheme of the Hymns, entitled Spirit, Sun, Earth, Air, Water, Fire, and Man: and the Epic Poem on what still appears to me the one only fit subject remaining for an Epic Poem, Jerusalem besieged & destroyed by Titus.—

And here comes, my dear Allsop!—here comes my sorrow and my weakness, my grievance and my confession. Anxious to perform the duties of the day arising out of the wants of the day, these wants too presenting themselves in the most painful of all forms, that of a debt owing to those who will not exact and yet need its payment—and the delay, the long (not live-long but *death*-long) BEHIND-HAND of my accounts to Friends whose utmost care and frugality on the one side and industry on the other, the wife's Management & the Husband's assiduity, are put in requisition to make both ends meet, I am at once forbidden to attempt and too perplext effectually to pursue, the *accomplishment* of the works worthy of me, those I mean above enumerated—even if, savagely as I have been injured by one of the two influencive Reviews & with more effective enmity undermined by the utter silence or occasional detractive compliments of the other,* I had the probable chance of disposing of them to the Booksellers so as even to liquidate my mere *Boarding* accounts during the time expended in the transcription, arrangement, and Proof-correction—and yet on the other hand my Heart & Mind are for ever recurring to them—Yes! my Conscience forces me to plead guilty—I have only by fits and starts even prayed, I have not even prevailed on myself to pray to God in sincerity and entireness, for the fortitude that might enable me to resign myself to the abandonment of all my Life's best Hopes—to say boldly to myself—'Gifted with powers confessedly above mediocrity, aided by an Education of which no less from almost unexampled Hardships & Sufferings than from manifold & peculiar advantages I have never yet found a Parallel, I have devoted myself to a Life of unintermitted Reading, Thinking, Meditating and Observing—I have not only sacrificed all wor[l]dly prospects of wealth & advancement but have in my inmost soul stood aloof even from temporary Reputation—in consequence of these toils & this self-dedication I possess a calm & clear consciousness that in many & most important departments of Truth & Beauty I have outstrode my

* Neither my Literary Life (2 Vol.) nor Sibylline Leaves (1 Vol.) nor Friend (3 Vol.) nor Lay-Sermons, nor Zapolya, nor Christabel, have ever been noticed by the Quarterly Review, of which Southey is yet the main support.—

Contemporaries, those at least of highest name—that the number of my printed works bears witness that I have not been idle, and the seldom acknowledged but yet strictly *proveable* effects of my labors appropriated to the immediate welfare of my Age, in the Morning Post before, and during the Peace of Amiens, in the Courier afterwards, and in the series & various subjects of my Lectures, at Bristol, and at the Royal, & Surry Institutions; in Fetter Lane; in Willis's Rooms; at the Crown & Anchor &c (add to which the unlimited freedom of my communications in colloquial life) may surely be allowed as evidence that I have not been useless in my generation; but from circumstances the *main* portion of my Harvest is still on the ground, ripe indeed and only waiting, a few for the sickle, but a large part only for the *sheaving*, and carting and housing—but from all this I must turn away, must let them rot as they lie, & be as tho' they never had been: for I must go to gather Blackberries, and Earth Nuts, or pick mushrooms & gild Oak-Apples for the Palates & Fancies of chance Customers.—I must abrogate the name of Philosopher, and Poet, and scribble as fast as I can & with as little thought as I can for Blackwood's Magazine, or as I have been employed for the last days, in writing MSS sermons for lazy Clergymen who stipulate that the composition must not be *more* than respectable, for fear they should be desired to publish the Visitation Sermon!'—This I have not yet had courage to do—My soul sickens & my Heart sinks—& thus oscillating between both I do neither—neither as it ought to be done, or to any profitable end. If I were to detail only the various, I might say, capricious interruptions that have prevented the finishing of this very scrawl, begun on the very day I received your last kind letter with the Hare, you would need no other illustrations—

Now I see but one possible plan of rescuing my permanent Utility. It is briefly this & plainly. For what we struggle with inwardly, we find at least easiest to *bolt out*—namely, that of engaging from the circle of those who think respectfully & hope highly of my powers & attainments a yearly sum, for three or four years, adequate to my actual Support with such comforts and decencies of appearance as my Health & Habits have made necessaries, so that my mind may be unanxious as far as the Present Time is concerned—that thus I should stand both enabled and pledged to begin with some one work of those above mentioned, and for two thirds of my whole Time to devote myself to this *exclusively* till finished—to take the chance of its success by the best mode of publication that would involve me in no risk—then to proceed with the next, & so on till the works above mentioned as already in full material existence should be reduced into *formal* and actual Being—while in the remaining third of my Time I might go on, maturing & compleating my great work, &—for if but easy in mind I have no doubt either of the re-awakening power or of the kindling inclination, my Christabel & what else the happier Hour might inspire–& without inspiration a Barrel organ may be played right deftly; but

> All otherwise the state of *Poet* stands:
> For lordly Want is such a tyrant fell,
> That where he rules, all power he doth expel.
> The vaunted verse a vacant head demands,
> Ne wont with crabbed Care the Muses dwell:
> *Unwisely weaves, who takes two webs* IN HAND!

Now Mr Green has offered to contribute from 30 to 40£ yearly for 3 or 4 years; my young Friend & Pupil, the Son of one of my dearest old friends, 50£; and I think that from 10 to 20£ I could rely on from another—the sum required would be about 250£—to be repaid, of course, should the disposal or sale & as far as the disposal & sale of my writings produce the means.—

I have thus placed before you at large & wanderingly as well as diffusely, the statement which I am inclined to send in a compressed form to a few of those, of whose kind dispositions towards me I have received assurances—& to their interest & influence I must leave it—anxious, however, before I do this, to learn from you your very very inmost feeling & judgement, as to the previous questions—Am I entitled, have I earned *a right*, to do this? Can I do it without *Moral* degradation? and lastly—Can it be done without loss of character in the eyes of my acquaintance & of my friends' acquaintance who may become informed of the circumstances?—That if attempted at all, it will be attempted in such a way and that such persons only will be spoken to, as will not expose me to indelicate rebuffs to be afterwards matter of Gossip, I know those, to whom I shall intrust the statement, too well, to be much alarmed about.—

Pray let me either see or hear from You as soon as possible. For indeed and indeed, it is no inconsiderable accession to the pleasure, I anticipate from disembarrassment, that *you* would have to contemplate in a more gracious form & in a more ebullient play of the inward Fountain, the mind & manners of,

My dear Allsop, | Your obliged & very affectionate Friend
S. T. Coleridge

[*To James Gillman*, 9 October 1825]

My dear Friend

It is a flat'ning Thought, that the more we have seen, the less we have to say. In Youth and early Manhood the Mind and Nature are, as it were, two rival Artists, both potent Magicians, and engaged, like the King's Daughter and the rebel Genie in the Arabian Nights' Enternts., in sharp conflict of Conjuration—each having for its object to turn the other into Canvas to paint on, Clay to mould, or Cabinet to contain. For a while the Mind seems to have the better in the contest, and makes of Nature what it likes; takes her Lichens and Weather-stains for Types & Printer's Ink and prints Maps & Fac Similes of Arabic and Sanscrit Mss. on her rocks; composes Country-Dances on her moon-shiny Ripples, Fandangos on her Waves and Walzes on her Eddy-pools; transforms her Summer Gales into Harps and Harpers, Lovers' Sighs and sighing Lovers, and her Winter Blasts into Pindaric Odes, Christabels & Ancient Mariners set to music by Beethoven, and in the insolence of triumph conjures her Clouds into Whales and Walrusses with Palanquins on their Backs, and chaces the dodging Stars in a Sky-hunt!—But alas! alas! that Nature is a wary wily long-breathed old Witch, tough-lived as a Turtle and divisible as the Polyp, repullulative in a thousand Snips and Cuttings, integra et in toto [whole and entire]! She is sure to get the better of Lady MIND in the long run, and to take her revenge too—transforms our To Day into a Canvass dead-colored to receive the dull featureless Portrait of Yesterday; not alone turns the mimic Mind, the ci-devant Sculptress with all her kaleidoscopic freaks and symmetries! into clay, but *leaves* it such a *clay*, to cast dumps or bullets in; and lastly (to end with that which suggested the beginning—) she mocks the mind with its own metaphors, metamorphosing the Memory into a lignum vitae Escrutoire [a Writing-desk made of the wood of life] to keep unpaid Bills & Dun's Letters in, with Outlines that had never been filled up, MSS that never went farther than the Title-pages, and Proof-Sheets & Foul Copies of Watchmen, Friends, Aids to Reflection & other *Stationary* Wares that have kissed the Publisher's Shelf with gluey Lips with all the tender intimacy of inosculation!—Finis!—And what is all this about? Why, verily, my dear Friend! the thought forced itself on me, as I was beginning to put down the first sentence of this letter, how impossible it would have been 15 or even ten years ago for me to have travelled & voyaged by Land, River, and Sea a hundred and twenty miles, with fire and water blending their souls for my propulsion, as if I had been riding on a Centaur with a Sopha for a Saddle—& yet to have nothing more to tell of it than that we had a very fine day, and ran aside the steps in Ramsgate Pier at ½ past 4 exactly, all having been well except poor Harriet, who during the middle Third of the Voyage fell into a reflecting melancholy, in the contemplation of successive specimens of her inner woman in a Wash-hand Basin. She looked pathetic; but I cannot affirm, that I observed any thing sympathetic in the countenances of her Fellow-passengers—which drew forth a sigh from me & a sage remark, how many of our virtues originate in the fear of Death—& that while we flatter ourselves that we are melting in Christian Sensibility over the sorrows of our human Brethren and Sisteren, we are in fact, tho' perhaps unconsciously, moved at the prospect of our own End—For who ever sincerely pities Sea-sickness, Toothache, or a fit of the Gout in a lusty Good-liver of 50?— . . .

[*To James Gillman, jun., aged eighteen*, 22 October 1826]

My dear James

There was a time—and indeed for the Many it still continues—when all the different departments of Literature and Science were regarded as so many different Plants, each springing up from a separate root, and requiring its own particular soil and aspect: and Mathematics and Classics, Philology, Philosophy, and Experimental Science, were treated as *indigenae* [natives] of different Minds—or of minds differently predisposed by their original constitution. Under this belief it was natural, that great stress was laid on the Student's having a *Turn*, or *Taste*, for this or that sort of Knowlege; and it was a valid excuse for reluctance in the study, and want of progress in the attainment, of any particular Branch, that the Individual had no *turn* that way.—But it is the Boast of genuine Philosophy to present a very different and far more hopeful View of the Subject. Without denying the importance or even the necessity of an original Tendency, or what is called a *Genius*, for the attainment of *Excellence* in any one Art or Science, but likewise without forgetting that even among the liberally educated Classes the fewest can or need be eminent Poets, Painters, or Naturalists—but that all ought to be well-informed and right-principled *Men*—the Philosopher considers the several knowleges and attainments, which it is the Object of a liberal Education to communicate or prepare for, as springing from one Root, and rising into one common Trunk, from the summit of which it diverges into the different Branches, and ramifies without losing its original unity into the minutest Twigs and Sprays of practical application: and so that in all alike it is but the same Principles unfolding into different Rules, and assuming different names & modifications, according to the different Objects, in which these Principles are to be realized. Now as in the present stage of your Studies, & indeed for the next two years of your life, you are engaged in forming the *Trunk* of the Tree of Knowlege, which *Trunk* belongs entire to each and every Branch of the Tree, singly as well as collectively—(the Clergyman must have the *whole*, the Lawyer the whole, the Physician the whole, yea, even the naval and the Military *Officer* must possess the *whole*, if either of these is to be more than a mere Tradesman and *Routinier*, a *hack* Parson, a *hack* Lawyer, &c, in short, a sapless *Stick*—for that is the right name for a Branch, in which the juices elaborated by the common trunk do not circulate—and for all the uses, that a stick can be applied to, such a man is good for—and good for nothing else!) it must be evident to you, that to have no *Taste*, no *Turn*, no Liking for *this* or for *that* is to confess an unfitness or dislike to a liberal Education *in toto* [as a whole]—And what is a liberal Education? That which *draws* forth and trains up the germ of free-agency in the Individual—Educatio, quae *liberum* facit [Education, which makes one *free*]: and the man, who has mastered all the conditions of *freedom*, is *Homo Liberalis* [a Freeborn, or Gentlemanly, Man]—the classical rendering of the modern term, *Gentle*man—because under the feudal system the *men of family* (Gentiles, generosi, quibus *gens* erat, et *gen*us°) alone possessed these conditions. I do not undervalue *Wealth*, but even if by descent or by Lottery (& since Mr Bish mourns in large Capitals, red, blue, and black, in every — corner over the Last, the downright *Last*,° you have but small chance, I suspect, of a snug £30,000 from this latter source, and your dear Father is too rational & upright a Practitioner and Highgate too healthy a Place for you to expect any one of the nine Cyphers as the Nominative of 000£ from the former)—but even if you had an independent fortune, it would not of itself suffice to make you an independent man, a free man, or a Gentleman. For believe me, my dear young Friend! it is no musty old Saw but a Maxim of Life, a medicinal Herb from the Garden of Experience that grows amid Sage, Thyme* and Heart's Ease, that He alone is *free* & entitled to the name

* This word reminds me of an Ode to PUNNING which I wrote at School, when I was of your age—& which began with 'SPELLING, avaunt!'!

of a Gentleman, who knows himself and walks in the light of his own consciousness.—

But for this reason, nothing can be rightly taken in, as a part of a liberal Education, that is not a Mean of acquainting the Learner with the nature and laws of his own Mind—as far as it is the representative of the Human Mind generally—. By knowing what it ought to be, it gradually becomes what it ought to be—and *this*, the Man's *ideal* Mind, serves for a Chronometer by which he can set his own pocket Watch and judge of his Neighbour's.—Most willingly, however, do I admit, that the far greater part of the process, which is called Education, and a classical, a liberal Education, corresponds most vilely to the character here given!—But that all knowlege, not merely mechanical and like a Carpenter's Ruler, having its whole value in the immediate outward use to which it is applied, without implying any portion of the science in the user himself, and which instead of re-acting on the mind tends to keep it in its original ignorance—all knowlege, I say, that enlightens and liberalizes, is a form and a means of Self-knowlege, whether it be grammar, or geometry, logical or classical. For such knowlege must be founded on *Principles*: and those Principles can be found only in the Laws of the Mind itself. Thus: the whole of Euclid's Elements is but a History and graphic Exposition of the powers and processes of the Intuitive Faculty—or a Code of the Laws, Acts and ideal Products of the pure Sense. We learn to *construe* our own perceptive power, while we *educe* into distinct consciousness its inexhaustible *constructive* energies. Every Diagram is a Construction of *Space*: and what is Space, but the universal antecedent Form and ground of all Seeing? Now that *Space* belongs to the mind itself, i.e. that it is but a *way* of contemplating objects, you may easily convince yourself by trying to imagine an outward space. You will immediately find that you imagine a space for *that* Space to exist in—in other words, that you turn this first space into a *thing* in space: or if you could succeed in abstracting from all thought of Color and Substance, & then shutting your eyes try to imagine it—it will be a mere Diagram, and no longer a construction *in* space but a construction *of* Space.—Not less certain and even more evident, is this position, in its application to *words* and language—to Grammar, Logic, Rhetoric, or the Art of Composition. For (as I have long ago observed to you) it is the fundamental Mistake of Grammarians and Writers on the philosophy of Grammar and Language [to assume] that words and their syntaxis are the immediate representatives of *Things*, or that they correspond to *Things*. Words correspond to

Thoughts; and the legitimate Order & Connection of words to the *Laws* of Thinking and to the acts and affections of the Thinker's mind.

In addition to the universal grounds on which I might rest the immense superiority of this method—i.e. Instruction by *insight*, and by the reduction of all Rules to their sources in the mind itself—over the ordinary plan (for *method that* cannot be called, in which there is no *μέθοδος* [*methodos*, a following after], no intelligible guiding principle of Transition and Progress) I recommend it to *you* as the most efficacious corroborant of the active Memory and the best Substitute for any defect or deficiency of the passive Memory.—Let us indulge our fancy for a moment and suppose that you or I or Sir Humphry Davy had discovered an easy and ready way of decomposing, rapidly and on a great scale, the Water of the Sea into its component Elements of Oxygen and Hydrogen, with a portion of Carbonic gas: so as to procure by an extemporaneous process a Fuel to any desirable extent, capable of boiling the undecompounded Water—How poor a thing would the most capacious Hold that ever Steam-Frigate could boast of, and the most abundantly stocked with Wood or Coal, and with the largest reservoirs to boot established at different Points, which the Vessel might stop at in the course of the Voyage from London to Bombay or Van Diemen's Land—how poor an affair would it all be, compared with the facilities given us by this discovery! The only danger would be, that the Colliers, Pitmen and Coal Merchants might waylay the Discoverer, and cut his throat.—Now, my dear James! not much unlike this is the difference between the reproductive power consequent on the full, clear and distinct Comprehension of PRINCIPLES, and a familiar acquaintance with the Rules of Involution and Evolution of Particulars in Generals, and the memory that results spontaneously from the impressions left on the Brain & Senses, and may be destroyed by a fit of Sickness, or be suspended at the very moment, you want it, by a Flutter or a dyspeptic Qualm, in favor of the former. In this Memory, to which we trust whatever we learn by rote and all insulated knowleges, seen each for itself without its relations and dependencies, the recollections stick together like the Dots in Frog-spawn by accidents of Place, Time, and Circumstance (Vide [See] Essays on Method, FRIEND, Vol. III: Essay the First) or like the eggs in a Caterpillar's Web, by threads of capricious and arbitrary Associations. At the best, the several knowleges are in the Mind as in a Lumber-garret: while *Principles* with the Laws, by which they are unfolded into their consequences, when they are once thoroughly mastered, become the mind itself and are living and constituent parts of it. We know what we want: and what we want, we reproduce—just as a neat-fingered Girl reconstructs the various figures on a Cat cradle. That this is no mere fancy, you have a proof in my own instance. In the gifts of passive or spontaneous Memory I am singularly deficient. Even of my own Poems I should be at a loss to repeat fifty lines. But then I make a point of speaking only on subjects, I understand on *principle* and *with insight*: and on these I set my Logic-engine and spinning jennies a going, and my friends, I suspect, more often complain of a superabundant, than of a deficient, supply of the Article in requisition.

To exemplify this method of instruction by applying it to the principal points of an academic or classical Education—and in the first instance, to the analysis (= parsing), hermeneusis (= construing) and synthesis (= composition) of the Learned Languages—the Greek in comparison and collation with our Mother Tongue affords, I think, the greatest facilities.—The Latin, at all events, is least suited to the experiment—both as being only a very scanty Dialect of the oldest and rudest Greek, and the derivation of its significant prefixes and affixes (or cases and tenses) requiring all the arts of the veteran Etymologist, τὰς γραμμάτων καὶ συλλαβῶν μεταμορφώσεις [the changes of letters and syllables].—Accordingly, with the Greek I have already begun with you, tho' but in a fragmentary way hitherto, and rather for the purpose of breaking down the *chevaux de frize* [barriers], which the *newness* & strangeness of the Subject throws round it, than in the expectation of leaving any distinct impression of the particular truths. For there is a state of mind the direct opposite to that which takes place in making an Irish *Bull.*° Miss Edgeworth has published a long Essay on *Bulls*—without understanding the precise meaning of the word which she makes synonimous with *Blunders*. But tho' all Bulls are Blunders, every Blunder is not a Bull. In a Bull there is always a *sensation* without a *sense* of connection between two incompatible thoughts.—The thoughts being incompatible, there cannot of course be any *sense* of—i.e. insight into—their connection or compatibility. But a *sensation*, a *feeling*, as if there was a connection, may exist, from various causes: as when for instance, the right order of the Thoughts would be thus—a b c d e f, of which *b* and *e* are incompatible ideas, tho' b is in just connection with c, c with d, and d with *e*—Now if from any heat & hurry of mind and temper such an extreme and undue vividness is given to b and e, as to bedim and practically extinguish the consciousness of (or distract the attention from) c and d, in this case b and e will appear next door neighbors, while the actual, tho' unconscious intervention of c and d, produces a *sensation* of the connection—just as in forgetting a name that was quite familiar to you—you have a *feeling* of remembering it, tho' the recollection is suspended.—Read this over till you understand it.—Now the opposite of this [stat]e is when we have the sense of the connection between any given series of Thoughts, but from their entire strangeness to the mind not the wonted, & therefore craved for, sensation. You understand them, & yet have the feeling of not understanding. The philosophy of the dead Languages I propose therefore to recommence systematically with you, beginning with the Greek Alphabet—then to the significant sounds or elementary *positions*, the Helms & Rudders of words, rather than *words*—and then to the different sorts of words. But for my present purpose, viz. of helping you to write Latin, I shall suppose all this done—& begin by pointing out the characteristic difference between Latin Sentences & English—& from this deduce the Rules & the Helps—This I defer to the next Post but one.—God bless you &

S. T. Coleridge.

[*To J. H. Green*,° 29 March 1832]

My very dear Friend

On Monday I had a sad trial of intestinal pain and restlessness; but thro' God's Mercy, without any craving for the Poison, which for more than 30 years has been the guilt, debasement, and misery of my Existence. I pray, that God may have mercy on me—tho' thro' unmanly impatiency of wretched sensations, that produced a disruption of my mental continuity of productive action I have for the better part of my life yearning [*sic*] towards God, yet having recourse to the evil Being—i.e. a continued act of thirty years' Self-poisoning thro' cowardice of pain, & without other motive—say rather without *motive* but solely thro' the goad *a tergo* [from behind] of unmanly and unchristian fear—God knows! I in my inmost soul acknowlege all my sufferings as the necessary effects of his Wisdom, and all the alleviations as the unmerited graces of his Goodness. Since Monday I have been tranquil; but still, placing the palm of my hand with its lower edge on the navel, I feel with no intermission a death-grasp, sometimes relaxed, sometimes tightened, but always present: and I am convinced, that if Medical Ethics permitted the production of a Euthanasia, & a Physician, convinced that at my time of Life there was no rational hope of revalescence to any useful purpose, should administer a score drops of the purest Hydro-cyanic Acid, & I were immediately after opened (as is my earnest wish) the state of the mesenteric region would solve the problem.

I trust, however, that I shall yet see you, as Job says, 'in this flesh'—& I write now tho' under an earnest conviction of the decay of my intellectual powers proportionate to the decay of the Organs, thro' which they are made manifest, & which you must have perceived, of late, more forcibly than myself—I write, my dear friend! first to acknowledge God's Goodness in my connection with you—secondly, to express my utter indifference, under whose *name* any truths are propounded to Mankind—God knows! it would be no pain to me, to foresee that my name should utterly cease—I have no desire for reputation—nay, no wish for *fame*—but I am truly thankful to God, that thro' you my labors of thought may be rendered not wholly unseminative. But in what last Sunday you read to me, I had a sort of Jealousy, probably occasioned by the weakened state of my intellectual powers, that you had in some measure changed your pole. My principle has ever been, that Reason is *subjective* Revelation, Revelation *objective* Reason—and that *our* business is not to *derive* Authority from the *mythoi* [myths] of the Jews & the first Jew-Christians (i.e. the O. and N. Testament) but to *give* it to them—never to assume their stories as facts, any more than you would Quack Doctors' affidavits on oath before the Lord Mayor—and verily in point of old Bailey Evidence this is a flattering representation of the Paleyian Evidence—but by *science* to confirm the *Facit* [(literally,) it makes; the solution of a problem given to make the arithmetical process easier to follow], kindly afforded to beginners in Arithmetic. If I lose my faith in *Reason*, as the perpetual revelation, I lose my faith altogether. I must deduce the objective from the subjective Revelation, or it is no longer a revelation for me, but a beastly fear, and superstition.

I hope, I shall live to see you next Sunday. God bless you, my dear Friend! We have had a sad sick House—& in consequence, I have seen but little of Mr Gillman, who has been himself ill—and likewise Miss Lucy Harding. For me, it is a great blessing & mercy, Life or Death, that I have been & still remain quiet, without any craving, but the contrary. Compared with this mercy, even the felt and doubtless by you perceived decay & languor of intellectual energy is a trifling counter-weight.—

Again, God bless you, my dear friend! | and
S. T. Coleridge—

Notebooks°

[October 1802] October 3—Night—My Dreams uncommonly illustrative of the non-existence of Surprize in sleep—I dreamt that I was asleep in the Cloyster at Christs Hospital & had awoken with a pain in my hand from some corrosion/boys & nurses daughters peeping at me/On their implying that I was not in the School, I answered yes I am/I am only twenty—I then recollected that I was thirty, & of course could not be in the School—& was perplexed—but not the least surprize that I could fall into such an error/So I dreamt of Dorothy, William & Mary—& that Dorothy was altered in every feature, a fat, thick-limbed & rather red-haired—in short, no resemblance to her at all—and I said, if I did not *know* you to be Dorothy, I never should *suppose* it/Why, says she—I have not a feature the same/& yet I was not surprized—

I was followed up & down by a frightful pale woman who, I thought, wanted to kiss me, & had the property of giving a shameful Disease by breathing in the face/

& again I dreamt that a figure of a woman of a gigantic Height, dim & indefinite & smokelike appeared—& that I was forced to run up toward it—& then it changed to a stool—& then appeared again in another place—& again I went up in great fright—& it changed to some other common thing—yet I felt no surprize.

[April 1803?] Language & all *symbols* give *outness* to Thoughts/& this the philosophical essence & purpose of Language/

[June 1803] Saturday, June 16—½ of a mile from John Stanley's toward Grasmere, whither I was going to stand Godfather to Wm Wordsworth's first Child—put the colored Glasses to my eyes as a pair of Spectacles, the red to the left, the yellow Glass to the right eye—saw only thro' the yellow—closed my right eye with my finger, without in the least altering the position of the left eye—& then I instantly saw thro' the red Glass. The right eye manifestly the stronger, tho what is curious, & to be explained by the greater Light of the yellow Glass, when I altered the Glasses, namely the yellow to the left eye, the red to the right, then I saw the Landscape as thro' the yellow—perhaps a very little reddish, while the clouds & skies were now as thro' the red./. These experiments must be tried over again & varied—

[August 1803] In extreme low Spirits, indeed it was downright despondency, as I was eating my morsel heartlessly, I thought of my Teeth of Teeth in general—the Tongue—& the manifest *means & ends* in nature/I cannot express what a manly comfort + religious resolves I derived from it—It was in the last Days of August, 1803.—I wish, I had preserved the very Day & hour.

[October–November, 1803] Seem to have made up my mind to write my metaphysical works, as *my Life*, & *in* my Life—intermixed with all the other events/or history of the mind & fortunes of S. T. Coleridge.

[October 1803] I am sincerely glad, that he [Wordsworth] has bidden farewell to all small Poems—& is devoting himself to his great work—grandly imprisoning while it deifies his Attention & Feelings within the sacred Circle & Temple Walls of great Objects & elevated Conceptions.—In these little poems his own corrections, coming *of necessity* so often, at the end of every 14, or 20 lines—or whatever the poem might chance to be—wore him out—difference of opinion with his best friends irritated him/& he wrote at times too much with a sectarian Spirit, in a sort of Bravado.—But now he is at the Helm of a noble Bark; now he sails right onward—it is all open Ocean, & a steady Breeze; and he drives before it, unfretted by short Tacks, reefing & unreefing the Sails, hawling & disentangling the ropes.—His only Disease is the having been out of his Element—his return to it is food to Famine, it is both the specific Remedy, & the condition of Health.

[October 1803] To return to the Question of Evil—woe to the man, to whom it is an uninteresting Question—tho' many a mind, overwearied by it, may shun it with Dread/and here, N.B. scourge with deserved & lofty Scorn those Critics who laugh at the discussion of old Questions—God, Right & Wrong, Necessity & Arbitrement—Evil, &c—No! forsooth!—the question must be new, *new spicy hot* Gingerbread, a French Constitution, a Balloon, change of Ministry, or which had the best of it in the Parliamentary Duel, Wyndham or Sheridan, or at the best, a chemical Theory, whether the new celestial Bodies shall be called Planets or Asteroids—&c—Something new, something *out* of themselves—for whatever is *in* them, is deep within them, must be *old as* elementary Nature. To find no contradiction in the union of old & novel—to contemplate the Ancient of Days with Feelings new as if they then sprang forth at his own Fiat—this marks the mind that feels the Riddle of the World, & may help to unravel it. But to return to the Question—the whole rests on the Sophism of imagining Change in a case of positive Substitution.—This, I fully believe, *settles* the Question/—The assertion that there is in the essence of the divine nature a necessity of omniform harmonious action, and that Order, & System/not number—in itself base & disorderly & irrational—/ define the creative Energy, determine & employ it—& that number is subservient to order, regulated, organized, made beautiful and rational, an object both of Imag. & Intellect, by Order—this is no mere assertion/it is strictly in harmony with the Fact, for the world appears so—& it is proved by whatever proves the Being of God—. Indeed, it is involved in the Idea of God.—

[October 1803] What is it, that I employ my Metaphysics on? To perplex our clearest notions, & living moral Instincts? To extinguish the Light of Love & of Conscience, to put out the Life of Arbitrement—to make myself & others *Worthless*, *Soul*-less, *God*less?—No! To expose the Folly & the Legerdemain of those, who have thus abused the blessed Organ of Language, to support all old & venerable Truths, to support, to kindle, to project, to make the Reason spread Light over our Feelings, to make our Feelings diffuse vital Warmth thro' our Reason—these are my Objects—& these my Subjects. Is this the metaphysics that bad Spirits in Hell delight in?

[December 1803] What a beautiful Thing Urine is, in a Pot, brown yellow, transpicuous, the Image, diamond shaped of the Candle in it, especially, as it now appeared, I have emptied the Snuffers into it, & the Snuff floating about, & painting all-shaped Shadows on the Bottom.

[December 1803] I will at least make the attempt to explain to myself the Origin of moral Evil from the *streamy* Nature of Association, which Thinking = Reason, curbs & rudders/how this comes to be so difficult/Do not the bad Passions in Dreams throw light & shew of proof upon this Hypothesis?—Explain those bad Passions: & I shall gain Light, I am sure—A Clue! A Clue!—an Hecatomb a la Pythagoras, if it unlabyrinths me.—Dec. 28, 1803—Beautiful luminous Shadow of my pencil point following it from the Candle—rather going before it & illuminating the word, I am writing. 11 °clock/—But take in the blessedness of Innocent Children, the blessedness of sweet

Sleep, &c &c &c: are these or are they not contradictions to the evil from *streamy* association?—I hope not: all is to be thought *over* and *into*—but what is the height, & ideal of mere association?—Delirium.—But how far is the state produced by Pain & Denaturalization? And what are these?—In short, as far as I can see any thing in this Total Mist, Vice is imperfect yet existing Volition, giving diseased Currents of association, because it yields on all sides & *yet* is—So think of Madness:—O if I live! Grasmere, Dec. 29. 1803.

[January 1804] All this evening, indeed all this day (Monday, Jan. 9) I ought to have [been] reading & filling the Margins of Malthus—I had begun & found it pleasant/why did I neglect it?—Because, I OUGHT not to have done this.—The same in reading & writing Letters, Essays, &c &c—surely this is well worth a serious Analysis, that understanding I may attempt to heal/for it is a deep & wide disease in my moral Nature, at once Elm-and-Oak-rooted.—Love of Liberty, Pleasure of Spontaneity, &c &c, these all express, not explain, the Fact.

[January 1804] Tuesd. Morn. Jan. 10. 1804.—After I had got into bed last night, I said to myself, that I had been pompously enunciating, as a difficulty, a problem of easy & common solution/viz. that it was the effect of Association, we from Infancy up to Manhood under Parents, Schoolmasters, Tutors, Inspectors, &c having had our pleasures & pleasant self-chosen Pursuits (self-chosen because pleasant, and not *originally* pleasant because self-chosen) interrupted, & we forced into dull unintelligible Rudiments or painful Labor/—*Now*, all Duty is felt as a *command*, commands *most often*, & therefore by Laws of Association felt as if *always*, from without & consequently, calling up the Sensations &c of the pains endured from Parents', Schoolmasters', &c &c—commands from without.—But I awoke with gouty suffocation this morning, ½ past one/& as soon as Disease permitted me to think at all, the shallowness & falsity of this Solution flashed on me at once/I saw, that the phænomenon occurred far far too early—in early Infancy, 2 & 3 months old, I have observed it/& have seen it in Hartley, turned up & lay'd bare to the unarmed Eye of merest common sense. That *Interruption* of itself is painful because & as far as it acts as Disruption/& then, without any reference to or distinct recollection of my former theory, I saw great Reason to attribute the effect wholly to the streamy nature of the associating Faculty and especially as it is evident that *they most* labor under this defect who are most reverie-ish & streamy—Hartley, for instance & myself/This seems to me no common corroboration of my former Thought on the origin of moral Evil in general.

[January 1804] Common people, and all men in common Life, notice only low & high *degrees*, & frame words accordingly/philosophers notice the *kind* & as it were *element*, & often finding no word to express this are obliged to invent a new word or to give a new sense to an old one/thus what Philosophers since the time of Mr Locke have called association was noticed by all men in certain *strong* instances & was called CUSTOM in common Life we know what *Pleasure* is, & what *Pain* is—*degrees* of the philosophical word Sensation/we have *Heat* & *Cold*, words of *degrees*, the kind of which late philosophers have agreed to express by the word Caloric, or Calorific/Now those words in common Life which seem the most general & elementary, as more, mind, Life, &c originally designated some one particular form or degree of these & became general, that is, philosophical, as men in common learnt to generalize, i.e. so far to become philosophers—/ now a philosopher, inventing a word or stamping with a new meaning an old word, in all civilized countries but especially since Printing & in proportion as we are Readers & form our Language from Books, these words slide into common use, generally much alloyed by the carelessness of common Life, or in the case of an old word new stamped, injured by a confusion of the two meanings, & Language degenerates—& more injured than benefited after a certain period, which it is indeed very difficult to assign, but which has assuredly passed by many centuries in this Country/Henry 8th the latest & yet if only out of Chaucer himself & his Contemporaries—taking in a space of 50 years—all the true glossary Synonyms were taken to explain the obscure words of Chaucer that might be found in those Books—I guess, we should be surprized at the close resemblance of that Language to our own/—Men of great Genius find in new words & new combinations the sin that most easily besets them/a strong feeling of originality seems to receive a gratification by new Terms—hence the all too often useless nomenclatures in the philosophical writings of all men of originality—some quite overlayed by it, as poor Behmen/ but if a *Daniel* whom Gower & Lydgate were so much below, & Chaucer so much above, had lived in the age of Chaucer, I doubt not, we should have been as much struck by the contrast as we now are by the contrast of Daniel's Language with Spenser's, & even the later Shakespeare—N.B.—The former Half of this Note I mean for my C.C. on the comforts to be derived from generalization & the natural

Tendency to generalize which Sickness & indeed all Adversity generates.

[December 1804] Idly talk they who speak of Poets as mere Indulgers of Fancy, Imagination, Superstition, &c—They are the Bridlers by Delight, the Purifiers, they that combine them with *reason* & order, the true Protoplasts, Gods of Love who tame the Chaos.

[December 1804] Saturday, Dec. 22nd, 1804. The duty of stating the power and in the very formation of the Letters, perceived during the formation, the meaning of the injured mind. The Best remains! Good God! wretched as I may be bodily, what is there good and excellent which I would not do—?—

But this is written in *involuntary* Intoxication. God bless all!

Sunday, Dec. 23rd/I do not understand the first sentence of the above—I wrote them after that convulsed or suffocated by a collection of wind in my stomach & alternately tortured by its colic pangs in my bowels, I in despair drank three glasses running of whisky & water/the violent medicine answered—I have been feeble in body during the next day, & active in mind—& how strange that with so shaken a nervous System I never have the Head ache!—I verily am a stout-headed, weak-bowelled, and O! most pitiably weak-*hearted* Animal! But I leave it, as I wrote it—& likewise have refused to destroy the stupid drunken Letter to Southey, which I wrote in the sprawling characters of Drunkenness/ If I should perish without having the power of destroying these & my other pocket books, the history of my own mind for my own improvement, O friend! Truth! Truth! but yet Charity! Charity! I have never loved Evil for its own sake; no! nor ever sought pleasure for its own sake, but only as the means of escaping from pains that coiled round my mental powers, as a serpent around the body & wings of an Eagle! My sole sensuality was *not* to be in pain!—

[January 1805] It is a most instructive part of my Life the fact, that I have been always preyed on by some Dread, and perhaps all my faulty actions have been the consequence of some Dread or other on my mind/from fear of Pain, or Shame, not from prospect of Pleasure/—So in my childhood & Boyhood the horror of being detected with a sorehead; afterwards imaginary fears of having the Itch in my Blood–/

then a short-lived Fit of Fears from sex—then horror of DUNS, & a state of struggling with madness from an incapability of hoping that I should be able to marry Mary Evans (and this strange passion of fervent tho' wholly imaginative and imaginary Love uncombinable by my utmost efforts with any regular Hope—/possibly from deficiency of bodily feeling, of tactual ideas connected with the image) had all the effects of direct Fear, & I have lain for hours together awake at night, groaning & praying—Then came that stormy time/and for a few months America really inspired Hope, & I became an exalted Being—then came Rob. Southey's alienation/my marriage—constant dread in my mind respecting Mrs Coleridge's Temper, &c—and finally stimulants in the fear & prevention of violent Bowel-attacks from mental agitation/then almost epileptic night-horrors in my sleep/& since then every error I have committed, has been the immediate effect of the Dread of these bad most shocking Dreams—any thing to prevent them/—all this interwoven with its minor consequences, that fill up the interspaces—the cherry juice running in between the cherries in a cherry pie/procrastination in dread of this—& something else in consequence of that procrast. &c/—and from the same cause the least languor expressed in a Letter from S.H. [Sara Hutchinson] drives me wild/& it is most unfortunate that I so fearfully despondent should have concentered my soul thus on one almost as feeble in Hope as myself. 11 Jan. 1805.— . . .

[1808–1818] Speaking of the original unific Consciousness, the primary Perception, & its extreme difficulty, to take occation to draw a lively picture of the energies, self-denials, sacrifices, toils, trembling knees, & sweat-drops on the Brow, of a philosopher who has really been sounding the depths of our being—& to compare it with the greatest & most perseverant Labors of Travellers, Soldiers, and whomever else Men honor & admire—how trifling the latter! And yet how cold our gratitude to the former—Say not, that they were vainly employed—compare the mind & motives of a vulgar Sensualist, or wild Savage with the mind of a Plato or an Epictetus—& then say if you dare that there is any more comparison between the *effects* of the toils of the Philosopher & the Worldling, than there is between the intensity of the toils themselves.

[May 1808] Ah! dear Book! Sole Confidant of a breaking Heart, whose social nature compels *some* Outlet. I write more unconscious that I am writing, than in my most earnest modes I *talk*—I am not then so unconscious of talking, as when I write in these dear, and only once profaned, Books, I am of the act of writing—So much so, that even in the last minute or two that I have been writing on my writing, I detected that the former Habit was predominant—I was only *thinking*. All minds must think by some *symbols*—the strongest minds possess the most vivid Symbols in the Imagination—yet this ingenerates a *want*, <u>ποθον</u>, *desiderium*, for vividness of Symbol: which something that is *without*, that has the property of *Outness* (a word which Berkeley preferred to 'Externality') can alone fully gratify/even that indeed not fully—for the utmost is only an approximation to that absolute *Union*, which the soul sensible of its imperfection in itself, of its *Halfness*, yearns after, whenever it exists free from meaner passions, as Lust, Avarice, love of worldly power, passion for distinction in all its forms of Vanity unmodified by those Instincts which often render it venial sometimes almost amiable, namely, ostentation of Wealth & Rank, as in Routs (1100 persons present, all of more or less distinction in one House!—The Prince of Wales, Townshend the Thief-catcher, & the Intelligencer from the Morning Post, all present!—what glory!—and the whole Street unpassable for those poor Creatures, who happen to have to drive thro' it, when their Wife is in labor, or their Mother dying, or some such Trifle) or in the more deceiving form of being the active man in a Patriotic Fund, or a Society for the Prevention of Vice, or for the Amelioration of the Poor, or any other of those dainty devices for turning the labouring *Nobles* of England into Neapolitan Lazzaroni [Beggars]—(I say, Nobles: for if to be a *Roman* Citizen was itself a Patent of Nobility, in the good Times of Rome, why should not the 'to be an Englishman' have the same claim, while he has hands to maintain himself, if Justice were done him, & both hand & heart to defend his Country?)—or lastly, in the pursuit of literary Reputation, which a few disguise to himselves as an honorable Love of *Fame*, & betray the truth to all sharp-sighted minds by their undue Irritation & vehemence of Language concerning every Review which attacks them, and every acquaintance, whose opinion of them does not soar so high as their own of themselves—(ex. gr. [Wordsworth (Coleridge used a code)] in relation to [Sharp] at [Sir George Beaumont's]—)—I say, every generous mind not already filled by some one of these passions feels its *Halfness*—it cannot *think* without a symbol—neither can it *live* without something that is to be at once its Symbol, & its *Other half*—(That phrase now so vulgar by the profane Use of it was most beautiful in its origin—)—Hence I deduce the habit, I have most unconsciously formed, of *writing* my inmost thoughts—I have not a soul on earth to whom I can reveal them—and yet

I am not a God, that I should stand alone

and therefore to you, my passive, yet sole true & kind, friends I reveal them. *Burn you I certainly shall, when I feel myself dying*; but in the Faith, that as the Contents of my mortal frame will rise again, so that your contents will rise with me, as a Phœnix from its pyre of Spice & Perfume.

[May 1808] If one thought leads to another, so often does it blot out another—This I find, when having lain musing on my Sopha, a number of interesting Thoughts having suggested themselves, I conquer my bodily indolence & rise to record them in these books, alas! my only Confidants.—The first Thought leads me on indeed to new ones; but nothing but the faint memory of having had them remains of the others, which had been even more interesting to me.—I do not know, whether this be an idiosyncracy, a peculiar disease, of *my* particular memory—but so it is with *me*—My Thoughts crowd each other to death.

[November 1809] If it were asked of me to justify the interest, which many good minds—what if I speak out, and say what I believe to be the truth—which the majority of the best and noblest minds feel in the great questions—Where am I? What and for what am I? What are the duties, which arise out of the relations of my Being to itself as heir of futurity, and to the World which is its present sphere of action and impression?—I would compare the human Soul to a Ship's Crew cast on an unknown Island (a fair Simile: for these questions could not suggest themselves unless the mind had previously felt convictions, that the present World was not its whole destiny and abiding Country) —What would be their first business? Surely, to enquire what the Island was? in what Latitude? what ships visited that Island? when? and whither they went?—and what chance that they should take off first one, & then another?—and after this—to think, how they should maintain & employ themselves during their stay—& how best stock themselves for the expected voyage, & procure the means of inducing the Captain to take them to the Harbour, which they wished to go to?—

The moment, when the Soul begins to be sufficiently self-conscious, to ask concerning itself, & its relations, is the first moment of its *intellectual* arrival into the World—Its *Being*—enigmatic as it must seem—is posterior to its *Existence*—. Suppose the shipwrecked man stunned, & for many weeks in a state of Ideotcy or utter loss of Thought & Memory—& then gradually awakened/

[November 1809] Important remark just suggests itself—13 Novr. 1809—That it is by a negation and voluntary Act of *no*-thinking that we think of earth, air, water &c as dead—It is necessary for our limited powers of Consciousness that we should be brought to this negative state, & that should pass into Custom—but likewise necessary that at times we should awake & step forward—& this is effected by Poetry & Religion/—. The Extenders of Consciousness—Sorrow, Sickness, Poetry, Religion—.—The truth is, we stop in the sense of Life just when we are not *forced* to go on—and then adopt a permission of our feelings for a precept of our Reason—

[May 1810] I do not like that presumptuous Philosophy which in its rage of explanation allows no xyz, no symbol representative of the vast Terra Incognita [Unknown Land] of Knowlege, for the Facts and Agencies of Mind and matter reserved for future Explorers/while the ultimate grounds of all must remain inexplorable or Man must cease to be progressive. Our Ignorance with all the intermediates of obscurity is the *condition* of our ever-increasing Knowlege. Consider for a moment the multitude and the importance of the phænomena now universally referred to electric agency—not to mention those now explained by the electro-chemical combination of the ponderable portion of oxygen Gas with metallic bases, or with hydrogen. (For I suppose Nitrogen to be included in the former, i.e. Ox. + Met. Base) Having made a numeration-table of these Phænomena, then consult the Works of those Physiologists who flourished & uttered oracles before the discovery of Electricity. Their several explanations will furnish a Lesson not only of modesty but of Logic—For, doubtless, by a more watchful, and austere, as well as more modest & for that cause *anticipative*, Logic, the false part of these explanations might even then have been detected—& such explanations given, as would have preserved our existing quantum [amount] of knowlege pure from positive error, by avowedly including our ignorance in our knowlege, under a common Symbol. Thus in the old explanation of Thunder & Lightning, that it was Fire by the *dashing together* of Clouds—Here was *presumption*/Had they been contented to say, that clouds contained Fire in a latent form, which under given circumstances passed from one to the other; but that these circumstances, that is, that the *law* of Fire as contained in Vapors, remained to be discovered—all would have been right—and Fire would have been a fair generic term, or Symbol, which thus limited would have represented as in a process of Algebra, that particular species of Fire, which in the conclusion would have come out as Electricity. But no!—the whole modus [means] must be explained & therefore the whole explanation must be seized from the scanty possessions behind us, not borrowed from the vast tract before us—and because the former pigmy Domain contained the fact of Fire generated by collision, as of Flint & Steel, or by Friction, as of the Wheel & Axle, therefore two *mists* in the air must be *dashed* against each other, or *rubbed* together. In like manner, I persuade myself, we might go thro' $\frac{9}{10}$ths of the old explanations, and abstracting our present discoveries detect their errors by the mere application of a sound & modest Logic—and thus disciplined we might venture to revise our present more fashionable explanations, and submit them to the same process. So would Logic not only be taught in the most delightful manner, enriching the memory, and stimulating the Imagination, while it both steadied & sharpened the Judgment, it would likewise become a *Scout* as well as a *Pioneer*, of Natural Philosophy—& gradually fill up that blank space in Bacon's grand Outline of our Desideratà, the *Ars Inveniendi* [Art of Discovering], I exceedingly wish, I could procure from Malta the Logica Venatrix of Giord. Bruno. What if I applied to Lord Mulgrave thro' Sir G.B.?—or to Mr P. thro' his Sister?

[May 1810] I wish, I dared use the Brunonian phrase°—& define Poetry—the Art of representing Objects in relation to the *excitability* of the human mind, &c—or what if we say—the communication of Thoughts and feelings so as to produce excitement by sympathy, for the purpose of immediate pleasure, the most pleasure from each part that is compatible with the largest possible sum of pleasure from the whole—?

The art of communicating whatever we wish to communicate so as to express and to produce excitement—or—in the way best fitted to express & to &c—or as applied to all the fine arts

A communication of mental excitement for the purposes of immediate pleasure, in which each part is fitted to afford as much pleasure as is compatible with the largest possible Sum from the whole—

Many might be the equally good definitions of Poetry, as metrical Language—I have given the former the preference as comprizing the essential of all the fine Arts, and applying to Raphael & Handel equally as to Milton/But of Poetry commonly so called we might justly call it—A mode of composition that calls into action & gratifies the largest number of the human Faculties in Harmony with each other, & in just proportions—at least, it would furnish a scale of merit if not a definition of *genus* [kind]—

Frame a numeration table of the primary faculties of Man, as Reason, *unified per Ideas* [by Ideas], Mater Legum [Mother of Laws], Arbitrement, Legibilitatis mater [mother of Amenability], Judgement, the discriminative, Fancy, the aggregative, Imagination, the modifying & *fusive*, the Senses & Sensations—and from these the different Derivatives of the Agreeable from the Senses, the Beautiful, the Sublime/the Like and the Different—the spontaneous and the receptive—the Free and the Necessary—And whatever calls into consciousness the greatest number of these in due proportion & perfect harmony with each other, is the noblest Poem.—

Not the mere quantity of pleasure received can be any criterion, for that is endlessly dependent on accidents of the Subject receiving it, his age, sensibility, moral habits, &c—but the worth, the permanence, and comparative Independence of the Sources, from which the Pleasure has been derived.

[June 1810] Mem.—If I should die without having destroyed this & my other Memorandum Books, I trust, that these Hints & first Thoughts, often too cogitabilia [things thinkable] rather than actual cogitata a *me* [things thought of by *me*], may not be understood as my fixed opinions—but merely as the suggestions of the disquisition; & acts of obedience to the apostolic command of Try all things: hold fast that which is good.

[May–July 1811] Why do you make a book? Because my Hands can extend but a few score Inches from my Body; because my poverty keeps those Hands empty when my Heart aches to empty them; because my Life is short, & my Infirmities; & because a Book, if it extends but to one Edition, will probably benefit three or four on whom I could not otherwise have acted; & should it live & deserve to live, will make ample Compensation for all the afore-stated Infirmities. O but think only of the thoughts, feelings, radical Impulses that have been implanted in how many thousands of thousands by the little Ballad of the Children in the Wood! The Sphere of Alexander the great's Agency is trifling compared with it.—

[October 1818] A Phænomenon in no connection with any other Phænomenon, as its immediate cause, is a Miracle: and what is believed to have been such is miraculous for the person so believing.—When it is *strange*, and surprizing (i.e. without any analogy in our former experience) it is *called* a Miracle. The *kind* defines the *Thing:* the circumstances the *word*. To stretch out my arm *is* a miracle, unless the Materialists should be more cunning than they have proved themselves hitherto—To re-animate a dead man by an act of the Will, no intermediate agency employed, not only *is*, but is *called*, a miracle. A Scripture Miracle therefore must not be so defined as to express not only its miracular essence but likewise the condition of its *appearing* miraculous—add therefore to the preceding the words, præter (omnem priorem) experientiam [beyond (all previous) experience].

It might be defined likewise—

An effect, not having its cause in any thing congenerous [of the same kind].— That Thought calls up Thought is no more miraculous, than that a Billiard Ball moves a billiard Ball—but that a Billiard Ball should excite a Thought, i.e. be perceived, *is* a Miracle—and were it strange, would be called such—For take the converse—that a Thought should call up a Billiard Ball.—Yet where is the difference, but that the one is a common experience—the other never yet experienced.—

It is not strictly accurate to affirm, that every thing would appear a Miracle if we were wholly uninfluenced by Custom & saw Things as they are—for *then* the very ground of all Miracle would probably vanish, namely, the *heterogeneity* of Spirit and Matter. For the Quid ulterius [What further]? of Wonder, we should have the Ne plus ultra [No going beyond this] of Adoration.

Again.

The word, Miracle, has an objective, a subjective, and a popular meaning.

Objective. The essence of a Miracle consists in the heterogeneity of the Consequent and its causative Antecedent.

Subjective—: in the assumption of the heterogeneity.

Add the wonder and surprize excited when the Consequent is out of the course of *Experience, and we have the Popular sense, and ordinary use, of the word.

* N.B. To have had a sight of a Thing does not justify us in saying, that we had *experience* of it. It must have had antecedents, from which we might anticipate it.—*Ex*-perior. Omne ex-pertum est *re*-pertum [I *ex*perience. Everything ex-perienced is *re*-experienced].

[1821] I should like to know, whether or how far the delight, I feel & have always felt, in adages or aphorisms of universal or very extensive application, is a general or common feeling with man, or a peculiarity of my own mind. I cannot describe how much pleasure I have derived from 'Extremes meet' for instance; or 'Treat every thing according to its Nature', and 'Be'! In the last I bring all inward Rectitude to its Test, in the former all outward Morality to its Rule, and in the first all problematic Results to their Solution, and reduce apparent Contraries to correspondent Opposites. How many hostile Tenets has it enabled me to contemplate, as Fragments of the Truth—false only by negation, and mutual exclusion—.

[February 1826] . . . It is a very difficult thing to return a sufficing answer, sufficing I mean & satisfactory to the Answerer's own Judgement, to the objection made as often as you complain of the irreligion of the present age—/In what age have not the same complaints been made in the same words?—But when I think of the great popularity with the Readers of the Period from the year 1635 (the year of Herbert's Death)° when he delivered the MSS to Mr Farrel to the [year] 1674, the date of the 10th Edition—& how large these Impressions must have been appears from Izaak Walton, who not long after the Death of Herbert states the number at 20,000 Copies—. When I read these poems, and compare my feelings & judgements respecting them at different periods of my own life—I cannot avoid drawing the Conclusion, that as the *Readers* did at that time comprize the most thinking & competent Members of the Community, there must have been more religious *experience*, more serious interest in the Christian Faith as a business-like Concern of each Individual, than there is at present—when even among serious people it is *Generals*, the broad total Truths only of Religion, that are interesting or intelligible to them—When a man begins to be interested *in detail*, *from hour to hour*, & feels Christianity as a Life, a Growth, a Pilgrimage thro' a hostile Country—then he will enjoy *The Temple*, for the same reason that men in general enjoy Franklin's Travels over the Frozen Zone or Parry's Voyages—

[July 1826] 19 July 1826. The noblest feature in the character of Germany I find in the so general tendency of the young men in all but the lowest ranks (N.b. and highest) to select for themselves some favorite study or object of pursuit, beside their *Brodt*-wissenschaft—their Bread-earner—and where circumstances allowed, to choose the latter with reference to the former. But this, I am told, is becoming less and less the fashion even in Germany; but in England it is the misery of our all-sucking all-whirling Money-Eddy—that in our universities those, who are not idle or mistaking Verses for Poetry and Poetry for the substitute instead of (as it should be) the corolla & fragrance of the austere and many Sciences, appreciate all knowlege as means to some finite and temporal end, the main value of which consists in its being itself a means to another finite & common end—Knowlege—Profession—Income—and consequently selecting their particular Profession in exclusive reference to the probability of their acquiring a good income & perhaps ultimate[ly] a Fortune thereby, then set about getting in the easiest way exactly that sort and that quantity of knowlege, which will pass them in their examinations for the Profession, and which is requisite to or likely to forward their views of making money in or by the Profession—Now as the Many are most often the Judges & Awarders of the Prize—or else, as in the Church, some Patron—Lord, Squire, Bishop, or State-minister, we may readily determine, a priori, what the qualifications will be that are most likely to be the means to such an end—But this is the worst sort of Slavery: for herein true Freedom consists, that the outward is determined by the inward, as the alone self-determinating Principle—what then must be the result, when in the vast majority of that class in which we are most entitled to expect the *conditions* of Freedom, and Freedom itself as manifested in the *Liberal* Arts and Sciences, all Freedom is stifled & overlaid from the very commencement of their career, as men—namely, in our Universities, Schools of Medicine, Law &c?

[August 1826] 7 August, 1826, ½ past 7—beautiful Sunset, as indeed for the last six months of this wonderful tho' to me overwhelming *Season* what evening has not been such—as I was pacing alongside the ivied wall that divides our Garden from Mr Nixon's Kitchen Garden, a dell or bottom to which ours is a Terrace, and musing on the ordinary exclusive attribution of Reality to the phænomena of the passive Sense (*See*, the *reverse* of a Verb Deponent: for as this is a Verb Active in the form of a Verb-passive, the other is a Verb Passive in the form of a Verb Active), and on the impossibility that a Mind in this sensual trance should attach any practical lively meaning to the Gospel Designation of a Christian as living *a life of Grace by Faith*, in the present state, and to pass to a life of Glory; as opposed to those, who *live* without God in the World; and (to comprise all in one) of the unmeaningness or dark superstition of the Eucharist to such men, and the consequent necessity of knowing and communing with an *other* world that now is, in order to an actual and lively Belief of another world *to come*—the Question started up in my mind—but must this knowlege be explicit, and be conveyed in distinct conceptions. If so, what shall I think of *such* a Woman *as Mrs Gillman*? Can I deny that she lives with God?—At this moment my eyes were dwelling on the lovely Lace-work of those fair fair Elm-trees, so richly so softly black between me and the deep red Clouds & Light of the Horizon, with their interstices of twilight *Air* made visible—and I received the solution of my difficulty, flashlike, in the word, BEAUTY! in the intuition of the *Beautiful*!—This too is *spiritual*—and [by] the Goodness of God this is the short-hand, Hieroglyphic of Truth—the mediator between Truth and Feeling, the Head & the Heart—The Sense [of] Beauty is *implicit* knowlege—a silent communion of the Spirit with the Spirit in Nature not without consciousness, tho' with the consciousness not successively unfolded!—The Beauty of Holiness! the Beauty of Innocence! the Beauty of Love! the Beauty of Piety! Far other is the pleasure which the refined Sensualist, the pure Toutos-kosmos [this-world] Man, receives from a fine Landscape—for him it is what a fine specimen of Calligraphy would be to an unalphabeted Rustic—To a spiritual Woman it is Music—the intelligible Language of Memory, Hope, Desiderium/the *rhythm* of the Soul's movements

To whatever we attribute reality, with that we claim kindred: for reality is a transfer of our own sense of Being.

[August 1826] The best plan, I think, for a man who would wish his mind to continue growing, is to find, in the first place, some means of ascertaining for himself whether it does or no.—And I can think of no better than early in life, say after 3 & 20, to procure gradually the works of some two or three great writers—say, for instance, Bacon, Jeremy Taylor, and Kant, with the De republicâ, de legibus, the Sophistes, and Politicus of Plato, and the Poetics, Rhetoric and περι πολιτειας [*Politics*] of Aristotle—and amidst all other reading, to make a point of reperusing some one, or some weighty part of some one, of these every four or five years—having for the beginning a separate note-book for each of these Writers, in which your impressions, suggestions, Conjectures, Doubts, and Judgements are to be recorded, with the date of each, & so worded as to represent most sincerely the exact state of your convictions at the time, such as they would be if you did not (which this plan will assuredly make you do, sooner or later) anticipate a change in them from increase of knowlege. 'It is probable that I am in the wrong—but so it now appears to me, after my best attempts—& I must therefore put it down, in order that I may find myself so, if so I am.—' It would make a little volume, to give in detail all the various moral as well as intellectual advantages, that would result from the systematic observation of this plan. Diffidence and Hope would reciprocally balance and excite each other. A continuity would be given to your Being, and its progressiveness ensured. All your knowlege otherwise obtained whether from Books, or Conversation, or Experience, would find centers round which it would organize itself. And lastly, the habit of confuting your past self, and detecting the causes & occasions of your having mistaken or overlooked the Truth, will give you both a quickness and a winning kindness resulting from sympathy in exposing the errors of others, as if you were persuading an alter ego [other self] of his mistake. And such indeed will your antagonist appear to you, another = your past Self—in all points on which the falsity is not too plainly a derivative from a corrupt heart & the predominance of bad passions or worldly interests over-laying the love of Truth *as* Truth. And even in this case the liveliness with which you will so often have expressed yourself in your private Note-books, in which the words unsought-for and untrimmed because intended for your own eye exclusively were the first-born of your first impressions, when you were either enkindled by admiration of your writer, or excited by a humble disputing with *him*, re-impersonated in his Book,—will be of no mean rhetorical advantage to you, especially in public & extemporary Debates, or animated conversation.

[December 1829] XXXII [i.e. Genesis 32].—The timid prudence, and foreboding yet discreet and provident Character of Jacob, are presented with exquisite truth and individuality of portraiture. In the constitution of Jacob's Mind there is an evident root, or radical mixture of feminine Qualities—The feminine fearfulness he inherited from his Father, Isaac, as likewise his devotional and *theopathic* temperament—likewise feminine. The other female qualities, of scheming, contrivance, and artful Management from his Mother, Rebecca—And it is delightful to observe the potenziation and maturation of Isaac's Qualities in Jacob. Of all the characters in the Old Covenant, Jacob is the most favored with Visions, Presentiments, Omens &c/And surely it is a very narrow spirit, that would consider this pre-aptitude of individual Nature as derogating from the divinity of the dispensation—Is it not rather an additional ground of probability—as shewing a pre-established Harmony. The same Providence that visited Jacob by Signs, Visions, and guiding Impulses, formed him with an original aptitude for & susceptibility of the same.—N. B. By *feminine* qualities I mean nothing detractory—no participation of the *Effeminate*. In the best and greatest of men, most eminently—and less so, yet still present in all but such [as] are below the average worth of Men, there is a feminine Ingredient.—There is the *Woman* in the Man—tho' not *perhaps* the Man in the Woman—Adam *therefore* loved *Eve*—and it is the Feminine in us even now, that makes every Adam love his Eve, and crave for an Eve—

Why, I have inserted the dubious 'perhaps'—why, it should be less accordant with truth to say, that in every good Woman there is the *Man* as an Under-song, than to say that in every true and manly Man there is a translucent Under-tint of the Woman—would furnish matter for a very interesting little Essay on sexual Psychology. At present, it is enough to say, that the Woman is to look up to the *Man*, not in herself but out of herself. The Man looks out of himself for the realization and totalization of that in himself, which in himself dare not be totalized or permitted to be on the surface.

[January 1830]

Important Memorandum.

Tuesday, 12 Jany. 1830.

Caveat Senescens [A warning to one who is growing old].

On this day I felt myself sufficiently recovered from the last severe Relapse to venture out and dine and spend the Evening at my old School and Form-fellow at Christ's Hospital, Mr Steele's, at the Bottom of the Hill, in Kentish Town. The Party,—Mr and Mrs Steele, their & in spirit and habitual affection *my*, Daughter, Susan Steele, Mr George Steele, Mr Steele's Son by a former Wife: and our family—i.e. Mr and James Gillman, Mrs Gillman, Miss Ingram and myself.—And an excellent Dinner, and a quiet sociable comfortable day we enjoyed—and I returned in the Carriage with Mrs G. and Miss Ingram (for I had, avoiding however the slipperiness of the snow-frozen boy-slidden Hill, walked *thither* down the Dutchess's Lane) by 11° clock.—After Tea and Coffee, Susan, Miss Ingram, James Gillman and George Steele being at the Piano, I sate down (—

for the first time since twelve months) to a Rubber of Whist, Mrs Gillman my Partner—against Mr Steele & Mr Gillman. And it [is] this that is the subject and occasion of this Memorandum. Tho' I gave all attention, that depended on my will, to the Game—I trumped Tricks which my Partner had already won—played a Trump, when the only other (I might have known) was in my Partner's Hand, played the last trump, so as to destroy the but for that certainty of my Partner's making a long sequence—in short, I both played and felt exactly as I have often observed very old Persons play, in the decay of their Faculties.—Now this, Tuesday Night, Jany 12th, 1830, was the first instance in which my own Mind has been rendered conscious of the Decay of mental powers coming on with decay of body and elanguescence of bodily Life—and it has begun just where it might have been expected to exhibit its first symptoms, i.e. in those directions of the mental powers in which they had always been the weakest—Now with me the peccant and weak part has always been in the application of the sense and judgement to outward things, in the moment of their presence and during the period of their Succession. In the application of my Mind to Thoughts, either in their renewal or their combination, or in the appreciative and inferential Acts, I have hitherto traced no decline. But the decay, no doubt, will eat inward from the surface, in which the caries or necrosis with me has commenced; even as in Worldlings, and sharp ready-witted men of the Senses, it begins from the centre and travels to the circumference—the knowlege of Things and the power of living to them remaining last—the interval between this and *the* last, if unhappily there should be any interval, being dotage and senile Idiocy.

Now when in my case the Decay thus commenced has penetrated inward to the power that converses with Thoughts, to the Subjective Faculties, it is only too probable that like the good Archbishop of Toledo in Gil Blas, I may, by the working of the decay itself, be unconscious of it, and incredulous. With this View, under the admonition of this fore-bodement I have made this Memorandum—and shall desire Mr Green and Mrs Gillman to take note that I have so done, and of the Number of the Fly-Catcher in which this, my 'Caveat Senescens' Memorandum, is to be found—in order that when the one or the other exerts, as I fervently beseech, hope and trust that they will not be deterred from doing, the trying but most needful duty of true and enduring Friends by admonishing me, that in such or such a composition, or that in such or such lengthy talkings, the decaying energy betrays itself—and I am becoming tiresome when I might tho' no longer able to amuse or instruct, remain neutral in the becoming and natural quietness of Old Age (which is not a matter of years)—then should I appear doubtful, this Memorandum may be recalled to my Memory—and the experiment may perhaps with this assistance prove less unavailing than happened in the instance of Gil Blas and the eloquent Archbishop of Toledo.—Well will it be if there should be ground for Thanksgiving as well as for Resignation/for the latter, in the fact that the Decay having pierced or crept inward so deep, is the consequence of moral infirmities, and an unhappy Habit, begun in ignorance and delusion, and sustained by an unmanly dread of pain and disruption/for Thanksgiving, if there still remain Reason, as a *Presence*, and FAITH as a *Power*.

[September 1833?] But for Christ, Christianity, Christendom, as center of convergence, I should utterly want the *historic* sense. Even as it is, I feel it very languid in all particular History—as far as it is History, and I do not detach and insulate (isolate, is the fashionable word) a part, and thus integrate it as a person, fact, or incident having its interest in itself. The Objects seem to me so mean, so transient, as to degrade the agents—and my predominant reflection is that which I have expressed in the Friend, & in a preceding page of this Number of the Fly-Catcher—the Brobdingnagism of the human Emotions, compared with the Lilliputian apparent Excitements, and immediate Aims.° Even the most impressive historic Associations add nothing for me to any locality, beyond its own abiding interest, as a Landscape. I was, of course, familiar with Smollett's Account of the Massacre of the MacGregor Clan; and at the head of Loch Lomond heard it talked of.—The Day after I passed thro' *Glenco*—was taken full possession of the peculiar characters of the Glen, above all, by the *sensation of height* abstracted from all *sense of comparative* height—I had passed thro' the Glen, without my thinking of the Massacre Nay I arrived at Fort William before I recalled the incident.—So standing before the door of No. 4. Spenser Place, Ramsgate, Mr Lockhart said to me—Surely you can not look at yonder point at yonder Bay without attaching a livelier interest to it, when you remember that there Julius Caesar first landed on this Island!—I replied with perfect truth—I attach a delightful interest to Julius Caesar on Shakespeare's account—(I meant, from having associated him with Shakespeare's Play so called), but no interest to Pegwell Bay on account of Julius Caesar.—Nor do I need it. You cannot look at it with more delight than I do.—

I do not mention this as a *privilege*. On the contrary, I hold it to be a defect, or at best a deficiency, in the Constitution of my Mind—& have often noticed the extreme contrast in this respect between myself and the two great Poets of our Age, Sir W. Scott and W. Wordsworth. I record it only as a peculiarity, whether accidental or psychognomenic [indicative of the mind].

Marginalia°

[*c*.1807. On Milton, *Paradise Lost* v, in 'Anderson's British Poets' (1793–5). Coleridge comments on lines 469–75:

> O Adam, one Almighty is, from whom
> All things proceed, and up to him return,
> If not deprav'd from good, created all
> Such to perfection, one first matter all,
> Indued with various forms, various degrees
> Of substance, and in things that live, of life.]

There is nothing wanting to render this a perfect enunciation of the only true System of Physics, but to declare the '*one first matter all*' to be a one Act or Power consisting in two Forces or opposite Tendencies, φυσις διπλοειδης potentialiter sensitiva [a nature with two forms, potentially sensitive]; and all that follows, the same in different Potencies. For matter can neither be *ground* or distilled into spirit. The Spirit is an Island harbourless, and every way inaccessible All its contents are its products: all its denizens indigenous. Ergo, as matter could exist only for the Spirit, and as for the Spirit it cannot exist, Matter as a *principle* does not exist at all—; but as a mode of Spirit, and derivatively, it may and does exist: it being indeed the intelligential act in its first Potency.

The most doubtful position in Milton's ascending Series is the Derivation of Reason from the Understanding—without a medium.

[After 1816. On Richard Baxter, *Reliquiae Baxterianae* (1696); on a flyleaf.]

Mem. Among the grounds for recommending the perusal of our elder writers, Hooker, Taylor, Baxter, in short almost any of the Folios composed from Edward VI to Charles II.

1. The overcoming the habit of deriving your whole pleasure passively from the Book itself, which can only be effected by excitement of Curiosity or of some Passion. Force yourself to reflect on what you said §ph by §ph, and in a short time you will derive your pleasure, an ample portion at least, from the activity of your own mind. All else is Picture Sunshine.

2. The conquest of party and sectarian Prejudices, when you have on the same table the works of a Hammond and a Baxter; and reflect how many & how momentous their points of agreement; how few and

almost childish the differences, which estranged and irritated these good men!

Let us but reflect, what their blessed Spirits now feel at the retrospect of their earthly frailties: and can we do other than strive to feel as they now *feel*, not as they once felt?—So will it be with the Disputes between good men of the present Day: and if you have no other reason to doubt your Opponent's Goodness than the point in Dispute, think of Baxter and Hammond, of Milton and Jer. Taylor, and let it be no reason at all!—

3. It will secure you from the narrow Idolatry of the Present Times and Fashions: and create the noblest kind of Imaginative Power in your Soul, that of living in past ages, wholly devoid of which power a man can neither anticipate the Future, nor even live a truly human life, a life of reason, in the Present.

4. In this particular work we may derive a most instructive Lesson that in certain points, as of Religion in relation to Law, the 'Medio tutissimus ibis [you will go most safely by the middle way]' is inapplicable. There is no *Medium* possible; and all the attempts, as those of Baxter, tho' no more were required than, 'I believe in God thro' Christ,' prove only the mildness of the Proposer's Temper, but as a rule would be either $= 0$, at least exclude only the two or three in a century that make it a matter of religion to declare themselves Atheists; or just as fruitful a rule for a Persecutor as the most complete set of Articles that could be framed by a Spanish Inquisition. For to 'believe' must mean to believe aright, and 'God' must mean the true God, and 'Christ' the Christ in the sense and with the attributes understood by Christians who are truly Xtians. An established Church with a Liturgy is the sufficient solution of the Problem de Jure Magistratus [concerning the Legal Authority of the Magistrate]. Articles of Faith are superfluous; for is it not too absurd for a man to hesitate at subscribing his name to doctrines which yet in the more aweful duty of Prayer & Profession, he dares affirm before his maker? They are therefore *merely* superfluous—not worth re-enacting, had they never been done away with—not worth removing now that they exist.

5. The characteristic Contra-distinction between the Speculative Reasoners of the Age before the Revolution and those since then is this: The former cultivated metaphysics without, or neglecting, empirical Psychology; the latter cultivate a mechanical Psychology to the neglect and contempt of Metaphysics. Both therefore almost equidistant from true *Philosophy*. Hence the belief in Ghosts, Witches,

sensible Replies to Prayer &c in Baxter, and a 100 others—See p. 81:° and look at Luther's Table Talk, &c &c.

6. The earlier part of this Volume is interesting as materials for medical History. The state of medical Science in the reign of Charles the First almost incredibly low!

[On Part 3, pp. 37–8.]

Baxter, like most Scholastic Logicians, had a sneaking affection for Puns. The cause is: the necessity of attending to the primary sense of words, i.e. the visual image or general relation exprest, & which remains common to all the after senses, however widely or even incongruously differing from each other in other respects.—For the same reason, School-masters are commonly Punsters. —'I have indorsed your Bill, Sir!' said a Pedagogue to a Merchant—meaning, that he had flogged his Son William. 'Nihil in intellectu quod non prius in sensu [There is nothing in the intellect that was not previously in the senses]', my old Master, Rd J. Boyer, the Hercules Furens of the phlogistic Sect,° but else an incomparable Teacher, used to translate—first reciting the Latin words & observing that they were the fundamental article of the Peripatetic School!—'You must flog a Boy before you can make him understand'—or 'You must lay it in at the Tail before you can get it into the Head.[']—

[1815–19. On Beaumont and Fletcher, *Dramatic Works* (1811), iii. 412–13 (the end of *The Queen of Corinth*).]

In respect of Style and Versification this Play and the Bonduca may be taken as the best, and yet as *characteristic*, specimens. Particularly, the first Scene of Bonduca—. Take the Richard the Second of Shakespeare, and having selected some one scene of about the same number of Lines, and consisting mostly of long Speeches, compare it with the first scene of Bonduca/ not for the idle purpose of finding out which is the better, but in order to see and understand the difference. The latter (B. and F.) you will find a well arranged bed of Flowers, each having its separate root, and its position determined aforehand by the *Will* of the Gardener—a fresh plant a fresh Volition. In the former an Indian Fig-tree, as described by Milton—all is growth, evolution, γενεσις [*genesis*, a coming into being]—each Line, each word almost, begets the following—and the Will of the Writer is an interfusion, a continuous agency, no series of separate Acts. —Sh. is the height, breadth, and depth of Genius: B. and F. the excellent mechanism, in juxta-position and succession, of Talent.

[On iv. 166–7 (*The Noble Gentleman*).]

Why have the Dramatists of the times of Elizabeth, James I. and the first Charles become almost obsolete? excepting Shakespeare?—Why do they no longer belong to the English People, being once so popular?—And why is Shakespeare an exception?—One thing among 50 necessary to the full solution is, that they all employed *poetry* and poetic diction on *unpoetic* Subjects, both Characters & Situations—especially, in their Comedy. Now Shakespeare's [are] all ideal—of no time, & therefore of all times.—Read for instance, *Marine's* panegyric on the Court, p. 168, Column 2nd—What can be more unnatural & inappropriate, (not only *is* but must be *felt* as such) than such poetry in the mouth of a silly Dupe.—In short, the scenes are mock dialogues, in which the Poet Solo plays the ventriloquist, but cannot suppress his own way of expressing himself. Heavy complaints have been made respecting the transprosing of the old plays by Cibber°—but it never occurred to these Critics to ask how it came that no one ever attempted to transprose a comedy of Shakespeare's./

[1833. On the Bible, the Book of Revelation in the edition by J. H. Heinrichs, *Apocalypsis Graece* . . ., ii (1821), 295, where Heinrichs has offered what seemed to Coleridge an especially obtuse interpretation.]

I have too clearly before me the idea of the poet's genius to deem myself other than a very humble poet; but in the very possession of the idea, I know myself so far a poet as to feel assured that I can understand and interpret a poem in the spirit of poetry, and with the poet's spirit. Like the ostrich, I cannot fly, yet I have wings that give me the feeling of flight; and as I sweep along the plain, can look up toward the bird of Jove, and can follow him, and say—

'Sovereign of the air, who descendest on thy nest in the cleft of the inaccessible rock, who makest the mountain pinnacle thy perch and halting-place, and scanning with steady eye the orb of glory right above thee, imprintest thy lordly talons in the impassive snows that shoot back and scatter round his glittering shafts,—I pay thee homage. Thou art my king. I give honour due to the vulture, the falcon, and all thy noble baronage; and no less to the lowly bird, the sky-lark, whom thou permittest to visit thy court, and chaunt her matin song within its cloudy curtains; yea, the linnet, the thrush, the swallow, are my brethren:—but still I am a bird, though but a bird of the earth.

Monarch of our kind, I am a bird, even as thou; and I have shed plumes, which have added beauty to the Beautiful,—grace to Terror, waving on the helmed head of the war-chief;—and majesty to Grief, drooping o'er the car of Death!'

[On Jacob Böhme, *Works* (1764–81) i; an inserted leaf.]

To the contemptuous and therefore unreasonable:
for what rea[son] for continuing to rea[d] what you despis[e]?
Reader

Plato and Aristotle both reprehend this point in Anaxagoras, that having, to his immortal honor in the history of Philosophy, first established the position of a supermundane *ΝΟΥΣ*, *ως το πρωτον, μεσον και ὑστατον* [MIND, as the first, midst and last], he yet never makes any regular or determinate Use of this supreme Agent in the course of his System, but does as the Poets do who introduce a God only when their invention fails them, or their Plot has been unskillfully laid—that is, he never introduces the NOUS (= Pure Intelligence) but as a Deus ex machinâ to cut a knot, he is not able to untie.

The fault of the great German Theosopher lies in the opposite extreme. But this ought not to excite thy scorn. For the *Attempt* is dictated by Reason, nay even by Consistency; and if he have failed by soaring too high, magnis tamen excidit ausis [yet he failed in a daring attempt]—and in no spirit of pride did he soar—but being a poor unlearned Man he contemplated Truth and the forms of Nature thro' a luminous Mist, the vaporous darkness rising from his Ignorance and accidental peculiarities of fancy and sensation, but the Light streaming into it from his inmost Soul. What wonder then, if in some places the Mist condenses into a thick smoke with a few wandering rays darting across it & sometimes overpowers the eye with a confused Dazzle? The true wonder is, that in so many places it thins away almost into a transparent Medium, and Jacob Behmen, *the Philosopher*, surprizes us in proportion as Behmen, the Visionary, had astounded or perplexed us. For Behmen was indeed a Visionary in two very different senses of that word. Frequently does he mistake the dreams of his own overexcited Nerves, the phantoms and witcheries from the cauldron of his own seething Fancy, for parts or symbols of a universal Process; but frequently likewise does he give incontestible proofs, that he possessed in very truth

The Vision and Faculty divine!

And even when he wanders in the shades, ast tenet umbra Deum [but the shadow contains a God].

Read then in meekness—lest to read him at all, which might be thy folly, should prove thy Sin.

[On Sir Thomas Browne, *Religio Medici* (1669); a page pasted in at pp. 146–7.]

Friendship satisfies the *highest* parts of our nature; but a wife, who is capable of friendship, satisfies *all*. The great business of real unostentatious Virtue is—not to eradicate any genuine instinct or appetite of human nature; but—to establish a concord and unity betwixt all parts of our nature, to give a Feeling & a Passion to our purer Intellect, and to intellectualize our feelings & passions. This a happy marriage, blessed with children, effectuates, in the highest degree, of which our nature is capable, & is therefore chosen by St Paul, as the symbol of the Union of the Church with Christ; that is, of the Souls of all good Men with God, the soul of the Universe—'I scarcely distinguish,' said once a good old man, 'the wife of my old age from the wife of my youth; for when we were both young, & she was beautiful, for *once* that I caressed her with a meaner passion I caressed her a thousand times with *Love*—& *these* caresses still remain to us!'

Besides, there is another Reason why Friendship is of somewhat less Value, than Love which includes Friendship—it is this—we may love many persons, all *very* dearly; but we cannot love many persons, all *equally* dearly. There will be *differences*, there will be *gradations*—our nature imperiously asks a *summit*, a *resting-place*—it is with the affections in Love, as with the Reason in Religion—we cannot *diffuse* & equalize—we must have a SUPREME—a *One the highest*. All languages express this sentiment. What is more common than to say of a man in love—'he idolizes her,' 'he makes *a god* of her.'—Now, in order that a person should *continue* to love another, better than all others, it seems necessary, that this feeling should be reciprocal. For if it be not so, Sympathy is broken off in the very highest point. A. (we will say, by way of illustration) loves B. above all others, in the best & fullest sense of the word, love: but B. loves C. above all others. Either therefore A. does not sympathize with B. in this most important feeling; & then his Love must necessarily be incomplete, & accompanied with a *craving* after something that *is not*, & yet *might be*; or he does sympathize with B. in loving C. above all others—& then, of course, he loves C. better than B. Now it is selfishness, at least it seems so to me, to desire that your *Friend* should love you better than all others—but not to wish that a *Wife* should.

[1811. On John Donne, *Poems* (1669); note on a flyleaf.]

To read Dryden, Pope &c, you need only count syllables; but to read Donne you must measure *Time*, & discover the *Time* of Each word by the Sense & Passion.—I would ask no surer Test of a Scotch-man's *Substratum* (for the Turf-cover of Pretension they all have) than to make him read Donne's Satires aloud. If he made manly Metre of them, & yet strict metre,—*then*—why, then he wasn't a Scotchman, or his Soul was geographically slandered by his Body's first appearing there.—

[On p. 10, commenting on either 'The Canonization'—which is more probable—or 'The Triple Fool'.]

One of my favorite Poems. As late as 10 years ago, I used to seek and find out grand lines and fine stanzas; but my delight has been far greater, since it has consisted more in tracing the leading Thought thro'out the whole. The former is too much like coveting your neighbour's Goods: in the latter you merge yourself in the Author—you *become He*.—

[On p. 125, 'Satyre III': 'Kind pity chokes my spleen'.]

If you would teach a Scholar in the highest form, how to *read*, take Donne, and of Donne this Satire. When he has learnt to read Donne, with all the force & meaning which are involved in the Words—then send him to Milton—& he will stalk on, like a Master, *enjoying* his Walk.

[On Henry Fielding, *Tom Jones* (1773), iv; on flyleaves.]

Manners change from generation to generation, and with manners morals appear to change,—actually change with some, but appear to change with all but the abandoned. A young man of the present day who should act as Tom Jones is supposed to act at Upton, with Lady Bellaston, &c. would not be a Tom Jones; and a Tom Jones of the present day, without perhaps being in the ground a better man, would have perished rather than submit to be kept by a harridan of fortune. Therefore this novel is, and, indeed, pretends to be, no exemplar of conduct. But, notwithstanding all this, I do loathe the cant which can recommend Pamela and Clarissa Harlowe as strictly moral, though they poison the imagination of the young with continued doses of *tinct. lyttæ*,° while Tom Jones is prohibited as loose. I do not speak of young women;—but a young man whose heart or feelings can be injured, or even his passions excited, by aught in this novel, is already thoroughly corrupt. There is a chearful, sun-shiny, breezy spirit that prevails everywhere, strongly contrasted with the close, hot, day-dreamy continuity of Richardson. Every indiscretion, every immoral act, of Tom Jones, (and it must be remembered that he is in every one taken by surprise—his inward principles remaining firm—) is so instantly punished by embarrassment and unanticipated evil consequences of his folly, that the reader's mind is not left for a moment to dwell or run riot on the criminal indulgence itself. In short, let the requisite allowance be made for the increased refinement of our manners—and then I dare believe that no young man who consulted his heart and conscience only, without adverting to *what the World* would say—could rise from the perusal of Fielding's Tom Jones, Joseph Andrews, and Amelia, without feeling himself a better man—at least, without an intense conviction that he *could* not be guilty of a *base* Act.

If I want a servant or mechanic, I wish to know what *he does*—but of a Friend, I must know what he *is*. And in no Writer is this momentous distinction so finely brought forward as by Fielding. We do not care what Blifil *does*—the *deed*, as separate from the agent, may be good or ill—but Blifil *is* a villain—and we feel him to be so, from the very moment he, the boy Blifil, restored Sophia's poor captive Bird to its native and rightful Liberty!

[On William Hayley, *Life of Milton* (1796), p. 75, where Hayley accepts the view that Milton's 'controversial merriment' is 'disgusting'.]

The man who reads a work meant for immediate effect on one age, with the notions & feelings of another, may be a refined gentleman, but must be a sorry Critic. He who possesses imagination enough to *live* with his forefathers, and leaving comparative reflection for an after moment, to give himself *up* during the first perusal to the feelings of a contemporary if not a partizan, will, I dare aver, rarely find any part of M.'s prose works *disgusting*.

[On Joannes Scotus Erigena, *De divisione Naturae* (1681), Part 2, p. iv.]

How is it to be explained that J. Erigena with so many other Christian Divines and Philosophers should not have perceived, that pious words and scriptural phrases may disguise but cannot transsubstantiate Pantheism—a handsome Mask that does not alter a single feature of the ugly Face, it hides?—How is it to be explained that so comprehensive and subtle an Intellect, as Scotus Erigena, should not have seen, that his 'Deus omnia et omnia Deus [God all things and all things God]' was incompatible with moral responsibility, and subverted all essential difference of Good and Evil, Right and Wrong?—I can suggest no other solution, but the Innocence of his Heart and the Purity of his Life—for the same reason, that so many young men in the unresisted buoyance of their Freedom embrace without scruple the doctrine of Necessity, and only at a later and less genial Period learn, and learn to value, their free-agency by its struggles to maintain itself against the increasing incroachments of Nature and Society. It is a great Mercy of God that a good Heart is often so effective an antidote to the heresies of the Head. I could name more than [one] learned, godly and religious Clergyman, who is a Pantheist thro' his Zeal for the TRINITY—without suspecting what nevertheless is demonstrably true, that Pantheism is but a painted Atheism and that the Doctrine of the Trinity is the great and only sure Bulwark against it. But these good men take up the venomous thing, and it hurteth them not.—

[On a flyleaf.]

The whole tremendous difficulty of a Creation ex nihilo [out of nothing]—and if ex aliquo [out of something], how could it be Creation?—and not in all propriety of language Formation or Construction?—this difficulty, I say, which appeared so gigantic to our Milton that he asserted the eternity of Matter to escape from it, and then to get rid of the offensive consequences reduces this matter to an Attribute of God, and plunges head over heels into Spinosism—this difficulty, I repeat for the third time (the sad necessity of all Philo-parenthesists [lovers of parentheses]!) arises wholly out of the Slavery of the Mind to the Eye and the visual Imagination (or Fancy), under the influence of which the Reasoner must have a *picture* and mistakes surface for substance—Such men—and their name is LEGION—consequently *demand Matter*, as a *Datum* [Given fact]. As soon as this gross Prejudice is cured by the appropriate discipline, and the Mind is familiarized to the contemplation of Matter as *a product*, in time, the resulting PHENOMENON of the equilibrium of the two antagonist Forces, Attraction and Repulsion, *that* the negative and *this* the Positive Pole of Gravity* (or the Power of DEPTH), the difficulty disappears—and the Idea of CREATION alone remains.—For to will causatively with foreknowlege is to *create*, in respect of all finite products. An absolute and coeternal Product (improperly so called) is either an Offspring, and the productive Act a *Begetting*, or a Procession. The WORD begetting, the Spirit proceeding.

[On Ben Jonson, *Dramatic Works* (1811), i; flyleaf. (This edition contains plays by Beaumont and Fletcher as well.)]

It would be amusing to collect from our Dramatists from Eliz. to Charles I. proofs of the manners of the Times. One striking symptom of general Coarseness (i.e., of *Manners*, which may co-exist with great refinement of Morals, as alas! vice versâ) is to be seen in the very frequent allusions to the Olfactories [and] their most disgusting Stimulants—and these too in the Conversation of virtuous Ladies. This would not appear so strange to one who had been on terms of familiarity with Sicilian and Italian Women of Rank: and bad as they may, too many of them, *actually be*, yet I doubt not, that the extreme grossness of their Language has imprest many an Englishman of the present Æra with far darker notions, than the same language would have produced in one of Eliz. or James Ist's Courtiers. Those who have read *Shakespeare only*, complain of occasional grossness in *his* plays—Compare him with his Contemporaries, & the inevitable conviction is that of the exquisite purity of his imagination—

[On i. 200–1.]

The more we reflect and examine, examine and reflect, the more

* Centrality, or Vis centrificalis [a centralizing power] would be the preferable term. It is the same with the Mosaic Darkness, in Hebrew, the withholder or Holder-in, *Inhibitor*,—[as opposed to (Coleridge uses a symbol for this phrase)] Light, as the distinctive *exhibitive* Power.

astonished are we at the immense Superiority of Shakespeare over his contemporaries—& yet what contemporaries! Giant Minds!—Think of Jonson's Erudition, & the force of learned Authority in that age—& yet in no genuine part of Shakespeare is to be found such an absurd rant & *ventriloquism* as this & too, too many other passages ferruminated from Seneca's Tragedies, & the later Romans by Jonson. Ventriloquism, because Sejanus is a Puppet out of which the Poet makes his own voice appear to come.—

[1827. On Manuel Lacunza, *The Coming of Messiah in Glory and Majesty*, translated and with an introduction by Coleridge's friend Edward Irving (1827), vol. i, p. lxxx.]

I can not forbear expressing my regret that Mr Irving has not adhered to the clear and distinct exposition of the understanding, *genere et gradu* [in kind and in degree], given in the Aids to Reflection. What can be plainer than to say: the understanding is the medial faculty or faculty of means, as reason on the other hand is the source of ideas or ultimate ends. By reason we determine the ultimate end: by the understanding we are enabled to select and adapt the appropriate means for the attainment of, or approximation to, this end, according to circumstances. But an ultimate end must of necessity be an idea, that is, that which is not represented by the sense, and has no entire correspondent in nature, or the world of the senses. For in nature there can be neither a first nor a last:—all that we can see, smell, taste, touch, are means, and only in a qualified sense, and by the defect of our language, entitled ends. They are only relatively ends in a chain of motives. B. is the end of A.; but it is itself a mean to C., and in like manner C. is a mean to D. and so on. Thus words are the means by which we reduce appearances, or things presented through the senses, to their several kinds, or *genera*; that is, we generalize, and thus think and judge. Hence the understanding, considered specially as an intellective power, is the source and faculty of words;—and on this account the understanding is justly defined, both by Archbishop Leighton, and by Immanuel Kant, the faculty that judges by, or according to, sense. However, practical or intellectual, it is one and the same understanding, and the definition, the medial faculty, expresses its true character in both directions alike. I am urgent on this point, because on the right conception of the same, namely, that understanding and sense (to which the sensibility supplies the material of outness, *materiam objectivam* [objective matter]) constitute the natural mind of man, depends the comprehension of St. Paul's whole theological system. And this natural mind, which is named the mind of the flesh, φρονημα σαρκός, as likewise ψυχικὴ συνεσις, the intellectual power of the living or animal soul, St. Paul everywhere contradistinguishes from the spirit, that is, the power resulting from the union and co-inherence of the will and the reason;—and this spirit both the Christian and elder Jewish Church named, *sophia*, or wisdom.

[On Luther, *Colloquia Mensalia*, translated into English by H. Bell (1652), pp. 206–7, on the temptation to despair: 'When Satan saith in thy heart, God will not pardon thy sins . . . how wilt thou then . . . comfort thyself[?]']

Oh! how true, how affectingly true is this! And when too Satan, the Tempter, becomes Satan, the Accuser, saying in my heart—This Sickness is the consequence of Sin or sinful infirmity—& thou hast brought thyself into a fearful dilemma; thou canst not hope for salvation as long as thou continuest in any sinful practice—and yet thou canst not abandon thy daily dose of this or that poison without suicide. For the Sin of thy Soul has become the necessity of thy Body—daily tormenting thee, without yielding thee any the least pleasurable sensation, but goading thee on by terror without Hope. Under such evidence of God's Wrath how can'st thou expect to be saved?—Well may the Heart cry out—Who shall deliver me from the *Body* of this Death! from this Death that lives and tyrannizes in my Body!—But the Gospel answers—There IS a Redemption from the Body promised—only cling to Christ. Call on him continually, with all thy heart and all thy Soul, to give thee Strength, to be strong in thy weakness—and what Christ doth not see good to relieve thee from, suffer *in hope*.

It may be better for thee to be kept humble and in self-abasement. The thorn in the flesh may remain—& yet the Grace of God thro' Christ prove sufficient for thee. Only cling to Christ, and do thy best. In all love, and well-doing gird thyself up to improve & use aright what remains free in thee, and if thou doest aught aright, say and thankfully believe, that Christ hath done it for thee. O what a miserable despairing Wretch should I become, if I believed the Doctrines of Bishop Jer. Taylor in his Treatise on Repentance—or those I heard preached by Dr Chalmers!! If I gave up the faith, that the Life of Christ would *precipitate* the remaining dregs of Sin in the crisis of Death, and that I shall rise a purer *capacity* of Christ, blind to be irradiated by his Light, empty to be possessed by his fullness, naked of merit to be cloathed with his Righteousness!—

[On pp. 230–1, a note addressed to Charles Lamb, from whom the book had been borrowed.]

In how many little escapes and corner-holes does the sensibility, the *fine*ness, (that of which refinement is but a counterfeit, at best but a Reflex) the geniality of nature appear in this Son of Thunder! O for a Luther in the present age! Why, Charles! with the very Handcuffs of his prejudices he would knock out the brains (nay, that is impossible—but) he would split the skulls, of our *Cristo-galli*—translate the word as you like—French Christians, or Coxcombs.

[On Henry More, *Theological Works* (1708); on flyleaves.]

There are three principal causes to which the imperfections and errors in the theological schemes and works of our elder Divines, the Glories of our Church, Men of almost unparalleled Learning, Genius, the rich and robust Intellects from the reign of Elizabeth to the death of Charles the Second, may, I think, be reasonably attributed. And striking, unusually striking instances of all three abound in this Volume—& in the works of no other Divine are they more worthy of being regretted. For hence has arisen a depreciation of Dr Henry More's theological writings, which yet contain more original, enlarged and elevating views of the Christian Dispensation, than I have met with in any other single Volume. For More had both the philosophic and poetic Genius, supported by immense erudition. But unfortunately, the two did not amalgamate. It was not his good fortune to discover, as in the preceding Generation William Shakespeare discovered, a mordaunt or common Base of both; and in which both, viz. the poetic and philosophic Power blended into one.

These causes are

Ist and foremost, the want of that logical προπαιδεια docimastica [examination as a preparation for learning], that Critique of the human intellect, which previous to the weighing and measuring of this or that begins by assaying the weights, measures, and scales themselves—that fulfilment of the heaven-descended, *Nosce teipsum* [Know thyself], in respect to the intellective part of Man, which was commenced in a sort of tentative *broad-cast* way by Lord Bacon in his Novum Organum, and brought to a systematic Completion by

Immanuel KANT in his Critik der reinen Vernunft, der Urtheilskraft, & der Metaphysische Anfangsgrunde der Naturwissenschaft. From the want of this searching Logic there is a perpetual confusion of the Subjective with the Objective, in the Arguments of our Divines, together with a childish or anile over-rating of Human Testimony, and an ignorance in the art of sifting it, which necessarily engendered Credulity.

2. The ignorance of Natural Science, their Physiography scant in fact and stuffed out with fables, their Physiology embrangled with an inapplicable Logic and a misgrowth of Entia Rationalia [Rational Entities], i.e. substantiated Abstractions; and their Physiogony a Blank, or Dreams of Tradition & such 'intentional Colors' as occupy space but cannot fill it. Yet if Christianity is to be the Religion of the World, if Christ be that Logos or Word that was in the beginning, by whom all things *became*; if it was the same, who said, Let there be *Light*; who in and by the Creation commenced that great redemptive Process, the history of LIFE which begins in its detachment from Nature and is to end in its union with God—if this be true, so true must it be, that the Book of Nature and the Book of Revelation with the whole history of Man as the intermediate Link must be the integral & coherent Parts of one great work. And the conclusion is: that a scheme of the Christian Faith which does not arise out of and shoot its beams downward into, the scheme of Nature, but stands aloof, as an insulated After-thought, must be false or distorted in all its particulars. In confirmation of this position, I may challenge any opponent to adduce a single instance in which the now exploded falsities of physical Science thro' all its revolutions from the second to the 17th Century of the Christian Æra did not produce some corresponding Warps in the theological systems and dogmas of the several periods.

III. The third and last cause, and especially operative in the writings of this Author, is the presence and *regnancy* of the false and fantastic Philosophy yet shot thro' with refracted Light from the not yet risen but rising Truth, a Scheme of Physics and Physiology compounded of Cartesian Mechanics, Empiricism (for it was the credulous Childhood of Experimentalism) and a corrupt mystical theurgical Pseudo-platonism, which infected the rarest minds under the Stuart Dynasty. The only not universal Belief in Witchcraft and Apparitions, and the vindication of such Monster follies by such Men, as Sir M. Hales, Glanville, Baxter, Henry More, and a host of others, are melancholy proofs of my position. Hence in the first Chapters of this Volume the most idle & fantastic Inventions of the ancients are sought to be made credible by the most fantastic hypotheses and analogies. See PAGE 67, first half of.

To the man who has habitually contemplated Christianity as interesting all rational finite Beings, as the very 'Spirit of Truth,' the application of the Prophecies, as so many *Fortune-tellings*, and Soothsayings, to particular Events & Persons, must needs be felt as childish—faces seen in the Moon, or the sediments of a Teacup. But reverse this—and a Pope, and a Bonaparte can never be wanting—the Mole-hill becomes an Andes.—On the other hand, there are few writers, whose works could be so easily defecated as More's. Mere Omission would suffice—& perhaps one half (an unusually large proportion) would come forth from the Furnace pure Gold; if but a 4th, how great a Gain!

[1810. On James Sedgwick, *Hints to the Public and the Legislature, on . . . Evangelical Preaching* (1808–10), Part 3, pp. 27–8, where Sedgwick, a barrister, with John Bunyan in mind, asserts that a tinker has no business preaching: 'Depend upon it such men were never sent by Providence to rule or regulate mankind.']

WHOO!!! Bounteous Providence, that always looks at the Baby Clothes & the Parents' Equipage before it picks out the proper Soul for the Baby. Ho! the Dutchess of Manchester is in labor—quick, Raphael or Uriel! bring a Soul out of the *Numa Bin*/ a young Lycurgus—or the Archbishop's Lady—Ho! a Soul from the Chrysostom or Athanasian Locker°!—But poor Moll Crispin is in the Throes with Twins—/ Well! there are plenty of Cobler & Tinker Souls in the Hold.—*John Bunyan*!—Why, thou miserable Barrister, it would take an Angel an eternity a post to tinker thee into a Skull of half his Capacity!

[On Part 3, pp. 31–4, an interpretation of a statement made in *Pilgrim's Progress*.]

False! We are told by Bunyan, that the Conscience can never find relief for its *disobedience* to the Law in the Law itself—and this is as true of the Moral as of the Mosaic Law. I am not defending Calvinism or Bunyan's Theology; but if victory, not truth, were my object, I could desire no easier task than to defend it against our doughty Barrister.—Well, but I *repent*—i.e. regret it.—Yes! and so you doubtless regret the loss of an Eye or Arm—will that make it grow again?—Think you this nonsense as applied to morality? Be it so! But yet nonsense most tremendously suited to human Nature it is, as the Barrister may find in the Arguments of the Pagan Philosophers against Christianity, who attributed a large portion of its success to its holding out an expiation, which no other Religion did.—Read but that most affecting & instructive anecdote selected from the Indostan Missionary Account by the Quart.Review. Again let me say, I am not giving my own opinion on this very difficult point; but of one thing I am convinced, that the *I am sorry for it—that's enough* men mean nothing but *regret* when they talk of repentance, and have consciences either so pure or so callous, as not to know what a direful and strange thing *Remorse* is! and how absolutely a fact sui generis [of its own kind, unique]! I have often remarked, & it cannot be too often remarked (vain as this may sound) that this essential *Heterogeneity* of Regret and Remorse is of itself a sufficient and the best, *proof* of Free *Will*, and *Reason*: the co-existence of which in man we call *Conscience*, and on this rests the whole Superstructure of human Religion—God, Immortality, Guilt, Judgment, Redemption. Whether another & different Superstructure may be raised on the same foundation, or whether the same Edifice is susceptible of important alteration, is another question—but such is the Edifice at present—and this its foundation: and the Barrister might as rationally expect to blow up Windsor Castle by breaking wind in one of its Cellars, as hope to demolish Calvinism by such arguments as his.

[On Part 3, pp. 35–40, on the text 'Thou shalt love the Lord thy God with all thy heart, with all thy soul, and with all thy strength, and with all thy mind; and thy neighbour as thyself', with Christ's comment, 'This do, and thou shalt live.']

So would Bunyan, and so would Calvin have preached: would both of them in the name of Christ have made this assurance to the Barrister—*This do*, and thou shalt live!—But what if he had not done it, but the very contrary? And what if the Querist should be a staunch Disciple of Dr Paley; & hold himself *morally obliged* not to hate his fellow-man, not because he is compelled by Conscience to see the exceeding sinfulness of Sin, & to abhor sin as sin, even as he eschews Pain as Pain—no, not even because God has forbidden it—but ultimately because the great Legislator is able and has threatened to put him to unspeakable Torture if he disobeys, and to give him all kind of pleasure if he does not—? Why, verily, in this case, I do foresee that both the Tinker and the Divine would wax warm, and rebuke the said Querist for vile Hypocrisy, and a most nefarious abuse of God's good gift, intelligible Language—What! do you call *this* loving the Lord your God with all your *heart*, with all your *soul*, &c &c—and your n. as yourself—/when in truth you *love* nothing, not even your own soul; but only set a superlative *value* on whatever will gratify your selfish lust of *Enjoyment*, and ensure you from Hell-fire at 1000 times the value of the dirty property. If you have the impudence to persevere in misnaming this *Love*, supply any one instance, in which you use the word in this sense? If your Son did not spit in your face, because he believed, you would disinherit him if he did, & this were his main *moral obligation*, would *you* allow, that your Son *loved* you—& with all his heart & mind, & strength, & soul?—Shame! Shame!

Now the power of *loving* God, of willing good (not of desiring the *agreeable*, and of preferring a larger tho' distant delight to an infinitely smaller immediate gratification = selfish prudence) Bunyan considers supernatural, & seeks its source in the free grace of the Creator thro' Christ the Redeemer/ this the Kantéan avers to be supersensual indeed, but not supernatural, but in the original and essence of human Nature, & forming its grand & aweful characteristic. Hence he calls it die Menschheit = the principle of Humanity—but yet no less than Calvin or the Tinker declares it a principle most mysterious, the undoubted Object of religious Awe, a perpetual witness of that God, whose Image (Icon) it is; a principle utterly *incomprehensible* by the discursive Intellect—and moreover teaches us, that the surest plan for stifling & paralyzing this divine *Birth* in the Soul (a phrase of Plato's as well as of the Tinker's) is by attempting to evoke it by, or substitute for it, the hopes & fears, the motives & calculations, of Prudence: which is an excellent & in truth indispensable *Servant*, but considered as Master & Primate of the moral Diocese precludes the possibility of Virtue (in Bunyan's phrase, HOLINESS OF SPIRIT) by introducing *Legality*: which is no cant-phrase of methodism, but of authenticated standing in the Ethics of the profoundest Philosophers—even those, who rejected Christianity, as a *miraculous* Event, & revelation itself as far as anything *supernatural* is implied in it.—I must not mention Plato, I suppose—he was a Mystic—nor Zeno—he & his were Visionaries—but Aristotle, the cold and dry Aristotle, has in a very remarkable passage in his lesser tract of Ethics asserted the same thing; and called it 'a divine principle, lying deeper than those things which can be explained, or enunciated discursively.'

[1817–19. On Shakespeare, *Dramatic Works* (1807). Coleridge wrote 200 notes in this copy of Shakespeare, which he had bound with blank pages interleaved especially for the purpose of annotation, as an aid to the preparation of his lecture series in 1818–19. The following note on the opening lines of *King Lear* is representative of the extended practical criticism found in these volumes. The text with which Coleridge begins is *King Lear* I. i. 1–7:

KENT. I thought the king had more affected the Duke of Albany, than Cornwall.

GLOUCESTER. It did always seem so to us: but now, in the division of his kingdom, it appears not which of the dukes he values most; for equalities are so weighed, that curiosity in neither can make choice of either's moiety.]

It was [not] without forethought, and it is not without its due significance, that the triple division is stated here as already determin'd, and in all its particulars, previously to the Trial of Professions, as the relative rewards of which the Daughters were to be made to consider their several portions. The strange yet by no means unnatural, mixture of Selfishness, Sensibility, and Habit of Feeling derived from & fostered by the particular rank and usages of the Individual—the intense desire to be intensely beloved, selfish and yet characteristic of the Selfishness of a loving and kindly nature—a feeble Selfishness, self-supportless and Leaning for all pleasure on another's Breast—the selfish Craving after a sympathy with a prodigal Disinterestedness, contradicted by its own ostentation and the mode and nature of its Claims—the anxiety, the distrust, the jealousy, which more or less accompany all selfish affections, and are among the surest contradiction of mere fondness from Love, and which originate Lear's eager wish to enjoy his Daughter's violent Professions, while the inveterate habits of Sovereignty convert the wish into claim and positive Right, and the incompliance with it into crime and treason—these facts, these passions, these moral verities, on which the whole Tragedy is founded, are all prepared for, and will to the retrospect be found implied in, these first 4 or 5 lines of the Play.—They let us know that the Trial is but a Trick—and that the grossness of the old King's rage is in part the natural result of a silly Trick suddenly and most unexpectedly baffled, and disappointed.* This having been provided in the fewest words, in a natural reply to as natural [a] question, which yet answers a secondary purpose of

* *Here* notice the improbability and nursery-tale character of the tale /prefixed as the *Porch* of the Edifice, not laid as its foundation—So Shylock's lb [pound] of Flesh—item, an old popular Ballad—with how great judgement what still remains is [?correlated] &c.

attracting our attention to the differences or diversity between the characters of Cornwall and Albany, the premises and *Data* [Given facts], as it were, having been thus afforded for our after-insight into the mind and mood of the Person, whose character, passions and sufferings are the main *subject-matter* of the Play—from Lear, the Persona PATIENS° of his Drama Shaksp. passes without delay to the second in importance, to the main *Agent*, and prime Mover—introduces Edmund to our acquaintance, and with the same felicity of Judgement in the same easy, natural way prepares us for his character in the seemingly casual communication of its origins and occasion.—From the first drawing up of the Curtain he has stood before us in the united strength and beauty of earliest Manhood. Our eyes have been questioning him. Gifted thus with high advantages of *person*, and further endowed by Nature with a powerful intellect and a strong energetic Will, even without any concurrence of circumstances and accident, Pride will be the Sin that most easily besets him/. But he is the known, and acknowledged Son of the princely Gloster—Edmund therefore has both the germ of Pride and the conditions best fitted to evolve and ripen it into a predominant feeling. Yet hitherto no reason appears why it should be other than the not unusual pride of Person, Talent and Birth, a pride auxiliary if not akin to many Virtues, and the natural ally of honorable [? impulses]. But, alas! in his own presence his own father takes shame to himself for the frank avowal that he is his Father—has blushed so often to acknowlege him that he is now braz'd to it. He hears his Mother and the circumstances of his Birth spoken of with a most degrading and licentious Levity—described as a Wanton by her own Paramour, and the remembrance of the animal sting, the low criminal gratifications connected with her Wantonness and prostituted Beauty assigned as the reason, why 'the Whoreson must be acknowledged.' This and the consciousness of its notoriety—the gnawing conviction that every shew of respect is an effort of courtesy which calls while it represses a contrary feeling—this is the ever-trickling flow of Wormwood and Gall into the wounds of Pride—the corrosive Virus which inoculates Pride with a venom not its own, with Envy, Hatred, a lust of that Power which in its blaze of radiance would hide the dark spots on his disk—pangs of shame, personally undeserved, and therefore felt as wrongs—and a blind ferment of vindictive workings towards the occasions and causes, especially towards a Brother whose stainless Birth and lawful Honors were the constant remembrancers of *his* debasement, and were ever in the way to prevent all chance of its being unknown or overlooked & forgotten.

Add to this that with excellent Judgement, and provident for the claims of the moral sense, for that which relatively to the Drama is called Poetic Justice; and as the fittest means for reconciling the feelings of the Spectators to the horrors of Gloster's after Sufferings—at least, of rendering them somewhat less unendurable—(for I will not disguise my conviction, that in this one point the Tragic has been urged beyond the outermost Mark and Ne plus Ultra ['Go no further'] of the Dramatic)—Shakespeare has precluded all excuse and palliation of the guilt incurred by both the Parents of the base-born Edmund by Gloster's confession, that he was at the time a married man and already blest with a lawful Heir of his fortunes. The mournful alienation of brotherly Love occasioned by Primogeniture in noble families, or rather by the unnecessary distinctions engrafted thereon, and this in Children of the same Stock, is still almost proverbial on the continent—especially as I know from my own observation in the South of Europe, and appears to have been scarcely less common in our own Island, before the Revolution of 1688, if we may judge from the characters and sentiments so frequent in our elder Comedies—the younger Brother, for instance, in B. and F's [Beaumont and Fletcher's] Scornful Lady, on one side, and the Oliver in Sh's own As You Like It, on the other. Need it be said how heavy an aggravation the stain of Bastardy must have been—were it only, that the younger Brother was liable to hear his own dishonor and his Mother's infamy related by his Father with an excusing shrug of the shoulders, and in a tone betwixt waggery and Shame.

By the circumstances here enumerated, as so many predisposing causes, Edmund's Character might well be deem'd already sufficiently explained and prepared for. But in this Tragedy the story or fable constrained Shakespeare to introduce wickedness in an outrageous form, in Regan and Gonerill. He had read Nature too heedfully not to know, that Courage, Intellect, and strength of Character were the most impressive Forms of Power: and that to Power in itself, without reference to any moral end, an inevitable Admiration & Complacency appertains, whether it be displayed in the conquests of a Napoleon or Tamurlane, or in the foam and thunder of a Cataract. But in the display of such a character it was of the highest importance to prevent the guilt from passing into utter *monstrosity*—which again depends on the presence or absence of causes and temptations sufficient to *account* for wickedness, without the necessity of recurring to a thorough fiendishness of nature for its origination—For such are the appointed relations of intellectual Power to Truth, and of Truth to Goodness, that it becomes both morally and poetic[ally] unsafe to present what is admirable—what our nature compels [us] to admire—in the mind, and what is most detestable in the Heart, as co-existing in the same individual without any apparent connection, or any modification of the one by the other. That Shakespeare has in one instance, that of Iago, approached to this, and that he has done it successfully, is perhaps the most astonishing proof of his Genius, and the opulence of its resources.—But in the present Tragedy, in which he [was] compelled to present a Goneril & Regan, it was most carefully to be avoided—and therefore the one only conceivable addition to the inauspicious influences on the preformation of Edmund's character is given in the information, that all the kindly counteractions to the mischievous feelings of Shame that might have been derived from co-domestication with Edgar & their common father, had been cut off by an absence from home and a foreign education from Boyhood to the present time—and the prospect of its continuance, as if to preclude all risk of his interference with the Father's Views for the elder and legitimate Son. 'He hath been out nine years, and away he shall again'—

[On Robert Southey, *The Life of Wesley* (1820), i. 218–19.]

O dear & honoured Southey! this, the favorite of my Library among many favorites, this the Book which I can read for the 20th time with delight, when I can read nothing else at all—this darling Book is nevertheless, an unsafe Book for all of unsettled minds. How many admirable young men do I know or have seen, whose mind would be a Shuttle cock, between the battledores which the bi-partite Author keeps in motion! A delightful Game, between you & your Duplicate—& for those, like you, harmless. But ah! what other Duplicate is there of R. Southey, but that of his own projection!—The same facts and incidents as those recorded in Scripture, and told in the same words—& the Workers—alas! in the next page these are Enthusiasts, Fanatics. But could this have been avoided, salvâ 'Veritate' [without injury to Truth]? Answer: The *Manner*, the *Way*, might have been avoided.

[On i. 345–6, where Wesley is described as reading aloud an extract from a treatise with which he did not agree in order to precipitate his separation from the Moravians.]

I strongly suspect either misquotation here, or misinterpretation. What can be more unfair, than to pinch out a bit of a book this way!

Most candid Critic! what if I By way of joke pinch your eye? then holding up the gobbet, cry—Ha! Ha! that Men should be such DOLTS! Behold this slimy Dab! & He who own'd it, dreamt that it could *see*! The Idea were mighty analytic—But should you like it, candid Critic? I cannot help thinking that the Biographer has been in several instances led by his venial partiality for his Hero into the neighbourhood of his Hero's faults. It was a fault common to Wesley and Swedenborg to limit the words of their opponents to the worst possible sense, instead of seeking as Leibniz did, the truest sense, & then finding the error in the insufficiency & exclusiveness of the position. The Moravian Leaders, being such as Southey himself has described them, could not be ignorant, methinks! that the act of restraining & withholding is as much a positive energy, if not more, than the act of *doing this or that*: and doubtless, for *many* minds the more profitable. Their error consisted in universalizing the position, and instead of 'many' putting 'all'.—

[On i. 356–8.]

R. Southey is an Historian worth his weight in diamonds, & were he (which Heaven forfend!) as fat as myself, and the Diamonds all as big as Birds' eggs, I should still repeat the appraisal. He may err in his own deductions from Facts; but he never deceives, by concealing any known part of the Grounds & Premises, on which he had formed his conclusions—or if there by any exception—and p. 318–321 are the only ground or occasion for this 'if'—yet it will be found to respect a complex Mass of Facts to be collected from jarring & motley narratives, all as accessible to his Readers as to himself. So *here*. That I am vexed with him for not employing stronger and more empassioned words of reprobation & moral Recoil in this *black blotch* of Wesley's Heart and Character, is in another point of view the highest honor to Southey as an Historian—since it is wholly and solely from his own statement of the Incidents that my impressions have been received.— The manner in which this most delightful of all Books of Biography has been received by the Wesleyan Methodists, demonstrates the justice of the main fault which judicious men charge against the work, viz. partiality toward the Sect & its Founder—a venial fault indeed, the liability to which is almost a desirable qualification in a Biographer.

[On i. 434.]

It is a matter of earnest thought and deep concernment to me—&

he little knows my heart who shall find the spirit of authorship in what I am about to say!—to think, that thousands will read this Chapter or the Substance of it in the Writings of Wesley himself, and never complain of obscurity, or that it is, as Hone called my Aids to Reflection, *a proper Brain-cracker*.—And why is this?—In the words, I use, or their collocation?—Not so: for no one has pointed out any passage of importance, which he having at length understood, he could propose any other & more intelligible words that would have conveyed precisely the same meaning!—No!—Wesley first *relates* his theory as a *history*—Ideas were for him & thro' him for his Readers so many *Proper Names*, the substratum of meaning being supplied by the general image or abstraction, of the human Form with the swarm of associations that cluster on it.—Wesley *takes for granted* that his Readers will all understand it all at once, and without effort—: The Readers are far too well pleased with this; or rather this procedure is far too much in accord both with their mental indolence & their Self-complacency, that they should think of asking themselves the question.—Reflect on the simple fact of the state of a Child's mind while with great delight he reads or listens to the story of Jack and the Bean Stalk!—How could this be, if in some sense he did not understand it?—Yea, the child does *under*stand each part of it, A and B and C; but not ABC = X. He understands it as we all understand our Dreams while we are dreaming, each Shape & Incident, or Group of Shapes & Incidents, by itself, unconscious of & therefore unoffended at, the absence of the logical copula or the absurdity of the Transitions. He understands it, in short, as the *Reading Public* understands this Exposition of Wesley's Theology.—Now compare this with the manner and even *obtruded* purpose of the Friend or the Aids to Reflection; in which the aim of every sentence is to solicit, nay, *tease* the Reader to ask himself, whether he *actually* does or does not, understand *distinctly*? Whether he has however reflected on the precise meaning of the word, however familiar it may be both to his ear & Mouth? Whether he has been hitherto aware of the mischief and folly of employing words on questions, to know the very truth of which is both his Interest & his Duty, without fixing the one meaning which on that question they are to represent?—Page after page for a Reader accustomed from Childhood either to learn by rote—i.e. without understanding at all, as Boys learn their Latin Grammar, or to content himself with the popular use of words, always wide and general—and expressing a whole County when perhaps the point in discussion concerns the difference between two Parishes of the same

County! (Ex. gr. MIND which in the popular use means sometimes Memory, sometimes Reason, sometimes Understanding, sometimes Sense (αισθησις), sometimes inclination—and sometimes all together confusedly.)—for such a Reader, I repeat, page after page is a process of mortification and awkward Straining. Will any one instruct me, how this is to be remedied? Will he refer me to any work already published, which has achieved the objects, at which I aim, without exciting the same complaints? But then I should wish my friendly Monitor to shew me at the same time one of these uncomplaining Readers, & convince me that he is actually Master of the Truths contained in that work—be it Plato's, Bacon's or Bull's or Waterland's. Alas! alas! with a poor illiterate but conscience-stricken, or soul-[?awakened] Hayme, or Pawson° I should find few difficulties, beyond those that are the price of all momentous knowlege. For while I was demonstrating the inner structure of our Spiritual Organismus [organism], he would have his mental eye fixed on the same Subject—i.e. his own mind—even as an Anatomist may be dissecting a human Eye, and the Pupils too far off to see this may yet be dissecting another eye, closely following the instructions of the Lecturer, & comparing his words with the shapes and textures which the knife discloses to them. But in the great majority of our Gentry and of our classically educated Clergy there is a fearful Combination of the *Sensuous* and the *Unreal*.—Whatever is *subjective*, the true & only proper *Noumenon*, or Intelligibile [object of intellectual apprehension], is unintelligible to them. But all *Sub*stance ipso nomine [by its very name] is necessarily *Subjective*, & what these men call reality is Object unsouled of all Subject—of course, an appearance only, which becomes connected with the sense of reality by its being common to any number of Beholders, present at the same moment—but an apparitio communis [shared apparition] is still but an *Apparition*, and can be substantiated for each Individual only by his attributing a Subject thereto, as its support and causa sufficiens [sufficient cause], even as the community of Appearance is the sign & presumptive proof, of its Objectivity. In short, I would fain bring the cause, I am pleading, to a short and simple yet decisive Test. Consciousness, *Ειμι* ['I am'; being], Mind, Life, Will, Body, Organ [as opposed to (Coleridge uses a symbol)] Machine, Nature, Spirit, Sin, Habit, Sense, Understanding, Reason—Here are fourteen words. Have you ever reflectively and quietly asked yourself the meaning of any one of these, and tasked yourself to return the answer in *distinct* terms, not applicable to any one of the other words? Or have you contented yourself with the vague floating meaning, that will just serve to save you from absurdity in the use of the word, just as the Clown's Botany would do who knew that Potatoes were Roots, and Cabbages Greens? Or, if you have the gift of Wit, shelter yourself under Augustine's Equivocation [']I know it perfectly well, till I am asked.'—Know? Aye, as an Oyster knows its Life; but do you know your knowlege? If the latter be your case, can you wonder that the Aids of Reflection are Clouds and Darkness for you?—

[On Jonathan Swift, *Works* (1768), v. The only note, on the endpaper, has to do with the fourth book of *Gulliver's Travels*.]

The great defect of the Houyhnhnms is not its misanthropy, and those who apply this word to it must really believe that the essence of human nature, that the anthropos misoumenos [hated man], consists in the shape of the Body. Now to shew the falsity of this was Swift's great object—he would prove to our feelings & imaginations and thereby teach *practically*, that it is Reason and Conscience which give all the loveliness and dignity not only to Man, but to the shape of Man—that deprived of these and retaining his Understanding he would be the most loathsome & hateful of all animals—that his Understanding would manifest itself only as malignant Cunning, his free will as obstinacy and unteachableness—and how true a picture this is, every Madhouse may convince any man/ a Brothel where Highwaymen meet, will convince every philosopher.—But the defect of this work is its inconsistency—the Houyhnhnms are not rational creatures, i.e. creatures of perfect reason—they are not progressive—they have servants without any reason for their natural inferiority or any explanation how the difference acted—and above all, they—i.e. Swift himself—has a perpetual affectation of being wiser* than his Maker,

* A case in point, and besides utterly inconsistent with the boasted Reason of the Houyhnhnm, may be seen p. 194, 195—where the Horse discourses on the human frame with the grossest prejudices that could possibly be inspired by vanity & self-opinion. That Reason which commands man to admire the fitness of the Horse & Stag for superior speed, of the Bird for Flight, &c &c—must it not have necessitated the rational Horse to have seen and acknowleged the admirable aptitude of the human Hand compared with his own fetlocks—of the human Limbs for climbing, for the management of Tools &c?—In short, compare the *effect* of the Satire, when it is founded in Truth & good sense (Chapt. V. for instance) with the wittiest of those passages which have their only support in spleen & want of reverence for the original

and of eradicating what God gave to be subordinated & used—ex. gr. the maternal & paternal affection (στοργὴ) [*storge*]—there is likewise a true Yahooism in the constant denial of the existence of Love, as not identical with Friendship, & yet distinct always & very often divided from Lust—. The best defence is that it is a Satyr—still it would have been felt a thousand times more deeply, if Reason had been truly pourtrayed—and a finer Imagination would have been evinced if the Author had shewn the effects of the possession of Reason & the moral sense in the outward form & gestures of the Horse/ In short, Critics in general complain of the Yahoos—I complain of the Houyhnhnms—

As to the *wisdom* of adopting this mode of proving the great Truths here exemplified, that is another question, which no feeling mind will find a difficulty in answering, who has read & understood the Paradise Scenes in Paradise Lost, & compared the moral effect on his heart: and the virtuous aspirations of Milton's Adam with Swift's Horses—but different men have different Turns of Genius—Swift's may be good, tho' very inferior to Milton's—they do not stand in each other's way.

[1818–27. On W. G. Tennemann, *Geschichte aer Philosophie* (1798–1817) viii, Part 2, p. 960 (where Tennemann summarizes Gerson's account of the faculties of the mind) and flyleaf.]

What G. calls Contemplation is what I call Positive Reason, Reason in her own Sphere, as distinguished from Negative or merely *formal* Reason, Reason in the sphere of the Understanding.

P. 960. Gerson's & St Victore's Contemplation is in my System = *Positive* Reason, or R. in her own Sphere as distinguished from the merely *formal* Negative Reason, R. in the lower sphere of the Understanding. The +R = Lux: −R = Lumen a Luce [positive Reason is Light: negative Reason is a Lamp from the Light]. By the one the Mind contemplates Ideas: by the other it meditates on Conceptions. Hence the distinction might be expressed by the names, Ideal Reason [in opposition to (Coleridge uses a symbol)] Conceptual Reason.

frame of Man—& the feelings of the Reader will be his faithful guides in the reperusal of the work—which I still think the highest effort of Swift's genius, unless we should except the Tale of the Tub—/ then I would put Lilliput—next Brobdignag—& Laputa I would expunge altogether/ It is a wretched abortion, the product of Spleen & Ignorance & Self-conceit—

The simplest yet practically sufficient Order of the Mental Powers is, beginning from the

lowest	highest
Sense	Reason
Fancy	Imagination
Understanding	Understanding
———	———
Understanding	Understanding
Imagination	Fancy
Reason	Sense.

Fancy and Imagination are Oscillations, *this* connecting R. and U; *that* connecting Sense and Understanding.

Table Talk°

[3 January 1823] Define a vulgar ghost with reference to all that is called ghostlike. It is visibility without tangibility; which is also the definition of a shadow. Therefore, a vulgar ghost and a shadow would be the same; because two different things properly have the same definition. A visible substance without susceptibility of impact, I maintain to be an absurdity. Unless there be an external substance, the bodily eye cannot see it; therefore, in all such cases, that which is supposed to be seen is, in fact, not seen, but is an image of the brain. External objects naturally produce sensation; but here, in truth, sensation produces, as it were, the external object.

In certain states of the nerves, however, I do believe that the eye, although not consciously so directed, may, by a slight convulsion, see a portion of the body, as if opposite to it. The part actually seen will by common association seem the whole; and the whole body will then constitute an external object, which explains many stories of persons seeing themselves lying dead. Bishop Berkeley once experienced this. He had the presence of mind to ring the bell, and feel his pulse; keeping his eye still fixed on his own figure right opposite to him. He was in a high fever, and the brain image died away as the door opened. I observed something very like it once at Grasmere; and was so conscious of the cause, that I told a person what I was experiencing, whilst the image still remained.

Of course, if the vulgar ghost be really a shadow, there must be some substance of which it is the shadow. These visible and intangible shadows, without substances to cause them, are absurd.

[1 May 1823] There is a great difference in the credibility to be attached to stories of dreams and stories of ghosts. Dreams have nothing in them which are absurd and nonsensical; and, though most of the coincidences may be readily explained by the diseased system of the dreamer, and the great and surprising power of association, yet it is impossible to say whether an inner sense does not really exist in the mind, seldom developed, indeed, but which may have a power of presentiment. All the external senses have their correspondents in the mind; the eye can see an object before it is distinctly apprehended—why may there not be a corresponding power in the soul? The power of prophecy might have been merely a spiritual excitation of this dormant faculty. Hence you will observe that the Hebrew seers sometimes seem to have required music. Everything in nature has a tendency to move in cycles; and it would be a miracle if, out of such myriads of cycles moving concurrently, some coincidences did not take place. No doubt, many such take place in the daytime; but then our senses drive out the remembrance of them, and render the impression hardly felt; but when we sleep, the mind acts without interruption. Terror and the heated imagination will, even in the daytime, create all sorts of features, shapes, and colours out of a simple object possessing none of them in reality.

But ghost stories are absurd. Whenever a real ghost appears—by which I mean some man or woman dressed up to frighten another—if the supernatural character of the apparition has been for a moment believed, the effects on the spectator have always been most terrible—convulsion, idiocy, madness, or even death on the spot. Consider the awful descriptions in the Old Testament of the effects of a spiritual presence on the prophets and seers of the Hebrews; the terror, the exceeding great dread, the utter loss of all animal power. But in our common ghost stories, you always find that the seer, after a most appalling apparition, as you are to believe, is quite well the next day. Perhaps, he may have a headache; but that is the outside of the effect produced. Allston, a man of genius, and the best painter yet produced by America, when he was in England told me an anecdote which confirms what I have been saying. It was, I think, in the University of Cambridge, near Boston, that a certain youth took it into his wise head to endeavour to convert a Tom Painish companion of his by appearing as a ghost before him. He accordingly dressed himself up in the usual way, having previously extracted the ball from the pistol which always lay near the head of his friend's bed. Upon first awaking, and seeing the apparition, the youth who was to be frightened, A., very coolly looked his companion the ghost in the face, and said, 'I know you. This is a good joke; but you see I am not frightened. Now you may vanish!' The ghost stood still. 'Come,' said A., 'that is enough. I shall get angry. Away!' Still the ghost moved not. 'By ——,' ejaculated A., 'if you do not in three minutes go away, I'll shoot you.' He waited the time, deliberately levelled the pistol, fired, and, with a scream at the immobility of the figure, became convulsed, and afterwards died. The very instant he believed it *to be* a ghost, his human nature fell before it.

[1 May 1830] A fall of some sort or other—the creation, as it were, of the non-absolute—is the fundamental postulate of the moral history of man. Without this hypothesis, man is unintelligible; with it, every phenomenon is explicable. The mystery itself is too profound for human insight.

Madness is not simply a bodily disease. It is the sleep of the spirit with certain conditions of wakefulness; that is to say, lucid intervals. During this sleep, or recession of the spirit, the lower or bestial states of life rise up into action and prominence. It is an awful thing to be eternally tempted by the perverted senses. The reason may resist—it does resist—for a long time; but too often, at length, it yields for a moment, and the man is mad for ever. An act of the will is, in many instances, precedent to complete insanity. I think it was Bishop Butler who said, that he was all his life struggling against the devilish suggestions of his senses, which would have maddened him, if he had relaxed the stern wakefulness of his reason for a single moment.

[12 May 1830] Shakespeare is the Spinozistic deity—an omnipresent creativeness. Milton is the deity of prescience; he stands *ab extra* [on the outside], and drives a fiery chariot and four, making the horses feel the iron curb which holds them in. Shakespeare's poetry is characterless; that is, it does not reflect the individual Shakespeare; but John Milton himself is in every line of the Paradise Lost. Shakespeare's rhymed verses are excessively condensed—epigrams with the point everywhere; but in his blank dramatic verse he is diffused, with a linked sweetness long drawn out. No one can understand Shakespeare's superiority fully until he has ascertained, by comparison, all that which he possessed in common with several other great dramatists of his age, and has then calculated the surplus which is entirely Shakespeare's own. His rhythm is so perfect, that you might be almost sure that you do not understand the real force of a line, if it does not run well as you read it. The necessary mental pause after every hemistich or imperfect line is always equal to the time that would have been taken in reading the complete verse.

[31 May 1830] Mrs Barbauld once told me that she admired the Ancient Mariner very much, but that there were two faults in it—it was improbable, and had no moral. As for the probability, I owned that that might admit some question; but as to the want of a moral, I told her that in my judgement the poem had too much; and that the only, or chief fault, if I might say so, was the obtrusion of the moral sentiment so openly on the reader as a principle or cause of action in a work of such pure imagination. It ought to have had no more moral than the Arabian Nights' tale of the merchant's sitting down to eat dates by the side of a well, and throwing the shells aside, and lo! a genie starts up, and says he *must* kill the aforesaid merchant, *because* one of the date shells had, it seems, put out the eye of the genie's son.

I took the thought of 'grinning for joy' in that poem from poor Burnett's remark to me, when we had climbed to the top of Plinlimmon, and were nearly dead with thirst. We could not speak from the constriction, till we found a little puddle under a stone. He said to me, 'You grinned like an idiot!' He had done the same.

[1 June 1830] There are three sorts of prayer: (1) public; (2) domestic; (3) solitary. Each has its peculiar uses and character. I think the Church ought to publish and authorize a directory of forms for the latter two. Yet I fear the execution would be inadequate. There is a great decay of devotional unction in the numerous books of prayers put out nowadays. I really think the hawker was very happy who blundered New Form of Prayer into New *former* Prayers.

I exceedingly regret that our Church pays so little attention to the subject of congregational singing. See how it is! In that particular part of the public worship in which, more than in all the rest, the common people might, and ought to, join—which, by its association with music, is meant to give a fitting vent and expression to the emotions—in that part we all sing as Jews; or, at best, as mere men, in the abstract, without a Saviour. You know my veneration for the Book of Psalms, or most of it; but with some half dozen exceptions, the Psalms are surely not adequate vehicles of Christian thanksgiving and joy! Upon this deficiency in our service, Wesley and Whitfield seized; and you know it is the hearty congregational singing of Christian hymns which keeps the humbler Methodists together. Luther did as much for the Reformation by his hymns as by his translation of the Bible. In Germany, the hymns are known by heart by every peasant: they advise, they argue from the hymns, and every soul in the Church praises God, like a Christian, with words which are natural and yet sacred to his mind. No doubt this defect in our service proceeded from the dread which the English Reformers had of being charged with introducing anything into the worship of God but the text of Scripture.

[2 July 1830] Every man is born an Aristotelian or a Platonist. I do not think it possible that anyone born an Aristotelian can become a Platonist; and I am sure no born Platonist can ever change into an

Aristotelian. They are the two classes of men, beside which it is next to impossible to conceive a third. The one considers reason a quality, or attribute; the other considers it a power. I believe that Aristotle never could get to understand what Plato meant by an idea. There is a passage, indeed, in the Eudemian Ethics which looks like an exception; but I doubt not of its being spurious, as that whole work is supposed by some to be. With Plato ideas are constitutive in themselves.

Aristotle was, and still is, the sovereign lord of the understanding—the faculty judging by the senses. He was a conceptualist, and never could raise himself into that higher state, which was natural to Plato, and has been so to others, in which the understanding is distinctly contemplated, and, as it were, looked down upon from the throne of actual ideas, or living, inborn, essential truths.

Yet what a mind was Aristotle's—only not the greatest that ever animated the human form!—the parent of science, properly so called, the master of criticism, and the founder or editor of logic! But he confounded science with philosophy, which is an error. Philosophy is the middle state between science, or knowledge, and *sophia*, or wisdom.

[8 September 1830] The fatal error into which the peculiar character of the English Reformation threw our Church, has borne bitter fruit ever since—I mean that of its clinging to Court and State, instead of cultivating the people. The Church ought to be a mediator between the people and the Government, between the poor and the rich. As it is, I fear the Church has let the hearts of the common people be stolen from it. See how differently the Church of Rome—wiser in its generation—has always acted in this particular. For a long time past the Church of England seems to me to have been blighted with prudence, as it is called. I wish with all my heart we had a little zealous imprudence.

[19 September 1830] It has never yet been seen, or clearly announced, that democracy, as such, is no proper element in the constitution of a State. The idea of a State is undoubtedly a Government ἐκ τῶν ἀρίστων [*ek tōn aristōn*, from among the best]—an aristocracy. Democracy is the healthful life-blood which circulates through the veins and arteries, which supports the system, but which ought never to appear externally, and as the mere blood itself.

A State, in idea, is the opposite of a Church. A State regards classes, and not individuals; and it estimates classes, not by internal merit, but external accidents, as property, birth, &c. But a Church does the reverse of this, and disregards all external accidents, and looks at men as individual persons, allowing no gradation of ranks, but such as greater or less wisdom, learning, and holiness ought to confer. A Church is, therefore, in idea, the only pure democracy. The Church, so considered, and the State, exclusively of the Church, constitute together the idea of a State in its largest sense.

[21 September 1830] I do not know whether I deceive myself, but it seems to me that the young men who were my contemporaries fixed certain principles in their minds, and followed them out to their legitimate consequences, in a way which I rarely witness now. No one seems to have any distinct convictions, right or wrong; the mind is completely at sea, rolling and pitching on the waves of facts and personal experiences. Mr —— is, I suppose, one of the rising young men of the day; yet he went on talking, the other evening, and making remarks with great earnestness, some of which were palpably irreconcilable with each other. He told me that facts gave birth to, and were the absolute ground of, principles; to which I said, that unless he had a principle of selection, he would not have taken notice of those facts upon which he grounded his principle. You must have a lantern in your hand to give light, otherwise all the materials in the world are useless, for you cannot find them, and if you could, you could not arrange them. 'But then,' said Mr ——, '*that* principle of selection came from facts!'—'To be sure!' I replied; 'but there must have been again an antecedent light to see those antecedent facts. The relapse may be carried in imagination backwards for ever—but go back as you may, you cannot come to a man without a previous aim or principle.' He then asked me what I had to say to Bacon's Induction: I told him I had a good deal to say, if need were; but that it was perhaps enough for the occasion, to remark, that what he was very evidently taking for the Baconian *In*duction, was mere *De*duction—a very different thing.

[8 October 1830] Galileo was a great genius, and so was Newton; but it would take two or three Galileos and Newtons to make one Kepler.° It is in the order of Providence, that the inventive, generative, constitutive mind—the Kepler—should come first; and then that the patient and collective mind—the Newton—should follow, and elaborate the pregnant queries and illumining guesses of the former. The laws of the planetary system are, in fact, due to Kepler. There is not a more glorious achievement of scientific genius upon record, than Kepler's guesses, prophecies, and ultimate apprehension of the law of the mean distances of the planets as connected with the periods of their revolutions round the sun. Gravitation, too, he had fully conceived; but, because it seemed inconsistent with some received observations on light, he gave it up, in allegiance, as he says, to Nature. Yet the idea vexed and haunted his mind; '*Vexat me et lacessit* [It annoys and irritates me],' are his words, I believe.

[24 August 1831] The English public is not yet ripe to comprehend the essential difference between the reason and the understanding—between a principle and a maxim—an eternal truth and a mere conclusion generalized from a great number of facts. A man, having seen a million moss-roses all red, concludes from his own experience and that of others that all moss-roses are red. That is a maxim with him—the *greatest* amount of his knowledge upon the subject. But it is only true until some gardener has produced a white moss-rose—after which the maxim is good for nothing. Again, suppose Adam watching the sun sinking under the western horizon for the first time; he is seized with gloom and terror, relieved by scarce a ray of hope that he shall ever see the glorious light again. The next evening, when it declines, his hopes are stronger, but still mixed with fear; and even at the end of a thousand years, all that a man can feel is, a hope and an expectation so strong as to preclude anxiety. Now compare this in its highest degree with the assurance which you have that the two sides of any triangle are together greater than the third. This, demonstrated of one triangle, is seen to be eternally true of all imaginable triangles. This is a truth perceived at once by the intuitive reason, independently of experience. It is, and must ever be so, multiply and vary the shapes and sizes of triangles as you may.

It used to be said that four and five *make* nine. Locke says, that four and five *are* nine. Now I say, that four and five *are not* nine, but that they will *make* nine. When I see four objects which will form a square, and five which will form a pentagon, I see that they are two different things; when combined, they will form a third different figure which we call nine. When separate they *are not* it, but will *make* it.

[12 September 1831] My system, if I may venture to give it so fine a name, is the only attempt I know ever made to reduce all knowledges into harmony. It opposes no other system, but shows what was true in each; and how that which was true in the particular, in each of them became error, because it was only half the truth. I have endeavoured to unite the insulated fragments of truth, and therewith to frame a perfect mirror. I show to each system that I fully understand and rightly appreciate what that system means; but then I lift up that system to a higher point of view, from which I enable it to see its former position, where it was, indeed, but under another light and with different relations—so that the fragment of truth is not only acknowledged, but explained. Thus, the old astronomers discovered and maintained much that was true; but, because they were placed on a false ground, and looked from a wrong point of view, they never did, they never could, discover the truth—that is, the whole truth. As soon as they left the earth, their false centre, and took their stand in the sun, immediately they saw the whole system in its true light, and their former station remaining, but remaining as a part of the prospect. I wish in short to connect by a moral copula natural history with political history; or, in other words, to make history scientific, and science historical—to take from history its accidentality, and from science its fatalism.

[10 June 1832] When I was a little boy at the Blue-coat School, there was a charm for one's foot when asleep; and I believe it had been in the school since its foundation, in the time of Edward VI. The march of intellect has probably now exploded it. It ran thus:

> Foot! foot! is fast asleep!
> Thumb! thumb! thumb! in spittle we steep:
> Crosses three we make to ease us,
> Two for the thieves, and one for Christ Jesus!

And the same charm served for a cramp in the leg, with the following substitution:

> The devil is tying a knot in my leg!
> Mark, Luke, and John, unloose it I beg!—
> Crosses three, etc.

And really, upon getting out of bed, where the cramp most frequently occurred, pressing the sole of the foot on the cold floor, and then repeating this charm with the acts configurative thereupon prescribed, I can safely affirm that I do not remember an instance in which the cramp did not go away in a few seconds.

I should not wonder if it were equally good for a stitch in the side; but I cannot say I ever tried it for that.

[21 July 1832] I have often wished that the first two books of the Excursion had been published separately, under the name of 'The Deserted Cottage'. They would have formed, what indeed they are, one of the most beautiful poems in the language.

Can dialogues in verse be defended? I cannot but think that a great philosophical poet ought always to teach the reader himself as from himself. A poem does not admit argumentation, though it does admit development of thought. In prose there may be a difference; though I must confess that, even in Plato and Cicero, I am always vexed that the authors do not say what they have to say at once in their own persons. The introductions and little urbanities are, to be sure, very delightful in their way; I would not lose them: but I have no admiration for the practice of ventriloquizing through another man's mouth.

I cannot help regretting that Wordsworth did not first publish his thirteen books on the growth of an individual mind—superior, as I used to think, upon the whole, to the Excursion. You may judge how I felt about them by my own poem upon the occasion. Then the plan laid out, and, I believe, partly suggested by me, was, that Wordsworth should assume the station of a man in mental repose, one whose principles were made up, and so prepared to deliver upon authority a system of philosophy. He was to treat man as man—a subject of eye, ear, touch, and taste, in contact with external nature, and informing the senses from the mind, and not compounding a mind out of the senses; then he was to describe the pastoral and other states of society, assuming something of the Juvenalian spirit as he approached the high civilization of cities and towns, and opening a melancholy picture of the present state of degeneracy and vice; thence he was to infer and reveal the proof of, and necessity for, the whole state of man and society being subject to, and illustrative of, a redemptive process in operation, showing how this idea reconciled all the anomalies, and promised future glory and restoration. Something of this sort was, I think, agreed on. It is, in substance, what I have been all my life doing in my system of philosophy.

I think Wordsworth possessed more of the genius of a great

philosophic poet than any man I ever knew, or, as I believe, has existed in England since Milton; but it seems to me that he ought never to have abandoned the contemplative position which is peculiarly—perhaps I might say exclusively—fitted for him. His proper title is *Spectator ab extra* [a Spectator from the outside].

[25 July 1832] The pith of my system is to make the senses out of the mind—not the mind out of the senses, as Locke did.

[6 August 1832] You will find this a good gauge or criterion of genius—whether it progresses and evolves, or only spins upon itself. Take Dryden's Achitophel and Zimri—Shaftesbury and Buckingham; every line adds to or modifies the character, which is, as it were, a-building up to the very last verse; whereas, in Pope's Timon etc. the first two or three couplets contain all the pith of the character, and the twenty or thirty lines that follow are so much evidence or proof of overt acts of jealousy, or pride, or whatever it may be that is satirized. In like manner compare Charles Lamb's exquisite criticisms on Shakespeare with Hazlitt's round and round imitations of them.

[16 August 1832] The discipline at Christ's Hospital in my time was ultra-Spartan—all domestic ties were to be put aside. 'Boy!' I remember Boyer saying to me once when I was crying the first day of my return after the holidays, 'Boy! the school is your father! Boy! the school is your mother! Boy! the school is your brother! the school is your sister! the school is your first cousin, and your second cousin, and all the rest of your relations! Let's have no more crying!'

No tongue can express good Mrs Boyer. Val. Le Grice and I were once going to be flogged for some domestic misdeed, and Boyer was thundering away at us by way of prologue, when Mrs B. looked in and said, 'Flog them soundly, sir, I beg!' This saved us. Boyer was so nettled at the interruption that he growled out, 'Away, woman! away!' and we were let off.

[4 May 1833] The wonderful powers of machinery can, by multiplied production, render the mere *arte facta* [artefacts, 'things made by art'] of life actually cheaper: thus, money and all other things being supposed the same in value, a silk gown is five times cheaper now than in Queen Elizabeth's time; but machinery cannot cheapen, in anything like an equal degree, the immediate growths of nature or the immediate necessaries of man. Now the *arte facta* are sought by the higher classes of society in a proportion incalculably beyond that in which they are sought by the lower classes; and therefore it is that the vast increase of mechanical powers has not cheapened life and pleasure to the poor as it has done to the rich. In some respects, no doubt, it has done so, as in giving cotton dresses to maidservants, and penny gin to all. A pretty benefit truly!

[3 July 1833] The definition of good prose is—proper words in their proper places; of good verse—the most proper words in their proper places. The propriety is in either case relative. The words in prose ought to express the intended meaning, and no more; if they attract attention to themselves, it is, in general, a fault. In the very best styles, as Southey's, you read page after page, understanding the author perfectly, without once taking notice of the medium of communication—it is as if he had been speaking to you all the while. But in verse you must do more; there the words, the *media*, must be beautiful, and ought to attract your notice—yet not so much and so perpetually as to destroy the unity which ought to result from the whole poem. This is the general rule, but, of course, subject to some modifications, according to the different kinds of prose or verse. Some prose may approach towards verse, as oratory, and therefore a more studied exhibition of the *media* may be proper; and some verse may border more on mere narrative, and there the style should be simpler. But the great thing in poetry is, *quocunque modo* [by some means or other], to effect a unity of impression upon the whole; and a too great fullness and profusion of point in the parts will prevent this. Who can read with pleasure more than a hundred lines or so of Hudibras at one time? Each couplet or quatrain is so whole in itself, that you can't connect them. There is no fusion—just as it is in Seneca.

[4 July 1833] Dr Johnson's fame now rests principally upon Boswell. It is impossible not to be amused with such a book. But his *bow-wow* manner must have had a good deal to do with the effect produced—for no one, I suppose, will set Johnson before Burke, and Burke was a great and universal talker, yet now we hear nothing of this except by some chance remarks in Boswell. The fact is, Burke, like all men of genius who love to talk at all, was very discursive and continuous; hence he is not reported; he seldom said the sharp short things that Johnson almost always did, which produce a more decided effect at the moment, and which are so much more easy to carry off. Besides, as to Burke's testimony to Johnson's powers, you must remember that Burke was a great courtier; and after all, Burke said and wrote more than once that he thought Johnson greater in talking than in writing, and greater in Boswell than in real life.

[20 August 1833] Hebrew is so simple, and its words are so few and near the roots, that it is impossible to keep up any adequate knowledge of it without constant application. The meanings of the words are chiefly traditional. The loss of Origen's Heptaglott Bible, in which he had written out the Hebrew words in Greek characters, is the heaviest which biblical literature has ever experienced. It would have fixed the sounds as known at that time.

Brute animals have the vowel sounds; man only can utter consonants. It is natural, therefore, that the consonants should be marked first, as being the framework of the word; and no doubt a very simple living language might be written quite intelligibly to the natives without any vowel sounds marked at all. The words would be traditionally and conventionally recognized as in shorthand—thus—*Gd crtd th Hvn nd th Rth.* I wish I understood Arabic; and yet I doubt whether to the European philosopher or scholar it is worth while to undergo the immense labour of acquiring that or any other Oriental tongue, except Hebrew.

[23 June 1834] You may conceive the difference in kind between the fancy and the imagination in this way—that if the check of the senses and the reason were withdrawn, the first would become delirium, and the last mania. The fancy brings together images which have no connection natural or moral, but are yoked together by the poet by means of some accidental coincidence; as in the well-known passage in Hudibras:

> The sun had long since in the lap
> Of Thetis taken out his nap,
> And like a lobster boyl'd, the morn
> From black to red began to turn.

The imagination modifies images, and gives unity to variety; it sees all things in one, *il più nell'uno.* There is the epic imagination, the perfection of which is in Milton; and the dramatic, of which Shakespeare is the absolute master. The first gives unity by throwing back into the distance; as after the magnificent approach of the Messiah to battle, the poet, by one touch from himself—

> far off their coming shone!—

makes the whole one image. And so at the conclusion of the description of the appearance of the entranced angels, in which every sort of image from all the regions of earth and air is introduced to diversify and illustrate, the reader is brought back to the single image by

> He call'd so loud that all the hollow deep
> Of Hell resounded.

The dramatic imagination does not throw back, but brings close; it stamps all nature with one, and that its own, meaning, as in Lear throughout.

[10 July 1834] I am dying, but without expectation of a speedy release. Is it not strange that very recently bygone images, and scenes of early life, have stolen into my mind, like breezes blown from the spice islands of Youth and Hope—those two realities of this phantom world! I do not add Love—for what is Love but Youth and Hope embracing, and so seen as one? I say realities; for reality is a thing of degrees, from the Iliad to a dream; *καὶ γάρ τ' ὄναρ ἐκ Δίος ἔστι* [for a dream too is from Zeus]. Yet, in a strict sense, reality is not predicable at all of aught below Heaven. 'Es enim *in coelis*, Pater noster, qui tu vere *es* [Thou art *in heaven*, Our Father, who *art* indeed]!' Hooker wished to live to finish his Ecclesiastical Polity—so I own I wish life and strength had been spared to me to complete my Philosophy. For, as God hears me, the originating, continuing, and sustaining wish and design in my heart was to exalt the glory of his name; and, which is the same thing in other words, to promote the improvement of mankind. But *visum aliter Deo* [God has decided otherwise], and his will be done.

A Moral and Political Lecture°

To calm and guide
The swelling democratic tide;
To watch the state's *uncertain* frame;
To baffle Faction's *partial* aim;
But chiefly with determin'd zeal
To quell the servile Band that kneel
To Freedom's jealous foes;
And lash that Monster, who is daily found
Expert and bold our country's peace to wound,
Yet dreads to handle arms, nor manly counsel knows.

Akenside

ADVERTISEMENT

They, who in these days of jealousy and party rage dare publicly explain the principles of freedom, must expect to have their intentions misrepresented, and to be entitled like the Apostles of Jesus, 'stirrers up of the people, and men accused of sedition.' The following lecture is therefore printed as it was delivered, the author choosing that it should be published with all the inaccuracies and inelegant colloquialisms of an hasty composition, rather than that he should be the object of possible calumny as one who had rashly uttered sentiments which he afterwards timidly qualified.

WHEN the wind is fair and the planks of the vessel sound, we may safely trust everything to the management of professional mariners; but in a tempest and on board a crazy bark, all must contribute their quota of exertion. The stripling is not exempted from it by his youth, nor the passenger by his inexperience. Even so in the present agitations of the public mind, everyone ought to consider his intellectual faculties as in a state of immediate requisition. All may benefit society in some degree. The exigencies of the times do not permit us to stay for the maturest years, lest the opportunity be lost, while we are waiting for an increase of power. Omitting therefore the disgusting egotisms of an affected humility, we shall briefly explain the design, and possible benefit, of the proposed political disquisitions.

Companies resembling the present will from a variety of circumstances consist chiefly of the zealous advocates for freedom. It will be therefore our endeavour, not so much to excite the torpid, as to regulate the feelings of the ardent: and above all, to evince the necessity of bottoming on fixed principles, that so we may not be the unstable patriots of passion or accident, or hurried away by names of which we have not sifted the meaning, and by tenets of which we have not examined the consequences. The times are trying: and in order to be prepared against their difficulties, we should have acquired a prompt facility of adverting in all our doubts to some grand and comprehensive truth. In a deep and strong soil must that blessing fix its roots, the height of which, like that of the tree in Daniel, is to 'reach to Heaven, and the sight of it to the ends of all the earth.'

The example of France is indeed a 'warning to Britain.' A nation wading to their rights through blood, and marking the track of freedom by devastation! Yet let us not embattle our feelings against our reason. Let us not indulge our malignant passions under the mask of humanity. Instead of railing with infuriate declamation against these excesses, we shall be more profitably employed in developing the sources of them. French freedom is the beacon, that while it guides us to equality should show us the dangers, that throng the road.

The annals of the French Revolution have recorded in letters of blood, that the knowledge of the few cannot counteract the ignorance of the many; that the light of philosophy, when it is confined to a small minority, points out the possessors as the victims, rather than the illuminators, of the multitude. The patriots of France either hastened into the dangerous and gigantic error of making certain evil the means of contingent good, or were sacrificed by the mob, with whose prejudices and ferocity their unbending virtue forbade them to assimilate. Like Samson, the people were strong—like Samson, the people were blind. Those two massy pillars of oppression's temple, monarchy and aristocracy

> With horrible convulsion to and fro
> They tugg'd, they shook—till down they came and drew
> The whole Roof after them with burst of Thunder
> Upon the heads of all who sat beneath,
> Lords, Ladies, Captains, Counsellors, and Priests,
> Their choice Nobility!
>
> Milton. Sam. Agon.

There was not a tyrant in Europe, that did not tremble on his throne. Freedom herself heard the crash aghast—yet shall she not have heard it unbenefited, if haply the horrors of that day shall have made other nations timely wise—if a great people shall from hence become adequately illuminated for a Revolution bloodless, like

Poland's, but not, like Poland's assassinated by the foul treason of tyrants against liberty.°

Revolutions are sudden to the unthinking only. Political disturbances happen not without their warning harbingers. Strange rumblings and confused noises still precede these earthquakes and hurricanes of the moral world. In the eventful years previous to a revolution, the philosopher as he passes up and down the walks of life, examines with an anxious eye the motives and manners, that characterize those who seem destined to be the actors in it. To delineate with a free hand the different classes of our present oppositionists to 'things as they are,' may be a delicate, but it is a necessary task—in order that we may enlighten, or at least beware of, the misguided men who have enlisted themselves under the banners of freedom from no principles or from bad ones, whether they be those 'who extol things vulgar' and

> admire they know not what,
> And know not whom, but as one leads the other—

or whether those,

> Whose end is private Hate, not help to Freedom,
> In *her* way to *Virtue* adverse and turbulent.

The first class among the professed friends of liberty is composed of men, who unaccustomed to the labour of thorough investigation and not particularly oppressed by the burthen of state, are yet impelled by their feelings to disapprove of its grosser depravities, and prepared to give an indolent vote in favour of reform. Their sensibilities unbraced by the co-operation of fixed principles, they offer no sacrifices to the divinity of active virtue. Their political opinions depend with weathercock uncertainty on the winds of rumour that blow from France. On the report of French victories they blaze into republicanism, at a tale of French excesses they darken into aristocrats; and seek for shelter those despicable adherents to fraud and tyranny, who ironically style themselves Constitutionalists. These dough-baked patriots may not however be without their use. This oscillation of political opinion, while it retards the day of revolution, may operate as a preventive to its excesses. Indecision of character, though the effect of timidity, is almost always associated with benevolence.

Wilder features characterize the second class. Sufficiently possessed of natural sense to despise the priest, and of natural feeling to hate the oppressor, they listen only to the inflammatory harangues of some mad-headed enthusiast, and imbibe from them poison, not food, rage, not liberty. Unillumined by philosophy and stimulated to a lust of revenge by aggravated wrongs, they would make the altar of freedom stream with blood, while the grass grew in the desolated halls of justice. These men are the rude materials from which a detestable minister manufactures conspiracies. Among these men he sends a brood of sly political monsters, in the character of sanguinary demagogues, and like Satan of old, 'the tempter ere the accuser' ensnares a few into treason, that he may alarm the whole into slavery. He who has dark purposes to serve must use dark means—light would discover, reason would expose him: he must endeavour to shut out both, or if this prove impracticable, make them appear frightful by giving them frightful names: for further than names the vulgar enquire not. Religion and reason are but poor substitutes for 'Church and Constitution'; and the sable-vested instigators of the Birmingham riots° well knew that a syllogism could not disarm a drunken incendiary of his firebrand, or a demonstration *helmet* a philosopher's head against a brickbat. But in the principles which this apostate has, by his emissaries, sown among a few blind zealots for freedom, he has digged a pit into which he himself may perhaps be doomed to fall. We contemplate those principles with horror. Yet they possess a kind of wild justice well calculated to spread them among the grossly ignorant. To unenlightened minds, there are terrible charms in the idea of retribution, however savagely it be inculcated. The groans of the oppressors make fearful yet pleasant music to the ear of him whose mind is darkness, and into whose soul the iron has entered.

This class, at present, is comparatively small—yet soon to form an overwhelming majority, unless great and immediate efforts are used to lessen the intolerable grievances of our poorer brethren, and infuse into their sorely wounded hearts the healing qualities of knowledge. For can we wonder that men should want humanity, who want all the circumstances of life that humanize? Can we wonder that with the ignorance of brutes they should unite their ferocity? Peace and comfort be with these! But let us shudder to hear from men of dissimilar opportunities sentiments of similar revengefulness. The purifying alchemy of education may transmute the fierceness of an ignorant man into virtuous energy—but what remedy shall we apply to him whom plenty has not softened, whom knowledge has not taught benevolence? This is one among the many fatal effects which result from the want of fixed principles. Convinced that vice is error, we shall entertain sentiments of pity for the vicious, not of indignation—and even with respect to that bad man,° to whom we have before alluded, although we are now groaning beneath the burthen of his misconduct, we shall harbour no sentiments of revenge; but rather condole with him that his chaotic iniquities have exhibited such a complication of extravagance, inconsistency, and rashness, as may alarm him with apprehensions of approaching lunacy!

There are a third class among the friends of freedom who possess not the wavering character of the first description, nor the ferocity last delineated. They pursue the interests of freedom steadily, but with narrow and self-centring views: they anticipate with exultation the abolition of privileged orders, and of acts that persecute by exclusion from the right of citizenship: they are prepared to join in digging up the rubbish of mouldering establishments and stripping off the taudry pageantry of Governments. Whatever is above them they are most willing to drag down; but alas! they use not the pulley! Whatever tends to improve and elevate the ranks of our poorer brethren, they regard with suspicious jealousy, as the dreams of the visionary; as if there were anything in the superiority of lord to gentleman, so mortifying in the barrier, so fatal to happiness in the consequences, as the more real distinction of master and servant, of rich man and of poor. Wherein am I made worse by my ennobled neighbour? do the childish titles of aristocracy detract from my domestic comforts, or prevent my intellectual acquisitions? but those institutions of society which should condemn me to the necessity of twelve hours daily toil, would make my soul a slave, and sink the rational being in the mere animal. It is a mockery of our fellow creatures' wrongs to call them equal in rights, when by the bitter compulsion of their wants we make them inferior to us in all that can soften the heart, or dignify the understanding. Let us not say that this is the work of time—that it is impracticable at present, unless we each in our individual capacities do strenuously and perseveringly endeavour to diffuse among our domestics those comforts and that illumination which far beyond all political ordinances are the true equalizers of men. But of the propriety and utility of holding up the distant mark of attainable perfection, we shall enter more fully towards the close of this address; we turn with pleasure to the contemplation of that small but glorious band, whom we may truly distinguish by the name of thinking and disinterested patriots. These are the men who have encouraged the sympathetic passions till they have become irresistible habits, and made their duty a necessary part of their self-interest, by the long continued cultivation of that moral taste which derives our most exquisite pleasures from the contemplation of possible perfection, and proportionate pain from the perception of existing depravation. Accustomed to regard all the affairs of man as a process, they never hurry and they never pause; theirs is not that twilight of political knowledge which gives us just light enough to place one foot before the other; as they advance, the scene still opens upon them, and they press right onward with a vast and various landscape of existence around them. Calmness and energy mark all their actions, benevolence is the silken thread that runs through the pearl chain of all their virtues. Believing that vice originates not in the man, but in the surrounding circumstances, not in the heart, but in the understanding, he is hopeless concerning no one —to correct a vice or generate a virtuous conduct he pollutes not his hands with the scourge of coercion, but by endeavouring to alter the circumstances removes, or by strengthening the intellect disarms, the temptation. The unhappy children of vice and folly, whose tempers are adverse to their own happiness as well as to the happiness of others, will at times awaken a natural pang; but he looks forward with gladdened heart to that glorious period when justice shall have established the universal fraternity of love. These soul-ennobling views bestow the virtues which they anticipate. He whose mind is habitually impressed with them soars above the present state of humanity, and may be justly said to dwell in the presence of the most high. Regarding every event even as he that ordains it, evil vanishes from before him, and he views with naked eye the eternal form of universal beauty.

Say why was Man so eminently raised
Amid the vast creation—why ordain'd
Thro' life and death to dart his piercing eye,
With thoughts beyond the limits of his frame,
But that the Omnipotent might send him forth
In sight of mortal and immortal powers,
As on a boundless theatre, to run
The great career of Justice—to exalt
His generous aim to all diviner deeds,
To chase each partial purpose from his breast
And thro' the tossing tide of chance and pain
To hold his course unfaltering? else why burns
In mortal bosoms this unquenched hope
That breathes from day to day sublimer things
And mocks possession?

would the forms
Of servile custom cramp the patriot's power,
Would sordid policies, the barbarous growth
Of ignorance and rapine bow him down
To tame pursuits, to Indolence and Fear?
Lo he appeals to Nature, to the winds
And rolling waves, the sun's unwearied course,
The elements and seasons—all declare
For what the Eternal Maker has ordain'd
The powers of Man: we feel within ourselves
His energy divine: he tells the heart
He meant, he made us to behold and love
What he beholds and loves, the general orb
Of Life and Being—to be great like him,
Beneficent and active.

Akenside

On such a plan has a Gerald formed his intellect.° Withering in the sickly and tainted gales of a prison, his healthful soul looks down from the citadel of his integrity on his impotent persecutors. I saw him in the foul and naked room of a jail—his cheek was sallow with confinement—his body was emaciated, yet his eye spoke the invincible purposes of his soul, and his voice still sounded with rapture the successes of freemen, forgetful of his own lingering martyrdom! Such too were the illustrious triumvirate* whom, as a Greek poet expresses it, it's not lawful for bad men even to praise. I will not say that I have abused your patience in thus indulging my feelings in these strains of unheard gratitude to men, who may seem to justify God in the creation of man. It is with pleasure that I am permitted to recite a yet unpublished tribute to their merit, the production of a man who has sacrificed all the energies of his heart and head—a splendid offering on the altar of Liberty.

TO THE EXILED PATRIOTS°

Martyrs of Freedom—ye who firmly good
  Stept forth the champions in her glorious cause,
Ye who against Corruption nobly stood
  For Justice, Liberty, and equal Laws.

* Muir, Palmer, and Margarot.°

Ye who have urged the cause of man so well
  Whilst proud Oppression's torrent swept along,
Ye who so firmly stood, so nobly fell,
  Accept one ardent Briton's grateful song.

For shall Oppression vainly think by Fear
  To quench the fearless energy of mind?
And glorying in your fall, exult it here
  As tho' no honest heart were left behind?

Thinks the proud tyrant by the pliant law
  The timid jury and the judge unjust,
To strike the soul of Liberty with awe,
  And scare the friends of Freedom from their trust?

As easy might the Despot's empty pride
  The onward course of rushing ocean stay;
As easy might his jealous caution hide
  From mortal eyes the orb of general day.

For like that general orb's eternal flame
  Glows the mild force of Virtue's constant light;
Tho' clouded by Misfortune, still the same,
  For ever constant and for ever bright.

Not till eternal chaos shall that light
  Before Oppression's fury fade away;
Not till the sun himself be lost in night;
  Not till the frame of Nature shall decay.

Go then secure, in steady virtue go,
  Nor heed the peril of the stormy seas—
Nor heed the felon's name, the outcast's woe;
  Contempt and pain, and sorrow and disease.

Tho' cankering cares corrode the sinking frame,
  Tho' sickness rankle in the sallow breast;
Tho' Death were quenching fast the vital flame,
  Think but for what ye suffer, and be blest.

So shall your great examples fire each soul,
  So in each free-born breast for ever dwell,
Till Man shall rise above the unjust controul—
  Stand where ye stood, and triumph where ye fell.

To accomplish the great object in which we are anxiously engaged to place liberty on her seat with bloodless hands, we have shown the necessity of forming some fixed and determinate principles of action to which the familiarized mind may at all times advert. We now proceed to that most important point, namely, to show what those principles must be. In times of tumult firmness and consistency are peculiarly needful, because the passions and prejudice of mankind are then more powerfully excited: we have shown in the example of France that to its want of general information, its miseries and its horrors may be attributed. We have reason to believe that a revolution in other parts of Europe is not far distant. Oppression is grievous—the oppressed feel and complain. Let us profit by the example of others; devastation has marked the course of most revolutions, and the timid assertors of freedom, equally with its clamorous enemies, have so closely associated the ideas that they are unable to contemplate the one disunited from the other. The evil is great, but it may be averted—it has been a general, but it is not therefore a necessary consequence. In order to avert it, we should teach ourselves and others habitually to consider that truth wields no weapon but that of investigation, we should be cautious how we indulge even the feelings of virtuous indignation. Indignation is the handsome brother of anger and hatred—benevolence alone beseems the philosopher. Let us not grasp even despotism with too abrupt a hand, lest, like the envenomed insect of Peru,* it infect with its poison the hand that removes it harshly. Let us beware that we continue not the evils of tyranny when the monster shall be driven from the earth. Its temple is founded on the ruins of mankind. Like the fane of Tescalipoca, the Mexican deity, it is erected with human skulls and cemented with human blood—let us beware that we be not transported into revenge while we are levelling the loathsome pile with the ground, lest when we erect the temple of freedom we but vary the style of architecture, not change the materials. Our object is to destroy pernicious systems, not their misguided adherents. Philosophy imputes not the great evil to the corrupted but to the system which presents the temptation to corruption. The evil must cease when the cause is removed, and the courtier who is enabled by State machinations to embroil or enslave a nation when levelled to the standard of men will be impotent of evil, as he is now unconscious of good. Humane from principle, not fear, the

* The Coya, an insect of so thin a skin, that on being incautiously touched, it bursts, and of so subtle a poison that it is immediately absorbed into the body, and proves fatal.

disciple of liberty shrinks not from his duty. He will not court persecution by the ill-timed obtrusion of truth, still less will he seek to avoid it by concealment or dereliction. J. H. Tooke on the morning of his trial wrote to a fellow sufferer in these words: 'Nothing will so much serve the cause of freedom as our *acquittal*, except our *execution*.'° He meant I presume to imply that whatever contributes to increase discussion must accelerate the progress of liberty. Let activity and perseverance and moderation supply the want of numbers. Convinced of the justice of our principles, let neither scorn nor oppression prevent us from disseminating them. By the gradual deposition of time, error has been piled upon error and prejudice on prejudice, till few men are tall enough to look over them, and they whose intellects surpass the common stature, and who describe the green vales and pleasant prospects beyond them, will be thought to have created images in vacancy and be honoured with the name of madman; but

It is the motive strong the conscience pure
That bids us firmly act or meek endure:
'Tis this will shield us when the storm beats hard
Content tho' poor had we no other guard!

Bowles

ESSAYS ON HIS TIMES

Pitt°

PLUTARCH, in his comparative biography of Rome and Greece, has generally chosen for each pair of Lives the two contemporaries who most nearly resembled each other. His work would perhaps have been more interesting, if he had adopted the contrary arrangement, and selected those rather, who had attained to the possession of similar influence, or similar fame, by means, actions, and talents the most dissimilar. For power is the sole object of philosophical attention in man, as in inanimate nature; and in the one equally as in the other, we understand it more intimately the more diverse the circumstances are with which we have observed it coexist. In our days the two persons who appear to have influenced the interests and actions of men the most deeply and the most diffusively are beyond doubt the Chief Consul of France, and the Prime Minister of Great Britain: and in these two are presented to us similar situations, with the greatest dissimilitude of characters.

William Pitt was the younger son of Lord Chatham, a fact of no ordinary importance in the solution of his character, of no mean significance in the heraldry of morals and intellect. His father's rank, fame, political connections, and parental ambition were his mould: he was cast, rather than grew. A palpable election, a conscious predestination controlled the free agency, and transfigured the individuality of his mind; and that, which he *might have been*, was compelled into that, which he *was to be*. From his early childhood it was his father's custom to make him stand up on a chair, and declaim before a large company; by which exercise, practised so frequently, and continued for so many years, he acquired a premature and unnatural dexterity in the combination of words, which must of necessity have diverted his attention from present objects, obscured his impressions, and deadened his genuine feelings. Not the thing on which he was speaking, but the praises to be gained by the speech, were present to his intuition; hence he associated all the operations of his faculties with words, and his pleasures with the surprise excited by them.

But an inconceivably large portion of human knowledge and human power is involved in the science and management of words; and an education of words, though it destroys genius, will often create, and always foster, talent. The young Pitt was conspicuous far beyond his fellows, both at school and at college. He was always full-grown: he had neither the promise nor the awkwardness of a growing intellect. Vanity, early satiated, formed and elevated itself into a love of power; and in losing this colloquial vanity, he lost one of the prime links that connect the individual with the species, too early for the affections, though not too early for the understanding. At college he was a severe student; his mind was founded and elemented in words and generalities, and these too formed all the superstructure. That revelry and that debauchery which are so often fatal to the powers of intellect would probably have been serviceable to him; they would have given him a closer communion with realities, they would have induced a greater presentness to present objects. But Mr Pitt's conduct was correct, unimpressibly correct. His after-discipline in the special pleader's office, and at the bar, carried on the scheme of his education with unbroken uniformity. His first political connections were with the Reformers; but those who accuse him of sympathizing or coalescing with their intemperate or visionary plans misunderstand his character, and are ignorant of the historical facts. Imaginary situations in an imaginary state of things rise up in minds that possess a power and facility in combining images. Mr Pitt's ambition was conversant with old situations in the old state of things, which furnish nothing to the imagination, though much to the wishes. In his endeavours to realize his father's plan of reform, he was probably as sincere as a being who had derived so little knowledge from actual impressions could be. But his sincerity had no living root of affection; while it was propped up by his love of praise and immediate power, so long it stood erect and no longer. He became a Member of the Parliament—supported the popular opinions, and in a few years, by the influence of the popular party, was placed in that high and awful rank in which he now is. The fortunes of his country, we had almost said the fates of the world, were placed in his wardship—we sink in prostration before the inscrutable dispensations of Providence, when we reflect in whose wardship the fates of the world were placed!

The influencer of his country and of his species was a young man, the creature of another's predetermination, sheltered and weather-fended from all the elements of experience; a young man whose feet had never wandered, whose very eye had never turned to the right or to the left, whose whole track had been as curveless as the motion of a fascinated reptile! It was a young man whose heart was solitary, because he had existed always amid objects of futurity, and whose imagination too was unpopulous, because those objects of hope to which his habitual wishes had transferred, and as it were *projected*, his existence, were all familiar and long-established objects!—A plant sown and reared in a hothouse, for whom the very air that surrounded him had been regulated by the thermometer of previous purpose; to whom the light of nature had penetrated only through glasses and covers; who had had the sun without the breeze; whom no storm had shaken; on whom no rain had pattered; on whom the dews of Heaven had not fallen!—A being who had had no feelings connected with man or nature, no spontaneous impulses, no unbiased and desultory studies, no genuine science, nothing that constitutes individuality in intellect, nothing that teaches brotherhood in affection! Such was the man—such, and so denaturalized, the spirit on whose wisdom and philanthropy the lives and living enjoyments of so many millions of human beings were made unavoidably dependent. From this time a real enlargement of mind became almost impossible. Preoccupations, intrigue, the undue passion and anxiety with which all facts must be surveyed; the crowd and confusion of those facts, none of them seen, but all communicated, and by that very circumstance, and by the necessity of perpetually classifying them, transmuted into words and generalities; pride; flattery; irritation; artificial power; these, and circumstances resembling these, necessarily render the heights of office barren heights, which command indeed a vast and extensive prospect, but attract so many clouds and vapours that most often all prospect is precluded. Still, however, Mr Pitt's situation, however inauspicious for his real being, was favourable to his fame. He heaped period on period; persuaded himself and the nation that extemporaneous arrangement of sentences was eloquence, and that eloquence implied wisdom. His father's struggles for freedom, and his own attempts, gave him an almost unexampled popularity; and his office necessarily associated with his name all the great events that happened during his administration. There were not, however, wanting men who saw through this delusion: and refusing to attribute the industry, integrity, and enterprising spirit of our merchants, the agricultural improvements of our landholders, the great inventions of our manufacturers, or the valour and skilfulness of our sailors to the merits of a Minister, they have continued to decide on his character from those acts and those merits which belong to him and to him alone. Judging him by this standard, they have been able to discover in him no one proof or symptom of a commanding genius. They have discovered him never controlling, never creating, events, but always yielding to them with rapid change, and sheltering himself from inconsistency by perpetual indefiniteness. In the Russian war, they saw him abandoning meanly what he had planned weakly, and threatened insolently. In the debates on the Regency, they detected the laxity of his constitutional principles, and received proofs that his eloquence consisted not in the ready application of a general system to particular questions, but in the facility of arguing for or against any question by specious generalities, without reference to any system. In these debates, he combined what is most dangerous in democracy with all that is most degrading in the old superstitions of monarchy, and taught an inherency of the office in the person in order to make the office itself a nullity, and the premiership, with its accompanying majority, the sole and permanent power of the State. And now came the French Revolution. This was a new event; the old routine of reasoning, the common trade of politics, were to become obsolete. He appeared wholly unprepared for it. Half favouring, half condemning, ignorant of what he favoured and why he condemned, he neither displayed the honest enthusiasm and fixed principle of Mr Fox, nor the intimate acquaintance with the general nature of man and the consequent *prescience* of Mr Burke.°

After the declaration of war, long did he continue in the common cant of office, in declamation about the Scheldt, and Holland, and all the vulgar causes of common contests! and when at last the immense genius of his new supporter had beat him out of these words (words, signifying places and dead objects, and signifying nothing more), he adopted other words in their places, other generalities—atheism and Jacobinism—phrases which he learnt from Mr Burke, but without learning the philosophical definitions and involved consequences with which that great man accompanied those words. Since the death of Mr Burke, the forms and the sentiments and the tone of the French have undergone many and important changes: how indeed is it possible, that it should be otherwise, while man is the creature of experience! But still Mr Pitt proceeds in an endless repetition of the same general phrases. This is his element; deprive him of general and abstract phrases, and you reduce him to silence. But you cannot deprive him of them. Press him to specify an individual fact of advantage to be derived from a war, and he answers, security! Call upon him to particularize a crime, and he exclaims—Jacobinism! Abstractions defined by abstractions! Generalities defined by generalities! As a Minister of Finance, he is still, as ever, the man of words and abstractions! Figures, custom-house reports, imports and exports, commerce and revenue—all flourishing, all splendid! Never was such a prosperous country, as England under his administration! Let it be objected, that the agriculture of the country is, by the overbalance of commerce, and by various and complex causes, in such a state, that the country hangs as a pensioner for bread on its neighbours, and a bad season uniformly threatens us with famine—this (it is replied) is owing to our prosperity—all prosperous nations are in great distress for food!—Still prosperity, still general phrases, unenforced by one single image, one single fact of real national amelioration, of any one comfort enjoyed where it was not before enjoyed, of any one class of society becoming healthier, or wiser, or happier. These are things, these are realities; and these Mr Pitt has neither the imagination to body forth, or the sensibility to feel for. Once indeed, in an evil hour, intriguing for popularity, he suffered himself to be persuaded to evince a talent for the real, the individual: and he brought in his Poor Bill!! When we hear the Minister's talents for finance so loudly trumpeted, we turn involuntarily to his Poor Bill—to that acknowledged abortion—that unanswerable evidence of his ignorance respecting all the fundamental relations and actions of property, and of the social union!

As his reasonings, even so is his eloquence. One character pervades his whole being. Words on words, finely arranged, and so dexterously consequent that the whole bears the semblance of argument, and still keeps awake a sense of surprise—but when all is done, nothing rememberable has been said; no one philosophical remark, no one image, not even a pointed aphorism. Not a sentence of Mr Pitt's has ever been quoted, or formed the favourite phrase of the day—a thing unexampled in any man of equal reputation. But while he speaks, the effect varies according to the character of his auditor. The man of no talent is swallowed up in surprise: and when the speech is ended, he remembers his feelings, but nothing distinct of that which produced them (how opposite an effect to that of nature and genius, from whose works the idea still remains, when the feeling is passed away—remains to connect itself with other feelings, and combine with new impressions!). The mere man of talent hears him with admiration, the mere man of genius with contempt—the philosopher neither admires nor contemns, but listens to him with a deep and solemn interest, tracing in the effects of his eloquence the power of words and phrases, and that peculiar constitution of human affairs in their present state which so eminently favours this power.

Such appears to us to be the Prime Minister of Great Britain, whether we consider him as a statesman, or as an orator. The same character betrays itself in his private life, the same coldness to realities, to images of realities, and to all whose excellence relates to reality. He has patronized no science, he has raised no man of genius from obscurity, he counts no one prime work of God among his friends. From the same source he has no attachment to female society, no fondness for children, no perceptions of beauty in natural scenery; but he is fond of convivial indulgences,° of that stimulation which, keeping up the glow of self-importance and the sense of internal power, gives feelings without the mediation of ideas.

These are the elements of his mind; the accidents of his fortune, the circumstances that enabled such a mind to acquire and retain such a power, would form a subject of a philosophical history, and that too of no scanty size. We can scarcely furnish the chapter of contents to a work which would comprise subjects so important and delicate as the causes of the diffusion and intensity of secret influence; the machinery and state-intrigue of marriages; the overbalance of the commercial interest; the panic of property struck by the late revolution; the short-sightedness of the careful; the carelessness of the far-sighted; and all those many and various events which have given to a decorous profession of religion, and a seemliness of private morals, such an unwonted weight in the attainment and preservation of public power. We are unable to determine whether it be more consolatory or humiliating to human nature that so many complexities of event, situation, character, age, and country should be necessary in order to the production of a Mr Pitt.

THE FRIEND°

[*Essay* II]

Sic oportet ad librum, praesertim miscellanei generis, legendum accedere lectorem, ut solet ad convivium conviva civilis. Convivator annititur omnibus satisfacere: et tamen si quid apponitur, quod hujus aut illius palato non respondeat, et hic et ille urbane dissimulant, et alia fercula probant, ne quid contristent convivatorem. Quis enim eum convivam ferat, qui tantum hoc animo veniat ad mensam, ut carpens quae apponuntur nec vescatur ipse, nec alios vesci sinat? et tamen his quoque reperias inciviliores, qui palam, qui sine fine damnent ac lacerent opus, quod nunquam legerint. Ast hoc plus quam *sycophanticum* est damnare quod nescias.

[*Translated freely by Coleridge in the footnote below*]

Erasmus

THE musician may tune his instrument in private, ere his audience have yet assembled: the architect conceals the foundation of his building beneath the superstructure. But an author's harp must be tuned in the hearing of those who are to understand its after-harmonies; the foundation-stones of his edifice must lie open to common view, or his friends will hesitate to trust themselves beneath the roof.

From periodical literature the general reader deems himself entitled to expect amusement, and some degree of information, and if the writer can convey any instruction at the same time and without demanding any additional thought (as the Irishman, in the hackneyed jest, is said to have passed off a light guinea between two good halfpence) this supererogatory merit will not perhaps be taken amiss. Now amusement in and for itself may be afforded by the gratification either of the curiosity or of the passions. I use the former word as distinguished from the love of knowledge, and the latter in distinction from those emotions which arise in well-ordered minds from the perception of truth or falsehood, virtue or vice—emotions which are always preceded by thought, and linked with improvement. Again, all information pursued without any wish of becoming wiser or better thereby, I class among the gratifications of mere curiosity, whether it be sought for in a light novel or a grave history. We may therefore omit the word information, as included either in amusement or instruction.

The present work is an experiment; not whether a writer may

honestly overlook the one, or successfully omit the other, of the two elements themselves which serious readers at least persuade themselves they pursue; but whether a change might not be hazarded of the usual order in which periodical writers have in general attempted to convey them. Having myself experienced that no delight either in kind or degree was equal to that which accompanies the distinct perception of a fundamental truth, relative to our moral being; having, long after the completion of what is ordinarily called a learned education, discovered a new world of intellectual profit opening on me—not from any new opinions, but lying, as it were, at the roots of those which I had been taught in childhood in my catechism and spelling-book; there arose a soothing hope in my mind that a lesser public might be found, composed of persons susceptible of the same delight, and desirous of attaining it by the same process. I heard a whisper too from within (I trust that it proceeded from conscience, not vanity) that a duty was performed in the endeavour to render it as much easier to them than it had been to me, as could be effected by the united efforts of my understanding and imagination.*

Actuated by this impulse, the Writer wishes, in the following essays, to convey not instruction merely, but fundamental instruction; not so much to show my reader this or that fact, as to kindle his own torch for him, and leave it to himself to choose the particular objects which he might wish to examine by its light. The Friend does not indeed exclude from his plan occasional interludes, and vacations of innocent entertainment and promiscuous information, but still in the main he proposes to himself the communication of such delight as rewards the march of truth, rather than to collect the flowers which diversify its

* In conformity with this anxious wish I shall make no apology for subjoining a translation of my motto to this essay.

(*Translation.*) A reader should sit down to a book, especially of the miscellaneous kind, as a well-behaved visitor does to a banquet. The master of the feast exerts himself to satisfy all his guests; but if after all his care and pains there should still be something or other put on the table that does not suit this or that person's taste, they politely pass it over without noticing the circumstance, and commend other dishes, that they may not distress their kind host, or throw any damp on his spirits. For who could tolerate a guest that accepted an invitation to your table with no other purpose but that of finding fault with everything put before him, neither eating himself, or suffering others to eat in comfort? And yet you may fall in with a still worse set than even these, with churls that in all companies and without stop or stay will condemn and pull to pieces a work which they had never read. But this sinks below the baseness of an *informer*, yea, though he were a false witness to boot! The man who abuses a thing of which he is utterly ignorant unites the infamy of both—and in addition to this makes himself the pander and sycophant of his own and other men's envy and malignity.

track, in order to present them apart from the homely yet foodful or medicinable herbs among which they had grown. To refer men's opinions to their absolute principles, and thence their feelings to the appropriate objects, and in their due degrees; and finally, to apply the principles thus ascertained to the formation of steadfast convictions concerning the most important questions of politics, morality, and religion—these are to be the objects and the contents of his work.

Themes like these not even the genius of a Plato or a Bacon could render intelligible without demanding from the reader thought sometimes, and attention generally. By thought I here mean the voluntary production in our own minds of those states of consciousness to which, as to his fundamental facts, the writer has referred us: while attention has for its object the order and connection of thoughts and images, each of which is in itself already and familiarly known. Thus, the elements of geometry require attention only; but the analysis of our primary faculties, and the investigation of all the absolute grounds of religion and morals, are impossible without energies of thought in addition to the effort of attention. The Friend will not attempt to disguise from his readers that both attention and thought are efforts, and the latter a most difficult and laborious effort; nor from himself, that to require it often or for any continuance of time is incompatible with the nature of the present publication, even were it less incongruous than it unfortunately is with the present habits and pursuits of Englishmen. Accordingly I shall be on my guard to make the numbers as few as possible which would require from a well-educated reader any energy of thought and voluntary abstraction.

But attention, I confess, will be requisite throughout, except in the excursive and miscellaneous essays that will be found interposed between each of the three main divisions of the work. On whatever subject the mind feels a lively interest, attention, though always an effort, becomes a delightful effort. I should be quite at ease could I secure for the whole work as much of it as a card party of earnest whist-players often expend in a single evening, or a lady in the making-up of a fashionable dress. But where no interest previously exists, attention (as every schoolmaster knows) can be procured only by terror: which is the true reason why the majority of mankind learn nothing systematically, except as schoolboys or apprentices.

Happy shall I be, from other motives besides those of self-interest, if no fault or deficiency on my part shall prevent the work from furnishing a presumptive proof that there are still to be found among us a respectable number of readers who are desirous to derive pleasure from the consciousness of being instructed or ameliorated: and who feel a sufficient interest as to the foundations of their own opinions in literature, politics, morals, and religion, to afford that degree of attention without which, however men may deceive themselves, no actual progress ever was or ever can be made in that knowledge which supplies at once both strength and nourishment.

[*The Landing-Place, or Essays Interposed for Amusement, Retrospect, and Preparation: Essay* II]°

Is it, I ask, most important to the best interests of mankind, temporal as well as spiritual, that certain works, the names and number of which are fixed and unalterable, should be distinguished from all other works, not in degree only but even in kind? And that these collectively should form the Book to which in all the concerns of faith and morality the last recourse is to be made, and from the decisions of which no man dare appeal? If the mere existence of a Book so called and charactered be, as the Koran itself suffices to evince, a mighty bond of union among nations whom all other causes tend to separate; if moreover the Book revered by us and our forefathers has been the foster-nurse of learning in the darkest, and of civilization in the rudest, times; and lastly, if this so vast and wide a blessing is not to be founded in a delusion, and doomed therefore to the impermanence and scorn in which sooner or later all delusions must end; how, I pray you, is it conceivable that this should be brought about and secured, otherwise than by a special vouchsafement to this one Book, exclusively, of that divine mean, that uniform and perfect middle way, which in all points is at safe and equal distance from all errors whether of excess or defect? But again if this be true (and what Protestant Christian worthy of his baptismal dedication will deny its truth), surely we ought not to be hard and over-stern in our censures of the mistakes and infirmities of those who, pretending to no warrant of extraordinary inspiration, have yet been raised up by God's providence to be of highest power and eminence in the reformation of his Church. Far rather does it behove us to consider, in how many instances the peccant humour native to the man had been wrought upon by the faithful study of that only faultless model, and corrected into an unsinning, or at least a venial, predominance in the writer or preacher. Yea, that not seldom the infirmity of a zealous soldier in the warfare of Christ has been made the very mould and groundwork of that man's peculiar gifts and virtues. Grateful too we should be, that the very faults of famous men have been fitted to the age on which they were to act: and that thus the folly of man has proved the wisdom of God, and been made the instrument of his mercy to mankind.

Anon.

WHOEVER has sojourned in Eisenach will assuredly have visited the Warteburg, interesting by so many historical associations, which stands on a high rock, about two miles to the south from the city gate. To this castle Luther was taken on his return from the imperial diet, where Charles the Fifth had pronounced the ban upon him, and limited his safe convoy to one and twenty days. On the last but one of these days, as he was on his way to Waltershausen (a town in the duchy of Saxe Gotha, a few leagues to the south-east of Eisenach) he was stopped in a hollow behind the Castle Altenstein, and carried to the Warteburg. The Elector of Saxony, who could not have refused to deliver up Luther, as one put in the ban by the Emperor and the Diet, had ordered John of Berleptsch, the governor of the Warteburg, and Burckhardt von Hundt, the governor of Altenstein, to take Luther to one or the other of these castles, without acquainting him which; in order that he might be able, with safe conscience, to declare that he did not know where Luther was. Accordingly they took him to the Warteburg, under the name of the Chevalier (Ritter) George.

To this friendly imprisonment the Reformation owes many of Luther's most important labours. In this place he wrote his works against auricular confession, against Jacob Latronum, the tract on the abuse of Masses, that against clerical and monastic vows, composed his Exposition of the 22, 27, and 68 Psalms, finished his Declaration of the Magnificat, began to write his Church Homilies, and translated the New Testament. Here too, and during this time, he is said to have hurled his inkstand at the Devil, the black spot from which yet remains on the stone wall of the room he studied in; which surely, no one will have visited the Warteburg without having had pointed out to him by the good Catholic who is, or at least some few years ago was, the warden of the castle. He must have been either a very supercilious or a very incurious traveller if he did not, for the gratification of his guide at least, inform himself by means of his penknife, that the said marvellous blot bids defiance to all the toils of the scrubbing-brush, and is to remain a sign for ever; and with this advantage over most of its kindred, that being capable of a double interpretation, it is equally flattering to the Protestant and the Papist, and is regarded by the wonder-loving zealots of both parties with equal faith.

Whether the great man ever did throw his inkstand at His Satanic Majesty, whether he ever boasted of the exploit, and himself declared the dark blotch on his study wall in the Warteburg to be the result and relict of this author-like hand-grenado (happily for mankind he used his inkstand at other times to better purpose, and with more effective hostility against the Arch-fiend), I leave to my reader's own judgement; on condition, however, that he has previously perused Luther's table talk, and other writings of the same stamp, of some of his most illustrious contemporaries, which contain facts still more strange and whimsical, related by themselves and of themselves, and accompanied with solemn protestations of the truth of their statements. Luther's table talk, which to a truly philosophic mind will not be less interesting than Rousseau's confessions, I have not myself the means of consulting at present, and cannot therefore say whether this inkpot adventure is, or is not, told or referred to in it; but many considerations incline me to give credit to the story.

Luther's unremitting literary labour and his sedentary mode of life during his confinement in the Warteburg, where he was treated with the greatest kindness, and enjoyed every liberty consistent with his own safety, had begun to undermine his former unusually strong health. He suffered many and most distressing effects of indigestion and a deranged state of the digestive organs. Melanchthon, whom he had desired to consult the physicians at Erfurth, sent him some deobstruent medicines, and the advice to take regular and severe exercise. At first he followed the advice, sat and laboured less, and spent whole days in the chase; but like the younger Pliny, he strove in vain to form a taste for this favourite amusement of the 'Gods of the earth,' as appears from a passage in his letter to George Spalatin, which I translate for an additional reason: to prove to the admirers of Rousseau (who perhaps will not be less affronted by this biographical parallel than the zealous Lutherians will be offended) that if my comparison should turn out groundless on the whole, the failure will not have arisen either from the want of sensibility in our great reformer, or of angry aversion to those in high places, whom he regarded as the oppressors of their rightful equals. 'I have been,' he writes, 'employed for two days in the sports of the field, and was willing myself to taste this bitter-sweet amusement of the great heroes: we have caught two hares, and one brace of poor little partridges. An employment this which does not ill suit quiet leisurely folks: for even in the midst of the ferrets and dogs, I have had theological fancies. But as much pleasure as the general appearance of the scene and the mere looking on occasioned me, even so much it pitied me to think of the mystery and emblem which lies beneath it. For what doth this symbol signify, but that the Devil, through his godless huntsman and dogs, the bishops and theologians to wit, doth privily chase and catch the innocent poor little beasts? Ah! the simple and credulous souls came thereby far too plain before my eyes. Thereto comes a yet more frightful mystery: as at my earnest entreaty we had saved alive one poor little hare, and I had concealed it in the sleeve of my greatcoat, and had strolled off a short distance from it, the dogs in the mean time found the poor hare. Such, too, is the fury of the Pope with Satan, that he destroys even the souls that had been saved, and troubles himself little about my pains and entreaties. Of such hunting then I have had enough.' In another passage he tells his correspondent, 'you know it is hard to be a prince, and not in some degree a robber, and the greater a prince the more a robber.' Of our Henry the Eighth he says, 'I must answer the grim lion that passes himself off for King of England. The ignorance in the book is such as one naturally expects from a king; but the bitterness and impudent falsehood is quite leonine.' And in his circular letter to the princes, on occasion of the Peasants' War, he uses a language so inflammatory, and holds forth a doctrine which borders so near on the holy right of insurrection, that it may as well remain untranslated.

Had Luther been himself a prince, he could not have desired better treatment than he received during his eight months' stay in the Warteburg; and in consequence of a more luxurious diet than he had been accustomed to, he was plagued with temptations both from the 'Flesh and the Devil.' It is evident from his letters* that he suffered under great irritability of his nervous system, the common effect of deranged digestion in men of sedentary habits, who are at the same time intense thinkers: and this irritability added to, and revivifying, the impressions made upon him in early life, and fostered by the theological systems of his manhood, is abundantly sufficient to explain all his apparitions and all his nightly combats with evil spirits. I see nothing improbable in the supposition that in one of those unconscious half-sleeps, or rather those rapid alternations of the sleeping with the half-waking state, which is the true witching time,

> the season
> Wherein the spirits hold their wont to walk,

the fruitful matrix of ghosts—I see nothing improbable, that in some

* I can scarcely conceive a more delightful volume than might be made from Luther's letters, especially from those that were written from the Warteburg, if they were translated in the simple, sinewy, idiomatic, hearty mother tongue of the original. A difficult task I admit—and scarcely possible for any man, however great his talents in other respects, whose favourite reading has not lain among the English writers from Edward the Sixth to Charles the First.

one of those momentary slumbers, in which the suspension of all thought in the perplexity of intense thinking so often passes, Luther should have had a full view of the room in which he was sitting, of his writing-table and all the implements of study, as they really existed, and at the same time a brain-image of the Devil, vivid enough to have acquired apparent *outness*, and a distance regulated by the proportion of its distinctness to that of the objects really impressed on the outward senses.

If this Christian Hercules, this heroic cleanser of the Augean Stable of apostasy, had been born and educated in the present or the preceding generation, he would, doubtless, have held himself for a man of genius and original power. But with this faith alone he would scarcely have removed the mountains which he did remove. The darkness and superstition of the age which required such a reformer had moulded his mind for the reception of ideas concerning himself better suited to inspire the strength and enthusiasm necessary for the task of reformation, ideas more in sympathy with the spirits whom he was to influence. He deemed himself gifted with supernatural influxes, an especial servant of heaven, a chosen warrior, fighting as the general of a small but faithful troop, against an army of evil beings headed by the Prince of the Air. These were no metaphorical beings in his apprehension. He was a poet indeed, as great a poet as ever lived in any age or country; but his poetic images were so vivid, that they mastered the poet's own mind! He was *possessed* with them, as with substances distinct from himself: Luther did not *write*, he *acted* poems. The Bible was a spiritual indeed but not a figurative armoury in his belief: it was the magazine of his warlike stores, and from thence he was to arm himself, and supply both shield and sword, and javelin, to the elect. Methinks I see him sitting, the heroic student, in his chamber in the Warteburg, with his midnight lamp before him, seen by the late traveller in the distant plain of Bischofsroda, as a star on the mountain! Below it lies the Hebrew Bible open, on which he gazes, his brow pressing on his palm, brooding over some obscure text, which he desires to make plain to the simple boor and to the humble artisan, and to transfer its whole force into their own natural and living tongue. And he himself does not understand it! Thick darkness lies on the original text: he counts the letters, he calls up the roots of each separate word, and questions them as the familiar spirits of an oracle. In vain! thick darkness continues to cover it! not a ray of meaning dawns through it. With sullen and angry hope he reaches for the Vulgate, his old and sworn enemy, the treacherous confederate of the

Roman Antichrist, which he so gladly, when he can, rebukes for idolatrous falsehoods, that had dared place

> Within the sanctuary itself their shrines,
> Abominations!

Now—O thought of humiliation—he must entreat its aid. See! there has the sly spirit of apostasy worked in a phrase which favours the doctrine of purgatory, the intercession of saints, or the efficacy of prayers for the dead. And what is worst of all, the interpretation is plausible. The original Hebrew might be forced into this meaning: and no other meaning seems to lie in it, none to hover above it in the heights of allegory, none to lurk beneath it even in the depths of Cabbala! This is the work of the Tempter! it is a cloud of darkness conjured up between the truth of the sacred letters and the eyes of his understanding by the malice of the evil one, and for a trial of his faith! Must he then at length confess, must he subscribe the name of Luther to an exposition which consecrates a weapon for the hand of the idolatrous hierarchy? Never! never!

There still remains one auxiliary in reserve, the translation of the seventy. The Alexandrine Greeks, anterior to the Church itself, could intend no support to its corruptions—the Septuagint will have profaned the altar of truth with no incense for the nostrils of the universal Bishop to snuff up. And here again his hopes are baffled! Exactly at this perplexed passage had the Greek translator given his understanding a holiday, and made his pen supply its place. O honoured Luther! as easily mightest thou convert the whole city of Rome, with the Pope and the conclave of cardinals inclusive, as strike a spark of light from the words, and *nothing but words*, of the Alexandrine version. Disappointed, despondent, enraged, ceasing to think, yet continuing his brain on the stretch in solicitation of a thought; and gradually giving himself up to angry fancies, to recollections of past persecutions, to uneasy fears and inward defiances and floating images of the evil Being, their supposed personal author; he sinks, without perceiving it, into a trance of slumber: during which his brain retains its waking energies, excepting that what would have been mere thoughts before, now (the action and counterweight of his senses and of their impressions being withdrawn) shape and condense themselves into things, into realities! Repeatedly half-wakening, and his eyelids as often reclosing, the objects which really surround him form the place and scenery of his dream. All at once he sees the Arch-fiend coming forth on the wall of the room, from the very spot, perhaps, on which his eyes had been fixed vacantly during the perplexed moments of his former meditation: the inkstand, which he had at the same time been using becomes associated with it: and in that struggle of rage, which in these distempered dreams almost constantly precedes the helpless terror by the pain of which we are finally awakened, he imagines that he hurls it at the intruder, or not improbably in the first instant of awakening, while yet both his imagination and his eyes are possessed by the dream, he *actually* hurls it. Some weeks after, perhaps, during which interval he had often mused on the incident, undetermined whether to deem it a visitation of Satan to him in the body or out of the body, he discovers for the first time the dark spot on his wall, and receives it as a sign and pledge vouchsafed to him of the event having actually taken place.

Such was Luther under the influences of the age and country in and for which he was born. Conceive him a citizen of Geneva, and a contemporary of Voltaire: suppose the French language his mother tongue, and the political and moral philosophy of English free-thinkers remodelled by Parisian *Forts Esprits* [Free-Thinkers], to have been the objects of his study—conceive this change of circumstances, and Luther will no longer dream of fiends or of Antichrist—but will he have no dreams in their place? His melancholy will have changed its drapery; but will it find no new costume wherewith to clothe itself? His impetuous temperament, his deep-working mind, his busy and vivid imaginations—would they not have been a trouble to him in a world where nothing was to be altered, where nothing was to obey his power to cease to be that which had been, in order to realize his preconceptions of what it ought to be? His sensibility, which found objects for itself and shadows of human suffering in the harmless brute, and even the flowers which he trod upon—might it not naturally, in an unspiritualized age, have wept, and trembled, and dissolved, over scenes of earthly passion, and the struggles of love with duty? His pity, that so easily passed into rage, would it not have found in the inequalities of mankind, in the oppressions of Governments and the miseries of the governed, an entire instead of a divided object? And might not a perfect constitution, a Government of pure reason, a renovation of the social contract, have easily supplied the place of the reign of Christ in the new Jerusalem, of the restoration of the visible Church, and the union of all men by one faith in one charity? Henceforward then, we will conceive his reason employed in building up anew the edifice of *earthly* society, and his imagination as pledging itself for the possible realization of the structure. We will lose the great reformer, who was born in an age which needed him, in the philosopher of Geneva [Rousseau], who was doomed to misapply his energies to materials the properties of which he misunderstood, and happy only that he did not live to witness the direful effects of his system.

[*The Landing-Place . . .*, *Essay* III]

Pectora cui credam? quis me lenire docebit
Mordaces curas, qui longas fallere noctes
Ex quo summa dies tulerit Damona sub umbras?
Omnia paulatim consumit longior aetas,
Vivendoque simul morimur, rapimurque manendo.
Ite tamen, lacrymae! purum colis aethera, Damon!
Nec mihi conveniunt lacrymae. Non omnia terrae
Obruta! vivit amor, vivit dolor! ora negatur
Dulcia conspicere: flere et meminisse relictum est.

[To whom shall I entrust my soul? Who will teach me to lighten | Consuming cares, who to beguile the long nights | Since his last day has borne Damon away within the shades? | Advancing age little by little devours everything, | While we live, we are dying, we stand firm and are being snatched away. | Go, tears! Damon dwells in the pure ether! | Tears fit not me. Not everything in the world | Has gone under! love lives, grief lives! we are forbidden | To look upon faces we love: but we can still weep and remember. (*CC*)]

THE two following essays I devote to elucidation, the first of the theory of Luther's apparitions stated perhaps too briefly in the preceding Number: the second for the purpose of removing the only difficulty which I can discover in the next section of the Friend to the reader's ready comprehension of the principles on which the arguments are grounded. First, I will endeavour to make my ghost theory more clear to those of my readers who are fortunate enough to find it obscure in consequence of their own good health and unshattered nerves. The window of my library at Keswick is opposite to the fireplace, and looks out on the very large garden that occupies the whole slope of the hill on which the house stands. Consequently, the rays of light transmitted *through* the glass (i.e. the rays from the garden, the opposite mountains, and the bridge, river, lake, and vale interjacent) and the rays reflected *from* it (of the fireplace etc.) enter the eye at the same moment. At the coming on of evening, it was my frequent amusement to watch the image or reflection of the fire, that seemed burning in the bushes or between the trees in different parts of the garden or the fields beyond it, according as there was more or less light; and which still arranged itself among the real objects of vision, with a distance and magnitude proportioned to its greater or lesser faintness. For still as the darkness increased, the image of the fire lessened and grew nearer and more distinct; till the twilight had deepened into perfect night, when all outward objects being excluded, the window became a perfect looking-glass: save only that my books on the side-shelves of the room were lettered, as it were, on their backs with stars, more or fewer as the sky was more or less clouded (the rays of the stars being at that time the only ones transmitted). Now substitute the phantom from Luther's brain for the images of *reflected* light (the fire for instance) and the forms of his room and its furniture for the *transmitted* rays, and you have a fair resemblance of an apparition, and a just conception of the manner in which it is seen together with real objects. I have long wished to devote an entire work to the subject of dreams, visions, ghosts, witchcraft, etc., in which I might first give and then endeavour to explain the most interesting and best-attested fact of each which has come within my knowledge, either from books or from personal testimony. I might then explain in a more satisfactory way the mode in which our thoughts, in states of morbid slumber, become at times perfectly *dramatic* (for in certain sorts of dreams the dullest wight becomes a Shakespeare) and by what law the form of the vision appears to talk to us its own thoughts in a voice as audible as the shape is visible; and this too oftentimes in connected trains, and not seldom even with a concentration of power which may easily impose on the soundest judgements, uninstructed in the optics and acoustics of the inner sense, for revelations and gifts of prescience. In aid of the present case, I will only remark, that it would appear incredible to persons not accustomed to these subtle notices of self-observation, what small and remote resemblances, what mere *hints* of likeness from some real external object, especially if the shape be aided by colour, will suffice to make a vivid thought consubstantiate with the real object, and derive from it an outward perceptibility. Even when we are broad awake, if we are in anxious expectations, how often will not the most confused sounds of nature be heard by us as articulate sounds? For instance, the babbling of a brook will appear for a moment the voice of a friend for whom we are waiting, calling out our own names etc. A short meditation, therefore, on the great law of the imagination, that a likeness in part tends to become a likeness of the whole, will make it not only conceivable but probable that the inkstand itself, and the dark-coloured stone on the wall, which Luther perhaps had never till then noticed, might have a considerable influence in the production of the Fiend, and of the hostile act by which his obtrusive visit was repelled.

A lady once asked me if I believed in ghosts and apparitions. I answered with truth and simplicity: *No, Madam! I have seen far too many myself.* I have indeed a whole memorandum book filled with records of these phenomena, many of them interesting as facts and data for psychology, and affording some valuable materials for a theory of perception and its dependence on the memory and imagination. 'In omnem actum Perceptionis imaginatio influit efficienter [The imagination effectively contributes to every act of Perception].'—Wolff. But he is no more, who would have realized this idea: who had already established the foundations and the law of the theory; and for whom I had so often found a pleasure and a comfort, even during the wretched and restless nights of sickness, in watching and instantly recording these experiences of the world within us, of the 'gemina natura, quae fit et facit, et creat et creatur [twin nature, which is product and maker, creator and creation]!' He is gone, my friend!° my munificent co-patron, and not less the benefactor of my intellect!—He who, beyond all other men known to me, added a fine and ever-wakeful sense of beauty to the most patient accuracy in experimental philosophy and the profounder researches of metaphysical science; he who united all the play and spring of fancy with the subtlest discrimination and an inexorable judgement; and who controlled an almost painful exquisiteness of taste by a warmth of heart, which in the practical relations of life made allowances for faults as quick as the moral taste detected them; a warmth of heart, which was indeed noble and pre-eminent, for alas! the genial feelings of health contributed no spark toward it! Of these qualities I may speak, for they belonged to all mankind. The higher virtues, that were blessings to his friends, and the still higher that resided in and for his own soul, are themes for the energies of solitude, for the awfulness of prayer!—virtues exercised in the barrenness and desolation of his animal being; while he thirsted with the full stream at his lips, and yet with unwearied goodness poured out to all around him, like the master of a feast among his kindred in the day of his own gladness! Were it but for the remembrance of him alone and of his lot here below, the disbelief of a future state would sadden the earth around me, and blight the very grass in the field.

[*The Landing-Place . . .*, *Essay* V]

Man may rather be defined a religious than a rational character, in regard that in other creatures there may be something of reason, but there is nothing of religion.

Harrington

IF the reader will substitute the word 'understanding' for 'reason,' and the word 'reason' for 'religion,' Harrington has here completely expressed the truth for which the Friend is contending. But that this was Harrington's meaning is evident. Otherwise instead of comparing two faculties with each other, he would contrast a faculty with one of its own objects, which would involve the same absurdity as if he had said, that man might rather be defined an astronomical than a seeing animal, because other animals possessed the sense of sight, but were incapable of beholding the satellites of Saturn, or the nebulae of fixed stars. If further confirmation be necessary, it may be supplied by the following reflections, the leading thought of which I remember to have read in the works of a continental philosopher. It should seem easy to give the definite distinction of the reason from the understanding, because we constantly imply it when we speak of the difference between ourselves and the brute creation. No one, except as a figure of speech, ever speaks of an animal *reason*;* but that many animals possess a share of understanding, perfectly distinguishable from mere instinct, we will allow. Few persons have a favourite dog without making instances of its intelligence an occasional topic of conversation.

* I have this moment looked over a translation of Blumenbach's Physiology by Dr Elliotson, which forms a glaring exception, p. 45. I do not know Dr Elliotson, but I *do* know Professor Blumenbach, and was an assiduous attendant on the lectures, of which this classical work was the textbook: and I know that that good and great man would start back with surprise and indignation at the gross materialism mortised on to his work: the more so because during the whole period in which the identification of man with the brute in *kind* was the *fashion* of naturalists Blumenbach remained ardent and instant in controverting the opinion, and exposing its fallacy and falsehood, both as a man of sense and as a naturalist. I may truly say, that it was uppermost in his heart and foremost in his speech. Therefore, and from no hostile feeling to Dr Elliotson (whom I hear spoken of with great regard and respect, and to whom I myself give credit for his manly openness in the avowal of his opinions), I have felt the present animadversion a duty of justice as well as gratitude. S. T. C. 8 April, 1817

They call for our admiration of the individual animal, and not with exclusive reference to the wisdom in nature, as in the case of the storge or maternal instinct of beasts; or of the hexangular cells of the bees, and the wonderful coincidence of this form with the geometrical demonstration of the largest possible number of rooms in a given space. Likewise, we distinguish various degrees of understanding there, and even discover from inductions supplied by the zoologists, that the understanding appears (as a general rule) in an inverse proportion to the instinct. We hear little or nothing of the instincts of 'the half-reasoning elephant,' and as little of the understanding of caterpillars and butterflies. (NB. Though reasoning does not in our language, in the lax use of words natural in conversation or popular writings, imply scientific conclusion, yet the phrase 'half-reasoning' is evidently used by Pope as a poetic hyperbole.) But reason is wholly denied, equally to the highest as to the lowest of the brutes; otherwise it must be wholly attributed to them, and with it therefore self-consciousness, and personality, or moral being.

I should have no objection to define reason with Jacobi, and with his friend Hemsterhuis, as an organ bearing the same relation to spiritual objects, the universal, the eternal, and the necessary, as the eye bears to material and contingent phenomena. But then it must be added, that it is an organ identical with its appropriate objects. Thus, God, the soul, eternal truth, etc. are the objects of reason; but they are themselves *reason*. We name God the Supreme Reason; and Milton says, 'Whence the soul Reason receives, and Reason is her Being.' Whatever is conscious self-knowledge is reason; and in this sense it may be safely defined the organ of the super-sensuous; even as the understanding wherever it does not possess or use the reason, as another and inward eye, may be defined the conception of the sensuous, or the faculty by which we generalize and arrange the phenomena of perception: that faculty, the functions of which contain the rules and constitute the possibility of outward experience. In short, the understanding supposes something that is understood. This may be merely its own acts or forms, that is, formal logic; but real objects, the materials of substantial knowledge, must be furnished, we might safely say revealed, to it by organs of sense. The understanding of the higher brutes has only organs of outward sense, and consequently material objects only; but man's understanding has likewise an organ of inward sense, and therefore the power of acquainting itself with invisible realities or spiritual objects. This organ is his reason. Again, the understanding and experience may exist* without reason. But reason cannot exist without understanding; nor does it or can it manifest itself but in and through the understanding, which in our elder writers is often called *discourse*, or the discursive faculty, as by Hooker, Lord Bacon, and Hobbes: and an understanding enlightened by reason Shakespeare gives as the contradistinguishing character of man, under the name *discourse of reason*. In short, the human understanding possesses two distinct organs, the outward sense, and 'the mind's eye' which is reason: wherever we use that phrase (the mind's eye) in its proper sense, and not as a mere synonym of the memory or the fancy. In this way we reconcile the promise of Revelation, that the blessed will see God, with the declaration of St John, God hath no one seen at any time.

We will add one other illustration to prevent any misconception, as if we were dividing the human soul into different essences, or ideal persons. In this piece of steel I acknowledge the properties of hardness, brittleness, high polish, and the capability of forming a mirror. I find all these likewise in the plate glass of a friend's carriage; but in addition to all these, I find the quality of transparency, or the power of transmitting as well as of reflecting the rays of light. The application is obvious.

If the reader therefore will take the trouble of bearing in mind these and the following explanations, he will have removed beforehand every possible difficulty from the Friend's political section. For there is another use of the word reason, arising out of the former indeed, but less definite, and more exposed to misconception. In this latter use it means the understanding considered as using the reason, so far as by the organ of reason only we possess the ideas of the necessary and the universal; and this is the more common use of the word, when it is applied with any attempt at clear and distinct conceptions. In this narrower and derivative sense the best definition of reason which I can give will be found in the third member of the following sentence, in which the understanding is described in its threefold operation, and

* Of this no one would feel inclined to doubt, who had seen the poodle dog, whom the celebrated Blumenbach, a name so dear to science, as a physiologist and comparative anatomist, and not less dear as a man, to all Englishmen who have ever resided at Göttingen in the course of their education, trained up, not only to hatch the eggs of the hen with all the mother's care and patience, but to attend the chicken[s] afterwards, and find the food for them. I have myself known a Newfoundland dog, who watched and guarded a family of young children with all the intelligence of a nurse, during their walks.

from each receives an appropriate name. The sense (vis sensitiva vel intuitiva [the sensitive or intuitive power]) *per*ceives: Vis regulatrix [the regulative Power] (the understanding, in its own peculiar operation) *con*ceives: Vis rationalis [the rational Power] (the reason or rationalized understanding) *comprehends*. The first is impressed through the organs of sense; the second combines these multifarious impressions into individual notions, and by reducing these notions to rules, according to the analogy of all its former notices, constitutes experience: the third subordinates both these notions and the rules of experience to absolute principles or necessary laws: and thus concerning objects which our experience has proved to have real existence, it demonstrates, moreover, in what way they are possible, and in doing this constitutes science. Reason therefore, in this secondary sense, and used not as a spiritual organ but as a faculty (namely, the understanding or soul enlightened by that organ)—reason, I say, or the scientific faculty, is the intellection of the possibility or essential properties of things by means of the laws that constitute them. Thus the rational idea of a circle is that of a figure constituted by the circumvolution of a straight line with its one end fixed.

Every man must feel, that though he may not be exerting different faculties, he is exerting his faculties in a different way, when in one instance he begins with some one self-evident truth (that the radii of a circle, for instance, are all equal) and in consequence of this being true sees at once, without any actual experience, that some other thing must be true likewise, and that, this being true, some third thing must be equally true, and so on till he comes, we will say, to the properties of the lever, considered as the spoke of a circle; which is capable of having all its marvellous powers demonstrated even to a savage who had never seen a lever, and without supposing any other previous knowledge in his mind, but this one, that there is a conceivable figure, all possible lines from the middle to the circumference of which are of the same length: or when, in the second instance, he brings together the facts of experience, each of which has its own separate value, neither increased nor diminished by the truth of any other fact which may have preceded it; and making these several facts bear upon some particular project, and finding some in favour of it, and some against it, determines for or against the project, according as one or the other class of facts preponderate: as, for instance, whether it would be better to plant a particular spot of ground with larch, or with Scotch fir, or with oak in preference to either. Surely every man will acknowledge that his mind was very differently employed in the first case from what it was in the second; and all men have agreed to call the results of the first class the truths of science such as not only are true, but which it is impossible to conceive otherwise: while the results of the second class are called *facts*, or things of experience: and as to these latter we must often content ourselves with the greater probability that they are so, or so, rather than otherwise—nay, even when we have no doubt that they are so in the particular case, we never presume to assert that they must continue so always, and under all circumstances. On the contrary, our conclusions depend altogether on contingent circumstances. Now when the mind is employed, as in the case first mentioned, I call it reasoning, or the use of the pure reason; but, in the second case, the understanding or prudence.

This reason applied to the motives of our conduct, and combined with the sense of our moral responsibility, is the conditional cause of conscience, which is a spiritual sense or testifying state of the coincidence or discordance of the free will with the reason. But as the reasoning consists wholly in a man's power of seeing whether any two ideas which happen to be in his mind are, or are not, in contradiction with each other, it follows of necessity, not only that all men have reason, but that every man has it in the same degree. For reasoning (or reason, in this its secondary sense) does not consist in the ideas, or in their clearness, but simply, when they are in the mind, in seeing whether they contradict each other or no.

And again, as in the determinations of conscience the only knowledge required is that of my own intention—whether in doing such a thing, instead of leaving it undone, I did what I should think right if any other person had done it; it follows that in the mere question of guilt or innocence, all men have not only reason equally, but likewise all the materials on which the reason, considered as conscience, is to work. But when we pass out of ourselves, and speak, not exclusively of the agent as meaning well or ill, but of the action in its consequences, then of course experience is required, judgement in making use of it, and all those other qualities of the mind which are so differently dispensed to different persons, both by nature and education. And though the reason itself is the same in all men, yet the means of exercising it, and the materials (i.e. the facts and ideas) on which it is exercised, being possessed in very different degrees by different persons, the practical result is, of course, equally different—and the whole groundwork of Rousseau's philosophy ends in a mere Nothingism. Even in that branch of knowledge, on which the ideas, on the congruity of which with each other the reason is to decide, are all possessed alike by all men, namely, in geometry (for all men in their senses possess all the component images, viz. simple curves and straight lines), yet the power of attention required for the perception of linked truths, even of *such* truths, is so very different in A and in B, that Sir Isaac Newton professed that it was in this power only that he was superior to ordinary men. In short, the sophism is as gross as if I should say—The souls of all men have the faculty of sight in an equal degree—forgetting to add that this faculty cannot be exercised without eyes, and that some men are blind and others short-sighted, etc.—and should then take advantage of this my omission to conclude against the use or necessity of spectacles, microscopes, etc.—or of choosing the sharpest-sighted men for our guides.

Having exposed this gross sophism, I must warn against an opposite error—namely, that if reason, as distinguished from prudence, consists merely in knowing that black cannot be white—or when a man has a clear conception of an enclosed figure, and another equally clear conception of a straight line, his reason teaches him that these two conceptions are incompatible in the same object, i.e. that two straight lines *cannot* include a space—the said reason must be a very insignificant faculty. But a moment's steady self-reflection will show us that in the simple determination 'Black is not White'—or, 'that two straight lines cannot include a space'—all the powers are implied that distinguish man from animals—first, the power of reflection—2d. of comparison—3d. and therefore of suspension of the mind—4th. therefore of a controlling will, and the power of acting from notions, instead of mere images exciting appetites; from motives, and not from mere dark instincts. Was it an insignificant thing to weigh the planets, to determine all their courses, and prophesy every possible relation of the heavens a thousand years hence? Yet all this mighty chain of science is nothing but a linking together of truths of the same kind, as, *the whole is greater than its part*:—or, if A and B = C, then A = B—or 3 + 4 = 7, therefore 7 + 5 = 12, and so forth. X is to be found either in A or B, or C or D: It is not found in A, B, or C, therefore it is to be found in D.—What can be simpler? Apply this to an animal—a dog misses his master where four roads meet—he has come up one, smells to two of the others, and then with his head aloft darts forward to the third road without any examination. If this was done by a conclusion, the dog would have reason—how comes it then, that he never shows it in his ordinary habits? Why does this story excite either wonder or incredulity?—If the story be a fact, and not a fiction, I should say—the breeze brought his master's scent down the fourth road to the dog's nose, and that *therefore* he did not put it down to the road, as in the two former instances. So awful and almost miraculous does the simple act of concluding, that *take* 3 *from* 4, *there remains one*, appear to us when attributed to the most sagacious of all animals.

LECTURES ON SHAKESPEARE°

Romeo and Juliet

[The seventh lecture, 9 December 1811]

IN a former lecture I endeavoured to point out the union of the poet and the philosopher, or rather the warm embrace between them, in the 'Venus and Adonis' and 'Lucrece' of Shakespeare. From thence I passed on to 'Love's Labours Lost,' as the link between his character as a poet and his art as a dramatist; and I showed that, although in that work the former was still predominant, yet that the germs of his subsequent dramatic power were easily discernible.

I will now, as I promised in my last, proceed to 'Romeo and Juliet,' not because it is the earliest, or among the earliest of Shakespeare's works of that kind, but because in it are to be found specimens, in degree, of all the excellences which he afterwards displayed in his more perfect dramas, but differing from them in being less forcibly evidenced, and less happily combined: all the parts are more or less present, but they are not united with the same harmony.

There are, however, in 'Romeo and Juliet' passages where the poet's whole excellence is evinced, so that nothing superior to them can be met with in the productions of his after years. The main distinction between this play and others is, as I said, that the parts are less happily combined, or to borrow a phrase from the painter, the whole work is less in keeping. Grand portions are produced: we have limbs of giant growth; but the production, as a whole, in which each part gives delight for itself, and the whole, consisting of these delightful parts, communicates the highest intellectual pleasure and satisfaction, is the result of the application of judgement and taste. These are not to be attained but by painful study, and to the sacrifice of the stronger pleasures derived from the dazzling light which a man of genius throws over every circumstance, and where we are chiefly struck by vivid and distinct images. Taste is an attainment after a poet has been disciplined by experience, and has added to genius that talent by which he knows what part of his genius he can make acceptable and intelligible to the portion of mankind for which he writes.

In my mind it would be a hopeless symptom, as regards genius, if I found a young man with anything like perfect taste. In the earlier works of Shakespeare we have a profusion of double epithets, and sometimes even the coarsest terms are employed, if they convey a more vivid image; but by degrees the associations are connected with the image they are designed to impress, and the poet descends from the ideal into the real world so far as to conjoin both—to give a sphere of active operations to the ideal, and to elevate and refine the real.

In 'Romeo and Juliet' the principal characters may be divided into two classes: in one class passion—the passion of love—is drawn and drawn truly, as well as beautifully; but the persons are not individualized farther than as the actor appears on the stage. It is a very just description and development of love, without giving, if I may so express myself, the philosophical history of it—without showing how the man became acted upon by that particular passion, but leading it through all the incidents of the drama, and rendering it predominant.

Tybalt is, in himself, a commonplace personage. And here allow me to remark upon a great distinction between Shakespeare and all who have written in imitation of him. I know no character in his plays (unless indeed Pistol be an exception) which can be called the mere portrait of an individual: while the reader feels all the satisfaction arising from individuality, yet that very individual is a sort of class character, and this circumstance renders Shakespeare the poet of all ages.

Tybalt is a man abandoned to his passions—with all the pride of family, only because he thought it belonged to him as a member of that family, and valuing himself highly, simply because he does not care for death. This indifference to death is perhaps more common than any other feeling: men are apt to flatter themselves extravagantly, merely because they possess a quality which it is a disgrace not to have, but which a wise man never puts forward but when it is necessary.

Jeremy Taylor in one part of his voluminous works, speaking of a great man, says that he was naturally a coward, as indeed most men are, knowing the value of life, but the power of his reason enabled him, when required, to conduct himself with uniform courage and hardihood. The good bishop perhaps had in his mind a story, told by one of the ancients, of a philosopher and a coxcomb on board the same ship during a storm: the coxcomb reviled the philosopher for betraying marks of fear: 'Why are you so frightened? I am not afraid of being drowned: I do not care a farthing for my life.'—'You are perfectly right,' said the philosopher, 'for your life is not worth a farthing.'

Shakespeare never takes pains to make his characters win your esteem, but leaves it to the general command of the passions, and to poetic justice. It is most beautiful to observe, in 'Romeo and Juliet,' that the characters principally engaged in the incidents are preserved innocent from all that could lower them in our opinion, while the rest of the personages, deserving little interest in themselves, derive it from being instrumental in those situations in which the more important personages develop their thoughts and passions.

Look at Capulet—a worthy, noble-minded old man of high rank, with all the impatience that is likely to accompany it. It is delightful to see all the sensibilities of our nature so exquisitely called forth; as if the poet had the hundred arms of the polypus, and had thrown them out in all directions to catch the predominant feeling. We may see in Capulet the manner in which anger seizes hold of everything that comes in its way, in order to express itself, as in the lines where he reproves Tybalt for his fierceness of behaviour, which led him to wish to insult a Montague, and disturb the merriment—

> Go to, go to;
> You are a saucy boy. Is't so, indeed?
> This trick may chance to scath you;—I know what.
> You must contrary me! marry, 'tis time.—
> Well said, my hearts!—You are a princox: go:
> Be quiet or—More light, more light!—For shame!
> I'll make you quiet.—What! cheerly, my hearts!
>
> Act I, Scene 5

The line

> This trick may chance to scath you;—I know what

was an allusion to the legacy Tybalt might expect; and then, seeing the lights burn dimly, Capulet turns his anger against the servants. Thus, we see that no one passion is so predominant but that it includes all the parts of the character, and the reader never has a mere abstract of a passion, as of wrath or ambition, but the whole man is presented to him—the one predominant passion acting, if I may so say, as the leader of the band to the rest.

It could not be expected that the poet should introduce such a character as Hamlet into every play; but even in those personages which are subordinate to a hero so eminently philosophical the passion is at least rendered instructive, and induces the reader to look with a keener eye and a finer judgement into human nature.

Shakespeare has this advantage over all other dramatists—that he has availed himself of his psychological genius to develop all the minutiae of the human heart: showing us the thing that, to common observers, he seems solely intent upon, he makes visible what we should not otherwise have seen: just as, after looking at distant objects through a telescope, when we behold them subsequently with the naked eye, we see them with greater distinctness, and in more detail, than we should otherwise have done.

Mercutio is one of our poet's truly Shakespearian characters; for throughout his plays, but especially in those of the highest order, it is plain that the personages were drawn rather from meditation than from observation, or to speak correctly, more from observation, the child of meditation. It is comparatively easy for a man to go about the world, as if with a pocket-book in his hand, carefully noting down what he sees and hears: by practice he acquires considerable facility in representing what he has observed, himself frequently unconscious of its worth, or its bearings. This is entirely different from the observation of a mind, which having formed a theory and a system upon its own nature, remarks all things that are examples of its truth, confirming it in that truth, and, above all, enabling it to convey the truths of philosophy, as mere effects derived from what we may call the outward watching of life.

Hence it is that Shakespeare's favourite characters are full of such lively intellect. Mercutio is a man possessing all the elements of a poet: the whole world was, as it were, subject to his law of association. Whenever he wishes to impress anything, all things become his servants for the purpose: all things tell the same tale, and sound in unison. This faculty, moreover, is combined with the manners and feelings of a perfect gentleman, himself utterly unconscious of his powers. By his loss it was contrived that the whole catastrophe of the tragedy should be brought about: it endears him to Romeo, and gives to the death of Mercutio an importance which it could not otherwise have acquired.

I say this in answer to an observation, I think by Dryden (to which indeed Dr Johnson has fully replied), that Shakespeare having carried the part of Mercutio as far as he could, till his genius was exhausted, had killed him in the third Act to get him out of the way. What shallow nonsense! As I have remarked, upon the death of Mercutio the whole catastrophe depends; it is produced by it. The scene in which it occurs serves to show how indifference to any subject but one, and aversion to activity on the part of Romeo, may be overcome and roused to the most resolute and determined conduct. Had not Mercutio been rendered so amiable and so interesting, we could not have felt so strongly the necessity for Romeo's interference, connecting it immediately, and passionately, with the future fortunes of the lover and his mistress.

But what am I to say of the Nurse? We have been told that her character is the mere fruit of observation—that it is like Swift's 'Polite Conversation,' certainly the most stupendous work of human memory, and of unceasingly active attention to what passes around us, upon record. The Nurse in 'Romeo and Juliet' has sometimes been compared to a portrait by Gerard Dow, in which every hair was so exquisitely painted, that it would bear the test of the microscope. Now, I appeal confidently to my hearers whether the closest observation of the manners of one or two old nurses would have enabled Shakespeare to draw this character of admirable generalization? Surely not. Let any man conjure up in his mind all the qualities and peculiarities that can possibly belong to a nurse, and he will find them in Shakespeare's picture of the old woman: nothing is omitted. This effect is not produced by mere observation. The great prerogative of genius (and Shakespeare felt and availed himself of it) is now to swell itself to the dignity of a god, and now to subdue and keep dormant some part of that lofty nature, and to descend even to the lowest character—to become everything, in fact, but the vicious.

Thus, in the Nurse you have all the garrulity of old age, and all its fondness; for the affection of old age is one of the greatest consolations of humanity. I have often thought what a melancholy world this would be without children, and what an inhuman world without the aged.

You have also in the Nurse the arrogance of ignorance, with the pride of meanness at being connected with a great family. You have the grossness, too, which that situation never removes, though it sometimes suspends it; and, arising from that grossness, the little low vices attendant upon it, which, indeed, in such minds are scarcely vices. Romeo at one time was the most delightful and excellent young man, and the Nurse all willingness to assist him; but her disposition soon turns in favour of Paris, for whom she professes precisely the same admiration. How wonderfully are these low peculiarities contrasted with a young and pure mind, educated under different circumstances!

Another point ought to be mentioned as characteristic of the ignorance of the Nurse: it is, that in all her recollections, she assists herself by the remembrance of visual circumstances. The great difference, in this respect, between the cultivated and the uncultivated mind is this—that the cultivated mind will be found to recall the past by certain regular trains of cause and effect; whereas, with the uncultivated mind, the past is recalled wholly by coincident images, or facts which happened at the same time. This position is fully exemplified in the following passages put into the mouth of the Nurse:

> Even or odd, of all days in the year,
> Come Lammas eve at night shall she be fourteen.
> Susan and she—God rest all Christian souls!—
> Were of an age.—Well, Susan is with God;
> She was too good for me. But, as I said,
> On Lammas eve at night shall she be fourteen;
> That shall she, marry: I remember it well.
> 'Tis since the earthquake now eleven years;
> And she was wean'd,—I never shall forget it,—
> Of all the days of the year, upon that day;
> For I had then laid wormwood to my dug,
> Sitting in the sun under the dove-house wall:
> My lord and you were then at Mantua.—
> Nay, I do bear a brain:—but, as I said,
> When it did taste the wormwood on the nipple
> Of my dug, and felt it bitter, pretty fool,
> To see it tetchy, and fall out with the dug!
> Shake, quoth the dove-house: 'twas no need, I trow,
> To bid me trudge.
> And since that time it is eleven years;
> For then she could stand alone.
>
> Act I, Scene 3

She afterwards goes on with similar visual impressions, so true to the character.—More is here brought into one portrait than could have been ascertained by one man's mere observation, and without the introduction of a single incongruous point.

I honour, I love, the works of Fielding as much, or perhaps more, than those of any other writer of fiction of that kind: take Fielding in his characters of postilions, landlords, and landladies, waiters, or indeed of anybody who had come before his eye, and nothing can be more true, more happy, or more humorous; but in all his chief personages, Tom Jones for instance, where Fielding was not directed by observation, where he could not assist himself by the close copying of what he saw, where it is necessary that something should take place, some words be spoken, or some object described, which he could not have witnessed (his soliloquies for example, or the interview between the hero and Sophia Western before the reconciliation), and I will venture to say, loving and honouring the man and his productions as I do, that nothing can be more forced and unnatural: the language is without vivacity or spirit, the whole matter is incongruous, and totally destitute of psychological truth.

On the other hand, look at Shakespeare: where can any character be produced that does not speak the language of nature? where does he not put into the mouths of his dramatis personae, be they high or low, kings or constables, precisely what they must have said? Where, from observation, could he learn the language proper to sovereigns, queens, noblemen or generals? yet he invariably uses it. Where, from observation, could he have learned such lines as these, which are put into the mouth of Othello when he is talking to Iago of Brabantio?

> Let him do his spite:
> My services, which I have done the signiory,
> Shall out-tongue his complaints. 'Tis yet to know,
> Which, when I know that boasting is an honour,
> I shall promulgate, I fetch my life and being
> From men of royal siege; and my demerits
> May speak, unbonneted, to as proud a fortune
> As this that I have reach'd: for know, Iago,
> But that I love the gentle Desdemona,
> I would not my unhoused free condition
> Put into circumscription and confine
> For the sea's worth.
>
> Act I, Scene 2

I ask where was Shakespeare to observe such language as this? If he did observe it, it was with the inward eye of meditation upon his own nature: for the time, he became Othello, and spoke as Othello, in such circumstances, must have spoken.

Another remark I may make upon 'Romeo and Juliet' is that in this tragedy the poet is not, as I have hinted, entirely blended with the dramatist—at least, not in the degree to be afterwards noticed in 'Lear,' 'Hamlet,' 'Othello,' or 'Macbeth.' Capulet and Montague not unfrequently talk a language only belonging to the poet, and not so characteristic of, and peculiar to, the passions of persons in the situations in which they are placed—a mistake, or rather an indistinctness, which many of our later dramatists have carried through the whole of their productions.

When I read the song of Deborah,° I never think that she is a poet, although I think the song itself a sublime poem: it is as simple a dithyrambic production as exists in any language; but it is the proper and characteristic effusion of a woman highly elevated by triumph, by the natural hatred of oppressors, and resulting from a bitter sense of wrong: it is a song of exultation on deliverance from these evils, a deliverance accomplished by herself. When she exclaims, 'The inhabitants of the villages ceased, they ceased in Israel, until that I, Deborah, arose, that I arose a mother in Israel,' it is poetry in the highest sense: we have no reason, however, to suppose that if she had not been agitated by passion, and animated by victory, she would have been able so to express herself; or that if she had been placed in different circumstances, she would have used such language of truth and passion. We are to remember that Shakespeare, not placed under circumstances of excitement, and only wrought upon by his own vivid and vigorous imagination, writes a language that invariably and intuitively becomes the condition and position of each character.

On the other hand, there is a language not descriptive of passion, nor uttered under the influence of it, which is at the same time poetic, and shows a high and active fancy, as when Capulet says to Paris,

> Such comfort as do lusty young men feel,
> When well-apparell'd April on the heel
> Of limping winter treads, even such delight
> Among fresh female buds, shall you this night
> Inherit at my house.
>
> Act I, Scene 2

Here the poet may be said to speak, rather than the dramatist; and it would be easy to adduce other passages from this play, where Shakespeare, for a moment forgetting the character, utters his own words in his own person.

In my mind, what have often been censured as Shakespeare's conceits are completely justifiable, as belonging to the state, age, or feeling of the individual. Sometimes, when they cannot be vindicated on these grounds, they may well be excused by the taste of his own and of the preceding age; as for instance, in Romeo's speech,

> Here's much to do with hate, but more with love:—
> Why then, O brawling love! O loving hate!
> O anything, of nothing first created!
> O heavy lightness! serious vanity!
> Misshapen chaos of well-seeming forms!
> Feather of lead, bright smoke, cold fire, sick health!
> Still-waking sleep, that is not what it is!
>
> Act I, Scene 1

I dare not pronounce such passages as these to be absolutely unnatural, not merely because I consider the author a much better judge than I can be, but because I can understand and allow for an effort of the mind, when it would describe what it cannot satisfy itself with the description of, to reconcile opposites and qualify contradictions, leaving a middle state of mind more strictly appropriate to the imagination than any other, when it is, as it were, hovering between images. As soon as it is fixed on one image, it becomes understanding; but while it is unfixed and wavering between them, attaching itself permanently to none, it is imagination. Such is the fine description of Death in Milton:

> The other shape,
> If shape it might be call'd, that shape had none
> Distinguishable in member, joint, or limb,
> Or substance might be call'd, that shadow seem'd,
> For each seem'd either: black it stood as night;
> Fierce as ten furies, terrible as hell,
> And shook a dreadful dart: what seem'd his head
> The likeness of a kingly crown had on.
>
> *Paradise Lost*, Book II

The grandest efforts of poetry are where the imagination is called forth, not to produce a distinct form, but a strong working of the mind, still offering what is still repelled, and again creating what is again rejected; the result being what the poet wishes to impress, namely, the substitution of a sublime feeling of the unimaginable for a mere image. I have sometimes thought that the passage just read might be quoted as exhibiting the narrow limit of painting, as compared with the boundless power of poetry: painting cannot go beyond a certain point; poetry rejects all control, all confinement. Yet we know that sundry painters have attempted pictures of the meeting between Satan and Death at the gates of Hell; and how was Death represented? Not as Milton has described him, but by the most defined thing that can be imagined—a skeleton, the dryest and hardest image that it is possible to discover; which, instead of keeping the mind in a state of activity, reduces it to the merest passivity—an image compared with which a square, a triangle, or any other mathematical figure, is a luxuriant fancy.

It is a general but mistaken notion that, because some forms of writing, and some combinations of thought, are not usual, they are not natural: but we are to recollect that the dramatist represents his characters in every situation of life and in every state of mind, and there is no form of language that may not be introduced with effect by a great and judicious poet, and yet be most strictly according to nature. Take punning, for instance, which may be the lowest, but at all events is the most harmless, kind of wit, because it never excites envy. A pun may be a necessary consequence of association: one man, attempting to prove something that was resisted by another, might, when agitated by strong feeling, employ a term used by his adversary with a directly contrary meaning to that for which that adversary had resorted to it: it might come into his mind as one way, and sometimes the best, of replying to that adversary. This form of speech is generally produced by a mixture of anger and contempt, and punning is a natural mode of expressing them.

It is my intention to pass over none of the important so-called conceits of Shakespeare, not a few of which are introduced into his later productions with great propriety and effect. We are not to forget that at the time he lived there was an attempt at, and an affectation of, quaintness and adornment, which emanated from the Court, and against which satire was directed by Shakespeare in the character of Osrick in Hamlet. Among the schoolmen of that age, and earlier, nothing was more common than the use of conceits: it began with the revival of letters, and the bias thus given was very generally felt and acknowledged.

I have in my possession a dictionary of phrases, in which the epithets applied to love, hate, jealousy, and such abstract terms are arranged; and they consist almost entirely of words taken from Seneca and his imitators, or from the schoolmen, showing perpetual antithesis, and describing the passions by the conjunction and combination of things absolutely irreconcilable. In treating the matter thus, I am aware that I am only palliating the practice in Shakespeare: he ought to have had nothing to do with merely temporary peculiarities: he wrote not for his own only, but for all ages, and so far I admit the use of some of his conceits to be a defect. They detract sometimes from his universality as to time, person, and situation.

If we were able to discover and to point out the peculiar faults as well as the peculiar beauties of Shakespeare, it would materially assist us in deciding what authority ought to be attached to certain portions of what are generally called his works. If we met with a play, or certain scenes of a play, in which we could trace neither his defects nor his excellences, we should have the strongest reason for believing that he had had no hand in it. In the case of scenes so circumstanced we might come to the conclusion that they were taken from the older plays, which, in some instances, he reformed or altered, or that they were inserted afterwards by some under-hand, in order to please the mob. If a drama by Shakespeare turned out to be too heavy for popular audiences, the clown might be called in to lighten the representation; and if it appeared that what was added was not in Shakespeare's manner, the conclusion would be inevitable that it was not from Shakespeare's pen.

It remains for me to speak of the hero and heroine, of Romeo and Juliet themselves; and I shall do so with unaffected diffidence, not merely on account of the delicacy, but of the great importance of the subject. I feel that it is impossible to defend Shakespeare from the most cruel of all charges—that he is an immoral writer—without entering fully into his mode of portraying female characters, and of displaying the passion of love. It seems to me that he has done both with greater perfection than any other writer of the known world, perhaps with the single exception of Milton in his delineation of Eve.

When I have heard it said, or seen it stated, that Shakespeare wrote for man, but the gentle Fletcher for woman, it has always given me something like acute pain, because to me it seems to do the greatest injustice to Shakespeare: when, too, I remember how much character is formed by what we read, I cannot look upon it as a light question, to be passed over as a mere amusement, like a game of cards or chess. I never have been able to tame down my mind to think poetry a sport, or an occupation for idle hours.

Perhaps there is no more sure criterion of refinement in moral character, of the purity of intellectual intention, and of the deep conviction and perfect sense of what our own nature really is in all its combinations, than the different definitions different men would give of love. I will not detain you by stating the various known definitions, some of which it may be better not to repeat: I will rather give you one of my own, which, I apprehend, is equally free from the extravagance of pretended Platonism (which, like other things which super-moralize, is sure to demoralize) and from its grosser opposite.

Considering myself and my fellow men as a sort of link between heaven and earth, being composed of body and soul, with power to reason and to will, and with that perpetual aspiration which tells us that this is ours for a while, but it is not ourselves; considering man, I say, in this twofold character, yet united in one person, I conceive that there can be no correct definition of love which does not correspond with our being, and with that subordination of one part to another which constitutes our perfection. I would say therefore that

> 'Love is a desire of the whole being to be united to something, or some being, felt necessary to its completeness, by the most perfect means that nature permits and reason dictates.'

It is inevitable to every noble mind, whether man or woman, to feel itself, of itself, imperfect and insufficient, not as an animal only, but as a moral being. How wonderfully, then, has Providence contrived for us, by making that which is necessary to us a step in our exaltation to a higher and nobler state! The Creator has ordained that one should possess qualities which the other has not, and the union of both is the most complete ideal of human character. In everything the blending of the similar with the dissimilar is the secret of all pure delight. Who shall dare to stand alone, and vaunt himself, in himself, sufficient? In poetry it is the blending of passion with order that constitutes perfection: this is still more the case in morals, and more than all in the exclusive attachment of the sexes.

True it is, that the world and its business may be carried on without marriage; but it is so evident that Providence intended man (the only animal of all climates, and whose reason is pre-eminent over instinct) to be the master of the world, that marriage, or the knitting together of society by the tenderest, yet firmest ties, seems ordained to render him capable of maintaining his superiority over the brute creation. Man alone has been privileged to clothe himself, and to do all things so as to make him, as it were, a secondary creator of himself, and of his own happiness or misery: in this, as in all, the image of the Deity is impressed upon him.

Providence, then, has not left us to prudence only; for the power of calculation, which prudence implies, cannot have existed, but in a state which presupposes marriage. If God has done this, shall we suppose that he has given us no moral sense, no yearning, which is something more than animal, to secure that without which man might form a herd, but could not be a society? The very idea seems to breathe absurdity.

From this union arise the paternal, filial, brotherly, and sisterly relations of life; and every state is but a family magnified. All the operations of mind, in short, all that distinguishes us from brutes, originate in the more perfect state of domestic life. One infallible criterion in forming an opinion of a man is the reverence in which he holds women. Plato has said that in this way we rise from sensuality to affection, from affection to love, and from love to the pure intellectual delight by which we become worthy to conceive that infinite in ourselves, without which it is impossible for man to believe in a God. In a word, the grandest and most delightful of all promises has been expressed to us by this practical state—our marriage with the Redeemer of mankind.

I might safely appeal to every man who hears me, who in youth has been accustomed to abandon himself to his animal passions, whether when he first really fell in love, the earliest symptom was not a complete change in his manners, a contempt and a hatred of himself for having excused his conduct by asserting that he acted according to the dictates of nature, that his vices were the inevitable consequences of youth, and that his passions at that period of life could not be conquered? The surest friend of chastity is love: it leads us, not to sink the mind in the body, but to draw up the body to the mind—the immortal part of our nature. See how contrasted in this respect are some portions of the works of writers, whom I need not name, with other portions of the same works: the ebullitions of comic humour have at times, by a lamentable confusion, been made the means of debasing our nature, while at other times, even in the same volume, we are happy to notice the utmost purity, such as the purity of love, which above all other qualities renders us most pure and lovely.

Love is not, like hunger, a mere selfish appetite: it is an associative quality. The hungry savage is nothing but an animal, thinking only of the satisfaction of his stomach: what is the first effect of love, but to associate the feeling with every object in nature? the trees whisper, the roses exhale their perfumes, the nightingales sing, nay the very skies smile in unison with the feeling of true and pure love. It gives to every object in nature a power of the heart, without which it would indeed be spiritless.

Shakespeare has described this passion in various states and stages, beginning, as was most natural, with love in the young. Does he open his play by making Romeo and Juliet in love at first sight—at the first glimpse, as any ordinary thinker would do? Certainly not: he knew what he was about, and how he was to accomplish what he was about: he was to develop the whole passion, and he commences with the first elements—that sense of imperfection, that yearning to combine itself with something lovely. Romeo became enamoured of the idea he had formed in his own mind, and then, as it were, christened the first real being of the contrary sex as endowed with the perfections he desired. He appears to be in love with Rosaline; but, in truth, he is in love only with his own idea. He felt that necessity of being beloved which no noble mind can be without. Then our poet, our poet who so well knew human nature, introduces Romeo to Juliet, and makes it not only a violent, but a permanent love—a point for which Shakespeare has been ridiculed by the ignorant and unthinking. Romeo is first represented in a state most susceptible of love, and then, seeing Juliet, he took and retained the infection.

This brings me to observe upon a characteristic of Shakespeare which belongs to a man of profound thought and high genius. It has been too much the custom, when anything that happened in his dramas could not easily be explained by the few words the poet has employed, to pass it idly over, and to say that it is beyond our reach, and beyond the power of philosophy—a sort of *terra incognita* [unknown land] for discoverers—a great ocean to be hereafter explored. Others have treated such passages as hints and glimpses of something now non-existent, as the sacred fragments of an ancient and ruined temple, all the portions of which are beautiful, although their particular relation to each other is unknown. Shakespeare knew the human mind, and its most minute and intimate workings, and he never introduces a word, or a thought, in vain or out of place: if we do not understand him, it is our own fault or the fault of copyists and typographers; but study, and the possession of some small stock of the knowledge by which he worked will enable us often to detect and explain his meaning. He never wrote at random, or hit upon points of character and conduct by chance; and the smallest fragment of his mind not unfrequently gives a clue to a most perfect, regular, and consistent whole.

As I may not have another opportunity, the introduction of Friar Laurence into this tragedy enables me to remark upon the different manner in which Shakespeare has treated the priestly character, as compared with other writers. In Beaumont and Fletcher priests are represented as a vulgar mockery; and, as in others of their dramatic personages, the errors of a few are mistaken for the demeanour of the many: but in Shakespeare they always carry with them our love and respect. He made no injurious abstracts: he took no copies from the worst parts of our nature; and, like the rest, his characters of priests are truly drawn from the general body.

It may strike some as singular that throughout all his productions he has never introduced the passion of avarice. The truth is, that it belongs only to particular parts of our nature, and is prevalent only in particular states of society; hence it could not, and cannot, be permanent. The Miser of Molière and Plautus is now looked upon as a species of madman, and avarice as a species of madness. Elwes,° of whom everybody has heard, was an individual influenced by an insane condition of mind; but, as a passion, avarice has disappeared. How admirably, then, did Shakespeare foresee, that if he drew such a character it could not be permanent! he drew characters which would always be natural, and therefore permanent, inasmuch as they were not dependent upon accidental circumstances.

There is not one of the plays of Shakespeare that is built upon anything but the best and surest foundation; the characters must be permanent—permanent while men continue men—because they stand upon what is absolutely necessary to our existence. This cannot be said even of some of the most famous authors of antiquity. Take the capital tragedies of Orestes, or of the husband of Jocasta: great as was the genius of the writers, these dramas have an obvious fault, and the fault lies at the very root of the action. In Oedipus a man is represented oppressed by fate for a crime of which he was not morally guilty; and while we read we are obliged to say to ourselves that in those days they considered actions without reference to the real guilt of the persons.

There is no character in Shakespeare in which envy is portrayed, with one solitary exception—Cassius, in 'Julius Caesar'; yet even there the vice is not hateful, inasmuch as it is counterbalanced by a number of excellent qualities and virtues. The poet leads the reader to suppose that it is rather something constitutional, something derived from his parents, something that he cannot avoid, and not something that he has himself acquired; thus throwing the blame from the will of man to some inevitable circumstance, and leading us to suppose that it is hardly to be looked upon as one of those passions that actually debase the mind.

Whenever love is described as of a serious nature, and much more when it is to lead to a tragical result, it depends upon a law of the mind which, I believe, I shall hereafter be able to make intelligible, and which would not only justify Shakespeare, but show an analogy to all his other characters.

Hamlet

[From the twelfth lecture, 2 January 1812]

. . . We will now pass to 'Hamlet,' in order to obviate some of the general prejudices against the author in reference to the character of the hero. Much has been objected to, which ought to have been praised, and many beauties of the highest kind have been neglected, because they are somewhat hidden.

The first question we should ask ourselves is—What did Shakespeare mean when he drew the character of Hamlet? He never wrote anything without design, and what was his design when he sat down to produce this tragedy? My belief is that he always regarded his story, before he began to write, much in the same light as a painter regards his canvas, before he begins to paint—as a mere vehicle for his thoughts—as the ground upon which he was to work. What then was the point to which Shakespeare directed himself in Hamlet? He intended to portray a person in whose view the external world, and all its incidents and objects, were comparatively dim, and of no interest in themselves, and which began to interest only when they were reflected in the mirror of his mind. Hamlet beheld external things in the same way that a man of vivid imagination, who shuts his eyes, sees what has previously made an impression on his organs.

The poet places him in the most stimulating circumstances that a human being can be placed in. He is the heir apparent of a throne; his father dies suspiciously; his mother excludes her son from his throne by marrying his uncle. This is not enough; but the ghost of the murdered father is introduced, to assure the son that he was put to death by his own brother. What is the effect upon the son?—instant action and pursuit of revenge? No: endless reasoning and hesitating—constant urging and solicitation of the mind to act, and as constant an escape from action; ceaseless reproaches of himself for sloth and negligence, while the whole energy of his resolution evaporates in these reproaches. This, too, not from cowardice, for he is drawn as one of the bravest of his time—not from want of forethought or slowness of apprehension, for he sees through the very souls of all who surround him, but merely from that aversion to action, which prevails among such as have a world in themselves.

How admirable, too, is the judgement of the poet! Hamlet's own disordered fancy has not conjured up the spirit of his father; it has been seen by others: he is prepared by them to witness its reappearance, and when he does see it, Hamlet is not brought forward as having long brooded on the subject. The moment before the Ghost enters, Hamlet speaks of other matters: he mentions the coldness of the night, and observes that he has not heard the clock strike, adding, in reference to the custom of drinking, that it is

> More honour'd in the breach than the observance.
>
> Act I, Scene 4

Owing to the tranquil state of his mind, he indulges in some moral reflections. Afterwards, the Ghost suddenly enters.

> HOR. Look, my lord! it comes.
> HAM. Angels and ministers of grace defend us!

The same thing occurs in 'Macbeth': in the dagger scene, the moment before the hero sees it, he has his mind applied to some indifferent matters; 'Go, tell thy mistress,' etc. Thus, in both cases, the preternatural appearance has all the effect of abruptness, and the reader is totally divested of the notion that the figure is a vision of a highly wrought imagination.

Here Shakespeare adapts himself so admirably to the situation—in other words, so puts himself into it—that, though poetry, his language is the very language of nature. No terms associated with such feelings can occur to us so proper as those which he has employed, especially on the highest, the most august, and the most awful subjects that can interest a human being in this sentient world. That this is no mere fancy, I can undertake to establish from hundreds, I might say thousands, of passages. No character he has drawn, in the whole list of his plays, could so well and fitly express himself, as in the language Shakespeare has put into his mouth.

There is no indecision about Hamlet, as far as his own sense of duty is concerned; he knows well what he ought to do, and over and over again he makes up his mind to do it. The moment the players, and the two spies set upon him, have withdrawn, of whom he takes leave with a line so expressive of his contempt,

> Ay so; good bye you.—Now I am alone,

he breaks out into a delirium of rage against himself for neglecting to perform the solemn duty he had undertaken, and contrasts the factitious and artificial display of feeling by the player with his own apparent indifference;

> What's Hecuba to him, or he to Hecuba,
> That he should weep for her?

Yet the player did weep for her, and was in an agony of grief at her sufferings, while Hamlet is unable to rouse himself to action, in order that he may perform the command of his father, who had come from the grave to incite him to revenge:

> This is most brave!
> That I, the son of a dear father murder'd,
> Prompted to my revenge by heaven and hell,
> Must, like a whore, unpack my heart with words,
> And fall a-cursing like a very drab,
> A scullion.
>
> Act II, Scene 2

It is the same feeling, the same conviction of what is his duty, that makes Hamlet exclaim in a subsequent part of the tragedy:

> How all occasions do inform against me,
> And spur my dull revenge! What is a man,
> If his chief good, and market of his time,
> Be but to sleep and feed? A beast, no more.
>
> * * * * *
>
> I do not know
> Why yet I live to say—'this thing's to do,'
> Sith I have cause and will and strength and means
> To do't.
>
> Act IV, Scene 4

Yet with all this strong conviction of duty, and with all this resolution arising out of strong conviction, nothing is done. This admirable and consistent character, deeply acquainted with his own feelings, painting them with such wonderful power and accuracy, and firmly persuaded that a moment ought not to be lost in executing the solemn charge committed to him, still yields to the same retiring from reality, which is the result of having what we express by the terms, a world within himself.

Such a mind as Hamlet's is near akin to madness. Dryden has somewhere said,

> Great wit to madness nearly is allied,

and he was right; for he means by 'wit' that greatness of genius which led Hamlet to a perfect knowledge of his own character, which, with all strength of motive, was so weak as to be unable to carry into act his own most obvious duty.

With all this he has a sense of imperfectness, which becomes apparent when he is moralizing on the skull in the churchyard. Something is wanting to his completeness—something is deficient which remains to be supplied, and he is therefore described as attached to Ophelia. His madness is assumed, when he finds that witnesses have been placed behind the arras to listen to what passes, and when the heroine has been thrown in his way as a decoy.

Another objection has been taken by Dr Johnson, and Shakespeare has been taxed very severely. I refer to the scene where Hamlet enters and finds his uncle praying, and refuses to take his life, excepting when he is in the height of his iniquity. To assail him at such a moment of confession and repentance, Hamlet declares,

> Why, this is hire and salary, not revenge.
>
> Act III, Scene 4

He therefore forbears, and postpones his uncle's death, until he can catch him in some act

> That has no relish of salvation in't.

This conduct, and this sentiment, Dr Johnson has pronounced to be so atrocious and horrible, as to be unfit to be put into the mouth of a human being. The fact, however, is that Dr Johnson did not understand the character of Hamlet, and censured accordingly: the determination to allow the guilty King to escape at such a moment is only part of the indecision and irresoluteness of the hero. Hamlet seizes hold of a pretext for not acting, when he might have acted so instantly and effectually: therefore, he again defers the revenge he was bound to seek, and declares his determination to accomplish it at some time,

> When he is drunk, asleep, or in his rage,
> Or in th' incestuous pleasures of his bed.

This, allow me to impress upon you most emphatically, was merely the excuse Hamlet made to himself for not taking advantage of this particular and favourable moment for doing justice upon his guilty uncle, at the urgent instance of the spirit of his father.

Dr Johnson farther states that in the voyage to England Shakespeare merely follows the novel as he found it, as if the poet had no other reason for adhering to his original; but Shakespeare never followed a novel because he found such and such an incident in it, but because he saw that the story, as he read it, contributed to enforce or to explain some great truth inherent in human nature. He never could lack invention to alter or improve a popular narrative; but he did not wantonly vary from it when he knew that, as it was related, it would so well apply to his own great purpose. He saw at once how consistent it was with the character of Hamlet that after still resolving, and still deferring, still determining to execute, and still postponing execution, he should finally, in the infirmity of his disposition, give himself up to his destiny, and hopelessly place himself in the power and at the mercy of his enemies.

Even after the scene with Osrick, we see Hamlet still indulging in reflection, and hardly thinking of the task he has just undertaken: he is all dispatch and resolution, as far as words and present intentions are concerned, but all hesitation and irresolution, when called upon to carry his words and intentions into effect; so that, resolving to do everything, he does nothing. He is full of purpose, but void of that quality of mind which accomplishes purpose.

Anything finer than this conception and working out of a great character is merely impossible. Shakespeare wished to impress upon us the truth that action is the chief end of existence—that no faculties of intellect, however brilliant, can be considered valuable, or indeed otherwise than as misfortunes, if they withdraw us from, or render us repugnant to action, and lead us to think and think of doing, until the time has elapsed when we can do anything effectually. In enforcing this moral truth, Shakespeare has shown the fullness and force of his powers: all that is amiable and excellent in nature is combined in Hamlet, with the exception of one quality. He is a man living in meditation, called upon to act by every motive human and divine, but the great object of his life is defeated by continually resolving to do, yet doing nothing but resolve.

Lay Sermons°

I

[From *The Statesman's Manual*, 1816]

BUT do you require some one or more particular passage from the Bible, that may at once illustrate and exemplify its applicability to the changes and fortunes of empires? Of the numerous chapters that relate to the Jewish tribes, their enemies and allies, before and after division into two kingdoms, it would be more difficult to state a single one, from which some guiding light might *not* be struck. And in nothing is Scriptural history more strongly contrasted with the histories of highest note in the present age than in its freedom from the hollowness of abstractions. While the latter present a shadow-fight of things and quantities, the former gives us the history of men, and balances the important influence of individual minds with the previous state of the national morals and manners, in which, as constituting a specific susceptibility, it presents to us the true cause both of the influence itself, and of the weal or woe that were its consequents. How should it be otherwise? The histories and political economy of the present and preceding century partake in the general contagion of its mechanic philosophy, and are the product of an unenlivened generalizing understanding. In the Scriptures they are the living *educts* of the imagination; of that reconciling and mediatory power, which incorporating the reason in images of the sense, and organizing (as it were) the flux of the senses by the permanence and self-circling energies of the reason, gives birth to a system of symbols, harmonious in themselves, and consubstantial with the truths of which they are the conductors. These are the wheels which Ezekiel beheld, when the hand of the Lord was upon him, and he saw visions of God as he sat among the captives by the river of Chebar. *Whithersoever the Spirit was to go, the wheels went, and thither was their spirit to go: for the spirit of the living creature was in the wheels also.* The truths and the symbols that represent them move in conjunction and form the living chariot that bears up (for *us*) the throne of the Divine Humanity. Hence, by a derivative, indeed, but not a divided, influence, and though in a secondary yet in more than a metaphorical sense, the Sacred Book is worthily entitled the Word of God. Hence too, its contents present to us the stream of time continuous as life and a symbol of eternity, inasmuch as the past and the future are virtually contained in the present. According therefore to our relative position on its banks the sacred history becomes prophetic, the sacred prophecies historical, while the power and substance of both inhere in its laws, its promises, and its comminations. In the Scriptures therefore both facts and persons must of necessity have a twofold significance, a past and a future, a temporary and a perpetual, a particular and a universal application. They must be at once portraits and ideals.

Eheu! paupertina philosophia in paupertinam religionem ducit:—A hunger-bitten and idea-less philosophy naturally produces a starveling and comfortless religion. It is among the miseries of the present age that it recognizes no medium between *literal* and *metaphorical.* Faith is either to be buried in the dead letter, or its name and honours usurped by a counterfeit product of the mechanical understanding, which in the blindness of self-complacency confounds symbols with allegories.° Now an allegory is but a translation of abstract notions into a picture-language which is itself nothing but an abstraction from objects of the senses; the principal being more worthless even than its phantom proxy, both alike unsubstantial, and the former shapeless to boot. On the other hand a symbol (*ὃ ἔστιν ἄει ταυτηγόρικον* [which is always tautegorical]) is characterized by a translucence of the special in the individual or of the general in the especial or of the universal in the general. Above all by the translucence of the eternal through and in the temporal. It always partakes of the reality which it renders intelligible; and while it enunciates the whole, abides itself as a living part in that unity of which it is the representative. The other are but empty echoes which the fancy arbitrarily associates with apparitions of matter, less beautiful but not less shadowy than the sloping orchard or hillside pasture-field seen in the transparent lake below. Alas! for the flocks that are to be led forth to such pastures! 'It shall even be as when the hungry dreameth, and behold! he eateth; but he waketh and his soul is empty: or as when the thirsty dreameth, and behold he drinketh; but he awaketh and is faint!' (Isaiah xxix. 8). O! that we would seek for the bread which was given from heaven, that we should eat thereof and be strengthened! O that we would draw at the well at which the flocks of our forefathers had living water drawn for them, even that water which, instead of mocking the thirst of him to whom it is given, becomes a well within himself springing up to life everlasting!

When we reflect how large a part of our present knowledge and civilization is owing, directly or indirectly, to the Bible; when we are compelled to admit, as a fact of history, that the Bible has been the main lever by which the moral and intellectual character of Europe has been raised to its present comparative height; we should be struck, methinks, by the marked and prominent difference of this book from the works which it is now the fashion to quote as guides and authorities in morals, politics, and history. I will point out a few of the excellencies by which the one is distinguished, and shall leave it to your own judgement and recollection to perceive and apply the contrast to the productions of highest name in these latter days. In the Bible every agent appears and acts as a self-subsisting individual: each has a life of its own, and yet all are one life. The elements of necessity and free will are reconciled in the higher power of an omnipresent Providence, that predestinates the whole in the moral freedom of the integral parts. Of this the Bible never suffers us to lose sight. The root is never detached from the ground. It is God everywhere: and all creatures conform to his decrees, the righteous by performance of the law, the disobedient by the sufferance of the penalty.

II

[From *A Lay Sermon*, 1817]

THE immediate occasions of the existing distress may be correctly given with no greater difficulty than would attend any other series of known historic facts; but toward the discovery of its true seat and sources, I can but offer a humble contribution. They appear to me, however, resolvable into the overbalance* of the commercial spirit in consequence of the absence or weakness of the counterweights; this overbalance considered as displaying itself, 1. In the commercial

* I entreat attention to the word *over*balance. My opinions would be greatly misinterpreted if I were supposed to think hostilely of the spirit of commerce to which I attribute the largest proportion of our actual freedom (i.e. as Englishmen, and not merely as landowners) and at least as large a share of our virtues as of our vices. Still more anxiously would I guard against the suspicion of a design to inculpate any number or class of individuals. It is not in the power of a minister or of a cabinet to say to the current of national tendency, stay here! or flow there! The excess can only be remedied by the slow progress of intellect, the influences of religion, and irresistible events guided by Providence. In the points even, which I have presumed to blame, by the word Government I intend all the directors of political power, that is, the great estates of the realm, temporal and spiritual, and not only the Parliament, but all the elements of Parliament.

world itself: 2. In the agricultural: 3. In the Government: and, 4. In the combined influence of all three on the more numerous and labouring classes.

Of the natural counterforces to the impetus of trade the first that presents itself to my mind is the ancient feeling of rank and ancestry, compared with our present self-complacent triumph over these supposed prejudices. Not that titles and the rights of precedence are pursued by us with less eagerness than by our forefathers. The contrary is the case; and for this very cause, because they inspire less reverence. In the old times they were valued by the possessors and revered by the people as distinctions of nature, which the Crown itself could only ornament, but not give. Like the stars in heaven, their influence was wider and more general, because for the mass of mankind there was no hope of reaching, and therefore no desire to appropriate, them. That many evils as well as advantages accompanied this state of things I am well aware: and likewise that many of the latter have become incompatible with far more important blessings. It would therefore be sickly affectation to suspend the thankfulness due for our immunity from the one in an idle regret for the loss of the other. But however true this may be, and whether the good or the evil preponderated, still it acted as a counterpoise to the grosser superstition for wealth. Of the efficiency of this counter-influence we can offer negative proof only: and for this we need only look back on the deplorable state of Holland in respect of patriotism and public spirit at and before the commencement of the French Revolution.

The limits and proportions of this address allow little more than a bare reference to this point. The same restraint I must impose on myself in the following. For under this head I include the general neglect of all the austerer studies; the long and omnious eclipse of philosophy; the usurpation of that venerable name by physical and psychological empiricism; and the non-existence of a learned and philosophic public, which is perhaps the only innoxious form of an *imperium in imperio* [empire within an empire], but at the same time the only form which is not directly or indirectly encouraged. So great a risk do I incur of malignant interpretation, and the assertion itself is so likely to appear paradoxical even to men of candid minds, that I should have passed over this point, most important as I know it to be; but that it will be found stated more at large, with all its proofs, in a work on the point of publication. The fact is simply this. We have—*lovers*, shall I entitle them? Or must I not rather hazard the introduction of their own phrases, and say, *amateurs* or *dilettanti*, as musicians, botanists, florists, mineralogists, and antiquarians. Nor is it denied that these are ingenuous pursuits, and such as become men of rank and fortune. Neither in these or in any other points do I complain of any excess in the pursuits themselves; but of that which arises from the deficiency of the counterpoise. The effect is the same. Every work which can be made use of either to immediate profit or immediate pleasure, every work which falls in with the desire of acquiring wealth suddenly, or which can gratify the senses, or pamper the still more degrading appetite for scandal and personal defamation, is sure of an appropriate circulation. But neither philosophy or theology in the strictest sense of the words can be said to have even a public existence among us. I feel assured that if Plato himself were to return and renew his sublime lucubrations in the metropolis of Great Britain, a handicraftsman from a laboratory who had just succeeded in disoxydating an earth° would be thought far the more respectable, nay, the more illustrious person of the two. Nor will it be the least drawback from his honours that he had never even asked himself what law of universal being nature uttered in this phenomenon: while the character of a visionary would be the sole remuneration of the man who from the insight into that law had previously demonstrated the necessity of the fact. As to that which passes with us under the name of metaphysics, philosophic elements, and the like, I refer every man of reflection to the contrast between the present times and those shortly after the restoration of ancient literature. In the latter we find the greatest men of the age, statesmen, warriors, monarchs, architects, in closest intercourse with philosophy. I need only mention the names of Lorenzo the Magnificent; Picus, Count Mirandula, Ficinus and Politian; the abtruse subjects of their discussion, and the importance attached to them, as the requisite qualifications of men placed by Providence as guides and governors of their fellow creatures. If this be undeniable, equally notorious is it that at present the more effective a man's talents are, and the more likely he is to be useful and distinguished in the highest situations of public life, the earlier does he show his aversion to the metaphysics and the books of metaphysical speculation which are placed before him: though they come with the recommendation of being so many triumphs of modern good sense over the schools of ancient philosophy. Dante, Petrarch, Spenser, Sir Philip Sidney, Algernon Sidney, Milton, and Barrow were Platonists. But all the men of genius with whom it has been my fortune to converse either profess to know nothing of the present systems, or to despise them. It would be equally unjust and irrational to seek the solution of this difference in the men; and if not, it can be found only in the philosophic systems themselves. And so in truth it is. The living of former ages communed gladly with a life-breathing philosophy. The living of the present age wisely leave the dead to take care of the dead.

But whatever the causes may be, the result is before our eyes. An excess in our attachment to temporal and personal objects can be counteracted only by a preoccupation of the intellect and the affections with permanent, universal, and eternal truths. Let no man enter, said Plato, who has not previously disciplined his mind by geometry. He considered this science as the first purification of the soul, by abstracting the attention from the accidents of the senses. We too teach geometry; but that there may be no danger of the pupil's becoming too abstract in his conceptions, it has been not only proposed, but the proposal has been adopted, that it should be taught by wooden diagrams! It pains me to remember with what applause a work that placed the inductions of modern chemistry in the same rank with the demonstrations of mathematical science was received even in a mathematical university. I must not permit myself to say more on this subject, desirous as I am of showing the importance of a philosophic class, and of evincing that it is of vital utility, and even an essential element in the composition of a civilized community. It must suffice that it has been explained in what respect the pursuit of truth for its own sake, and the reverence yielded to its professors, has a tendency to calm or counteract the pursuit of wealth; and that therefore a counterforce is wanting wherever philosophy is degraded in the estimation of society. What are *you* (a philosopher was once asked) in consequence of your admiration of these abstruse speculations? He answered: What I am, it does not become me to say; but what thousands are who despise them, and even pride themselves on their ignorance, I see—and tremble!

AIDS TO REFLECTION°

[From the *Preface*]

READER!—You have been bred in a land abounding with men, able in arts, learning, and knowledges manifold, this man in one, this in another, few in many, none in all. But there is one art of which every man should be master, the art of reflection. If you are not a *thinking* man, to what purpose are you a *man* at all? In like manner, there is one knowledge, which it is every man's interest and duty to acquire, namely, self-knowledge: or to what end was man alone, of all animals, indued by the Creator with the faculty of self-consciousness? Truly said the pagan moralist, E coelo descendit, *Γνῶθι Σέαυτον* [From heaven descended 'Know Thyself'].

But you are likewise born in a Christian land: and revelation has provided for you new subjects for reflection, and new treasures of knowledge, never to be unlocked by him who remains self-ignorant. Self-knowledge is the key to this casket; and by reflection alone can it be obtained. Reflect on your own thoughts, actions, circumstances, and—which will be of especial aid to you in forming a *habit* of reflection,—accustom yourself to reflect on the words you use, hear, or read, their birth, derivation, and history. For if words are not things, they are living powers, by which the things of most importance to mankind are actuated, combined, and humanized. Finally, by reflection you may draw from the fleeting facts of your worldly trade, art, or profession, a science permanent as your immortal soul; and make even these subsidiary and preparative to the reception of spiritual truth, 'doing as the dyers do, who having first dipt their silks in colours of less value, then give them the last tincture of crimson in grain.'

S. T. COLERIDGE

[From *Introductory Aphorisms*]

APHORISM I

IN philosophy equally as in poetry it is the highest and most useful prerogative of genius to produce the strongest impressions of novelty, while it rescues admitted truths from the neglect caused by the very circumstance of their universal admission. Extremes meet. Truths, of all others the most awful and interesting, are too often considered as *so* true, that they lose all the power of truth, and lie bedridden in the dormitory of the soul, side by side with the most despised and exploded errors.

APHORISM II

THERE is one sure way of giving freshness and importance to the most commonplace maxims—that of reflecting on them in direct reference to our own state and conduct, to our own past and future being.

LIFE is the one universal soul, which by virtue of the enlivening Breath, and the informing Word, all organized bodies have in common, each after its kind. This, therefore, all animals possess, and man as an animal. But, in addition to this, God transfused into man a higher gift, and specially imbreathed—even a living (that is, self-subsisting) soul, a soul having its life in itself. 'And man became a living soul.' He did not merely possess it, he became it. It was his proper being, his truest self, *the* man in the man. None then, not one of human kind, so poor and destitute, but there is provided for him, even in his present state, *a house not built with hands*. Aye, and spite of the philosophy (falsely so called) which mistakes the causes, the conditions, and the occasions of our becoming conscious of certain truths and realities for the truths and realities themselves—a house gloriously furnished. Nothing is wanted but the eye, which is the light of this house, the light which is the eye of this soul. This seeing light, this enlightening eye, is reflection. It is more, indeed, than is ordinarily meant by that word; but it is what a Christian ought to mean by it, and to know too, whence it first came, and still continues to come—of what light even this light is *but* a reflection. This, too, is thought; and all thought is but unthinking that does not flow out of this, or tend towards it.

[From *Prudential Aphorisms*]

FELICITY, in its proper sense, is but another word for fortunateness, or happiness; and I can see no advantage in the improper use of words, when proper terms are to be found, but, on the contrary, much mischief. For, by familiarizing the mind to equivocal expressions, that is, such as may be taken in two or more different meanings, we introduce confusion of thought, and furnish the sophist with his best and handiest tools. For the juggle of sophistry consists, for the greater part, in using a word in one sense in the premiss, and in another sense in the conclusion. We should accustom ourselves to think, and reason, in precise and steadfast terms; even when custom, or the deficiency, or the corruption of the language will not permit the same strictness in speaking. The mathematician finds this so necessary to the truths which he is seeking that his science begins with, and is founded on, the definition of his terms. The botanist, the chemist, the anatomist, etc., feel and submit to this necessity at all costs, even at the risk of exposing their several pursuits to the ridicule of the many, by technical terms, hard to be remembered, and alike quarrelsome to the ear and the tongue. In the business of moral and religious reflection, in the acquisition of clear and distinct conceptions of our duties, and of the relations in which we stand to God, our neighbour, and ourselves, no such difficulties occur. At the utmost we have only to rescue words, already existing and familiar, from the false or vague meanings imposed on them by carelessness, or by the clipping and debasing misusage of the market. And surely happiness, duty, faith, truth, and final blessedness, are matters of deeper and dearer interest for all men, than circles to the geometrician, or the characters of plants to the botanist, or the affinities and combining principle of the elements of bodies to the chemist, or even than the mechanism (fearful and wonderful though it be!) of the perishable tabernacle of the soul can be the anatomist. Among the aids to reflection, place the following maxim prominent: let distinctness in expression advance side by side with distinction in thought. For one useless subtlety in our elder divines and moralists, I will produce ten sophisms of equivocation in the writings of our modern preceptors: and for one error resulting from excess in distinguishing the indifferent, I would show ten mischievous delusions from the habit of confounding the diverse.

[From *Moral and Religious Aphorisms*]

LEIGHTON

Your blessedness is not—no, believe it, it is not where most of you seek it, in things below you. How can that be? It must be a higher good to make you happy.

COMMENT

EVERY rank of creatures, as it ascends in the scale of creation, leaves death behind it or under it. The metal at the height of being seems a mute prophecy of the coming vegetation, into a mimic semblance of which it crystallizes. The blossom and flower, the acme of vegetable life, divides into correspondent organs with reciprocal functions, and by instinctive motions and approximations seems impatient of that fixture, by which it is differenced in kind from the flower-shaped psyche [butterfly], that flutters with free wing above it. And wonderfully in the insect realm doth the irritability, the proper seat of instinct, while yet the nascent sensibility is subordinated thereto—most wonderfully, I say, doth the muscular life in the insect, and the musculo-arterial in the bird, imitate and typically rehearse the adaptive understanding, yea, and the moral affections and charities, of man. Let us carry ourselves back, in spirit, to the mysterious week, the teeming work-days of the Creator: as they rose in vision before the eye of the inspired historian of 'the generations of the heaven and the earth, in the days that the Lord God made the earth and the heavens'. And who that hath watched their ways with an understanding heart could, as the vision, evolving, still advanced towards him, contemplate the filial and loyal bee; the home-building, wedded, and divorceless swallow; and above all the manifoldly intelligent ant tribes, with their commonwealths and confederacies, their warriors and miners, the husbandfolk, that fold in their tiny flocks on the honeyed leaf, and the virgin sisters, with the holy instincts of maternal love, detached and in selfless purity—and not say to himself, Behold the shadow of approaching humanity, the sun rising from behind, in the kindling morn of creation! Thus all lower natures find their highest good in semblances and seekings of that which is higher and better. All things strive to ascend, and ascend in their striving. And shall man alone stoop? Shall his pursuits and desires, the reflections of his inward life, be like the reflected image of a tree on the edge of a pool, that grows downward, and seeks a mock heaven in the unstable element beneath it, in neighbourhood with the slim water-weeds and oozy bottom-grass that are yet better than itself and more noble, in as far as substances that appear as shadows are preferable to shadows mistaken for substance! No! it must be a higher good to make you happy. While you labour for anything below your proper humanity, you seek a happy life in the region of death. Well saith the moral poet—

> Unless above himself he can
> Erect himself, how mean a thing is man!

[From *Aphorisms on Spiritual Religion*]

I WILL now suppose the reader to have thoughtfully reperused the paragraph containing the tenets peculiar to Christianity, and if he have his religious principles yet to form, I should expect to hear a troubled murmur: How can I comprehend this? How is this to be proved? To the first question I should answer: Christianity is not a theory, or a speculation; but a life. Not a philosophy of life, but a life and a living process. To the second: Try it. It has been eighteen hundred years in existence: and has one individual left a record, like the following? 'I tried it; and it did not answer. I made the experiment faithfully according to the directions; and the result has been, a conviction of my own credulity.' Have you, in your experience, met with anyone in whose words you could place full confidence, and who has seriously affirmed, 'I have given Christianity a fair trial. I was aware, that its promises were made only conditionally. But my heart bears me witness, that I have to the utmost of my power complied with these conditions. Both outwardly and in the discipline of my inward acts and affections, I have performed the duties which it enjoins, and I have used the means which it prescribes. Yet my assurance of its truth has received no increase. Its promises have not been fulfilled: and I repent me of my delusion!' If neither your own experience nor the history of almost two thousand years has presented a single testimony to this purport; and if you have read and heard of many who have lived and died bearing witness to the contrary: and if you yourself have met with some *one*, in whom on any other point you would place unqualified trust, who has on his own experience made report to you, that 'he is faithful who promised, and what he promised he has proved himself able to perform': is it bigotry, if I fear that the unbelief, which prejudges and prevents the experiment, has its source elsewhere than in the uncorrupted judgement; that not the strong free mind, but the enslaved will, is the true original infidel in this instance? It would not be the first time that a treacherous bosom-sin had suborned the understandings of men to bear false witness against its avowed enemy, the right though unreceived owner of the house, who had long warned it out, and waited only for its ejection to enter and take full possession of the same.

I have elsewhere in the present work, though more at large in the 'Elements of Discourse' which, God permitting, will follow it, explained the difference between the understanding and the reason, by reason meaning exclusively the speculative or scientific power so called, the *nous* or *mens* of the ancients. And wider still is the distinction between the understanding and the spiritual mind. But no gift of God does or can contradict any other gift, except by misuse or misdirection. Most readily therefore do I admit that there can be no contrariety between revelation and the understanding; unless you call the fact that the skin, though sensible of the warmth of the sun, can convey no notion of its figure, or its joyous light, or of the colours it impresses on the clouds, a contrariety between the skin and the eye; or infer that the cutaneous and the optic nerves *contradict* each other.

But we have grounds to believe that there are yet other rays or effluences from the sun, which neither feeling nor sight can apprehend, but which are to be inferred from the effects. And were it so even with regard to the spiritual sun, how would this contradict the understanding or the reason? It is a sufficient proof of the contrary, that the mysteries in question are not *in the direction* of the understanding or the (speculative) reason. They do not move on the same line or plane with them, and therefore cannot contradict them. But besides this, in the mystery that most immediately concerns the believer, that of the birth into a new and spiritual life, the common sense and experience of mankind come in aid of their faith. The analogous facts, which we know to be true, not only facilitate the apprehension of the facts promised to us, and expressed by the same words in conjunction with a distinctive epithet; but being confessedly not less incomprehensible, the certain knowledge of the one disposes to the belief of the other. It removes at least all objections to the truth of the doctrine derived from the mysteriousness of its subject. The life we seek after is a mystery; but so both in itself and in its origin is the life we have. In order to meet this question, however, with minds duly prepared, there are two preliminary enquiries to be decided; the first respecting the purport, the second respecting the language of the Gospel.

First then of the purport, viz. what the Gospel does *not*, and what it *does* profess to be. The Gospel is not a system of theology, nor a syntagma of theoretical propositions and conclusions for the enlargement of speculative knowledge, ethical or metaphysical. But it is a history, a series of facts and events related or announced. These do indeed involve, or rather, I should say they at the same time *are*, most important doctrinal truths; but still facts and declarations of facts.

Secondly, of the language. This is a wide subject. But the point to which I chiefly advert is the necessity of thoroughly understanding the distinction between analogous and metaphorical language. Analogies are used in aid of conviction: metaphors as a means of illustration. The language is analogous wherever a thing, power, or principle in a higher dignity is expressed by the same thing, power, or principle in a lower but more known form. Such, for instance, is the language of John iii. 6. *That which is born of the flesh, is flesh; that which is born of the spirit, is spirit.* The latter half of the verse contains the fact asserted; the former half the analogous fact, by which it is rendered intelligible. If any man choose to call this metaphorical or figurative, I ask him whether with Hobbes and Bolingbroke he applies the same rule to the attributes of the Deity? Whether he regards the divine justice, for instance, as a metaphorical term, a mere figure of speech? If he disclaims this, then I answer, neither do I regard the words *born again*, or *spiritual life*, as figures or metaphors. I have only to add that these analogies are the material, or (to speak chemically) the *base*, of symbols and symbolical expressions; the nature of which is always *taut*egorical (i.e. expressing the *same* subject but with a difference) in contradistinction from metaphors and similitudes, that are always *alle*gorical (i.e. expressing a *different* subject but with a resemblance) will be found explained at large in the Statesman's Manual, p. 35–38.°

Of metaphorical language, on the other hand, let the following be taken as an instance and illustration. I am speaking, we will suppose, of an act which in its own nature, and as a producing and efficient cause, is transcendent; but which produces sundry effects, each of which is the same in kind with an effect produced by a cause well known and of ordinary occurrence. Now when I characterize or designate this transcendent act, in exclusive reference to these its effects, by a succession of names borrowed from their ordinary causes; not for the purpose of rendering the act itself, or the manner of the agency, conceivable, but in order to show the nature and magnitude of the benefits received from it, and thus to excite the due admiration, gratitude, and love in the receivers—in this case I should be rightly described as speaking metaphorically. And in this case to confound the similarity in respect of the effects relatively to the recipients with an identity in respect of the causes or modes of causation relatively to the transcendent act or the divine agent, is a confusion of metaphor with analogy, and of figurative with literal; and has been and continues to be a fruitful source of superstition or enthusiasm in believers, and of objections and prejudices to infidels and sceptics.° . . . Forgiveness of sin, the abolition of guilt, through the redemptive power of Christ's love, and of his perfect obedience during his voluntary assumption of humanity, is expressed, on account of the resemblance of the consequences in both cases, by the payment of a debt for another, which debt the payer had not himself incurred. Now the impropriation of this metaphor—(i.e. the taking it literally) by transferring the sameness from the consequents to the antecedents, or inferring the identity of the causes from a resemblance in the effects—this is the point on which I am at issue: and the view or scheme of redemption grounded on this confusion I believe to be altogether un-Scriptural.

Indeed, I know not in what other instance I could better exemplify the species of sophistry noticed in p. 215,° as the Aristotelean *μετάβασις εἰς ἄλλο γένος*, or clandestine passing over into a diverse kind. The purpose of a metaphor is to illustrate a something less known by a partial identification of it with some other thing better understood, or at least more familiar. Now the article of redemption may be considered in a twofold relation—in relation to the antecedent, i.e. the redeemer's act, as the efficient cause and condition of redemption; and in relation to the consequent, i.e. the effects in and for the redeemed. Now it is the latter relation in which the subject is treated of, set forth, expanded, and enforced by St Paul. The mysterious act, the operative cause is transcendent—factum est [It has been done, It is a fact]: and beyond the information contained in the enunciation of the fact, it can be characterized only by the consequences. It is the consequences of the act of redemption, that the zealous Apostle would bring home to the minds and affections both of Jews and Gentiles. Now the Apostle's opponents and gainsayers were principally of the former class. They were Jews: not only Jews unconverted, but such as had partially received the Gospel, and who, sheltering their national prejudices under the pretended authority of Christ's original apostles and the Church in Jerusalem, set themselves up against Paul as followers of Cephas. Add too, that Paul himself was 'a Hebrew of the Hebrews'; intimately versed 'in the Jew's religion above many, his equals, in his own nation, and above measure zealous of the traditions of his fathers.' It might, therefore, have been anticipated that his reasoning would receive its outward forms and language, that it would take its predominant colours, from his own past, and his opponents' present, habits of thinking; and that his figures, images, analogies, and references would be taken preferably from objects, opinions, events, and ritual observances ever uppermost in the imaginations of his own countrymen. And such we find them: yet so judiciously selected, that the prominent forms, the figures of most frequent recurrence, are drawn from points of belief and practice, from laws, rites, and customs, that then prevailed through the whole Roman world, and were common to Jew and Gentile.

Now it would be difficult if not impossible to select points better suited to this purpose, as being equally familiar to all, and yet having a special interest for the Jewish converts, than those are from which the learned Apostle has drawn the four principal metaphors, by which he illustrates the blessed consequences of Christ's redemption of mankind. These are: 1. Sin-offerings, sacrificial expiation. 2. Reconciliation, atonement, καταλλαγὴ [*katallagē*].* 3. Ransom from slavery, redemption, the buying back again, or being bought back, from *re* and *emo*. 4. Satisfaction of a creditor's claims by a payment of the debt. To one or other of these four heads all the numerous forms and exponents of Christ's mediation in St Paul's writings may be referred. And the very number and variety of the words or periphrases used by him to express one and the same thing furnish the strongest presumptive proof that all alike were used metaphorically. (In the following

* This word occurs but once in the New Testament, viz. Romans v. 11, the marginal rendering being, reconciliation. The personal noun καταλλακτὴς, is still in use with the modern Greeks for a money-changer, or one who takes the debased currency, so general in countries under despotic or other dishonest Governments, in exchange for sterling coin or bullion; the purchaser paying the *catallage*, i.e. the difference. In the elder Greek writers the verb means *to exchange for an opposite*, as καταλλάσσετο τὴν ἔχθρην τοῖς στασιώταις.—He exchanged within himself enmity for friendship (that is, he reconciled himself) with his party—or as we say, *made it up* with them, an idiom which (with whatever loss of dignity) gives the exact force of the word. He made up the difference. The Hebrew word of very frequent occurrence in the Pentateuch, which we render by the substantive, atonement, has its radical or visual image in *copher*, pitch. Gen. vi. 14, *thou shalt pitch it within and without with pitch*. Hence, to unite, to fill up a breach, or leak, the word expressing both the *act*, viz. the bringing together what had been previously separated, and the *means*, or material, by which the reunion is effected, as in our English verbs, *to caulk*, *to solder*, *to poy* or *pay* (from *poix*, pitch), and the French, *suiver*. Thence, metaphorically, *expiation*, the *piacula* having the same root, and being grounded on another property or use of gums and rosins, the supposed *cleansing* powers of their fumigation. Numbers viii. 21: 'made *atonement* for the Levites to *cleanse* them.' Lastly (or if we are to believe the Hebrew Lexicons, properly and most frequently) *ransom*. But if by *proper* the interpreters mean *primary* and *radical*, the assertion does not need a confutation: all radicals belonging to one or other of three classes, 1. Interjections, or sounds expressing sensations or passions. 2. Imitations of sounds, as splash, roar, whizz, etc. 3. and principally, visual images, objects of sight. But as to frequency, in all the numerous (fifty, I believe) instances of the word in the Old Testament, I have not found one in which it can, or at least need, be rendered by *ransom*: though beyond all doubt *ransom* is used in the Epistle to Timothy, as an equivalent term.

notation, let the small letters represent the effects or consequences, and the capitals the efficient causes or antecedents. Whether by causes we mean acts or agents, is indifferent. Now let X signify a transcendent, i.e. a cause beyond our comprehension and not within the sphere of sensible experience: and on the other hand, let A, B, C, and D represent each some one known and familiar cause in reference to some single and characteristic effect: viz. A in reference to k, B to l, C to m, and D to n. Then I say X + k l m n is in different places expressed by (or as =) A + k; B + l; C + m; D + n. And these I should call metaphorical exponents of X.)

Now John, the beloved disciple, who leant on the Lord's bosom, the Evangelist *κατὰ πνεῦμα*, i.e. according to the spirit, the inner and substantial truth of the Christian creed—John, recording the Redeemer's own words,° enunciates the fact itself, to the full extent in which it is enunciable for the human mind, simply and without any metaphor, by identifying it in kind with a fact of hourly occurrence—expressing it, I say, by a familiar fact the same in kind with that intended, though of a far lower dignity—by a fact of every man's experience, known to all, yet not better understood than the fact described by it. In the redeemed it is a regeneration, a birth, a spiritual seed impregnated and evolved, the germinal principle of a higher and enduring life, of a *spiritual* life—that is, a life, the actuality of which is not dependent on the material body, or limited by the circumstances and processes indispensable to its organization and subsistence. Briefly, it is the differential of immortality, of which the assimilative power of faith and love is the integrant, and the life in Christ the integration.

But even this would be an imperfect statement, if we omitted the aweful truth, that besides that dissolution of our earthly tabernacle which we call death, there is another death, not the mere negation of life, but its positive opposite. And as there is a mystery of life and an assimilation to the principle of life, even to him who is *the* life; so is there a mystery of death and an assimilation to the principle of evil *ἄμφιθαλης θανάτῳ* [abounding in death]! a fructifying of the corrupt seed, of which death is the germination. Thus, the regeneration to spiritual life is at the same time a redemption from the spiritual death. . . .

IN wonder all philosophy began: in wonder it ends: and admiration fills up the interspace. But the first wonder is the offspring of ignorance: the last is the parent of adoration. The first is the birth-throe of our knowledge: the last is its euthanasy and apotheosis.

SEQUELAE [(THOUGHTS) FOLLOWING]: OR THOUGHTS SUGGESTED BY THE PRECEDING APHORISM

As in respect of the first wonder we are all on the same level, how comes it that the philosophic mind should in all ages be the privilege of a few? The most obvious reason is this: the wonder takes place before the period of reflection, and (with the great mass of mankind) long before the individual is capable of directing his attention freely and consciously to the feeling, or even to its exciting causes. Surprise (the form and dress which the wonder of ignorance usually puts on) is worn away, if not precluded, by custom and familiarity. So is it with the objects of the senses, and the ways and fashions of the world around us: even as with the beat of our own hearts, which we notice only in moments of fear and perturbation. But with regard to the concerns of our inward being, there is yet another cause that acts in concert with the power in custom to prevent a fair and equal exertion of reflective thought. The great fundamental truths and doctrines of religion, the existence and attributes of God, and the life after death, are in Christian countries taught so early, under such circumstances, and in such close and vital association with whatever makes or marks reality for our infants minds, that the words ever after represent sensations, feelings, vital assurances, sense of reality—rather than thoughts, or any distinct conception. Associated, I had almost said identified, with the parental voice, look, touch, with the living warmth and pressure of the mother, on whose lap the child is first made to kneel, within whose palms its little hands are folded, and the motion of whose eyes its eyes follow and imitate—(yea, what the blue sky is to the mother, the mother's upraised eyes and brow are to the child, the type and symbol of an invisible Heaven!)—from within and from without, these great first truths, these good and gracious tidings, these holy and humanizing spells, in the preconformity to which our very humanity may be said to consist, are so infused, that it were but a tame and inadequate expression to say, we all take them for granted. At a later period, in youth or early manhood, most of us, indeed (in the higher and middle classes at least) read or hear certain proofs of these truths—which we commonly listen to, when we listen at all, with much the same feelings as a popular prince on his coronation day, in the centre of a fond and rejoicing nation, may be supposed to hear the champion's challenge to all the non-existents that deny or dispute his rights and royalty. In fact, the order of proof is most often reversed or transposed. As far, at least, as I dare judge from the goings on in my own mind, when with keen delight I first read the works of Derham, Niewentiet, and Lyonnet, I should say that the full and lifelike conviction of a gracious Creator is the proof (at all events, performs the office and answers all the purpose of a proof) of the wisdom and benevolence in the construction of the creature.

Do I blame this? Do I wish it to be otherwise? God forbid! It is only one of its accidental, but too frequent, consequences, of which I complain, and against which I protest. I regret nothing that tends to make the Light become the life of men, even as the life in the eternal Word is their only and single true light. But I do regret, that in after years—when by occasion of some new dispute on some old heresy, or any other accident, the attention has for the first time been distinctly attracted to the superstructure raised on these fundamental truths, or to truths of later revelation supplemental of these and not less important—all the doubts and difficulties, that cannot but arise where the understanding, 'the mind of the flesh,' is made the measure of spiritual things; all the sense of strangeness and seeming contradiction in terms; all the marvel and the mystery that belong equally to both, are first thought of and applied in objection exclusively to the latter. I would disturb no man's faith in the great articles of the (falsely so called) religion of nature. But before the man rejects, and calls on other men to reject, the revelations of the Gospel and the religion of all Christendom, I would have him place himself in the state and under all the privations of a Simonides, when on the fortieth day of his meditation the sage and philosophic poet abandoned the problem in despair. Ever and anon he seemed to have hold of the truth; but when he asked himself what he *meant* by it, it escaped from him, or resolved itself into meanings that destroyed each other. I would have the sceptic, while yet a sceptic only, seriously consider whether a doctrine, of the truth of which a Socrates could obtain no other assurance than what he derived from his strong *wish* that it should be true; or that which Plato found a mystery hard to discover, and when discovered, communicable only to the fewest of men; can consonantly with history or common sense, be classed among the articles, the belief of which is ensured to all men by their mere common sense? Whether, without gross outrage to fact, they can be said to constitute a religion of nature, or a natural theology antecedent to revelation or superseding its necessity? Yes! in prevention (for there is little chance, I fear, of a cure) of the pugnacious dogmatism of partial reflection, I would prescribe to every man who feels a commencing alienation from the catholic faith,° and whose studies and attainments authorize him to argue on the subject at all, a patient and thoughtful perusal of the arguments and representations which Bayle supposes to have passed through the mind of Simonides.° Or I should be fully satisfied if I could induce these eschewers of mystery to give a patient, manly, and impartial perusal to the single Treatise of Pomponatius, De Fato.*

When they have fairly and satisfactorily overthrown the objections and cleared away the difficulties urged by this sharp-witted Italian against the doctrines which they profess to retain, then let them commence their attack on those which they reject. As far as the supposed irrationality of the latter is the ground of argument, I am much deceived if, on reviewing their forces, they would not find the ranks woefully thinned by the success of their own fire in the preceding engagement—unless, indeed, by pure heat of controversy, and to storm the lines of their antagonists, they can bring to life again the arguments which they had themselves killed off in the defence of their own positions. In vain shall we seek for any other mode of meeting the broad facts of the scientific Epicurean, or the requisitions and queries of the all-analysing Pyrrhonist, than by challenging the tribunal to which they appeal as incompetent to try the question. In order to non-suit the infidel plaintiff, we must remove the cause from the faculty that judges according to sense, and whose judgements, therefore, are valid only on objects of sense, to the superior courts of conscience and intuitive reason! 'The words I speak unto you are Spirit,' and such only 'are life,' i.e. have an inward and actual power abiding in them.

But the same truth is at once shield and bow. The shaft of atheism glances aside from it to strike and pierce the breastplate of the heretic. Well for the latter, if plucking the weapon from the wound he recognizes an arrow from his own quiver, and abandons a cause that connects him with such confederates! Without further rhetoric, the sum and substance of the argument is this: an insight into the proper functions and subaltern rank of the understanding may not, indeed, disarm the psilanthropist of his metaphorical glosses, or of his *Versions* fresh from the forge and with no other stamp than the private mark of

* The philosopher whom the Inquisition would have burnt alive as an atheist, had not Leo X and Cardinal Bembo decided that the work might be formidable to those semi-pagan Christians who regarded revelation as a mere makeweight to their boasted religion of nature; but contained nothing dangerous to the Catholic Church or offensive to a true believer.

the individual manufacturer; but it will deprive him of the only rational pretext for having recourse to tools so liable to abuse, and of such perilous example.

WITH these expositions I hasten to contrast the Scriptural article respecting original sin, or the corrupt and sinful nature of the human will, and the belief which alone is required of us, as Christians. And here the first thing to be considered, and which will at once remove a world of error, is: that this is no tenet first introduced or imposed by Christianity; and which, should a man see reason to disclaim the authority of the Gospel, would no longer have any claim on his attention. It is no perplexity that a man may get rid of by ceasing to be a Christian, and which has no existence for a philosophic deist. It is a fact affirmed, indeed, in the Christian Scriptures alone with the force and frequency proportioned to its consummate importance; but a fact acknowledged in every religion that retains the least glimmering of the patriarchal faith in a God infinite yet personal! A fact assumed or implied as the basis of every religion, of which any relics remain of earlier date than the last and total apostasy of the pagan world, when the faith in the great I AM, the Creator, was extinguished in the sensual polytheism which is inevitably the final result of pantheism or the worship of nature; and the only form under which the pantheistic scheme—that, according to which the world is God, and the material universe itself the one only absolute being—can exist for a people, or become the popular creed. Thus, in the most ancient books of the Brahmins, the deep sense of this fact, and the doctrines grounded on obscure traditions of the promised remedy, are seen struggling, and now gleaming, now flashing, through the mist of pantheism, and producing the incongruities and gross contradictions of the Brahmin mythology: while in the rival sect— in that most strange phenomenon, the religious atheism of the Buddhists! with whom God is only universal Matter considered abstractly from all particular forms—the fact is placed among the delusions natural to man, which, together with other superstitions grounded on a supposed essential difference between right and wrong, the sage is to decompose and precipitate from the menstruum of his more refined apprehensions! Thus, in denying the fact, they virtually acknowledge it.

From the remote East turn to the mythology of Minor Asia, to the descendants of Javan *who dwelt in the tents of Shem, and possessed the Isles.*° Here again, and in the usual form of an historic solution, we find the same fact, and as characteristic of the human race, stated in that earliest and most venerable mythus (or symbolic parable) of Prometheus—that truly wonderful fable, in which the characters of the rebellious spirit and of the divine friend of mankind (*Θέος φιλάνθρωπος*) are united in the same person: and thus in the most striking manner noting the forced amalgamation of the patriarchal tradition with the incongruous scheme of pantheism. This and the connected tale of Io, which is but the sequel of the Prometheus,° stand alone in the Greek mythology, in which elsewhere both gods and men are mere powers and products of nature. And most noticeable it is, that soon after the promulgation and spread of the Gospel had awakened the moral sense, and had opened the eyes even of its wiser enemies to the necessity of providing some solution of this great problem of the moral world, the beautiful parable of Cupid and Psyche° was brought forward as a rival fall of man: and the fact of a moral corruption connatural with the human race was again recognized. In the assertion of original sin the Greek mythology rose and set.

But not only was the fact acknowledged of a law in the nature of man resisting the law of God. (And whatever is placed in active and direct oppugnancy to the Good is, *ipso facto* [by that very fact], positive Evil.) It was likewise an acknowledged mystery, and one which by the nature of the subject must ever remain such—a problem, of which any other solution than the statement of the fact itself was demonstrably impossible. That it is so, the least reflection will suffice to convince every man who has previously satisfied himself that he is a responsible being. It follows necessarily from the postulate of a responsible will. Refuse to grant this, and I have not a word to say. Concede this, and you concede all. For this is the essential attribute of a will, and contained in the very idea, that whatever determines the will acquires this power from a previous determination of the will itself. The will is ultimately self-determined, or it is no longer a will under the law of perfect freedom, but a nature under the mechanism of cause and effect. And if by an act, to which it had determined itself, it has subjected itself to the determination of nature (in the language of St Paul, the law of the flesh), it receives a nature into itself, and so far it becomes a nature: and this is a corruption of the will and a corrupt nature. It is also a fall of man, inasmuch as his will is the condition of his personality; the ground and condition of the attribute which constitutes him Man. And the groundwork of personal being is a capacity of acknowledging the moral law (the law of the spirit, the law of freedom, the divine will) as that which should, of itself, suffice to determine the will to a free obedience of the law, the law working thereon *by its own exceeding lawfulness.** This, and this alone, is *positive* good: good in itself, and independent of all relations. Whatever resists and, as a positive force, opposes this in the will is therefore evil. But an evil in the will is an evil will; and as all moral evil (i.e. all evil that is evil without reference to its contingent physical consequences) is *of* the will, this evil will must have its source in the will. And thus we might go back from act to act, from evil to evil, *ad infinitum* [to infinity], without advancing a step.

We call an individual a *bad* man, not because an action is contrary to the law, but because it has led us to conclude from it some principle opposed to the law, some private maxim or by-law in the will contrary to the universal law of right reason in the conscience, as the ground of the action. But this evil principle again must be grounded in some other principle which has been made determinant of the will by the will's own self-determination. For if not, it must have its ground in some necessity of nature, in some instinct or propensity imposed, not acquired, another's work, nor our own. Consequently, neither act nor principle could be imputed; and relatively to the agent, not *original*, not *sin*.

Now let the ground on which the fact of an evil inherent in the will is affirmable in the instance of any one man be supposed equally applicable in every instance, and concerning all men: so that the fact is asserted of the individual, not because he has committed this or that crime, or because he has shown himself to be this or that man, but simply because he is a man. Let the evil be supposed such as to imply the impossibility of an individual's referring to any particular time at which it might be conceived to have commenced, or to any period of his existence at which it was not existing. Let it be supposed, in short, that the subject stands in no relation whatever to time, can neither be called *in* time or *out of* time; but that all relations of time are as alien and heterogeneous in this question as the relations and attributes of space (north or south, round or square, thick or thin) are to our affections and moral feelings. Let the reader suppose this, and he will have before him the precise import of the scriptural doctrine of original sin: or rather of the fact acknowledged in all ages, and recognized, but not originating in the Christian Scriptures.

In addition to this memento it will be well to remind the inquirer

* If the law worked *on* the will, it would be the working of an extrinsic, alien force, and as St Paul profoundly argues, prove the will sinful.

that the steadfast conviction of the existence, personality, and moral attributes of God is presupposed in the acceptance of the Gospel, or required as its indispensible preliminary. It is taken for granted as a point which the hearer had already decided for himself, a point finally settled and put at rest: not by the removal of all difficulties, or by any such increase of insight as enabled him to meet every objection of the Epicurean or the sceptic with a full and precise answer; but because he had convinced himself that it was folly as well as presumption in so imperfect a creature to expect it; and because these difficulties and doubts disappeared at the beam, when tried against the weight and convictive power of the reasons in the other scale. It is, therefore, most unfair to attack Christianity, or any article which the Church has declared a Christian doctrine, by arguments which, if valid, are valid against all religion. Is there a disputant who scorns a mere postulate, as the basis of any argument in support of the faith; who is too high-minded to beg his ground, and will take it by a strong hand? Let him fight it out with the atheists, or the Manichaeans; but not stoop to pick up their arrows, and then run away to discharge them at Christianity or the Church!

The only true way is to state the doctrine, believed as well by Saul of Tarsus, 'yet breathing out threatenings and slaughter against' the Church of Christ, as by Paul the Apostle 'fully preaching the Gospel of Christ.' A moral evil is an evil that has its origin in a will. An evil common to all must have a ground common to all. But the actual existence of moral evil we are bound in conscience to admit; and that there is an evil common to all is a fact; and this evil must therefore have a common ground. Now this evil ground cannot originate in the divine Will: it must therefore be referred to the will of man. And this evil ground we call original sin. It is a mystery, that is, a fact, which we see, but cannot explain; and the doctrine a truth which we apprehend, but can neither comprehend nor communicate. And such by the quality of the subject (viz. a responsible will) it must be, if it be truth at all. . . .

[From the *Conclusion*]

MATERIALISM, conscious and avowed materialism, is in ill repute: and a confessed materialist therefore a rare character. But if the faith be ascertained by the fruits; if the predominant, though most often unsuspected, persuasion is to be learnt from the influences under which the thoughts and affections of the man move and take their direction; I must reverse the position. Only not all are materialists. Except a few individuals, and those for the most part of a single sect, everyone who calls himself a Christian holds himself to have a soul as well as a body. He distinguishes mind from matter, the subject of his consciousness from the objects of the same. The former is his mind: and he says, it is immaterial. But though *subject* and *substance* are words of kindred roots, nay, little less than equivalent terms, yet nevertheless it is exclusively to sensible objects, to bodies, to modifications of matter, that he habitually attaches the attributes of reality, of substance. Real and tangible, substantial and material, are synonyms for him. He never indeed asks himself, what he means by mind? But if he did, and tasked himself to return an honest answer—as to what, at least, he had hitherto meant by it—he would find that he had described it by negatives, as the opposite of bodies, ex. gr. as a somewhat opposed to solidity, to visibility, etc. as if you could abstract the capacity of a vessel, and conceive of it as a somewhat by itself, and then give to the emptiness the properties of containing, holding, being entered, and so forth. In short, though the proposition would perhaps be angrily denied in words, yet in fact he thinks of his mind, as a property or accident of a something else, that he calls a *soul* or *spirit*: though the very same difficulties must recur, the moment he should attempt to establish the difference. For either this soul or spirit is nothing but a thinner body, a finer mass of matter: or the attribute of self-subsistency vanishes from the soul on the same grounds on which it is refused to the mind.

I am persuaded, however, that the dogmatism of the corpuscular school, though it still exerts an influence on men's notions and phrases, has received a mortal blow from the increasingly *dynamic* spirit of the physical sciences now highest in public estimation. And it may safely be predicted that the results will extend beyond the intention of those who are gradually effecting this revolution. It is not chemistry alone that will be indebted to the genius of Davy, Oersted, and their compeers: and not as the founder of Physiology and philosophic anatomy alone will mankind love and revere the name of John Hunter. These men have not only *taught*, they have compelled us to admit that the immediate objects of our senses, or rather the grounds of the visibility and tangibility of all objects of sense, bear the same relation and similar proportion to the intelligible object—i.e. to the object which we actually mean when we say, 'It is such or such a thing,' or 'I have seen this or that'—as the paper, ink, and differently combined straight and curved lines of an edition of Homer bear to what we understand by the words, Iliad and Odyssey. Nay, nothing would be more easy than so to construct the paper, ink, painted capitals, etc. of a printed disquisition on the eye, or the muscles and cellular texture (i.e. the flesh) of the human body, as to bring together every one of the sensible and ponderable stuffs or elements that are sensuously perceived in the eye itself, or in the flesh itself. Carbon and nitrogen, oxygen and hydrogen, sulphur, phosphorus, and one or two metals and metallic bases, constitute the whole. It cannot be these, therefore, that we mean by an *eye*, by our *body*. But perhaps it may be a particular combination of these? But here comes a question: In this term do you or do you not include the principle, the operating cause, of the combination? If not, then detach this eye from the body! Look steadily at it—as it might lie on the marble slab of a dissecting room. Say it were the eye of a murderer, a Bellingham: or the eye of a murdered patriot, a Sidney!—Behold it, handle it, with its various accompaniments or constituent parts, of tendon, ligament, membrane, blood-vessel, gland, humours; its nerves of sense, of sensation, and of motion. Alas! all these names, like that of the organ itself, are so many anachronisms, figures of speech, to express that which has been: as when the guide points with his finger to a heap of stones, and tells the traveller, 'That is Babylon, or Persepolis.' Is this cold jelly 'the light of the body?' Is this the micranthropos in the marvellous microcosm? Is this what you mean when you well define the eye as the telescope and the mirror of the soul, the seat and agent of an almost magical power?

Pursue the same inquisition with every other part of the body, whether integral or simply ingredient; and let a Berzelius or a Hatchett° be your interpreter, and demonstrate to you what it is that in each actually meets your senses. And when you have heard the scanty catalogue, ask yourself if these are indeed the living flesh, the blood of life? Or not far rather—I speak of what, as a man of common sense, you really *do*, not what, as a philosopher, you *ought* to believe—is it not, I say, far rather the distinct and individualized agency that by the given combinations utters and bespeaks its presence? Justly and with strictest propriety of language may I say, *speaks*. It is to the coarseness of our senses, or rather to the defect and limitation of our percipient faculty, that the visible object appears the same even for a moment. The characters, which I am now shaping on this paper abide. Not only the forms remain the same, but the particles of the colouring stuff are fixed, and, for an indefinite period at least, remain the same. But the particles that constitute the *size*, the visibility of an organic structure [. . .] are in perpetual flux. They are to the combining and constitutive power as the pulses of air to the voice of a discourser; or of one who sings a roundelay. The same words may be repeated; but in each second of time the articulated air hath passed away, and each act of articulation appropriates and gives momentary form to a new and other portion. As the column of blue smoke from a cottage chimney in the breathless summer noon, or the steadfast-seeming cloud on the edge-point of a hill in the driving air-current, which momently condensed and recomposed is the common phantom of a thousand successors—such is the flesh which our bodily eyes transmit to us; which our palates taste; which our hands touch.

But perhaps the material particles possess this combining power by inherent reciprocal attractions, repulsions, and elective affinities; and are themselves the joint artists of their own combinations? I will not reply, though well I might, that this would be to solve one problem by another, and merely to shift the mystery. It will be sufficient to remind the thoughtful querist, that even herein consists the essential difference, the contradistinction, of an organ from a machine; that not only the characteristic shape is evolved from the invisible central power, but the material mass itself is acquired by assimilation. The germinal power of the plant transmutes the fixed air and the elementary base of water into grass or leaves; and on these the organific principle in the ox or the elephant exercises an alchemy still more stupendous. As the unseen agency weaves its magic eddies, the foliage becomes indifferently the bone and its marrow, the pulpy brain, or the solid ivory. That what you see *is* blood, *is* flesh, is itself the work, or shall I say, the translucence, of the invisible energy which soon surrenders or abandons them to inferior powers (for there is no pause nor chasm in the activities of nature), which repeat a similar metamorphosis according to their kind. These are not fancies, conjectures, or even hypotheses, but facts; to deny which is impossible, not to reflect on which is ignominious. And we need only reflect on them with a calm and silent spirit to learn the utter emptiness and unmeaningness of the vaunted mechanico-corpuscular philosophy, with both its twins, materialism on the one hand, and idealism, rightlier named *subjective idolism*, on the other: the one obtruding on us a world of spectres and apparitions; the other a mazy dream!

Let the mechanic or corpuscular scheme, which in its absoluteness and strict consistency was first introduced by Descartes, be judged by the results. *By its fruits shall it be known.*

On the Constitution of the Church and State, according to the Idea of Each°

CHAPTER II

The idea of a State in the larger sense of the term, introductory to the constitution of the State in the narrower sense, as it exists in this country

A CONSTITUTION is the attribute of a State, i.e. of a body politic, having the principle of its unity within itself, whether by concentration of its forces, as a constitutional pure monarchy, which, however, has hitherto continued to be *ens rationale* [entity of (i.e. created by) the reason], unknown in history (*B. Spinozae Tract. Pol. cap. VI. De Monarchiâ ex rationis praescripto*)—or—with which we are alone concerned—by equipoise and interdependency: the *lex equilibrii* [law of balance], the principle prescribing the means and conditions by and under which this balance is to be established and preserved, being the constitution of the State. It is the chief of many blessings derived from the insular character and circumstances of our country, that our social institutions have formed themselves out of our proper needs and interests; that long and fierce as the birth-struggle and the growing-pains have been, the antagonist powers have been of our own system, and have been allowed to work out their final balance with less disturbance from external forces than was possible in the Continental States.

O ne'er enchain'd nor wholly vile,
O Albion! O my Mother Isle!
Thy valleys fair as Eden's bowers
Glitter green with sunny showers!
Thy grassy uplands' gentle swells
Echo to the bleat of flocks;
Those grassy hills, those glittering dells,
Proudly ramparted with rocks:
And Ocean 'mid his uproar wild
Speaks safety to his Island-child!
Hence thro' many a fearless Age
Has social Freedom lov'd the Land,
Nor Alien Despot's jealous rage
Or warp'd thy growth or stamp'd the servile Brand.

Ode to the Departing Year, Dec. 1796

Now, in every country of civilized men acknowledging the rights of property, and by means of determined boundaries and common laws united into one people or nation, the two antagonist powers or opposite interests of the State, under which all other State interests are comprised, are those of permanence and of progression.*

It will not be necessary to enumerate the several causes that combine to connect the permanence of a State with the land and the landed property. To found a family, and to convert his wealth into land, are twin thoughts, births of the same moment, in the mind of the opulent merchant, when he thinks of reposing from his labours. From the class of the *Novi Homines* [New Men] he redeems himself by becoming the staple ring of the chain by which the present will become connected with the past; and the test and evidence of permanency afforded. To the same principle appertain primogeniture and hereditary titles, and the influence which these exert in accumulating large masses of property, and in counteracting the antagonist and dispersive forces which the follies, the vices, and misfortunes of individuals can scarcely fail to supply. To this, likewise, tends the proverbial obduracy of prejudices characteristic of the humbler tillers of the soil, and their aversion even to benefits that are offered in the form of innovations. But why need I attempt to explain a fact which no thinking man will deny, and where the admission of the fact is all that my argument requires?

* Permit me to draw your attention to the essential difference between *opposite* and *contrary*. Opposite powers are always of the same kind, and tend to union, either by equipoise or by a common product. Thus the + and − poles of the magnet, thus positive and negative electricity are opposites. Sweet and sour are opposites; sweet and bitter are contraries. The feminine character is *opposed* to the masculine; but the effeminate is its *contrary*. Even so in the present instance, the interest of permanence is opposed to that of progressiveness; but so far from being contrary interests, they, like the magnetic forces, suppose and require each other. Even the most mobile of creatures, the serpent, makes a *rest* of its own body, and drawing up its voluminous train from behind on this fulcrum, propels itself onward. On the other hand, it is a proverb in all languages, that (relatively to man at least) what would stand still must retrograde. You, my dear Sir,° who have long known my notions respecting the power and value of words, and the duty as well as advantage of using them appropriately, will forgive this.

Many years ago, in conversing with a friend, I expressed my belief that in no instance had the false use of a word become current without some practical ill consequence, of far greater moment than would *primo aspectu* [at first sight] have been thought possible. That friend, very lately referring to this remark, assured me that not a month had passed since then, without some instance in proof of its truth having occurred in his own experience; and added, with a smile, that he had more than once amused himself with the thought of a verbarian Attorney-General, authorized to bring information ex officio against the writer or editor of any work in extensive circulation, who, after due notice issued, should persevere in misusing a word.

On the other hand, with as little chance of contradiction, I may assert that the progression of a State in the arts and comforts of life, in the diffusion of the information and knowledge, useful or necessary for all; in short, all advances in civilization, and the rights and privileges of citizens, are especially connected with and derived from the four classes of the mercantile, the manufacturing, the distributive, and the professional. To early Rome, war and conquest were the substitutes for trade and commerce. War was their trade. As these wars became more frequent, on a larger scale, and with fewer interruptions, the liberties of the plebeians continued increasing: for even the sugar plantations of Jamaica would (in their present state, at least), present a softened picture of the hard and servile relation, in which the plebeian formerly stood to his patrician patron.

Italy is supposed at present to maintain a larger number of inhabitants than in the days of Trajan or in the best and most prosperous of the Roman empire. With the single exception of the ecclesiastic State, the whole country is cultivated like a garden. You may find there every gift of God—only not freedom. It is a country rich in the proudest records of liberty, illustrious with the names of heroes, statesmen, legislators, philosophers. It hath a history all alive with the virtues and crimes of hostile parties, when the glories and the struggles of ancient Greece were acted over again in the proud republics of Venice, Genoa, and Florence. The life of every eminent citizen was in constant hazard from the furious factions of their native city, and yet life had no charm out of its dear and honoured walls. All the splendours of the hospitable palace, and the favour of princes, could not soothe the pining of Dante or Machiavel, exiles from their free, their beautiful Florence. But not a pulse of liberty survives. It was the profound policy of the Austrian and the Spanish Courts by every possible means to degrade the profession of trade; and even in Pisa and Florence themselves to introduce the feudal pride and prejudice of less happy, less enlightened countries. Agriculture, meanwhile, with its attendant population and plenty, was cultivated with increasing success; but from the Alps to the Straits of Messina the Italians are slaves.

We have thus divided the subjects of the State into two orders, the agricultural or possessors of land; and the merchant, manufacturer, the distributive, and the professional bodies, under the common name of citizens. And we have now to add that by the nature of things common to every civilized country, at all events by the course of events in this country, the first is subdivided into two classes, which, in imitation of our old law-books, we may entitle the major and minor barons; both these, either by their interests or by the very effect of their situation, circumstances, and the nature of their employment, vitally connected with the permanency of the State, its institutions, rights, customs, manners, privileges—and as such, opposed to the inhabitants of ports, towns, and cities, who are in like manner and from like causes more especially connected with its progression. I scarcely need say that in a very advanced stage of civilization the two orders of society will more and more modify and leaven each other, yet never so completely but that the distinct character remains legible, and to use the words of the Roman Emperor, even in what is struck out the erasure is manifest. At all times the lower of the two ranks, of which the first order consists, or the franklins, will, in their political sympathies, draw more nearly to the antagonist order than the first rank. On these facts, which must at all times have existed, though in very different degrees of prominence or maturity, the principle of our constitution was established. The total interests of the country, the interests of the State, were entrusted to a great council or Parliament, composed of two Houses. The first consisting exclusively of the major barons, who at once stood as the guardians and sentinels of their several estates and privileges, and the representatives of the common weal. The minor barons, or franklins, too numerous, and yet individually too weak, to sit and maintain their rights in person, were to choose among the worthiest of their own body representatives, and these in such number as to form an important though minor proportion of a second House—the majority of which was formed by the representatives chosen by the cities, ports, and boroughs; which representatives ought on principle to have been elected not only by, but from among, the members of the manufacturing, mercantile, distributive, and professional classes.

These four classes, by an arbitrary but convenient use of the phrase, I will designate by the name of the Personal Interest, as the exponent of all moveable and personal possessions, including skill and acquired knowledge, the moral and intellectual stock-in-trade of the professional man and the artist, no less than the raw materials, and the means of elaborating, transporting, and distributing them.

Thus, in the theory of the constitution it was provided that even though both divisions of the Landed Interest should combine in any legislative attempt to encroach on the rights and privileges of the Personal Interest, yet the representatives of the latter forming the clear and effectual majority of the lower House, the attempt must be abortive: the majority of votes in both Houses being indispensable, in order to the presentation of a bill for the Completory Act—that is, to make it a law of the land. By force of the same mechanism must every attack be baffled that should be made by the representatives of the minor landholders, in concert with the burgesses, on the existing rights and privileges of the peerage, and of the hereditary aristocracy, of which the peerage is the summit and the natural protector. Lastly, should the nobles join to invade the rights and franchises of the franklins and the yeomanry, the sympathy of interest, by which the inhabitants of cities, towns, and seaports are linked to the great body of the agricultural fellow commoners who supply their markets and form their principal customers, could not fail to secure a united and successful resistance. Nor would this affinity of interest find a slight support in the sympathy of feeling between the burgess senators and the county representatives, as members of the same House; and in the consciousness which the former have of the dignity conferred on them by the latter. For the notion of superior dignity will always be attached in the minds of men to that kind of property with which they have most associated the idea of permanence: and the land is the synonym of country.

That the burgesses were not bound to elect representatives from among their own order, individuals bonâ fide belonging to one or other of the four divisions above enumerated; that the elective franchise of the towns, ports, etc., first invested with borough rights, was not made conditional, and to a certain extent at least dependent on their retaining the same comparative wealth and independence, and rendered subject to a periodical revisal and readjustment; that in consequence of these and other causes the very weights intended for the effectual counterpoise of the great landholders, have, in the course of events, been shifted into the opposite scale; that they now constitute a large proportion of the political power and influence of the very class whose personal cupidity, and whose partial views of the landed interest at large they were meant to keep in check; these are no part of the constitution, no essential ingredients in the idea, but apparent defects and imperfections in its realization—which, however, we will neither regret nor set about amending, till we have seen whether an equivalent force had not arisen to supply the deficiency—a force great enough to have destroyed the equilibrium, had not such a transfer taken place previously to, or at the same time with, the operation of the new forces. Roads, canals, machinery, the press, the periodical and daily press, the might of public opinion, the consequent increasing desire of popularity among public men and functionaries of every description, and the increasing necessity of public character, as a means or condition of political influence—I need but mention these to stand acquitted of having started a vague and naked possibility in extenuation of an evident and palpable abuse.

But whether this conjecture be well or ill grounded, the principle of the constitution remains the same. That harmonious balance of the two great correspondent, at once supporting and counterpoising, interests of the State, its permanence, and its progression: that balance of the landed and the personal interests was to be secured by a legislature of two Houses; the first consisting wholly of barons or landholders, permanent and hereditary senators; the second of the knights or minor barons, elected by, and as the representatives of, the remaining landed community, together with the burgesses, the representatives of the commercial, manufacturing, distributive, and professional classes—the latter (the elected burgesses) constituting the major number. The king, meanwhile, in whom the executive power is vested, it will suffice at present to consider as the beam of the constitutional scales. A more comprehensive view of the kingly office must be deferred, till the remaining problem (the idea of a national Church) has been solved.

I must here entreat the reader to bear in mind what I have before endeavoured to impress on him, that I am not giving an historical account of the legislative body; nor can I be supposed to assert that such was the earliest mode or form in which the national council was constructed. My assertion is simply this, that its formation has advanced in this direction. The line of evolution, however sinuous, has still tended to this point, sometimes with, sometimes without, not seldom, perhaps, against, the intention of the individual actors, but always as if a power, greater, and better, than the men themselves, had intended it for them. Nor let it be forgotten that every new growth, every power and privilege, bought or extorted, has uniformly been claimed by an antecedent right; not acknowledged as a boon conferred, but both demanded and received as what had always belonged to them, though withheld by violence and the injury of the times. This too, in cases where, if documents and historical records, or even consistent traditions, had been required in evidence, the monarch would have had the better of the argument. But, in truth, it was no more than a practical way of saying: this or that is contained in the idea of our Government, and it is a consequence of the 'Lex, Mater Legum [Law, Mother of Laws],' which, in the very first law of

State ever promulgated in the land, was presupposed as the ground of that first law.

Before I conclude this part of my subject, I must press on your attention, that the preceding is offered only as the constitutional idea of the State. In order to correct views respecting the constitution, in the more enlarged sense of the term, viz. the constitution of the nation, we must, in addition to a grounded knowledge of the State, have the right idea of the national Church. These are two poles of the same magnet; the magnet itself, which is constituted by them, is the constitution of the nation.

CHAPTER V

Of the Church of England, or National Clergy, according to the constitution: its characteristic ends, purposes, and functions: and of the persons comprehended under the clergy, or the functionaries of the National Church

AFTER these introductory preparations, I can have no difficulty in setting forth the right idea of a national Church as in the language of Elizabeth the *third* great venerable estate of the realm. The first being the estate of the landowners or possessors of fixed property, consisting of the two classes of the barons and the franklins; the second comprising the merchants, the manufacturers, free artisans, and the distributive class. To comprehend, therefore, this third estate, in whom the reserved nationalty° was vested, we must first ascertain the end, or national purpose, for which it was reserved.

Now, as in the former state the permanency of the nation was provided for; and in the second estate its progressiveness, and personal freedom; while in the king the cohesion by interdependence, and the unity of the country, were established; there remains for the third estate only that interest, which is the ground, the necessary antecedent condition, of both the former. Now these depend on a continuing and progressive civilization. But civilization is itself but a mixed good, if not far more a corrupting influence, the hectic of disease, not the bloom of health, and a nation so distinguished more fitly to be called a varnished than a polished people; where this civilization is not grounded in cultivation, in the harmonious development of those qualities and faculties that characterize our humanity. We must be men in order to be citizens.

The nationalty, therefore, was reserved for the support and maintenance of a permanent class or order, with the following duties. A certain smaller number were to remain at the fountain-heads of the humanities, in cultivating and enlarging the knowledge already possessed, and in watching over the interests of physical and moral science; being, likewise, the instructors of such as constituted, or were to constitute, the remaining more numerous classes of the order. This latter and far more numerous body were to be distributed throughout the country, so as not to leave even the smallest integral part or division without a resident guide, guardian, and instructor; the objects and final intention of the whole order being these—to preserve the stores, to guard the treasures, of past civilization, and thus to bind the present with the past; to perfect and add to the same, and thus to connect the present with the future; but especially to diffuse through the whole community, and to every native entitled to its laws and rights, that quantity and quality of knowledge which was indispensable both for the understanding of those rights, and for the performance of the duties correspondent. Finally, to secure for the nation, if not a superiority over the neighbouring States, yet an equality at least in that character of general civilization which equally with, or rather more than, fleets, armies, and revenue, forms the ground of its defensive and offensive power. The object of the two former estates of the realm, which conjointly form the State, was to reconcile the interests of permanence with that of progression—law with liberty. The object of the national Church, the third remaining estate of the realm, was to secure and improve that civilization without which the nation could be neither permanent nor progressive.

That in all ages individuals who have directed their meditations and their studies to the nobler characters of our nature, to the cultivation of those powers and instincts which constitute the man, at least separate him from the animal, and distinguish the nobler from the animal part of his own being, will be led by the *supernatural* in themselves to the contemplation of a power which is likewise super-*human*; that science, and especially moral science, will lead to religion, and remain blended with it—this, I say, will, in all ages, be the course of things. That in the earlier ages, and in the dawn of civility, there will be a twilight in which science and religion give light, but a light refracted through the dense and the dark, a superstition—this is what we learn from history, and what philosophy would have taught us to expect. But we affirm that in the spiritual purpose of the word, and as understood in reference to a future State, and to the abiding essential interest of the individual as a person, and not as the citizen, neighbour, or subject, religion may be an indispensable ally, but is not the essential constitutive end of that national institute which is unfortunately, at least improperly, styled a Church—a name which in its best sense is exclusively appropriate to the Church of Christ. If this latter be ecclesia, the communion of such as are called out of the world, i.e. in reference to the especial ends and purposes of that communion; this other might more expressively have been entitled *enclesia*, or an order of men, chosen in and of the realm, and constituting an estate of that realm.° And in fact, such was the original and proper sense of the more appropriately named *clergy*. It comprehended the learned of all names, and the clerk was the synonym of the man of learning. Nor can any fact more strikingly illustrate the conviction entertained by our ancestors, respecting the intimate connection of this clergy with the peace and weal of the nation, than the privilege formerly recognized by our laws, in the well-known phrase, 'benefit of clergy.'°

Deeply do I feel, for clearly do I see, the importance of my theme. And had I equal confidence in my ability to awaken the same interest in the minds of others, I should dismiss as affronting to my readers all apprehension of being charged with prolixity, while I am labouring to compress in two or three brief chapters the principal sides and aspects of a subject so large and multilateral as to require a volume for its full exposition. With what success will be seen in what follows, commencing with the churchmen, or (a far apter and less objectionable designation) the national clerisy.

The clerisy of the nation, or national Church, in its primary acceptation and original intention comprehended the learned of all denominations—the sages and professors of the law and jurisprudence; of medicine and physiology; of music; of military and civil architecture; of the physical sciences; with the mathematical as the common organ of the preceding; in short, all the so-called liberal arts and sciences, the possession and application of which constitute the civilization of a country, as well as the theological. The last was, indeed, placed at the head of all; and of good right did it claim the precedence. But why? Because under the name of theology, or divinity, were contained the interpretation of languages; the conservation and tradition of past events; the momentous epochs, and revolutions of the race and nation; the continuation of the records;

logic, ethics, and the determination of ethical science, in application to the rights and duties of men in all their various relations, social and civil; and lastly, the ground-knowledge, the *prima scientia* [first knowledge] as it was named—philosophy, or the doctrine and discipline* of ideas.

Theology formed only a part of the objects, the theologians formed only a portion of the clerks or clergy of the national Church. The theological order had precedency indeed, and deservedly; but not because its members were priests, whose office was to conciliate the invisible powers, and to superintend the interests that survive the grave; not as being exclusively, or even principally, sacerdotal or templar, which, when it did occur, is to be considered as an accident of the age, a mis-growth of ignorance and oppression, a falsification of the constitutive principle, not a constituent part of the same. No! The theologians took the lead, because the science of theology was the root and the trunk of the knowledge that civilized man, because it gave unity and the circulating sap of life to all other sciences, by virtue of which alone they could be contemplated as forming, collectively, the living tree of knowledge. It had the precedency, because under the name theology were comprised all the main aids, instruments, and materials of national education, the *nisus formativus* [formative impulse] of the body politic, the shaping and informing spirit, which *educing*, i.e. eliciting, the latent *man* in all the natives of the soil, trains them up to citizens of the country, free subjects of the realm. And lastly, because to divinity belong those fundamental truths which are the common groundwork of our civil and our religious duties, not less

* That is, of knowledges immediate, yet real, and herein distinguished in kind from logical and mathematical truths, which express not realities, but only the necessary forms of conceiving and perceiving, and are therefore named the *formal* or *abstract* sciences. Ideas, on the other hand, or the truths of philosophy, properly so called, correspond to substantial beings, to objects whose actual subsistence is implied in their idea, though only by the idea revealable. To adopt the language of the great philosophic apostle, they are 'spiritual realities that can only spiritually be discerned,' and the inherent aptitude and moral preconfiguration to which constitutes what we mean by ideas, and by the presence of ideal truth, and of ideal power, in the human being. They, in fact, constitute his humanity. For try to conceive a man without the ideas of God, eternity, freedom, will, absolute truth, of the good, the true, the beautiful, the infinite. An animal endowed with a memory of appearances and of facts might remain. But the man will have vanished, and you have instead a creature, 'more subtle than any beast of the field, but likewise cursed above every beast of the field; upon the belly must it go and dust must it eat all the days of its life.' But I recall myself from a train of thoughts little likely to find favour in this age of sense and selfishness.

indispensable to a right view of our temporal concerns, than to a rational faith respecting our immortal wellbeing. (Not without celestial observations can even terrestrial charts be accurately constructed.) And of especial importance is it to the objects here contemplated that only by the vital warmth diffused by these truths throughout the many, and by the guiding light from the philosophy, which is the basis of divinity, possessed by the few, can either the community or its rulers fully comprehend, or rightly appreciate, the permanent distinction, and the occasional contrast, between cultivation and civilization; or be made to understand this most valuable of the lessons taught by history, and exemplified alike in her oldest and her most recent records—that a nation can never be a too cultivated, but may easily become an over-civilized race.

NOTES

ABBREVIATIONS

The following abbreviations have been used in the notes and apparatus:

CC *The Collected Works of Samuel Taylor Coleridge* (Princeton and London: Princeton University Press and Routledge & Kegan Paul Ltd., 1969–).

CL *The Collected Letters of Samuel Taylor Coleridge*, ed. Earl Leslie Griggs (Oxford and New York: Oxford University Press, 1956–71).

CM S. T. Coleridge, *Marginalia*, ed. George Whalley (Princeton and London: Princeton University Press and Routledge & Kegan Paul Ltd., 1980–) [in *CC*].

CN *The Notebooks of Samuel Taylor Coleridge*, ed. Kathleen Coburn (New York, Princeton, and London: Princeton University Press and Routledge & Kegan Paul Ltd., 1957–).

DNB *Dictionary of National Biography*.

LCL Loeb Classical Library.

OED *The Oxford English Dictionary*, 12 vols. (Oxford: Clarendon Press, 1933; repr. 1961, 1970, 1978).

POEMS

1–153 Coleridge's first published work was a poem that appeared in the *Cambridge Intelligencer* while he was still a student; his last, a revised edition of his *Poetical Works* in three volumes. Although his really fruitful period as a poet lasted only ten years, with the great year, the 'annus mirabilis' of his close association with Wordsworth, at the centre of it, he wrote poetry all his life. The poems are here arranged chronologically by date of composition, as far as that date can at present be ascertained; the dates appear with the titles in the Contents. The footnotes that Coleridge, in an age of annotated verse, occasionally included with his poems, have been omitted unless he invariably used them or they appear to be especially helpful to the understanding of the poem.

The question of the text of the poems is complicated, because most of them appeared in variant forms in Coleridge's lifetime. The text here is based on the standard edition, the *Complete Poetical Works*, ed. E. H. Coleridge (1912). Since I believe that E. H. Coleridge was correct in taking the edition of 1834, the last supervised by Coleridge himself, as the best edition for the poems included in it, I have restored accidentals of the 1834 text that E. H. Coleridge chose to emend according to his own standards of punctuation. E. H. Coleridge made the poems look more 'poetic' and, to our eyes, more Victorian than they did in 1834, by greatly increasing the amount of capitalization, by introducing italics occasionally for emphasis, and by distinguishing between sounded and silent '-ed' endings by accenting the former ('becalméd') and replacing the 'e' with an apostrophe in the latter ('sooth'd'). In this matter the 1834 text, which itself varies between '-ed', the apostrophe, and occasionally 'é', has been followed. E. H. Coleridge's text, admirable in other ways, was unfortunately printed so as to obscure stanza divisions, which I have therefore also restored. Exceptions to this textual policy are indicated in the notes.

1 *Monody on the Death of Chatterton.* Thomas Chatterton (1752–70) was a symbol of neglected genius for the poets of Coleridge's generation. Despairing of literary success, he had taken arsenic and died in London at the age of seventeen. His fame now rests on the 'Rowley Poems', compositions in a deliberately antiquated style which he tried to pass off as the work of a fifteenth-century monk, Thomas Rowley.

3 l. 47. *Dacyan foe.* An allusion to Chatterton's mythology, in which Aella figures as an Anglo-Saxon hero against the Danes, e.g. in *Aella; A Tragycal Enterlude* and the 'Songe to Aella'.

11 *Sonnets on Eminent Characters.* Coleridge published in all twelve sonnets, addressed mostly to political figures, in a newspaper, the *Morning Chronicle*, in December 1794 and January 1795. He came later to repudiate the politics or the poetry or both (as is the case with 'Pitt') of several of the series. The subjects of the selection given here may be briefly identified. Edmund Burke (1729–97), the great statesman and orator, champion of the American Revolution, had at the time of the composition of this sonnet just retired from politics after establishing himself as a leader of opposition to the principles of the French Revolution. Joseph Priestley (1733–1804) was a well-known scientist, political radical, and leader of the Unitarians in England; but anti-Jacobin rioters set fire to his house in Birmingham and in 1794 he was driven to emigrate to America. William Pitt the Younger (1759–1806), second son of the great Whig statesman, the Earl of Chatham, was Prime Minister of England from 1783 to 1801 and again from 1804 to 1806. William Lisle Bowles (1762–1850), author of *Fourteen Sonnets* (1789), was at this time one of the gods of Coleridge's idolatry: see *Biographia Literaria*, pp. 161–9.

12 *Pitt.* Text of 1803, after which date Coleridge omitted this poem.

l. 12. *Sister.* Justice.

To the Rev. W. L. Bowles. The text published here is not the original version of 1794, but the revised text that Coleridge included in collections of his poetry from 1796.

13 *Religious Musings. Argument.* Adopted from the early editions of the poem, although Coleridge himself omitted it after 1803.

17 l. 159. *Even now.* Coleridge included in the later editions of this poem a footnote that explains its topical reference:

January 21st, 1794, in the debate on the Address to his Majesty, on the speech from the Throne, the Earl of Guildford [*sic*] moved an amendment to the following effect: 'That the House hoped his Majesty would seize the earliest opportunity to conclude a peace with France,' etc. This motion was opposed by the Duke of Portland, who 'considered the war to be merely grounded on one principle—the preservation of the Christian religion'. May 30th, 1794, the Duke of Bedford moved a number of resolutions, with a view to the establishment of a peace with France. He was opposed (among others) by Lord Abingdon in these remarkable words: 'The best road to peace, my Lords, is war! and war carried on in the same manner in which we are taught to worship our Creator, namely, with all our souls, and with all our minds, and with all our hearts, and with all our strength.'

l. 172. *wedded lord*. Catherine the Great (1729–96), Tsarina of Russia, was thought to have been responsible for the death of her husband Peter III after deposing him. The 'connatural mind' of the following line is Frederick William II of Prussia, Catherine's ally in the shameful partition of Poland in 1793.

18 l. 223. *blind Ionian*. Homer. The reference is to gods of the *Iliad* especially.

19 l. 234. *patriot Sage*. Benjamin Franklin (1706–90), the American journalist, diplomat, patriot, and scientist, whose famous experiment in electricity, involving a kite in a thunderstorm, is alluded to here.

20 l. 269. *Simoom*. Coleridge's lines are meant to evoke Africa: the simoom, a hot, suffocating, lethal desert wind, had been described by a contemporary traveller as coming over the desert like a 'purple haze'.

l. 275. *Behemoth*. The Hebrew word for a huge animal is meant here to refer specifically to the elephant.

l. 292. *Lazar-house*. A hospital, particularly a hospital for lepers. Coleridge's lines designate specific social evils that are consequences of war: poverty that leads to crime; prostitution; neglect and starvation; forced enlistment; increasing numbers of widows and orphans.

l. 304. *fifth seal*. An allusion to Rev. 6: 9–10: 'And when he had opened the fifth seal, I saw under the altar the souls of them that were slain for the word of God, and for the testimony which they held; and they cried with a loud voice, saying, "How long, O Lord, holy and true, dost thou not judge and avenge our blood on them that dwell on the earth?"'

21 l. 315. *storm begins*. With the French Revolution.

l. 323. *abhorred Form*. The whore of Babylon—again an allusion to the apocalypse (Rev. 17). Coleridge remarked of the passage here, 'I am convinced that the Babylon of the Apocalypse does not apply to Rome exclusively; but to the union of Religion with Power and Wealth, wherever it is found.'

22 l. 359. *Thousand Years.* The Millennium (literally, a thousand years), the period during which Christ is to rule on earth (Rev. 20).

l. 370. *sentient brain.* The philosopher David Hartley, whose doctrine of association Coleridge was to discuss and reject in the seventh chapter of *Biographia Literaria.*

l. 380. *Jasper Throne.* The throne of God in Rev. 4: 2–3.

24 l. 2. *To an Infant. unclasped knife.* An open clasp-knife.

Lines. The text is from *Poetical Works* (1828); the poem was not included in the collection of 1834.

27 *The Eolian Harp.* An aeolian harp or wind-harp (Aeolus is the name of the god of the winds) is a rectangular box strung on one side and placed in an open window to make musical sounds when the wind crosses it.

29 *Reflections on Having Left a Place of Retirement.* l. 12. *Bristowa's.* Bristol's.

30 l. 49. *Howard's.* John Howard (1726–90), philanthropist and prison reformer, died in Russia on one of several surveys of prisons, of camp fever caught while tending the sick.

31 *Ode to the Departing Year.* l. 40. *Northern Conqueress.* Catherine the Great (1729–96), Tsarina of Russia, scandalous in her private life (Coleridge alludes to the murder of her husband in 'Religious Musings') and ruthlessly ambitious. The capture of the Turkish town of Izmail, on the Danube, in 1790, and the sack of Warsaw in 1794 were recent atrocities associated with her reign.

34 l. 76. *Lampads seven.* The seven 'lamps of fire burning before the throne, which are the seven Spirits of God' in Rev. 4: 5.

36 *To the Rev. George Coleridge.* In a letter written just two months earlier to Thomas Poole, the 'friend' of this poem, Coleridge had described his brother George (1764–1828) as 'worth the whole family in a Lump' (see p. 496). George had been a second father to Coleridge after their father's death, and it later gave Coleridge great pain to be rejected by him at the time of his (Coleridge's) separation from his wife.

37 l. 26. *Manchineel.* A West Indian tree with attractive fruit but poisonous sap. Coleridge makes fun of the literary abuse of this metaphor in the first chapter of the *Biographia* (p. 160).

41 *The Wanderings of Cain. Death of Abel.* To write, that is, a work modelled on Salomon Gessner's prose epic *The Death of Abel* (1758; trans. 1761).

46 *The Rime of the Ancient Mariner.* Although it appears in sequence at the date of composition, the 'Ancient Mariner' is presented not in its original form (as it appeared in *Lyrical Ballads*, 1798) but in the revised version, with epigraph and marginal gloss, that Coleridge published in *Sibylline Leaves* (1817) and in all subsequent editions of his poetry. The opening stanzas of the earlier version convey something of its different character:

It is an ancyent Marinere
And he stoppeth one of three:
'By thy long grey beard and thy glittering eye
Now wherefore stoppest me?

The Bridegroom's doors are open'd wide,
And I am next of kin;
The Guests are met, the Feast is set,—
May'st hear the merry din.'

But still he holds the wedding-guest—
There was a Ship, quoth he—
'Nay, if thou'st got a laughsome tale,
Marinere! come with me.'

For Coleridge's account of the genesis of the volume that opened with this poem, see *Biographia Literaria*, ch. XIV (pp. 314–16). He discusses the 'Ancient Mariner' specifically in *Table Talk*, p. 593.

66 *Christabel. Preface. imitated*. Byron and Scott. 'Christabel' had circulated in manuscript long before it was published in 1816, and Scott's *Lay of the Last Minstrel* (1805), the first of a series of popular historical poems, had been influenced directly by it.

85 *Fire, Famine, and Slaughter. A War Eclogue*. Coleridge dramatizes in the mode of *Macbeth* recent and past events of the war carried on between England (under Pitt, 'letters four do form his name') and France since 1792. The English Government was known to have supported unsuccessful Royalist uprisings in the Vendée in France in 1793; the harsh suppression of the French people at that time was likened by anti-war and anti-Ministerial writers to the suppression of the Irish Rebellion that broke out, with support from France, early in 1798.

87 *Frost at Midnight*. l. 15. *that film*. In early editions of this poem, Coleridge included a footnote that clarifies the popular superstition, associated with coal fires, that is alluded to here: 'In all parts of the kingdom these films are called *strangers* and are supposed to portend the arrival of some absent friend.' The same superstition is referred to in Cowper's poem, *The Task*.

89 *France: An Ode. Argument*. This statement, published by Coleridge in only one early edition of the poem, clarifies the political context: Britain had declared war on France in February 1792; news of the invasion of Switzerland (Helvetia) by France reached England in March 1798.

99 *The Nightingale*. l. 13. *melancholy bird*. The well-known epithet is quoted from Milton's 'Il Penseroso', but Coleridge included in most editions of 'The Nightingale' the following explanatory footnote: 'This passage in Milton possesses an excellence far superior to that of mere description; it is spoken in the character of the melancholy Man, and has therefore a dramatic propriety. The Author makes this remark, to rescue himself

from the charge of having alluded with levity to a line in Milton; a charge than which none could be more painful to him, except perhaps that of having ridiculed his Bible.'

102 *Kubla Khan*. The circumstantial details of the prefatory note have been called into question, and it seems in fact that they were never intended to be taken as literally true, but were meant to suggest the frame of mind in which this poem (which Coleridge called a 'psychological curiosity') was to be read, as is the case with the note prefixed to 'This Lime-Tree Bower My Prison'.

104 *Recantation*. A comic political allegory in which the course of the French Revolution becomes the career of a frightened ox, Louis XVI is represented by the farmer Lewis, and the 'sage' who changes his view of events stands for the English politicians Tierney and Sheridan. *Text*: *Sibylline Leaves* (1817).

106 l. 78. *fasting-spittle*. A local allusion explained by Coleridge's note in *Sibylline Leaves*: 'According to the superstition of the West-Countries, if you meet the Devil, you may either cut him in half with a straw, or force him to disappear by spitting over his horns.' Fasting-spittle is the saliva that is in the mouth before one's fast is broken.

113 *Dejection: An Ode*. This ode exists in several versions, and there has been considerable critical debate not only about which version is to be preferred but also about the sequence of composition. The version given here is the one used exclusively by Coleridge after 1817, but readers may be interested in comparing *CL* ii. 790–8 and 815–19.

l. 7. *Eolian lute*. See note to p. 27.

123 *Constancy to an Ideal Object*. l. 30. *round its head*. The famous 'Brocken Spectre', an eerie phenomenon with a natural explanation, was used as an image by several Romantic writers. It is a shadow created by the rising sun, cast upon and magnified by morning mists.

129 *A Tombless Epitaph*. l. 1. *Satyrane*. The name Coleridge took for himself in 'Satyrane's Letters' in the *Biographia*. 'Idoloclastes' means 'breaker of idols', and Sir Satyrane, though the son of a satyr, was the champion of Una in Spenser's *Faerie Queen*, Bk. I, canto vi.

130 l. 22. *Parnassian forest*. Both the mountain Parnassus and the fountain Hippocrene (on Mount Helicon) were sacred to the Muses, and are conventional figures for poetry and poetic inspiration.

131 *Limbo*. The original notebook entries (*CN* iii. 4073–4) will repay study for the context (and the textual variants) they provide for this poem and 'Ne Plus Ultra' following.

132 *Ne Plus Ultra*. The title, a phrase supposed to have been carved on the mountains called the Pillars of Hercules at the western end of the Mediterranean Sea, is a command to go no further, '[Let there be] no more [sailing] beyond.' It may therefore suggest either an insuperable barrier or an ultimate boundary with nothing beyond it.

l. 18. *Lampads Seven.* As in the 'Ode to the Departing Year', this apocalyptic image refers to the 'seven Spirits of God' in Rev. 4: 5.

133 *On Donne's Poetry.* Coleridge never published this poem, which appeared posthumously in 1836, having been taken from among his marginalia.

134 *Fancy in Nubibus.* l. 11. *Chian strand.* The island of Chios, reputed birthplace of Homer.

A Character. Coleridge's political self-defence, in the spirit of Swift's 'Verses on the Death of Dr Swift'.

135 l. 16. *Phoebus' breed.* Phoebus Apollo, the sun-god, patron of poetry.

l. 46. *Sic vos non vobis.* Coleridge uses the Virgilian tag that occurs also at the end of the second chapter of *Biographia Literaria* (p. 181): 'Thus you labour, not for yourselves.'

136 l. 64. *Goose and Goody.* Probably the champion of reform, Sir Francis Burdett, who had been caricatured by Gillray as a goose, and the Chancellor, Lord Eldon, 'an exceedingly good-natured man' according to Hazlitt's *Spirit of the Age* (1825), which praised both men. The attack on Coleridge in Hazlitt's work prompted the writing of 'A Character'.

l. 72. *Ἔστησε.* A bilingual play on words. Coleridge has converted the initials of his name, S. T. C., into a Greek verb. 'Punic' is pronounced with a short 'u', as in 'pun'.

138 *Lines. Berengarius.* The theologian Berengar of Tours (*c*.999–1088), excommunicated in 1050 for his views on the Eucharist, recanted and made his peace with the Church by a formal retraction, but is thought never quite to have given up his position.

139 *Work Without Hope.* An interesting context for this sonnet is to be found in the early draft in *CL* v. 414–16.

140 *The Improvisatore. my Jo, John.* The line is the title of a song by Burns, a love song for an ageing couple.

148 *Alice Du Clos.* l. 91. *Ellen.* Alice.

153 *Epitaph.* l. 7. *forgiven for fame.* Preferring, that is, mercy to praise and forgiveness to fame.

PROSE

BIOGRAPHIA LITERARIA

Although Coleridge had conceived as early as 1803 the project of writing 'my metaphysical works, as *my Life*, & in my Life', it was not until 1815 that circumstances presented an opportunity of making the project a reality. At that time, he planned to issue a volume of poems (*Sibylline Leaves*) with a preface; before long, the preface came to be thought of as a companion volume that would take up a number of the critical controversies in which Coleridge had been involved since his collaboration with Wordsworth in *Lyrical Ballads* (1798). Particularly important were the question of the relationship between poetry and criticism, and the issue of poetic diction. Coleridge and his friends had been the victims of savage reviews for some years; Wordsworth's *Excursion* (1814) had recently been attacked by Francis Jeffrey in the influential *Edinburgh Review*; so Coleridge took the opportunity offered by 'a sort of introduction to a volume of poems' to analyse the weaknesses of the reviewing system, and to propose as a model of fair criticism his own counterbalancing assessment of Wordsworth's work. In the distinction between fancy and imagination and in the debate about poetic diction—the proposition that English poetry had cut itself off from real life by an increasing dependence upon a sort of dialect of the language, found only in books of verse—Coleridge's opponent was, however, Wordsworth himself, whose recent volume of *Poems* (1815) revived a controversy that had been started by the 1800 Preface to *Lyrical Ballads*, in a way that made Coleridge feel he must clarify the limitations of his agreement with his friend's theories.

Most of the *Biographia* was dictated by Coleridge to a friend, John Morgan, during the summer of 1815. Among the many complications that held up publication of the work from the time when the manuscript was delivered to the printer until July 1817 were problems of length. The fear that it would be too long contributed to Coleridge's decision to cut short the philosophical argument in the centre of the book; then the publisher's discovery that it would be too short for the two volumes that had eventually been agreed upon led Coleridge hastily to add 'Satyrane's Letters' and chapter XXIII, the critique on *Bertram*, to fill the gap. The extent to which these sections have been successfully integrated in the *Biographia* is still debated.

Text: Coleridge himself observed that the *Biographia* had been 'wildly printed', and he had no opportunity to correct it in a second edition. The text used here is the first edition of 1817, modernized according to the principles described in the Note on the Text, pp. xvii–xviii, but retaining, in the absence of full modernization, which would involve more loss than gain, some of the peculiarities of the original. The punctuation, particularly, has been very little altered although it is in some cases eccentric by modern standards, since, in this dictated text, it may reflect the cadences of the author's speech, and is frequently a guide to his meaning. Titles, which are often rather approximate, have also been left as they appear in the 1817 text.

160 *index expurgatorius*. The Expurgatory Index is the list of authors and works forbidden by the Church of Rome to be read until they shall have been expurgated.

161 (*CC*). See Abbreviations (p. 697) for full identification of source.

Mr Bowles's. William Lisle Bowles (1762–1850), famous in his time as the author of topographical lyrics, notably the sonnets mentioned by

Coleridge, *Fourteen Sonnets* (1789), which Coleridge probably read in the enlarged second edition, under the title *Sonnets, Written Chiefly in Picturesque Spots, During a Tour* (1789).

165 *Darwin's Botanic Garden. The Botanic Garden*, published in two parts in 1789 and 1792 by Erasmus Darwin (1731–1802), was an influential didactic poem introducing the Linnaean system of botanical classification by means of heroic verse and ample footnotes.

166 *In the Nutricia. . . .* C's point is that both lines have the same meaning, 'The pure stream wanders on among coloured pebbles'.

176 *the rude Syrinx*. Ovid relates the Greek myth of the nymph Syrinx who, pursued by Pan, prayed for assistance and was turned into a reed which Pan, however, cut down to make himself a pipe to play. The point here is that only a Pan or Apollo—the patron of music—could make a fine instrument out of such materials.

177 *couched*. In the sense of having been operated on for a cataract in the eye; hence, one with renewed vision.

178 *Friend No. 10*. Coleridge's own periodical, published 1809–10, issued in volumes in 1812, and extensively revised for publication in 1818.

182 *Camera obscura*. See note to p. 393.

184 *Mr Southey*. Coleridge's relations with his brother-in-law Robert Southey (1774–1843) are outlined in the 'Chronology' and are touched on in letters and notebook entries included in this volume. The first readers of the *Biographia* knew him as the author of exotic epic poems (*Thalaba*, 1801; *Madoc*, 1805; *The Curse of Kehama*, 1810; *Roderick; the Last of the Goths*, 1814; etc.) and as a regular contributor to the Tory *Quarterly Review*. He was appointed Poet Laureate in 1813.

194 *Psilosophy*. Coleridge's coinage, 'slender wisdom', i.e. false philosophy.

196 *a bull*. 'A self-contradictory proposition' (*OED*), popular as a type of joke in Coleridge's time. The statement, 'My mother was unfortunately barren', is an example. Coleridge's own illustration hinges on the idea of the changeling, an ugly creature substituted for a normal baby carried off by fairies.

203 *animalcula infusoria*. One-celled organisms which took their name from their being found in infusions of decaying animal and vegetable matter. Coleridge draws attention to the smallest possible form of life, the nucleus of such creatures.

206 *Hartley*. The philosopher David Hartley (1705–57) expounded his influential doctrine of the association of ideas in *Observations on Man* (1749), which traces the development of moral ideas, including faith in God, to the physical impact of sense-impressions. Coleridge as a young man was an enthusiastic disciple of Hartley's, but in his thirties he deliberately rejected both Hartley's and other materialist systems.

211 *suffictions*. Coleridge has coined two words, the Greek transliterated

'hypopoiesis' and the Latin-based 'suffiction', both signifying something *made up* underlying something else, as distinguished from 'hypothesis' and 'supposition', words that signify something (already existing) *put under* something else.

218 *Spectator.* There is some confusion in the reference here. 'Grimalkins' are she-cats, and Coleridge seems to be recalling a story about cats walking about *inside* a harpsichord; the only paper in the *Spectator* on such a topic, however, is no. 361 (24 April 1712), which has to do with whistles ('cat-calls') used by members of the audience in London theatres.

219 *armed vision.* That is, to one equipped, for example, with a magnifying glass. 'Armed' is a term adopted from magnetism.

226 *Price.* In *A Free Discussion of the Doctrines of Materialism and Necessity* (1778), Joseph Priestley (1733–1804), scientist, political radical, leader of the Unitarian movement in England and for some time in the 1790s an idol of Coleridge's, had defended his necessitarian position against arguments in favour of free will by Richard Price.

228 *Locke's Essay.* A reference to the famous discussion of innate ideas in the *Essay Concerning Human Understanding* (1690).

in. Perhaps a printer's error for 'on'.

229 *illustrious Florentine.* Marsilio Ficino (1433–99).

philosopher of Nola. Giordano Bruno (*c.*1548–1600).

Jacob Behmen. Coleridge's spelling of the name was common in the period and has been retained, but it is customary now to see Boehme or Böhme given as the name of the German mystic (1575–1624) whose work found a champion in Coleridge.

writer of the Continent. F. W. J. von Schelling (as indicated in the chapter-heading), to whose works Coleridge was indebted for several passages in the *Biographia*, notably in ch. XII.

236 *Braunonian system.* Also called Brunonian, this was the practice associated with Dr John Brown (1735–88), who believed that illnesses were caused by a deficiency or surplus of 'excitability' in the body, and could be corrected by appropriate doses of stimulants or narcotics.

240 *add to . . . caloric.* A deliberate comic use of the jargon of chemistry, signifying 'as much as is sufficient of China tea' to which is added 'the oxide of hydrogen' that has been made as hot as possible—that is, boiling water.

fox-brush . . . incumbrance of tails. There is a proverbial story of a fox who, having lost his own tail, tried to persuade the other foxes to cut theirs off as well. Coleridge's moral might be paraphrased, 'It is better to be a little vain of a fine tail than to go about with a bare bottom.' The revolutionaries of France, called *sansculottes* because they wore long trousers and not the breeches of gentlemen, were depicted in English political cartoons as literally without anything to cover their bottoms (*culs*).

249 *gagging bills*. Popularly so called, the two Acts that became law on 18 December 1795 prohibited seditious speeches and gave magistrates the power to disperse political meetings of fifty or more people.

275 *Darwin and Roscoe*. Erasmus Darwin, whose poetry has been mentioned earlier in the *Biographia*, was a celebrated physician at Derby; William Roscoe, historian, biographer, and poet, also had careers in law and banking.

277 *Trullibers*. Parson Trulliber, a boorish and uncharitable parson-farmer in Fielding's novel *Joseph Andrews*, stands for clergymen who have completely lost sight of their professional responsibilities.

298 *eternal I AM*. The Hebrew name for God, Yahweh or Jehovah, signifies 'I am' (Exodus 3: 14); as Byron neatly if frivolously says in *Don Juan*, ''Tis strange—the Hebrew noun which means "I am," | The English always use to govern "damn".'

313 *Your affectionate, etc*. Coleridge himself was the author of this letter.

The Ancient Mariner. A fragment of this essay survives, but Coleridge decided not to publish it with 'The Ancient Mariner' in *Sibylline Leaves* or elsewhere.

316 *recent collection*. For *Poems* (1815), Wordsworth wrote a new Preface and 'Supplementary' Essay, and printed the old 1800 Preface as an appendix.

317 *Alexis of Virgil*. That is, in a more perfect society readers would not take pleasure in immoral subjects (such as Coleridge considered the homosexual love-affairs celebrated in Anacreon's twenty-ninth Ode and Virgil's second Eclogue to be), and the primary end of poetry would therefore be defeated.

329 '*Alonzo and Imogen*'. M. G. Lewis's popular Gothic ballad, 'Alonzo the Brave and the Fair Imogine', published in *The Monk* in 1796, described grisly events in a jaunty anapaestic metre.

330 *Sir J. Reynolds*. First President of the Royal Academy, Sir Joshua Reynolds (1723–92) is the eminent artist whose advice (from the 1769 *Discourses*) Coleridge has been quoting.

335 *forms of nature*. Here, as throughout this chapter, Coleridge is quoting the 1800 Preface, and commenting on Wordsworth's statements point by point.

343 *Sir Roger L'Estrange*. Coleridge means to illustrate a spectrum of prose styles from the learned dignity of Hooker's *Ecclesiastical Polity* (1594–7) to the political pamphlets of L'Estrange and the slangy satire of Tom Brown's *Amusements, Serious and Comical* (1700).

347 *invaluable system*. An allusion to a current educational controversy, in which Coleridge supported the 'Madras system' of Andrew Bell, in

which older pupils assisted in the teaching, against what he saw as a corruption of the system in Lancaster's methods.

352 *my judgement*. In the 1800 Preface, Wordsworth quoted a simple and moving stanza from 'The Babes in the Wood' to prove that simplicity is not in itself contemptible, as long as 'the *matter*' or subject is serious. He contrasted it with the parody of such a style in a quatrain by Dr Johnson, quoted by Coleridge, with a paraphrase of Wordsworth's comment on it, on p. 358.

Sterne. Three much-admired chapters in Laurence Sterne's *Sentimental Journey* (1768).

354 *mordant*. In the dyeing process, the mordant is the substance used to fix colours.

379 *Edinburgh Review*. Founded in 1802 by Francis Jeffrey, Sydney Smith, and others, the Whiggish *Edinburgh Review* distinguished itself from other periodical reviews by its policy of selectivity (which meant that reviews could be much longer than they were elsewhere) and by its practice of asperity. It was extraordinarily influential. The famous review of Wordsworth's *Excursion* (1814), which began 'This will never do', is repeatedly referred to by Coleridge in this chapter, and is a fair specimen of the treatment Coleridge and his friends had at the hands of anonymous reviewers.

380 *viva sectio*. Cutting up of living things; vivisection.

Lessing. Coleridge means that he has adopted most of the previous paragraph from the German dramatist and critic G. E. Lessing, whose work he greatly admired.

386 *wares of his pack*. Coleridge alludes again to the review of the *Excursion* in the *Edinburgh Review*, which ended with complaints about the choice of a pedlar as a central figure in the poem.

393 *camera obscura*. Literally 'dark chamber', the camera obscura was a box or darkened room admitting light through one small hole only; the hole contained a lens which would project an inverted image of the external scene on a screen or wall or piece of paper set at its focal point.

the Egyptian statue. A colossal statue in Egypt was said to produce a sound when struck by the rays of the rising sun; it was identified by the Greeks with the legend of the goddess Eos (Dawn) greeting her son Memnon, Prince of Ethiopia. Coleridge's point seems to be that as light, a visual sensation, produces sound in the statue, so the description of a sound in Milton's lines conjures up a visual image.

415 *white of their eye*. Proverbial: who can find fault with them?

417 *un philosophe*. Literally 'philosopher'; but as Coleridge's distaste suggests, the French term was reserved more particularly for the free-thinking satirists of the eighteenth century—Voltaire, Diderot, etc.

422 *Corresponding Society*. An English Jacobinical working-class movement

of the 1790s, estimated at its peak to consist of over a hundred societies (in communication or 'correspondence' with French Jacobins) in provincial centres as well as in London. In 1792 the Sheffield association distributed 1,600 copies of the first part of Thomas Paine's inflammatory *Rights of Man*.

426 *extinguisher*. A cone-shaped metal cap for putting out the flame of a lamp or candle (in the latter case, the extinguisher was often attached as part of the candlestick); before gas lighting was used in cities, houses used to have large extinguishers hung on their railings to be used on links or torches.

431 *Surinam toad*. This toad, in frequent use as an image among the Romantic poets, carries its eggs stuck in the soft skin of its back until they hatch, and therefore seems to sprout little toads. Window taxes had been in use as a form of property tax in England since 1697, but in 1792, under the Government of Pitt the Younger, houses with fewer than seven windows were made exempt. The *additionals* are presumably clauses added to existing legislation by Pitt.

433 *Alien Act*. By Alien Bills first passed in 1792–3 and periodically renewed, the Crown was empowered to banish foreigners from England.

poison. Venetian glass was credited with the power of detecting poisons.

443 *my friend*. Some of Coleridge's original readers would have realized that this friend was Wordsworth.

452 *twelve months*. Charles Robert Maturin's tragedy *Bertram* had a very successful run at Drury Lane in 1816, and went through seven editions in that year. Coleridge's comments on the actors and staging show that he had seen as well as read the play.

453 *morals and taste*. In the 1800 Preface to *Lyrical Ballads*, Wordsworth had also attacked the 'degrading thirst after outrageous stimulation' which led audiences to turn from 'the invaluable works of our elder writers' to 'frantic novels, sickly and stupid German Tragedies, and deluges of idle and extravagant stories in verse'. Coleridge here turns his attention to the decline of the drama, as revealed by the use of animals on stage (Polito's, mentioned later, was a menagerie) and in the popularity of adaptations of the sentimental plays of August von Kotzebue, whose *Stranger*, *Lovers' Vows*, and *Pizarro* (Sheridan's version of *The Spaniards in Peru*) were all performed at Drury Lane in 1798–9.

457 *Carrier*. Sent to suppress an uprising in the Vendée, the French revolutionary Jean Baptiste Carrier was responsible for the deaths of thousands through mass drownings and similar brutalities in October 1793, but went to the guillotine himself in 1794.

459 *ghost*. As he reveals later on, Coleridge is following the text of Shadwell's *Libertine* (1675), a version of the Don Juan legend still popular on the English stage.

462 *I observed*. In the letters published as 'Satyrane's Letters', p. 440.

467 *Macheath.* The dashing highwayman hero of Gay's comic opera, *The Beggar's Opera* (1728).

472 *quere wake.* The parenthetical remark is Coleridge's own questioning of the text: 'Is "make" a printer's error for "wake"?'

476 *animal magnetism.* Animal magnetism was a topic of general interest at the time of the *Biographia.* An extension of mesmerism, later gaining respectability as 'hypnotism' in restricted medical use, 'animal magnetism' was believed by Coleridge and others to exhibit the will of the magnetizer working through the imagination of the patient to affect both mind and body.

482 *Word.* The Logos, Jesus Christ.

483 *Letters. Text: CL,* subject to the minor changes made consistently throughout this volume; see Note on the Text, pp. xvii–xviii. Where the letter is not given in full, points of ellipsis are used to indicate the omissions. Recipients for whom no special note is given may be identified by referring to the Chronology or Index to the Prose.

[*To Robert Southey,* 18 September 1794]. *Pantisocracy.* A form of government in which all are equal in power. For a year after they met in the summer of 1794, Coleridge and Southey discussed plans and attempted to raise money to found a classless Utopian society in America. Shad, mentioned in this letter, was the servant of Southey's rich aunt. At one time, the plan was to involve twelve couples: Southey was engaged to Edith Fricker, and Coleridge became engaged to her sister Sara in August 1794. After being scaled down to farming in Wales, the scheme was eventually abandoned because of Southey's reluctance.

485 [*To 'Citizen' John Thelwall,* 19 November 1796]. John Thelwall (1764–1834), son of a silk mercer, with some training in divinity, law, and medicine, was by this time an atheist and notorious political radical. Along with John Horne Tooke and Thomas Hardy (founder of the pro-revolutionary Corresponding Society of London), he had been imprisoned in the Tower on charges of sedition in 1794. Their acquittal was the occasion of public rejoicing, and all three were popular heroes. Thelwall published *Poems written in Close Confinement in the Tower and Newgate* in 1795.

487 [*To Benjamin Flower,* 11 December 1796]. Benjamin Flower (1755–1829), a Cambridge printer who had published some of Coleridge's poems in his *Cambridge Intelligencer.*

488 *Godwin.* William Godwin's *Enquiry Concerning Political Justice,* a rationalist argument concerning the perfectibility of man and of human society, had great impact on the intellectuals of Coleridge's generation. Godwin was unfortunately (from Coleridge's point of view) an atheist.

491 [*To John Thelwall,* 17 December 1796]. *Della Cruscan.* A name applied in the nineties to feeble sentimental verse, derived from the name of a school of poetry associated with Robert Merry, who had lived in

Florence, was a member of its Della Cruscan Academy, and signed his contributions to the periodical *The World* 'Della Crusca'.

act of Uniformity. Coleridge alludes to the Act of 1662 which barred both Thelwall and himself from careers in the university or the Church of England by requiring that all clergymen and schoolmasters accept the Book of Common Prayer as the only guide for public worship.

496 [*To Thomas Poole*, March 1797]. *Parson Adams*. The much-loved unworldly clergyman of Fielding's novel *Joseph Andrews*.

497 *Shenstone's*. The affectionately comic portrait in William Shenstone's poem *The Schoolmistress* (1742).

498 [*To Joseph Cottle*, April 1797]. Joseph Cottle (1770–1853), a minor poet and bookseller of Bristol, published several of Coleridge's early works, including the first edition of *Lyrical Ballads*.

510 [*To William Sotheby*, 13 July 1802]. William Sotheby (1757–1833), an author without financial anxiety, shared many of Coleridge's interests: he had, for example, taught himself German in order to translate Wieland's *Oberon* in 1798. Coleridge was later to act informally as critic of Sotheby's plays, poems, and translations before publication.

erste Schiffer. A prose poem (*The First Navigator*) by Salomon Gessner, whose *Death of Abel* influenced Coleridge and Wordsworth (see Coleridge's 'Prefatory Note' to 'The Wanderings of Cain', p. 41).

526 [*To Thomas Allsop*, 30 March 1820]. Thomas Allsop (1795–1880), a young man who had introduced himself to Coleridge after attending his lectures in 1818.

536 [*To James Gillman, jun., aged eighteen*, 22 October 1826]. *Gentiles . . . genus*. Coleridge brings together a number of words significantly related to one another by their derivation from the Latin *gens*, 'race' or 'clan'; by the odd italicization in the last word he may be implying further that all are derived ultimately from the Greek word for 'earth', *gē*. The line may be translated 'Men of family, well-born, who belong to a particular tribe, and kind'.

downright Last. Lotteries were abolished in England in 1826, the last being held on 18 October that year.

539 *Irish* Bull. See note to p. 196n.

540 [*To J. H. Green*, 29 March 1832]. Joseph Henry Green (1791–1863), an eminent surgeon interested in German literature, who studied with Coleridge on a regular basis 1818–34 and became his literary executor.

543 *Notebooks*. The transcribing, dating, and annotating of Coleridge's chaotic and now widely dispersed notebooks has been one of the scholarly triumphs of this century, and readers are urged to spend some time with Kathleen Coburn's edition (*CN*) in order to understand the sorts of problems presented by the original documents. The small selection of notes here is intended to cover a long span of Coleridge's life;

to suggest the variety of the notebooks, and the varied uses to which they were put; to convey some of the qualities of mind that make Coleridge's informal prose especially attractive; and to illuminate aspects of the poetry and the *Biographia*. It is unrepresentative, however, in that there are no very long notes, and that large areas of interest to Coleridge—scientific and theological speculation, biblical exegesis, the close study of philosophical texts—have been omitted. Furthermore, the notes appear necessarily out of context and sketchily annotated; for fuller explication, see *CN*.

Text: *CN*, subject to the minor changes described in the Note on the Text, pp. xvii–xviii. In all but two cases (where points of ellipsis are used), the complete entry is given.

553 *Brunonian phrase*. See note to p. 236.

556 *Herbert's Death*. Coleridge is reflecting on the religious poetry of George Herbert (1593–1633), one of his favourites among English poets, and frequently mentioned in the *Biographia*.

562 *immediate Aims*. An allusion to the Brobdingnagian giants and Lilliputian miniatures of Swift's *Gulliver's Travels*. 'Fly-Catcher' was Coleridge's name for a series of small notebooks, catchers of stray thoughts.

564 *Marginalia*. Quite early in life, and with the encouragement of his friends, Coleridge adopted the practice of writing notes in the margins of books as he read them, and of using the flyleaves of books for notes that overran the marginal space, or for general notes summing up his opinion of the work in question. After his death, his literary executors collected and published many of the notes in miscellaneous volumes of criticism—*Notes on Shakespeare*, *Notes on English Divines*, etc.—and all the marginalia are now being re-edited (*CM*) as part of the Bollingen edition of his *Collected Works*. About 400 annotated books have survived, and the marginalia, with the texts to which they are attached and with suitable scholarly annotation, are expected to fill five volumes. It will therefore be obvious that the selection here is a most inadequate sample: it does not represent the books that were most important to Coleridge; it does not systematically present all the marginalia in any volume; it does not include the passages from the printed text that stimulated the marginalia in the first place; it does not permit the reader, as *CM* does, in some measure to accompany Coleridge in the reading of a particular work. It is frankly a miscellany of detached observations that it is hoped will illuminate the major works represented in this volume, and arouse the reader's appetite for more.

Text: *CM*, subject to the minor changes described in the Note on the Text, pp. xvii–xviii. It should be noted that for material that is as yet unpublished, a draft of the *CM* text was checked against originals where possible, but that there may be slight differences in the final *CM* form of the text. The marginalia are arranged as in *CM*, alphabetically by author, and where no new author's name is given at the beginning of an entry, it is to be understood that Coleridge's note appears in the same work as the preceding note. A date is given as the first part of the head-note in square brackets, when the note can be confidently dated.

566 *p. 81*. An instance of physical cure attributed to God.

phlogistic Sect. 'Phlogistic' is not used in its scientific sense, but as a punning reference to Boyer's skill in flogging. *Hercules Furens*, 'The Madness of Hercules' (but with overtones of rage or fury) is the title of a tragedy by Seneca.

567 *Cibber*. Colley Cibber (1671–1757) adapted many older plays, including Shakespeare's *Richard III* and several of the comedies of Beaumont and Fletcher.

571 *tinct. lyttæ*. Pharmaceutical jargon: a solution ('tincture') supposed to be aphrodisiac.

578 *Athanasian Locker*. The archangels, 'Raphael or Uriel', are to bring the soul of a ruler such as Numa or Lycurgus for the duchess's son, and of a theologian as great Chrysostom or Athanasius for the child of the archbishop.

582 *Persona* PATIENS. Word-play with *dramatis persona* (the actor in a drama), and the opposition between *ago* 'I act' and *patior* 'I suffer, I am passive'. Lear is the suffering character in this play.

587 *Hayme, or Pawson*. John Haime, a soldier, and John Pawson, who trained to be an architect, were Wesleyan converts and became Methodist ministers.

591 *Table Talk*. With a view perhaps to becoming Coleridge's Boswell, his nephew and son-in-law Henry Nelson Coleridge kept records of Coleridge's conversation in the last ten years of his life and published an edited version as *Specimens of the Table Talk of the Late Samuel Taylor Coleridge* in 1835. Although the published text is, like the text of the *Lectures on Shakespeare* below, at least two removes from Coleridge's famous talk, and although some of Coleridge's friends complained that it did not do him justice, the *Table Talk* provides valuable confirmation and clarification of some of Coleridge's ideas, and conveys something of the charm of his character in maturity.

Text: the first edition of 1835, modernized according to the principles outlined in the Note on the Text, pp. xvii–xviii.

596 *Kepler*. Johann Kepler (1571–1630), German astronomer contemporary with Galileo. His three laws of planetary motion provided a basis for much of Newton's work. Coleridge is exercising his preference for theoretical over experimental science.

604 *A Moral and Political Lecture*. This lecture, printed shortly after it was delivered at Bristol in 1795, provides interesting evidence of Coleridge's early political views—not as radical as they were later represented—as well as of his style of oratory.

Text: *CC*, subject to the modernization described in the Note on the Text, pp. xvii–xviii.

606 *liberty*. In 1794, after seeing their country twice diminished in its partition by Prussia and Russia, the Polish people rebelled, led by Kosciusko. This rebellion, a subject of some of Coleridge's most ambitious poetry at the time (e.g. 'Ode to the Departing Year'), was suppressed in October 1794, and the country was again divided between the conquerors in October 1795.

607 *Birmingham riots*. Coleridge implies that the Anglican clergy incited the anti-Jacobin rioters in Birmingham in the early nineties, who burnt down buildings, notably (in July 1791) the home of Joseph Priestley. Coleridge alludes to the Priestley incident specifically, in 'Religious Musings'.

608 *that bad man*. William Pitt the Younger (1759–1806), Prime Minister 1783–1801, and again 1804–6.

610 *intellect*. Joseph Gerrald, one of the defendants in the infamous trials for sedition held in Edinburgh in 1793 and 1794, was sentenced to transportation for fourteen years in March 1794, but spent a year in Newgate before dying in exile in 1796.

Muir, *Palmer*, *and Margarot*. Thomas Muir, Thomas Fysshe Palmer, and Maurice Margarot were, like Gerrald, unsuccessful defendants in the Edinburgh trials for sedition, and therefore in the eyes of liberals were martyrs to Pitt's policy of suppression.

To the Exiled Patriots. The poem is by Robert Southey.

613 *execution*. John Horne Tooke was tried for treason in London in November 1794. In a reversal of the events in Edinburgh, the first three defendants in the London trials—Hardy, Horne Tooke, and Thelwall (later Coleridge's friend)—were acquitted, and similar charges against others were dropped.

614 *Essays on His Times*. *Pitt*. This 'character' of the Prime Minister, William Pitt the Younger (1759–1806), was first published in the *Morning Post* of 19 March 1800, and was intended to be paired with a similar sketch of Bonaparte, though the latter article never appeared. Pitt, who held office 1783–1801 and again 1804–6, had been regularly attacked in Coleridge's political poetry and journalism.

Text: *CC*, subject to modernization as outlined in the Note on the Text, pp. xvii–xviii.

617 *Burke*. The long-standing friendship between the two great politicians, Charles James Fox (1749–1806) and Edmund Burke (1749–97), had been shattered by their publicly taking opposite stands on the French Revolution—Fox enthusiastically in favour, Burke opposed—in 1790.

619 *indulgences*. i.e. drink and drinking companions.

620 *The Friend*. Coleridge's main contribution to the tradition of the periodical essay that includes the *Tatler* and *Spectator* of Addison and Steele and Johnson's *Rambler* and *Idler* suffered various Coleridgean irregularities from the start. The work was largely dictated by Coleridge to Sara Hutchinson, and essays as a consequence frequently violated the

conventions of the genre by running over from one number to the next. Publication was delayed by printing and postal problems; it proved difficult to extract payment from the subscribers; and after twenty-eight numbers the *Friend* ceased to appear. It was, however, sold in sets, reissued in 1812, and reorganized and reprinted in three volumes in 1818. The selection here follows the order of the 1818 version.

Text: CC, subject to modernization as described in the Note on the Text, pp. xvii–xviii.

623 *The Landing-Place . . ., Essay II.* This essay on Martin Luther (1483–1546), the leader of the Reformation in Germany, has as an undertheme the paradoxical comparison—introduced in a preceding essay in the 1818 collection—of Luther and the free-thinking Jean-Jacques Rousseau (1712–78).

632 *The Landing-Place . . ., Essay III. my friend.* Thomas Wedgwood (1771–1805).

640 *Lectures on Shakespeare.* It should be said at once that this text is not Coleridge's composition. Although he gave several series of literary lectures, the lecture is an ephemeral form and Coleridge himself never published his notes. Much of what has been presented as his literary criticism is in fact a patchwork, patiently assembled by Coleridge's children, his first editors, of entries from the notebooks, manuscript remains, and marginalia. In 1856, however, John Payne Collier published Coleridge's *Seven Lectures on Shakespeare and Milton*, reproducing the highlights (at least) of seven lectures in the 1811–12 series from shorthand notes that he had taken at the time; and although what he printed is at two removes from Coleridge's speech, being a slightly polished version of the notes, it is still the best record available of the opinions expressed by Coleridge and of the sequence of his thought on those occasions. (Collier's later career was shadowed by charges of literary forgery, and for a time his version of these lectures was under suspicion, but the evidence of its authenticity is fully convincing, and it has been accepted as part of the Coleridge canon.) Coleridge was forever resolving, and forever neglecting, to lecture from a fully prepared text. He was a notoriously digressive lecturer and the selection here may not therefore be quite typical, but it represents the two most successful lectures in the successful series of 1811–12.

Text: John Payne Collier, ed., *Seven Lectures on Shakespeare and Milton* (1856), modernized according to the principles outlined in the Note on the Text, pp. xvii–xviii. A new transcription of Collier's shorthand notes, *Coleridge on Shakespeare. The Text of the Lectures of 1811–12*, ed. R. A. Foakes (London: Routledge & Kegan Paul, 1971), provides a slightly different text, but since it is soon to be superseded by Professor Foakes's edition of the *Literary Lectures* in the Bollingen edition, the more familiar version is used here.

646 *Romeo and Juliet. Deborah.* In Judges 5.

654 *Elwes*. John Elwes (1714–89), 'a byword for sordid penury' according to the *DNB*.

660 *Lay Sermons*. The full titles of the two works joined under this name may serve as an introduction: they are *The Statesman's Manual or The Bible the Best Guide to Political Skill and Foresight. A Lay Sermon addressed to the Higher Classes of Society* (1816) and *A Lay Sermon addressed to the Higher and Middle Classes on the Existing Distresses and Discontents* (1817). Published in a post-war period of economic difficulty, the 'lay sermons' in a characteristically Coleridgean way analyse the *malaise* of English society and prescribe for its cure.

Text: CC, modernized according to the principles outlined in the Note on the Text, pp. xvii–xviii.

661 *I. allegories*. This distinction is also discussed in *Aids to Reflection*, pp. 672–3.

664 *II. an earth*. Coleridge added an explanatory note to this phrase in one copy: 'i.e. detecting and separating the metallic base of.'

666 *Aids to Reflection*. The *Aids* began modestly as a proposed anthology of extracts (with editorial commentary) from the works of Archbishop Leighton (1611–84), and even when the book achieved its final form, much more Coleridge than Leighton, passages by him and other seventeenth-century divines were retained as focuses for meditation. This extraordinary work, addressed to the questioning Christian and intended to encourage rather than to put to rest serious enquiry, leads the reader up a moral scale from matters of worldly prudence to the great issues of spiritual religion. Coleridge brought to the *Aids* the fruits of his study of the Bible, and the methods and theories developed by him are of considerable interest to students of literature, notably the emphasis on the power of language; the distinction between metaphor and symbol and the illustration of the value of that distinction in the study of texts concerned with redemption; and the approach to myth. All these are included here. A fundamental part of Coleridge's argument, the distinction between reason and understanding, however, is only touched on in this selection; readers are therefore directed to other expositions of the distinction in this volume, e.g. pp. 597 and 633–9.

Text: the first edition of 1825, but incorporating substantive emendations from the second edition of 1831, and modernized in accidentals according to the principles outlined in the Note on the Text, pp. xvii–xviii.

672 *Aphorisms on Spiritual Religion. Statesman's Manual*. Included above, p. 661.

sceptics. The application of the principle follows directly here, although more than a hundred pages divide the two parts of the 1825 text. The doctrine of redemption is Coleridge's subject; the texts are from St Paul, expecially Rom. 5 and Heb. 9, 10.

673 *p. 215*. On p. 215 of the first edition of 1825, Coleridge alluded to the important logical distinction between difference of degree and difference of kind, quoted the Aristotelean phrase as he does here, and glossed it thus: 'Transition into a new kind, or the falsely applying to X what had been truly asserted of A, and might have been true of X, had it differed from A in its degree only.'

675 *own words*. In John 3, especially 3: 3, ('Verily, verily I say unto thee, Except a man be born again, he cannot see the kingdom of God') and 3: 6 as quoted above, p. 672.

678 *catholic faith*. Not the doctrines of Roman Catholicism specifically, but essential articles of Christian faith in general.

Simonides. The article on 'Simonides' in the *Dictionary* of Pierre Bayle tells the story of Simonides' failure to produce a definition of 'God', and devotes part of a long footnote (note F) to a speculative reconstruction of the reasons for his refusal, as he anticipates objections to every part of his statement.

679 *who dwelt . . . the Isles*. The Greeks.

680 *the Prometheus*. Prometheus, the 'rebellious Spirit', stole fire from heaven to assist mankind, and was punished by Zeus by being fastened to a rock, where an eagle came every day to feed on his liver. Io, a maiden loved by Zeus but turned by Hera into a heifer and driven by a gadfly from country to country, encounters Prometheus in Aeschylus' *Prometheus Bound*; Prometheus prophesies that one of her descendants will set him free.

Cupid and Psyche. Psyche, a mortal whose name means 'soul', was loved by Cupid, but prohibited from looking at him. When she disobeyed, he left her, and it was only after undergoing a series of trials that she could be reunited with him.

684 *Conclusion. a Berzelius or a Hatchett*. Eminent contemporary chemists.

686 *On the Constitution of the Church and State, according to the Idea of Each*. First published in December 1829, this work of political theory was prompted by the movements for Catholic emancipation and Parliamentary reform, but has been influential beyond its immediate causes for its analysis of the State and for the idea of the clerisy, both represented in the two chapters extracted here.

Text: *CC*, modernized according to the principles outlined in the Note on the Text, pp. xvii–xviii.

687 *dear Sir*. John Hookham Frere (1769–1846), translator of Aristophanes, friend and supporter of Coleridge, for whom the first draft of *Church and State* had been written about 1825.

692 *nationalty*. Coleridge's term for that part of the wealth of a nation that is not divided among individuals but is set aside for the needs of the nation as a whole.

694 *that realm*. Reduced to their roots, 'ecclesia' ('church') and 'enclesia' signify 'called out' and 'called in'.

benefit of clergy. Originally, exemption from punishment for crime, extended to the learned—in practice, those who could read and write.

FURTHER READING

MAJOR EDITIONS

The best course for those who wish to know more of Coleridge, or to improve their understanding of what they have read, is to read more of Coleridge's own work and, after that, to read the works of his contemporaries, particularly of those whose lives intersected with his—Wordsworth, Lamb, Southey, De Quincey, Hazlitt, and to a lesser extent Scott and Byron. Coleridge's own writings will eventually all be available in modern editions, but to date only the *Collected Letters*, ed. Earl Leslie Griggs, is complete, in six volumes. The edition of the *Notebooks*, by Kathleen Coburn, is almost finished: three double volumes (text and notes) of five have appeared, covering Coleridge's life to 1819. Of sixteen titles in the massive edition of the *Collected Works*, published for the Bollingen Foundation by Princeton University Press, and under the general editorship of Kathleen Coburn, eight have appeared and one—the important edition of the *Marginalia*, by George Whalley—has been published in part. The published titles are *Lectures 1795: On Politics and Religion*, *The Watchman*, *Essays on His Times*, *The Friend*, *Lay Sermons*, *Biographia Literaria*, *On the Constitution of the Church and State*, *Marginalia*, vol. i, and *Logic*. Others are well advanced: *Lectures 1808–19: On Literature*, *Lectures 1818–1819: On the History of Philosophy*, *Aids to Reflection*, *Shorter Works and Fragments*, *Table Talk*, *Opus Maximum*, and *Poetical Works*. Until the collected edition is complete, however, readers must turn to older editions for these texts. A useful set is the *Complete Works*, ed. W. G. T. Shedd, in seven volumes (New York: Harper, 1853; repr. 1871, 1875, 1884), which includes Coleridge's chief scientific text, the 'Theory of Life' or, to give it its full title, *Hints Towards the Formation of a More Comprehensive Theory of Life* (1848). The standard edition of the poetical and dramatic works is still the *Complete Poetical Works*, ed. E. H. Coleridge (Oxford: Oxford University Press, 1912; frequently reprinted). Several of Coleridge's essays on aesthetics are included and annotated in the second volume of *Biographia Literaria*, ed. J. Shawcross (Oxford: Oxford University Press, 1907; frequently repr.; rev. 1954). His literary criticism is collected in T. M. Raysor, ed., *Coleridge's Shakespearean Criticism* (1930; rev. edn. London: Everyman's Library, 1960) and *Coleridge's Miscellaneous Criticism* (Cambridge, Mass.: Harvard University Press, 1936). A stimulating selection of extracts from Coleridge's notebooks, marginalia, unpublished manuscripts, and other out-of-the-way places is Kathleen Coburn's *Inquiring Spirit* (London: Routledge & Kegan Paul, 1951; repr. Toronto: University of Toronto Press, 1979).

BIOGRAPHIES

There are many specialized biographies dealing with particular aspects or sections of Coleridge's life—his relationship with Wordsworth, his addiction to opium, his Malta period—but two works can be recommended as broader general studies. The first has yet to be supplanted as a condensed, unadorned account: it is James Dykes Campbell's *Coleridge: A Narrative of the Events of His Life* (London: Macmillan, 1894), originally published as the biographical preface to his edition of *Poetical Works* (London and New York: Macmillan, 1893). A fuller study is Walter Jackson Bate's *Coleridge* (New York: Macmillan, 1968). A delightful collection of accounts of Coleridge by his contemporaries is *Coleridge the Talker*, eds. Richard W. Armour and Raymond F. Howes (1940; rev. edn. New York and London: Johnson Reprint, 1969).

BIBLIOGRAPHIES

The entry on Coleridge in the third volume of the *New Cambridge Bibliography of English Literature* is the most convenient guide to works by Coleridge as well as to articles and monographs about him published up to 1967. For works published since that date, see *The Romantic Movement*, ed. David Erdman, which appeared first as an annual supplement to *English Language Notes* and then, since 1980, as a separate publication produced by Garland Publishing of New York. An elaborate catalogue of Coleridge manuscripts (with the exception of letters) has recently appeared, compiled by Barbara Rosenbaum, in the *Index of English Literary Manuscripts*, vol. iv, Part 1 (London and New York: Mansell, 1982).

SPECIAL STUDIES

This brief alphabetical list attempts to include some work that can be recommended in each area of concern to students of Coleridge, but it is necessarily incomplete and selective. Where the title is not self-explanatory, the contents may be indicated in parentheses.

J. A. Appleyard, *Coleridge's Philosophy of Literature: The Development of a Concept of Poetry 1791–1819*, Cambridge, Mass.: Harvard University Press, 1965.

Owen Barfield, *What Coleridge Thought*, 2nd edn., London: Oxford University Press, 1972.

J. Robert Barth, SJ, *Coleridge and Christian Doctrine*, Cambridge, Mass.: Harvard University Press, 1969.

John Beer, *Coleridge's Poetic Intelligence*, London: Macmillan, 1977.

Roberta F. Brinkley, *Coleridge on the Seventeenth Century*, Durham, NC: Duke University Press, 1955. (A collection of Coleridge's comments on writers of the seventeenth century, with comment.)

Richard Holmes, *Coleridge*, Past Masters Series, Oxford: Oxford University Press, 1982. (A short comprehensive account; paperback.)

Humphry House, *Coleridge*, London: Rupert Hart-Davis, 1953. (A sympathetic introduction to Coleridge, through the letters and notebooks, with commentary on the major poems.)

T. H. Levere, *Poetry Realized in Nature: Samuel Taylor Coleridge and Early Nineteenth-Century Science*, Cambridge: Cambridge University Press, 1981.

J. L. Lowes, *The Road to Xanadu: A Study in the Ways of the Imagination*, London: Constable, 1927. (A reconstruction of the genesis of the major poems; a classic for its method.)

Thomas McFarland, *Coleridge and the Pantheist Tradition*, Oxford: Clarendon Press, 1969. (Coleridge and Spinozism; Coleridge's 'plagiarism'.)

John H. Muirhead, *Coleridge as Philosopher*, London: Allen & Unwin, 1930; repr. 1954.

I. A. Richards, *Coleridge on Imagination*, London: Kegan Paul, 1934.

Carl Woodring, *Politics in the Poetry of Coleridge*, Madison: University of Wisconsin Press, 1961.

INDEX OF POEM TITLES AND FIRST LINES

Titles are set in italics. In titles (but not first lines) the definite and indefinite articles are ignored when either is the first word; otherwise, organization is strictly alphabetical so that 'I stood' follows 'In Xanadu'.

INDEX TO PROSE

It is often difficult to recover passages one remembers reading in Coleridge's prose. An exhaustive index could not be considered for this volume, but the following selective index of topics and key words has been designed to meet the everyday needs of the reader. Proper names appear only when they represent the main subject of a passage or a key word by which the passage may be found. Titles of works are given under their authors' names. To distinguish them from topics proper, instances of figurative language and illustrations are followed in the index headings by the notation '(fig.)'. Coleridge's own words have been used as far as possible in preparing the index entries, and cross-references have been introduced for such closely related terms as 'soul', 'spirit', and 'mind'. When the index does involve an editorial interpretation, part of the text is given in parentheses beside the page number to allow the reader to locate the passage. The abbreviation 'fn.' designates one of Coleridge's own footnotes, 'n.' an editorial note.